Fatal Misconception

Fatal Misconception

The Struggle to Control World Population

MATTHEW CONNELLY

The Belknap Press of Harvard University Press

Cambridge, Massachusetts / London, England   2008

Library of Congress Cataloging-in-Publication Data

Connelly, Matthew James.
Fatal misconception : the struggle to control world population / Matthew Connelly.
    p.   cm.
Includes bibliographical references and index.
ISBN 978-0-674-02423-6 (hardcover : alk. paper)   1. Population policy—History.
2. Population—Economic aspects.   3. Overpopulation.   4. Birth control—History.
5. International relations.   I. Title.

HB883.5.C65   2008
363.9—dc22   2007040553

To my parents, for having so many children

CONTENTS

PREFACE

This has been a hard book to write, but the dedication really did write itself. When I first set down to tell the story of the population control movement, I realized that it was already a tribute to my parents. After all, I am the youngest of eight children. Just mentioning this fact strikes most people with amazement. When they hear that my parents are Catholic, they seize on it as a simple explanation. In fact, by 1967, the year I was conceived and born, American Catholics were practicing contraception at virtually the same rate as everyone else. My grandmother, who was particularly devout, greeted news of each new child with dismay. When they grew up to make her proud, her son would ask which of her grandchildren she wished had never been born—the only sharp words anyone can remember passing between them.

The late 1960s and early 1970s were a tough time to raise a large family, which may be why mine considered toughness an essential trait. It was not just the uncertainties of the era, though I well remember waiting in long lines at gas stations as the tenth passenger in a nine-passenger station wagon. We forget that one of the few certainties of that troubled time, endorsed in virtually identical planks in the Democratic and Republican platforms of 1968, was that population control should be an urgent priority. A best-seller published that year, Paul Ehrlich's *The Population Bomb,* insisted that "the battle to feed all of humanity is over. In the 1970's the world will undergo famines—hundreds of millions of people are going to starve to death in spite of any crash programs embarked on now." On the very day I was born, the cabinet of India under Indira Gandhi considered for the first time compulsory sterilization for parents with more than three children. And Ehrlich believed that "the birth of each American child is 50 times the disaster for the world as the birth of a child in India."

This book is critical of Gandhi, Ehrlich, and others who argued that only population control could save the world. Is it because, somehow, I've chosen to take it personally? In fact, I did not know any of this when I started my research. I had become a historian in order to study the rise and fall of the great powers. My mentor, Paul Kennedy, had literally written the book on the subject. I became interested in the rise of world population only because Kennedy insisted that we needed to look beyond great-power rivalries to understand the new, post–Cold War era. We eventually co-authored a cover story for the *Atlantic Monthly* that warned that population growth in poor countries, the increasing awareness of gross economic in-equality, and the prospect of mass migration might lead to clashes between "the West" and "the rest." We called for more and better development aid, technology transfers, and additional funding for contraception—but not, Kennedy made clear, abortion. My mentor drew a moral distinction be-tween them. I, on the other hand, like many people who grew up in the wake of *Roe v. Wade,* did not feel any particular passion about either contra-ception or abortion. The right to decide whether and when to have chil-dren seemed like "settled law."

I was unsettled by what I discovered when, years later, I started research-ing a book on the subject, still thinking it was just a way to broaden our understanding of international security. Much to my chagrin, I found that other authors had issued similar warnings decades earlier that swelling numbers of poor people had begun to understand their plight and posed an imminent threat to themselves and others. A decade later, this stack grows ever higher. The *Atlantic* article was just one in a long line of works that re-duce differences in wealth and power to a question of differential fertility—too often in terms of "us" and "them."

Poor countries have long had high rates of fertility, and it has seemed obvious that this must be part of their problem, and that having fewer chil-dren can provide a solution. In fact, this view is not supported by the data. Obviously, a society with large numbers of working people and relatively few children tends for a time to have more disposable income (until, that is, this generation tries to collect its pensions). But even the most sophisti-cated quantitative research cannot resolve a question that really comes down to values. Most people are quite happy to reduce their "per capita GNP" by having children, and not all of them regret it. Programs to distribute contraceptives in poor countries have not had more than a marginal effect on population growth. Far more important is whether people actually want to have smaller families. When India, China, and other countries have tried to change these preferences, whether through cash payments or outright compulsion, the results have been disastrous. We will be living with them for generations to come.

As I went from archive to archive, pored through thousands of documents, and interviewed some of the people who made this history, I began to realize that much more is at stake than how we might redefine national security. This is a story of how some people have tried to control others without having to answer to anyone. They could be ruthless and manipulative in ways that were, and are, shocking. Perhaps we would expect no less of nativists or eugenicists, who assumed that people unlike themselves must be "beaten men from beaten races." Yet many more actually had the best of intentions, hoping to reduce poverty and prevent conflict, not unlike Kennedy and me. Of course, we did not call for coercive measures, only more contraception. But we also knew that population growth was slowing. The people I write about, on the other hand, were facing something utterly unprecedented in human history: world population was doubling and doubling again at an accelerating rate.

When contemplating the seemingly inexorable rise in human numbers, the most thoughtful observers have eventually asked themselves: "What are people for?" One cannot read the debates that ensued without starting to take them personally, because they ultimately concern the meaning and purpose of life itself. This book will not try to settle existential questions. But a work of history can at least show what happens when some people believe they can answer on behalf of others because they think they know best. In effect, they diagnosed political problems as pathologies that had a biological basis. At its most extreme, this logic has led to sterilization of the "unfit" or ethnic cleansing. But even family planning could be a form of population control when proponents aimed to plan *other people's* families, demeaning those "targeted" as "acceptors," including tens of millions of poor people who were paid money to agree to sterilization. No less manipulative were those who denied hundreds of millions more people access to contraceptives and abortion because they wanted them to have more babies.

This book is about the most ambitious population control schemes of all, which aimed to remake humanity by controlling the population of the world, typically by reducing the fertility of poor people and poor countries. But all population control projects shared the premise that societies should consciously reproduce themselves by design, even if that meant controlling how people disposed of their own bodies. And all looked at human beings not as individuals but as populations that could be shaped through the combined force of faith and science. That is why nativism, eugenics, pronatalism, and coercive or manipulative forms of "family planning" share a common history, one that can help us understand how they developed, how they diverged, and how the cause of reproductive rights was finally redeemed.

Nowadays those writing on these issues are expected to identify themselves as "pro-life" or "pro-choice." These two camps are locked in confrontation, espousing principles that have come to seem irreconcilable. But this is a history of how some people systematically devalued *both* the sanctity of life *and* the autonomy of the individual. Because I am late to this fight, this book reflects the passion of a convert—not to one camp or the other, but rather to the belief that we must make common cause if we are to stop what may be even more dangerous experiments of the future. By confronting them together, the different sides might recognize in population control something that all of us should reject, and in that way find new ways to renew a dialogue about the meaning of life, and the meaning of freedom.

ABBREVIATIONS

ABCL	American Birth Control League
AES	American Eugenics Society
AIWC	All India Women's Conference
AMA	American Medical Association
BCIIC	Birth Control International Information Center
CMH	Committee on Maternal Health
FAO	Food and Agriculture Organization
FHF	Family Health Foundation
FPA	Family Planning Association
IEC	information-education-communication
IMS	Indian Medical Service
IPPF	International Planned Parenthood Federation
IUD	intrauterine device
IUSIPP	International Union for the Scientific Investigation of Population Problems
NCWC	National Catholic Welfare Conference
Norad	Norwegian Aid Agency
OEO	Office of Economic Opportunity
OPR	Office of Population Research
PAA	Population Association of America
PCC	Population Crisis Committee
PID	pelvic inflammatory disease
PPFA	Planned Parenthood Federation of America

SIDA	Swedish International Development Authority
SSRC	Social Science Research Council
UNDP	UN Development Program
UNESCO	UN Educational, Scientific, and Cultural Organization
UNFPA	UN Fund for Population Activities
USAID	U.S. Agency for International Development
WHO	World Health Organization
WPEC	World Population Emergency Campaign
ZPG	Zero Population Growth

Fatal Misconception

Introduction

HOW BIOLOGY BECAME HISTORY

Thinking about how populations grow and change has long provided a means to imagine the future. While political contests, culture wars, and technological revolutions continually surprise us, the procession of generations appears to provide one of life's few certainties. Most babies born today will, like their parents, mature and bear children of their own. It is but a mathematical exercise, one that UN demographers perform with great regularity, to specify assumptions about fertility, mortality, and migration and summon into view a world of people fifty or a hundred years hence. The numbers can be broken down continent by continent, country by country, such that each one of us feels part of some collective fate. Even when the United Nations is careful to explain that it is providing projections, not predictions, journalists report them as statistical prophecy.

But just imagine, for a moment, the world and its people in a vision far darker than anything the UN has ever anticipated. Imagine a world with an average life expectancy of less than thirty years. Many babies do not live to see their first birthday. Subject to chronic malnutrition, children are vulnerable to disease, grow slowly, and find it harder to learn. Those who survive to adulthood seem stunted, with an average body mass a third smaller than our own. The great majority live off the land. The few who inhabit cities—dwelling with their own waste and drinking water alive with microbes—are even more likely to die early deaths. Altogether, there are not even a billion people living on earth, less than a sixth as many as there are today.

This is not some post-apocalyptic future. It is the world we left behind two hundred years ago. It was then that Thomas Malthus wrote his *Essay on*

1

the Principle of Population, as It Affects the Future Improvement of Society. For Malthus, privation not only cut down human populations but diminished the bodies—even the souls—of those who survived:

> The children are sickly from insufficient food. The rosy flush of health gives place to the pallid cheek and hollow eye of misery. Benevolence yet lingering in a few bosoms makes some faint expiring struggles, till at length self-love resumes his wonted empire and lords it triumphant over the world.[1]

Malthus was actually living among the best-fed and healthiest people in Europe. Averaging five foot six and 136 pounds, British men stood a full three inches taller than their French counterparts, who subsisted on just 1,800 calories a day. Considering the world in which he wrote, it is understandable that Malthus thought it folly to stand in the way of merciless nature when it dispatched the indigent, and risk being dragged down to share the same fate. In the years that followed, he continually revised and extended his essay. The tone of unremitting gloom never lifted. "Misery and the fear of misery" were, for Malthus, "the necessary and inevitable results of the laws of nature in the present stage of man's existence."[2]

In fact, humanity was entering an entirely new stage of existence. In the nineteenth century the peoples of northwestern Europe would experience steady improvement in life expectancy. Rather than dispelling the Malthusian nightmare, the resulting population growth—still slow by recent standards—only made it seem more compelling. It was Malthus who inspired Charles Darwin to argue in 1859 that the struggle for existence could give rise to new, better-adapted species. This gave Darwin's cousin, Francis Galton, the idea that humans could be bred like racehorses. But others worried that it would be the poorest and most fecund examples of humanity who would overrun all the rest.[3]

Instead, the European peoples continued multiplying until, by the early twentieth century, more than a third of humanity hailed from this one overgrown peninsula and swarmed across every other continent. Though they were starting to have fewer children, they lived longer than any of their contemporaries. When Asians, Africans, and Amerindians also began to survive in greater numbers, the growth of world population sharply accelerated. In the last century, humanity has experienced more than twice as great a gain in longevity as in the previous two thousand centuries, and more than four times the growth in population.[4]

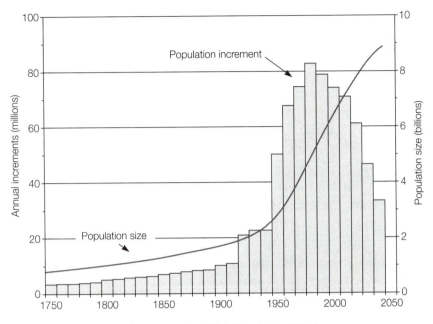

World Population Growth, 1750–2050. U.N. Population Division.

The more world population expanded, the more collapse appeared inevitable. By the 1980s, the earth was gaining some eighty million inhabitants every year. Yet population growth, along with the improvements in nutrition and public health that made it possible, continued nonetheless. At last, at the beginning of a new millennium, both the rate of increase and the annual increments have diminished, even while gains in per capita food consumption and average life expectancy show no sign of ceasing. Malthus considered it unconscionable, but now it has become a commonplace: people the world over have sex without having children.

This history has already transformed humanity. To the extent that sheer numbers of workers and consumers contribute to climate change, it will raise the seas and alter the very air we breathe. It is a signal event, not merely in human history, but in the history of life on earth. Yet it is fundamentally different from any other episode in natural history. For the first time the future of a species—not only its numbers, but its very nature—has become the object of its own design. As people eradicated diseases, regulated migration, and manipulated fertility rates, the quantity and "quality" of human populations—counted and categorized with unprecedented precision—became a subject of scientific experiment and political struggle.

Some scientists argue that we should consider humanity like any other species, because natural selection and "selfish genes" provide the basis for human behavior, whether sexuality, or aggression, or altruism. History will eventually be revealed as nothing but a specialized branch of biology, explaining particular wars or sexual revolutions but not war or sexism as such. Scientists even suggest that they might be able to apply their insights to mold human behavior for the better.[5]

In fact, such arguments only provide further proof that "sociobiologists" have it backward: our biology is becoming a branch of history, subject to human will and human error. Our choice of whether to understand or to ignore this history, especially eugenics and other attempts to improve human populations, will help determine how this happens. If humanity tries to remake itself once again, repeating the errors of the past will prove all the more unforgivable.

Some of the greatest historians agree that changes in population have changed the world. Geoffrey Barraclough once observed that "the demographic revolution of the half-century between 1890 and 1940 was the basic change marking the transition from one era of history to another." Population growth in the half century that followed was, if anything, even more revolutionary. For Eric Hobsbawm, it was so spectacular that it constituted perhaps the single most profound development of this "age of extremes." "The global growth of population," William H. McNeill agreed, "is the most fundamental and pervasive disturber of human society in modern times." Even Francis Fukuyama has conceded that the end of the Cold War did not, after all, signify the end of history, because biotechnological advances may portend a new age of social engineering.[6]

Yet few scholars of world politics have paid any serious attention, and most still devote their careers to studying territorial and ideological conflicts. Has population change been so much less significant? The worldwide influenza pandemic that started in the waning days of World War I killed more people than all the fighting on all the fronts in the preceding four years. Conversely, because of improved public health and nutrition, World War II scarcely slowed the relentless growth in world population. And surely nothing was so important in the history of the Cold War as what did *not* happen: global nuclear war. Such a war might have meant tens, if not hundreds, of millions of deaths, the emptying of cities—especially in Europe and North America—and refugee flows from the Northern to the Southern Hemisphere. Yet the Cold War era witnessed changes that, while unfolding over decades, were no less profound: World population more than doubled; the proportion of North Americans and Europeans shrank by more than a third; cities in Asia, Africa, and Latin America became the largest in the world; and migration flows from South to North began to make "Anglos" a minority in the United States and Islam the second-largest religion in France. Indeed, leaders of the great conflicts of the last century like Hitler, Eisenhower, and Mao defined victory or defeat not just in terms of territory gained or lost, but in the size or "quality" of population that remained on each side.

Struggles over population should concern anyone who wants to understand the changing nature of international relations. For centuries, state sovereignty has provided the organizing principle of world politics. If a government can exercise exclusive authority over a specified territory, it is usually because it claims to represent the nation that inhabits it—regardless of whether the people concerned come from different places, speak different languages, and have different loyalties. As a matter of law, the borders of states and nations are assumed to be coterminous, which is why the two terms have become synonymous. But a moment's reflection makes apparent how few countries conform to the ideal type of "nation-state," and how often boundary disputes, cross-border migration, and claims to minority rights have roiled international relations. At the same time, the increasingly free flows of capital, goods, ideas, and elites have made the exclusive control of territory both more difficult and less decisive.

Controlling how a nation reproduces itself has provided an alternative approach to policing a nation's borders, one measured in time as much as space. The regulation of public health, reproduction, and migration—what this book calls the politics of population—created an arena in which people fought over such questions as, Who shall inherit America? Barring entry to "beaten men from beaten races," promoting the fertility of the native-born, and sterilizing the "unfit" made a nation seem like more than just a political construct, but a biological reality, one that could be purified, enlarged, or even "improved." The most ambitious example of population control, one that became a global campaign that encompassed most of humanity, aimed to convince or coerce people to plan smaller families. Eventually it took on a momentum all its own. But part of this campaign's initial appeal becomes apparent when we realize that all of the governments that first adopted family planning programs in the 1950s and 1960s had contentious borders, including India, Pakistan, South Korea, Taiwan, Turkey, and Honduras.

States have struggled to control populations and call them nations, but fertility, mortality, and migration obviously predate the nineteenth-century invention of the nation-state. In what may be the first proposal for a policy to control world population, Benjamin Franklin noted in 1751 how rapidly the colonies of North America were growing year by year, which is what gave Malthus the idea that such growth could not possibly be sustainable. Franklin was already worried that if the next wave of settlers proved more industrious and frugal, "they will gradually eat the Natives out." He was especially concerned about "swarthy" German immigrants, but derived a more global conclusion:

> The Number of purely white People in the World is proportionably very small. . . . I could wish their Numbers were increased. And while we are, as I may call it, *Scouring* our Planet, by *clearing America* of Woods, and so making this Side of our Globe reflect a brighter Light to the Eyes of Inhabitants in *Mars* or *Venus*, why should we, in the Sight of Superior Beings, darken its People? Why increase the Sons of *Africa*, by planting them in *America*, where we have so fair an Opportunity, by excluding all Blacks and Tawneys, of increasing the lovely White and Red? But perhaps I am partial to the Complexion of my Country, for such Kind of Partiality is natural to Mankind.

Those concerned about population trends would often adopt this planetary perspective—seeing the earth as a whole—even when, as in this case, it was to imagine how it should be divided between different races. And not all were as forthright as Franklin in admitting their "partiality" in pursuing such schemes. He was only the first of many who had local and parochial reasons to begin "thinking globally."[7]

Population change, and the recognition that it might be possible to direct it, thus prompted a more profound question: Who shall inherit the earth? State sovereignty seemed to stand in the way of meeting existential threats, such as degeneration, global famine, and uncontrolled migration. These fears provided new reasons and new ways to divide nations and divide the world, inciting ethnic conflict and raising the specter of racial, religious, or class war on a global scale. But this challenge to the principle of nation-state sovereignty—and especially the concern that the earth could not support swelling human numbers—also inspired visions of a "global family." Scientists and activists organized across borders to press for common norms of reproductive behavior. International and nongovernmental organizations spearheaded a worldwide campaign to reduce fertility. Together they created a new kind of global governance, in which proponents tried to control the population of the world without having to answer to anyone in particular.

Even now, long after the demise of population control as an organized movement, fear of the fertility and mobility of particular groups continues to spark ethnic strife. Demographic statistics are fodder for conflict among Americans worried about Hispanics, Europeans worried about Muslims, and Israelis worried about Palestinians, to name just a few. At the same time, concerns that we have grown beyond the earth's carrying capacity spur demands for new global norms and institutions. Ethnic conflict and global governance present the greatest challenges to an inter*national* system based on the principle of state sovereignty. A history of the population control movement can therefore show how local political fragmentation and the impulse to organize globally are intimately related, and thus help us understand a process as powerful as it is paradoxical: our world is both coming together and coming apart.

But where do we begin? The idea of population control is at least as ancient as Plato's *Republic,* which described how a "Guardian" class could be bred to rule, the unfit left to die, and everyone sold the same myth that political inequality reflected the natural order of things. It has been argued that some kind of population policy is common to every culture. Most have been pro-natalist, in that they taught people by means more or less subtle—from tax breaks to witch hunts—to "be fruitful and multiply."

The idea of controlling the population of the world, by contrast, is a modern phenomenon. By the end of the nineteenth century, as rival empires seized the few remaining regions that had withstood European settlers, observers began to see trends in fertility, mortality, and migration as interconnected. Concerns about "race suicide," pandemics, and "the yellow peril" inspired the first concerted effort to regulate migration worldwide, which by the start of the twentieth century had begun to contain Asians within their own continent. Immigration restrictions, in turn, provided models for eugenicists in the 1920s and 1930s who wanted to regulate fertility among the "fit" and the "unfit." Campaigns to improve public health, by contrast, promised to regenerate whole nations, especially imperial powers that faced population decline. But when, by the 1940s and 1950s, new public health techniques had begun to bring dramatic gains in life expectancy to colonized peoples as well, population growth began to appear like a global crisis. Advocates of eugenics and birth control consciously emulated public health campaigns when they developed a common program for "family planning." It appealed to leaders of newly independent nations as well as international and nongovernmental organizations as a means to achieve "modernization" in a single generation.

Very capable scholars have followed different threads in this history. If we take a step back, we can see how they were intertwined and where they ultimately led: a worldwide movement to plan population growth, which culminated in massive campaigns that swept across East and South Asia, Africa, and the Americas from the 1960s to the 1980s. This movement was an arena as much as an agenda. Feminists, environmentalists, and a host of others all demanded a place within it, even while pushing in different directions. But all together aimed to change the way people considered their sexuality, their families, their place in the world, and their collective future.

Authors have filled libraries with books about the most sinister episodes in the larger, longer history of population control, such as forced migrations and genocide. Certain of these are so notorious that it is difficult to discuss some of the aforementioned connections without appearing to indict everyone involved, if only for guilt by association. The very word *eugenics,* for instance, immediately summons to mind the horror of the Holocaust. Even sophisticated population researchers nowadays assume it is synonymous with racism. But the idea of improving the genetic makeup of humankind counted adherents all over the world, including everyone from W. E. B. Du Bois to John Maynard Keynes. Eugenics was invoked to justify everything from free day care to forced sterilization. Similarly, though seldom used today, the term *population control* for most proponents signified a way to lift people out of poverty, even to save the earth. Many were no less well intentioned or well received than those who nowadays speak of human capital, sustainable development, and the quality of life.

The challenge for historians—and the opportunity for people who want to learn from this history—is to discover how, specifically, such protean concepts evolve into norms, practices, and institutions that empower people or manipulate them, enrich or impoverish, give life or take it away, sometimes all at the same time. This means looking past the slogans, as important as slogans are in telling us which ideas resonate. It requires investigating how money was raised and how it was spent, who was hired and how they were trained, how programs were supposed to work and why they usually failed. When people set out to save the world, the devil is in the details.

Along with tracing the parallels and connections between different kinds of population control, we need to explain why they spread and struck root in so many places. Most histories have been national in nature, even when they compare two or more countries. They have shown how ideas like eugenics and family planning could mean different things in different contexts, and meant very little if they were not made relevant to local circumstances. But even while acknowledging this diversity, we need to explain why divergent tendencies emerged almost simultaneously and how they came together in a movement to control the population of the world. This means exploring commonalities among population activists, the ways they communicated and mobilized across borders, and the key features in the global context that facilitated their spread.

If we understand population control as a movement, one that inherited different tendencies but acquired its own direction and momentum, a new and surprising history starts to unfold. We begin to notice, for instance, that neo-Malthusians, eugenicists, pro-natalists, and nativists all began to organize at the same time—a time in which the world had begun to seem small while population trends appeared out of control. We can recognize how cordoning off settlement colonies constituted the first policy that was expected to shape world population. We discern pro-natalist campaigns of the 1920s and 1930s as the home front of an undeclared war between new and old empires over lebensraum, a front in which women's lives were being sacrificed years before the official outbreak of hostilities. We can also trace the diverging trajectories of these tendencies, and ponder how they might have developed differently—for instance, if eugenicists had managed to work together worldwide, or if a nascent sex reform movement had not been strangled in occupied Europe during World War II.

Fortuitous events helped Margaret Sanger, the foremost proponent of birth control as panacea for the world's problems, to move forward with a consolidated program of family planning. It appeared to provide a global solution to a crisis in colonial rule occurring from Nairobi to New Delhi, from Algiers to Hanoi. Rather than addressing political and economic inequality, imperial administrators blamed divergent population trends, even if they preferred to leave it to international and nongovernmental organizations to pick up the pieces. A transnational network of population experts took up where empires left off. By not specifying who would actually do the planning, they could adapt family planning to different conditions. For rich people and rich countries, it meant helping couples plan larger families, thus reversing the fertility declines that appeared to portend both the "Twilight of Parenthood" and the "Twilight of the West." For poor people and poor countries, it meant engineering incentives and disincentives to induce people to stop having so many babies and start being "modern."

A global history can also show that what we thought we knew about population control is not really true. For instance, eugenics did not end with what is now understood as the Holocaust, the most notorious episode of all. Eugenic sterilizations continued across several American states, Scandinavia, Japan, India, and China. What most contemporaries interpreted as the Nazis' persecution of all their opponents—failing to recognize how Jews were targeted for total elimination—did not seem relevant to those intent on improving "the global family" through both voluntarist and coercive forms of family planning (or what some referred to privately as "crypto-eugenics"). In the 1960s, as some began to argue that humanity's survival depended on sterilization camps, a belated recognition of this history reminded people that population control had a dubious provenance and dangerous potential.

Examining population control as a global movement can both prompt and answer questions that might not even occur to us otherwise. For instance, why did the international diplomacy of population control feature such strange coalitions, in which communist and Catholic countries were arrayed opposite socialist and capitalist states? Why were population control programs so similar in countries that were otherwise so different—with the same technologies and techniques deployed almost simultaneously everywhere from Taiwan to India to Tunisia to Haiti—all too often with terrible results? And why did organizations that proclaimed family planning as a human right fail to oppose increasingly coercive policies in India and China? Why, instead, did they provide friendly advice and support, and defend them to all the world?

This is the first global history of population control, but not because it covers every country. To begin with, some counted more than others. When people first began trying to remake humankind, Britain and France ruled over most of the world's population. Even their domestic debates therefore had wider implications. Fascist Italy, Nazi Germany, and the USSR were particularly ruthless in controlling population, inspiring others to develop more appealing alternatives. Sweden showed how a welfare state could instead promote family planning both at home and abroad. Japan was very nearly the first country to adopt it as a means to reduce population growth during the Allied occupation, and later became second only to the United States as a source of international assistance.

But this account gives far and away the most attention to Americans. They were the first to pursue policies intended to shape world population. They played a leading role in institutionalizing both the science of demography and the political strategy of family planning, at the same time mentoring protégés around the world. They were disproportionately represented in the international and nongovernmental organizations that created standardized population control programs, which were largely funded by public and private sources in the United States. Conversely, American Catholics were particularly influential within the Church and spearheaded its campaign to keep birth control and abortion illegal everywhere.

Intramural struggles among Americans proved crucial in Puerto Rico, a proving ground for both the birth control pill and state-supported sterilization, and Japan, where occupation officials and local proponents managed to legalize abortion, but left women with few other options. In Taiwan and South Korea, American consultants developed programs using incentive payments to motivate medical workers to insert IUDs in more women, a technique that caused untold misery when exported to countries with far too few clinics to treat all of those who suffered side effects. Latin America, on the other hand, was resistant to such efforts, in part because of transnational cooperation among clerical and lay Catholic elites. Because they were so effective in limiting access to birth control and abortion, this region continues to have some of the highest rates of unsafe abortion in the world.

No other country attempted to control its fertility over such a long period, or with such wide-ranging influence, as India. India attracted the first sustained effort by British and American birth control activists to establish clinics abroad. It invited the first United Nations advisory missions in demography and family planning. It hosted the founding conference of the International Planned Parenthood Federation (IPPF). India was the first country to adopt policies to reduce fertility. Pakistan followed shortly. Together with Bangladesh and Sri Lanka, they were testing grounds for new contraceptive techniques and absorbed the lion's share of international assistance. Indeed, the phrase "countries like India" became shorthand for poor countries with high fertility. In contemporary debates China looms larger, even though it owes comparatively little to foreign aid. The pro-life movement made the one-child policy a poster child to oppose family planning everywhere. But only by tracing the historical trajectory of the population control movement—driven and defined by a series of campaigns across South Asia during the 1960s and 1970s—can we understand why organizations like the IPPF and the UN Fund for Population Activities (UNFPA) landed in China, and why they found it so difficult to escape.

To inventory population control country-by-country helps us know where the story begins and ends, but it is not the best way to narrate the rise and demise of a global movement. How, for instance, would one assess the role of the Vatican? Its influence obviously had little to do with a few hundred acres of Roman real estate, but also varied—not just from country to country, but from diocese to diocese. When Pius XI condemned Malthusianism and eugenics in 1930, insisting that the patriarchal family was more sacred than the state, he, like every pope who followed him, asserted an authority that superseded state sovereignty. Many of the leaders of the population control movement, for their part, did not identify themselves primarily in national terms. Margaret Sanger said that she "never had a country," and devoted her life to the cause. For eugenicists as well, beginning with Galton, population control was a secular religion, a faith handed down to save humanity from degeneration.[8]

All of these leaders tended to treat governments as means to an end, and claimed to represent constituencies that would not fit in a national framework, including universal sisterhood, future generations, or the community of the faithful both living and dead (including "infants hidden in the mother's womb," as Pius XI insisted).[9] Networks of scientists and activists set the agenda for the global population control campaign of the 1960s and 1970s, nongovernmental organizations pioneered the most influential projects, and it was all carried forward under the auspices of the United Nations. Anyone who has spent more than a few hours reading the papers of the leading figures will notice that these were people on the move, writing on postcards and stationery from steamships, hotels, and airlines.

Not all population controllers were cosmopolitans. And if the history they made was global in scope, it was experienced by particular people in particular places. But like more and more of the history of our times, it was not organized and divided among different countries, like separate containers to be poured out one at a time. If one succeeded in covering every country, a study organized in this way would miss much of the story, and the most essential part of it. Even to explain the policies of governments like those of the United States and India, it can be a mistake to consider them as two cases, "the U.S." and "India," much less "Washington" and "Delhi." As federal systems, it sometimes makes more sense to focus on connections and parallels between North Carolina and Madras, and the mediating role of groups like the Association for Voluntary Sterilization. Similarly, scientists and fund-raisers promised a global solution to population problems, with contraceptive technologies and techniques aimed at both slum dwellers and subsistence farmers. The resulting backlash, articulated in terms of Black Power, the Bihar movement, a new international economic order, and much more, was heard around the world.

As messy as it may seem, we must recognize that population change and struggles to control it are interconnected and transnational phenomena, and should be studied as such. The decisions couples make can unmake nations, and historically governments have had little say in the matter. The mere possibility of shaping demographic trends encouraged people to conceive of alternative ways of organizing politics, whether in terms of religions, races, generations, or civilizations. Now it can help us understand the changing nature of international relations and assess more critically the prospects of global governance.

To come to grips with both the history and future of population control, we need to recognize that it is made up of networks—networks of ideas, of individuals, and of institutions—which were organized to control humankind. A narrative about networks places certain demands on readers, because they will not find one or a few protagonists who advance all the action, or a center from which it emanates, or even a fixed target that focused all their efforts. After all, the objects of population control—whether migrants, or persecuted minorities, or working mothers—tended to be moving targets. What readers will see are new ideas emerging, catching fire, and sparking political movements. People organize across borders to advance their agendas, convening conferences, raising money, and founding new institutions. Sometimes these movements lose momentum and die out, only to reemerge when conditions become more favorable, giving rise to new ideas, new policies, and new programs. The often unsavory work of discriminating against unwanted immigrants or inducing people to have more or fewer children, in turn, provoked schisms and countermobilizations, discrediting population control and driving it underground. Networks, by

their very nature, are resilient. This helps explain why population controllers persisted despite the no-less-formidable opposition they aroused.

It would have been easier—certainly for the author—to pull all these
strands together into a neat package and present the history of population
control as a global conspiracy. At the least, some readers might have preferred a tidier story limited to the campaign to stop world population
growth. Both critics on the left and "pro-life" forces on the right have
long attacked population control as something that white, wealthy elites—
especially in the United States—perpetrated on the rest of the world. The
evidence they offer for such claims tends to be fragmentary, even if it is now
all over the Internet (as a search for the terms *eugenics* and *birth control*
instantly reveals). Sometimes this evidence has been manipulated to serve
an ideological attack. In no case has anyone explored all the archives that
might reveal such a conspiracy—especially those of the Population Council, the International Planned Parenthood Federation, the Ford and Rockefeller foundations, the key UN agencies, the World Bank, and so on.

This book can demonstrate that some of the leading protagonists did, in
fact, act in underhanded ways, pretending their organizations were dedicated to one agenda while secretly harboring another. Some readers might
jump to the conclusion that nothing has changed—that when these same
institutions now argue for reproductive rights and health, they really mean
population control. If so, it is hard to explain why they would have granted
the author remarkably unfettered access to their archives. This openness to
outside scrutiny suggests that the people who now lead these institutions
have nothing to hide.

The Catholic Church, on the other hand, has been more reluctant to
provide meaningful access to its documentary record. It was the most powerful force opposed to birth control and abortion worldwide, but it was not
necessarily opposed to population control per se—especially when it merely
required denying more than half the world's population control over their
own bodies. Popes, bishops, and priests fought some forms of control only
to defend others, leaving women to their own devices in situations that
were manifestly unjust. Men reserved unto themselves the power to decide
when life begins and which methods of family planning are "natural"—as
it turned out, a method that required permission from a priest, agreement
of husbands, strict discipline and record keeping, and still failed most of
those who tried it. "Natural law" as defined by popes and enforced by governments had the effect and sometimes the intent of making people breed.

But pro-natalist policies invariably, and predictably, drove contraception and abortion underground, where its toll became incalculable. Seemingly opposed forms of population control thus ran parallel, leading to suffering and death for those caught in the middle.

No matter what may still lie smoldering in the archives, population control will never be explicable if it is reduced to a conspiracy. Those who wanted to control world population could not have gotten anywhere if they did not find allies abroad. Such far-flung movements encompassed people inspired by different ideas, who then pulled in different directions. Ideas can take on lives of their own, and people are even more unpredictable. Individuals changed this history when they changed their minds and decided to work for different goals. Most importantly, the fate of population control was ultimately decided by people who would not be controlled. They included immigrants who subverted institutionalized racism, litigants who challenged compulsory sterilization, officials who risked their careers by declining to coerce others, and countless millions who refused to reproduce according to some global—or divine—plan.

Rather than a conspiracy theory, this book presents a cautionary tale. It is a story about the future, and not just the past. It therefore takes the form of a narrative unfolding over time, including very recent times. It describes the rise of a movement that sought to remake humanity, the reaction of those who fought to preserve patriarchy, and the victory won for the reproductive rights of both women and men—a victory, alas, Pyrrhic and incomplete, after so many compromises, and too many sacrifices. It asks readers to empathize, if not sympathize, with the protagonists as they experience triumphs and disappointments. They developed an appealing set of ideas, rallied "the great and the good," and mobilized a phalanx of supporting institutions—including the world's richest foundations, one of the largest nongovernmental organizations, the biggest foreign aid programs, the World Bank, and a purpose-built UN agency. They were opposed by an even more formidable apparatus, the Roman Catholic Church, which rallied Christian and Muslim conservatives and answers only to God.

A narrative makes us think harder about how all of these institutions and individuals were interconnected and changed over time. And because key decision-makers on both sides made decentralization a deliberate strategy and sometimes covered their tracks, a narrative is required to reveal their identity and uncover the critical turning points. It can show how, upon meeting determined resistance, both sides faltered, changed direction, and were finally transformed. Most supporters of family planning are now troubled, if not puzzled, at the mere mention of "population control." Most Catholics defy church teaching on the subject and think nothing of it. Few people are even aware of the struggles and sacrifices that were required to secure reproductive rights, such that most of us now believe them to be inalienable.

World population growth is slowing, and the heroic age of population control appears to be over, at least for now. People now worry that there are too many pensioners in the West, too few girls in the East, and that too much information is provided to prospective parents, who may be tempted to redesign their offspring. We can therefore begin to see the end of a remarkable story, which is another reason it can now be told. The history of how population control first began, and took off in such frightening directions, can help us prepare for what may be the even more ominous history to come.

NOTE

In the text, names have been rendered according to North American usage—that is, surname last, so Shidzue Ishimoto rather than Ishimoto Shidzue. The exceptions are heads of state so well known that reversing the order would be baffling rather than helpful, such as Mao Zedong. Place names are from the nomenclature of the era, thus Ceylon until 1972, and thereafter Sri Lanka. Whenever the value of a sum of money is adjusted for inflation, both figures are provided, with the latter denoted as "today's money," circa 2006, according to the calculator available at www.measuringworth.com.

Some readers might wonder why many other terms and values contained in this book have not been translated and updated, beginning with population control. But the term *family planning,* in the sense of promoting reproductive rights, means the opposite of *population control.* Some of the protagonists used the two terms interchangeably. When it is clear that they were mainly concerned with achieving population targets, this book does not refer to it as family planning. To do so would confuse the distinction that the great majority of people in the field now strive to uphold.

More generally, this book must include many terms employed by population controllers to convey their ideas and work. Describing human beings as "unfit," referring to them as "acceptors," or counting "IUDs inserted" will strike some readers as offensive, and it should. But it faithfully reflects how, for many decades, population policies and programs were designed and conducted—so much so that using "scare quotes" in every instance would become distracting. If this book is properly understood as a critique of population control, the scare quotes should be assumed.

1

POPULATIONS OUT OF CONTROL

The population of the world is no more and no less than the sum total of billions of acts large and small that together create the conditions of life and death. No one can say which one in particular marked the beginning of what would one day become a globe-spanning movement to shape demographic trends. But the summer of 1877 was more momentous than most. It was a time in which the world's population seemed out of control, with appalling famines in India and China, great waves of migration across the Atlantic and Asia, and the beginning of large-scale, organized violence against Chinese immigrants in the United States. It was also when the world first heard of a remarkable family that had grown to some twelve hundred people in seven generations. Through criminality, pauperism, and disease, the pseudonymous Jukes were alleged to have cost the American people over $20 million in today's dollars—"to say nothing of the money paid for whiskey," as noted by the *New York Times*.[1]

No less remarkable, in its way, was a drama that began to unfold in the Queen's Bench Court in London on June 18, 1877. After three months of delays, two freethinkers and socialists, Charles Bradlaugh and Annie Besant, rose in their own defense against the charge of publishing an obscene pamphlet. It was to be a test case for a novel cause: the need to help people have sex without having children. The Home Office was reluctant to provide them with a platform. The pamphlet in question, Charles Knowlton's *Fruits of Philosophy*, had circulated quietly in small editions for decades. But by advertising the pamphlet together with their defiance of the state, Bradlaugh and Besant sold 133,000 copies by the time they went on trial.[2]

The public was clearly most interested in Knowlton's description of sexual anatomy and contraceptive technique, including condoms, sponges, douching, and withdrawal. But Besant rested her defense on their social value. And it was Besant, 29 years old, beautiful and brilliant, who would attract national and even international attention. Citing Malthus and the prospect of world population doubling and doubling again, Knowlton had warned that "the time will come when the earth cannot support its inhabitants." Besant brought him up to date, detailing with evidence from Dickensian London the danger that Britain was becoming as overpopulated as China and would suffer the same "barbarous means" of limiting further growth: "war, famine, disease, misery, starvation, overcrowding, preventable disease, infanticide, baby-farming, and all other horrors." Besant warned that if Britons were to prevent such "natural" or "positive" checks on population, as they surely must, then they would confront a new danger. Citing Darwin and Galton, she described how interfering with nature without scientifically controlling it would lead to the proliferation of the physically, mentally, and morally weakest members of humanity. It was a "crime" against society to bring into the world children for whom one could not provide.[3]

Besant impressed the all-male jury but did not persuade them. They found the two defendants guilty, though all of them seemed reluctant to do so and some made a donation for the appeal. In the meantime, the first reports of the famines in India and China began to appear in London newspapers. Besant used her next court appearance as an opportunity to renew her calls for population control. She insisted that the famines were "caused entirely by over-population." Besant was finally released together with Bradlaugh because of a legal technicality. She went on to publish a new pamphlet of her own, *The Law of Population*. It sold hundreds of thousands of copies and was translated into German, French, Italian, and Dutch.[4]

Meanwhile, something quite unexpected had started to happen. All across England, the birth rate began to fall. The decline was most sudden and precipitous among professional couples and their domestic servants. But people of every social class and occupation began striving to limit the size of their families, albeit with different degrees of determination and success. And in one area after another, 1877 marked the turning point. It is impossible to prove that the publicity surrounding the trial was a precipitating cause, but the statistics leave no doubt that a profound change was occurring in the most intimate history of the era. Men and women were talking about sex in ways that would have astonished their parents. Succeeding generations would know that their very existence had been deliberated and decided upon by mortals like themselves.[5]

Although it was nowhere more rapid, the decline in the birth rate of the European peoples had not actually begun in England. A defense witness, Charles Drysdale, cited evidence for widespread fertility limitation in France. As with French cuisine, the genius of French birth control was making do with seasonal ingredients that were readily at hand, including prolonged nursing, coitus interruptus, and every conceivable alternative to conventional intercourse—even abstinence. Farmers feared that otherwise estate laws mandating that property be distributed equally would hopelessly fragment their fields. When his legions marched into Moscow, Napoleon could recruit from a population as large as any in Europe, perhaps even Russia's. The generation that followed his nephew to defeat at Sedan was already outnumbered by their German foes.[6]

Yet aside from the French exception, and a few forerunners among the nobility, it was these last two decades of the nineteenth century that saw the ideal and reality of smaller families first sweep across Europe, spurred on by the development of new or improved diaphragms, condoms, and abortion techniques. Marital fertility rates in Denmark, Norway, Finland, Sweden, Germany, Belgium, Britain, Switzerland, the Netherlands, and Austria had all peaked by 1880, and once they began to fall they never recovered. Indeed, when plotted on a graph, they appear to have plunged off a cliff, with the countries of southern and eastern Europe following like lemmings.[7]

At virtually the same time as European fertility began to fall, there was a "fast, massive, and irreversible" continent-wide decline in infant mortality. The countries with the worst rates witnessed the most marked improvement. In 1867, nearly one in four Germans perished before their first birthday, but the survivors saw less than a tenth of their grandchildren suffer the same fate. Austria and the Netherlands had similarly steep declines, whereas the improvement in infants' health came earlier and more gradually in Norway and Sweden. The mere fact that mothers had fewer children and longer breaks between them reduced risks for all. But middle- and upper-class women with smaller families and the domestic help to care for them were also leading campaigns to improve public health and hygiene. By the 1890s, an estimated half million British women were volunteering for work in social welfare.[8]

Paying closer attention to infants' health, nutrition, and schooling generally meant parents had to pay more for their upbringing. At the same time new labor laws barred children from the workplace. These laws, together with regulations that purported to protect mothers from the hazards of working outside the home, both reflected and reinforced new social norms promoting a nurturing childhood and romantic marriage. In 1891 Pope Leo XIII provided his imprimatur in the landmark encyclical *Rerum Novarum,* which endorsed men's demand for a "bread-winner's wage" to better rule their families. But if this domestic sphere was to form a virtuous circle, most people began to realize that it would have to become smaller. Parents who were more hopeful that their offspring would survive and prosper, and sacrificed to ensure that they would, were also more inclined to wonder, and worry, about the world their children would inherit twenty or thirty years hence. This created a constituency for social movements that sought to shape long-term demographic trends and not simply trust that God would provide.[9]

Of course, many Europeans continued to have large families, and many still struggled to survive from day to day. For women, bringing new life into the world represented a significant risk of death—roughly 1 in 200 births were fatal in England in the 1880s, and each succeeding pregnancy worsened the odds. While rates of infant mortality were generally improving, they remained appallingly high. Until the turn of the century, 1 in 10 babies perished before their first birthday among the English upper classes. Infant mortality was 50 percent higher in working-class homes. Young mothers who did not go to work sometimes did not eat, but to discontinue breast-feeding deprived babies of their hard-won immunity to such diseases as tuberculosis, typhoid, and smallpox. Leaving infants at home in crowded tenements could be lethal during their first vulnerable months. The bacillus that caused TB spread rapidly through the coughing, vomiting, and convulsions of those struggling to survive. Typhoid was marked by telltale rose spots on the belly but also, less subtly, high fever, vomiting, and diarrhea. Preventing deadly dehydration required drinking from the same filthy water or milk supplies that had carried the pathogen in the first place. As one observer indelicately put it, "Every case of typhoid fever means a short circuit between the alvine discharges of one person and the mouth of another." Smallpox, by contrast, could spread through the air, though more commonly through contact with the hideous lesions it left across the face and limbs, scarring survivors ever afterward.[10]

Yet the same twenty years that witnessed rapid and irreversible declines in fertility were also marked by revolutionary scientific advances that raised the prospect of eradicating epidemic diseases. In the early 1880s researchers managed to isolate the bacilli that caused tuberculosis and typhoid. By the end of the decade, the first TB case notification systems had been created in Edinburgh and Manchester, enabling authorities to isolate the afflicted and thus halt new outbreaks. The triumph of "germ theory" did not by itself immediately save lives, and improvements in nutrition, housing, and sanitation were more important in mortality declines. But following the trails of pathogens, rather than scattershot attacks on "miasmas" or bad habits, did make possible the most dramatic advances in the decades to come.[11]

In the 1930s and 1940s, when new public health techniques brought even more rapid declines in mortality to the very poorest countries, social scientists pored over European demographic statistics in the hope of discovering the conditions that caused a corresponding decline in fertility. They developed a theory called "the demographic transition," in which population trends were posited to advance in three stages. First, couples had many children and many of them died, producing little, if any, growth. Urbanization and industrialization reduced mortality without, at first, affecting fertility, such that population soared. Population stabilized once again as people recognized the advantages in reducing fertility. At first this model appeared to describe and even explain both the intimate and the political-economic history of a whole continent, providing "development" lessons for all humanity.[12]

But when demographers interested in exporting "modernization" began to study more closely how it actually advanced in individual provinces and smaller communities, they came to realize that there had not been one demographic transition, but many. For instance, in half the districts of Germany, parents had fewer babies *before* their prospects for survival improved. England and Belgium showed the same, seemingly inverse, relationship, while analyses of local data on fertility and infant mortality in many other countries revealed no correlation at all. Similarly, in northern and central Italy, families tended to be smaller in large towns and cities compared with the countryside—though whether it was because of more constricted quarters, increased opportunities to learn about avoiding pregnancy, or some other factor altogether is unclear even now. The same is true of Lisbon and its environs, but not in the northern part of Portugal, or southern Italy for that matter. In fact, in most of Europe changes in reproductive behavior came nearly simultaneously in urban and rural areas, such that farmers had families more similar to those of their cousins in a nearby city than to those of farmers in a different region. This raises the question of whether rural or urban living really made a difference.[13]

Other demographers have devised models less dependent on simplistic notions of development and modernization. When many infants do not survive to adolescence, and those who do survive can pay their own way, it is sensible for couples to have many children, especially when they depend on their offspring for support in old age. When the "wealth flow" is downward rather than upward—with college savings accounts being only the most onerous example—parents may see one or two children as more than sufficient. Though harder to quantify, "cultural diffusion" through social networks could convey this message either explicitly, as in the Besant-Bradlaugh trial, or by example. The only factor that has consistently and convincingly been found to correlate with lower fertility is increasing women's education. The birth rate varies among societies with near-universal school enrollment, but within a much narrower range.[14]

While trends in mortality and fertility progressed in more ways than one, transition theory did convey a larger historical truth. The most profound changes generally happen slowly, and this is especially true of the history of populations. The continent-wide fall in fertility eventually slowed growth, but the momentum that had built up earlier, together with greater life expectancy and higher rates of marriage, meant it would continue for decades. Initially, improved health can actually increase fertility, since healthier mothers are more likely to conceive and to survive childbirth. Even when a generation of young people resolved to have smaller families, so many had been born when fertility levels were still high and mortality was falling that they could not help but sire an unprecedentedly large new cohort. And the exceptions to these trends are counted in the millions. Just a handful could make all the difference. After all, even profound historical changes sometimes have small beginnings.

In 1879 an Irish Catholic stonemason named Michael Higgins and his long-suffering wife conceived their sixth child, almost certainly unplanned, and decided to name her Margaret. Carving headstones provided steady business in Corning, New York, but Higgins drank and talked too much to be a good businessman. Anne Higgins, frequently ill, had five more children and wasted away by the age of 50. Her daughter, Margaret Sanger, would dedicate her first book to this woman who, unusual even in her own time, "gave birth to eleven living children."[15]

Elise Ottesen-Jensen was born seven years after Sanger in Norway. Her father, a Lutheran minister, was posted well north of the arctic circle. Her mother lost six of their first nine children in infancy, and went on to have five more. A younger sister was sent away after becoming pregnant, scarcely understanding her condition, and committed suicide after being forced to give up the baby. Ottesen-Jensen would go on to become a pioneer of sex education. Marie Stopes, on the other hand, the leading birth control advocate in Britain, was one of only two daughters born to a bookish, upper-middle-class couple. One of the first Englishwomen to earn a doctorate in paleobotany, she claimed to still be a virgin at age 34 because no one had ever explained how to consummate her marriage, and her husband proved incapable. Birth control pioneers of Asia like Baroness Shidzué Ishimoto and Lady Rama Rau had more privileged upbringings. Their elite status enabled them to travel to Europe and North America and join with Sanger, Ottesen-Jensen, and Stopes to lead a movement that would leave no family, no matter how rich or poor, untouched or indifferent.

These exceptional women broke all the rules and made many new ones—rules to regulate reproduction, rules to reform relations between the sexes, rules that reshaped attitudes about the very meaning, purpose, and value of human life. Yet in the last decades of the nineteenth century, long before these women entered public life, trends in fertility, mortality, and migration had begun to provoke an intellectual and political reaction that aimed to control people as "populations." Any woman—or man—who imagined a movement that might instead liberate and empower individuals by giving them control of their own bodies would have to contend with these more conservative forces, or compromise with them.

We know about falling rates of fertility and mortality precisely because governments of the day were concerned to monitor these trends more closely through birth and death registration, census taking, and statistical analysis. States were intent on taking the measure of the nation because nationality was beginning to entail a whole panoply of rights and duties. Between 1872 and 1885, nearly all of the European states plus Japan adopted military conscription. Comparing the number of births each year now provided a means to estimate the future correlation of forces. Women who petitioned for recognition of their rights in this period tended to argue that it was a prerequisite for fulfilling their duties as mothers of the nation. In Paris, looking after expectant mothers came to be seen as a matter of national security. In 1893, France began to provide free medical care for childbirth. Germany was already mandating maternity leave with pay.[16]

Before, states made little distinction among those residing on its territory. But institutionalizing citizenship provided protection not only from rival states, but also from the mobile poor—who might otherwise demand and thereby undermine the privileges citizenship provided, and thus the solidarity it ensured. With their increasing responsibilities in social welfare, taking on and further expanding functions once fulfilled by municipalities and charities, states took on the corollary role of registering aliens and expelling those considered a public charge or public danger. This process accelerated in the last decades of the nineteenth century, which witnessed an upsurge in the number of poor people migrating across Europe and with it growing concern about crime and infectious disease. In this new age of "official nationalisms," with governments promoting a single identity for all inhabitants, the boundaries of states and nations were expected to be one and the same.[17] As Alphonse Bertillon wrote in 1878, every country needed a "social accounting." For Bertillon, "a Nation is similar to a factory":

> Whether it is people or things that are produced, the keeping of books is subject to the same rules and obligations: One must record exactly what *enters,* what *exits,* establish the *balance* of this two-way movement and *verify,* according to the *state* of the register and the products in the store (inventory or counting), the accuracy of the account of *movements* (what comes in and what goes out).[18]

Napoleon had once mocked England as "a nation of shopkeepers." Ironically, it was now a Frenchman who first imagined nation-building as a form of inventory control.

Defining and delimiting the membership of the nation was only the beginning of a process of making societies more "legible," as James Scott describes it, and thus more amenable to policy interventions. The new science of demography—which had its first international congress in 1878—would be a handmaiden to this process. The development of reliable means of identifying particular individuals, such as the system of physical measurements Bertillon proposed in 1880, was no less essential. People would be fixed within interlocking fields of surveillance by information-gathering agencies, from police forces to public health authorities to poor-relief workers. Aggregating and analyzing this information could reveal unexpected and useful facts, such as excess mortality in a particular district or the carrier of a contagious disease. "Where the premodern state was content with a level of intelligence sufficient to allow it to keep order, extract taxes, and raise armies," Scott argues, "the modern state increasingly aspired to 'take in charge' the physical and human resources of the nation and make them more productive."[19]

In the late nineteenth century, however, this often remained only an aspiration. In demanding a more exact social accounting, Bertillon was decrying the poor state of French statistics on immigration, which he was certain were the worst in Europe. But even Bismarck's Prussia was poorly informed about the increasing numbers of immigrants arriving from the east. The growth of state surveillance only made the remaining blind spots all the more intolerable. And the observation of regularities in rates of infant mortality and suicide increased demands for intervention. Seemingly random tragedies were transformed into statistics, which could make them seem mundane, but also predictable and therefore amenable to better management. Thus, for England's registrar-general, William Farr, it was intolerably wasteful that 4,604 of every 10,000 children born in Liverpool in the 1860s died by their fifth birthday. Statisticians in Germany calculated the value of individual lives, suggesting that premature death was a loss of national wealth. In both countries reformers pressed for better sanitation and improved training for midwives. Neo-Malthusians, on the other hand, argued that poor people already had too many offspring to care for them properly. Their surviving children were therefore a net liability. The debate was just beginning, but critical elements of population control—especially this calculus of human worth—were already falling into place.[20]

If people had begun to look like populations, and appeared to require control, one reason is that they were moving about the globe in greater numbers, at greater speed, and with more freedom, than ever before. In earlier centuries, most long-distance migrants arrived at their destinations in shackles or burdened with debt contracts. But the slave trade, exile, and indenture were gradually disappearing, and people had more choice in where they went. The development of steam power and metal ships slashed the price of ocean passage. Europeans who emigrated to the Americas and Australia generally found conditions more favorable to large families. That "fatal shore," Australia, now had one of the lowest mortality rates in the world. The period between 1880 and 1915 marked the greatest mass emigration of Europeans ever recorded, some thirty-two million all told.[21]

The same technologies that reduced the costs of migration also cut rates for shipping freight. Together with relatively low tariffs, telegraph cables, and refrigerated cargo holds, this helped to create an intercontinental trade in meat, fish, and grains. At the same time, the establishment of the gold standard in 1878 and a global capital market financed further expansion. Never before had such a large share of the world's wealth crossed borders, and Europeans found that their wages had begun to go farther at food markets. Even the poorest Londoner, surviving on ten shillings a week, began to depend on wheat from Madras, tea from Ceylon, and sugar from Queensland and Natal.[22]

The population boom of the European peoples that had begun in the eighteenth century and continued to accelerate in the last decades of the nineteenth was like one of the great transoceanic steamers that now linked them together. It was divided into different classes of service, but all were headed in the same direction. Both rich and poor experienced declining mortality and declining fertility, though not necessarily in that order. Eventually death rates and birth rates would come into balance and stabilize population levels, yet it took time just to slow down, much less turn around.

Moreover, while statisticians had begun to plot the course of population trends, no one was in command. Instead, a host of different agencies, charities, and pressure groups jostled at the wheel. Health visitors, sanitary inspectors, school nurses, doctors, ministers, and priests could all exert influence and sometimes power over individual families. Their collective efforts affected key demographic determinants such as nutrition, public health, and labor law. Leo XIII felt obliged to insist that children belonged to their fathers and every family belonged to God. But despite the increasingly obvious fact that families were growing smaller, even Rome preferred to remain silent about contraception. The official policy for confessors amounted to "Don't ask, don't tell." In 1886 they were instructed to inquire only when they had "founded suspicion" that penitents were trying to avoid childbirth. If governments, for their part, were becoming more intrusive, no state actually set demographic goals or created programs to achieve them. Even in France, which would have witnessed population decline were it not for immigration, law naturalizing new citizens was still largely determined by ideological debates about rights and duties, not demography.[23]

Aside from a few, generally ineffectual Malthusians, no one was yet organized or even inclined to tell people how to plan their families. A Malthusian League created by Besant, Bradlaugh, and their disciples limited itself to social critique. The only prescription offered by its longtime president, Charles Drysdale, was to fine families with more than four children, something a sympathetic MP warned would not garner a single vote—even his own. In Holland a *neo*-Malthusian League showed the potential of a different approach. It flourished by training doctors and midwives, publishing how-to pamphlets, and aiding Dr. Aletta Jacobs and her Amsterdam birth control clinic, the world's first. Its French counterpart, La Ligue de la Régénération Humaine, published one of these Dutch pamphlets in translation as its first act, and it also openly distributed contraceptives. But while Malthusians called for international cooperation in population reduction and published their propaganda in multiple languages—even Esperanto—most failed to respond to parents' evident demand for practical means to have smaller families. The Malthusian message disregarded the very things that Europeans increasingly had in common: declines in fertility and mortality and expanding migration and trade.[24]

Such dramatic and uncoordinated changes were bound to arouse anxieties, even if the Malthusians were not the ones to capitalize on them. An additional reason for apprehension was the spectacle presented by the rest of the world. For in the very period when Europeans were eating better, living longer, and traveling with ever greater ease and speed, Asians, Africans, and Latin Americans were enduring some of the most catastrophic population losses of the modern era. The crises that coincided with the Besant-Bradlaugh trial were just the beginning, and by 1902 India and China alone suffered thirty to sixty million deaths from famine. The telegraph, rising literacy rates, and increasingly popular newspapers and periodicals ensured that people the world over would read all about it, with illustrative photographs and engravings adding to the impact.[25]

Most contemporary observers simply assumed that non-Europeans lacked the foresight and discipline to avoid overpopulation, pestilence, and famine. Drysdale presented China as an object lesson in Malthusian excess. In fact, Asia's demographic history was no less varied than Europe's, including many instances of relatively low rates of marital fertility. Moreover, until the nineteenth century, Europeans did not generally have higher standards of living than their Asian counterparts. Examining what was happening by the end of that century reveals how Europeans had begun to live longer partly *because* people in other parts of the world were suffering deprivation and dying young.[26]

The first of three waves of acute famine and epidemic disease began in 1876, when what we now understand as the El Niño effect drastically reduced rainfall in a broad swath of territories, including parts of Brazil, North Africa, India, China, and the Philippines. On Negros, for instance, Filipino sharecroppers were literally decimated after three years of drought and locusts. As much as a quarter of Morocco's rural population perished. In Peking imperial authorities belatedly struggled to organize relief for one of the worst-hit provinces. When help finally arrived, Shanxi was deserted, with the dead sometimes bricked into their homes by fleeing relatives in order to deter animal and human scavengers.[27]

Before circumstances reached this point, people would employ a range of strategies to cope with the disaster—slaughtering livestock, liquidating assets, and organizing raids on profiteers. But hunger of this intensity, scope, and duration often proved too much to bear. Most casualties occurred when, as in battle, people defeated by famine broke ranks and fled. In Ceará, a northeastern province of Brazil, for instance, an exodus of starving peasants began suddenly when, after winter rains failed once again, they saw sprouts from their carefully husbanded seeds shrivel and die. With no food to take with them, a descent to the coast turned into a rout. The dead and dying littered the roadside. Even seventy years later, Brazilians hearing the single word *drought* thought only of "one place and one date," when half a million perished, their bodies mummified under the unrelenting sun.[28]

Rather than summoning relief, global communications sometimes encouraged elites to enforce free-market orthodoxy. Brazilians emulated the example set by British officials in India, where this famine struck first. There had recently been abundant harvests, and mountains of grain were heaped along the waterfront in Madras. But the hungry people of the hinterland, like Bombay Presidency and the North Western Provinces, now had to compete in a global market. Poor harvests in England made consumers there insistent that these stores be well guarded and rapidly put to sea. While Malthus was increasingly a prophet without honor as far as his home country was concerned, the British rulers of India were among his

most devoted disciples. He had personally taught a generation of these officials at Haileybury, the administrative college of the East India Company.
Their successors shared the assumption that "every benevolent attempt
made to mitigate the effects of famine and defective sanitation serves but to
enhance the evils resulting from overpopulation," as Minister of Finance Sir
Evelyn Baring put it.[29]

The British created relief projects that proved more deadly than the conditions they were meant to relieve. They set them up at such distances from
famine-struck districts that many of the starving died struggling to reach
the sites. Once there, men, women, and children were made to build roads
and dig irrigation ditches for rations inferior to those given inmates at
Buchenwald. Moreover, many brought with them epidemic disease, which
spread rapidly in their crowded conditions. Consequently, a relief project
might look more like "a battlefield," as an English official observed, "its
sides being strewn with the dead, the dying and those recently attacked."[30]

In other ways as well, and continuing long after the crisis had passed, infrastructure projects created the very conditions that caused early death.
Improved transportation provided new and rapid paths for migration and
trade, but also for cholera, which in this way came to central India and the
Punjab. It was among the most terrifying diseases because its onset was so
sudden, striking even the apparently healthy all at once with vomiting and
uncontrollable diarrhea. Stool soon turned clear except for the milky-white
cholera vibrios themselves, thus finding their way into new water sources,
and new victims. Most would die after losing more than a tenth of their
body weight, their eyes and cheeks sunken into clammy, expressionless
faces. It would all be over in a few excruciating hours.[31]

Relief projects also provided breeding grounds for an even greater killer:
the malaria-bearing anopheline mosquito. Areas with poorly designed irrigation deteriorated into waterlogged swamps. Forest clearing for new road
and rail links caused erosion that exposed a rock-hard subsurface riddled
with cracks and potholes and inevitably gave rise to stagnant pools of water.
The roads and tracks themselves often blocked normal drainage routes and
thus created more breeding sites. While Indians might have had immunity
to the local variety of malaria, relief projects exposed both migrant and local populations to new strains against which they had no defense, especially in their famine-weakened condition. Intermittent fevers and chills
left them further exhausted, and death often followed as the parasite seized

control of their bodies. When malaria attacked the brain, for instance, victims suffered fits and confusion before descending into a coma.[32]

Even though colonial authorities did not yet understand that malaria was carried by mosquitoes, as early as the 1840s they associated it with waterlogging from irrigation and infrastructure projects. They knew enough to avoid their own handiwork and construct their camps in hill stations. Similarly, even before Robert Koch discovered the vibrio that caused cholera, medical authorities suspected that it was linked to poor sanitation, and they steadily improved the water supply of the cantonments, cities, and towns where their compatriots lived. But little, if any, effort was expended in providing new water supplies to accommodate the population flows that followed new transport links. Improvements in housing and sanitation for British troops and officials paralleled developments at home, while mortality soared outside these favored enclaves. Even when, in 1897, Dr. Ronald Ross discovered that mosquitoes transmit malaria—and hence the importance of proper drainage—sanitation officers were pressured to downplay or dismiss these findings lest they interfere with the enormous profits to be made from irrigated agriculture. India was the most irrigated imperial possession in the world.[33]

Perhaps increased farm production and improved transport would have justified the terrible price paid by those who constructed irrigation canals, roads, and rails if they had actually made possible emergency food shipments. Instead, they exposed previously inaccessible regions to the hard back of the invisible hand. Those areas best served by rails and roads were worse off because grain reserves were more easily transported out of them and on to the highest bidders. Shipping rates were much lower for goods going to ports for export than for those going to destinations within India. As authorities observed in Madras, "Facilities for moving grain by rail were rapidly raising prices everywhere."[34]

All in all, London rejected the notion that "England ought to pay tribute to India for having conquered her," as Lord Salisbury put it. Instead, in good years as well as bad, India transferred several million pounds of taxes to London in payment of debts contracted for the construction of new infrastructure. Most of the new rail lines did not even make a profit, but the government in Calcutta had guaranteed a 5 percent return. Indian taxpayers had to make up the difference. Officials did not want to call it tribute, just as they were loathe to acknowledge the number of people who were dying of starvation, lacking the courage of their imperialist and Malthusian convictions. By insisting that India was both overpopulated and a paying investment, they could have their cake and eat it too.[35]

India provides the most striking illustration of the direct relationship between how Europeans and their progeny prospered from global markets that left other people with no means of subsistence. It would be an exaggeration to suggest that all of those who died in late-nineteenth-century famines were victims of an emerging world system inspired by Adam Smith and abetted by Malthus. Yet the global exchange of foodstuffs was always dangerous, if not deadly, to those left with nothing to trade. Famine refugees could be forced into work that only killed them more slowly, such as extracting rubber from the malaria-infested forests of the Amazon. Even this was easier than the slave labor that continued to be inflicted on many Africans during the same era. In 1885 an international conference in Berlin handed the Congo to King Leopold II of Belgium as his personal property—the principle that Europeans could dispossess any "uncivilized" people was not even debated. Millions would be killed or have their limbs hacked off when they did not cooperate in harvesting latex rubber to export overseas. Some of it would be shaped into the condoms and diaphragms that enabled other people to have smaller, more prosperous families. To the extent that the food in one's belly and the clothes on one's back were produced and consumed through colonial expropriation and market integration, increasing numbers of people the world over were locked in a struggle for survival.[36]

Nowhere was this clearer than in the fastest-growing settlement colonies, where the "white deluge" had driven some indigenous peoples to the brink of extinction. During the last decades of the nineteenth century, Asian migrants began to arrive as well. It was here that the first mass political movements emerged that demanded that governments control world population—not limiting its growth, as the Malthusians urged, but rather controlling its movement, composition, and "quality."[37]

Asian labor made vital contributions to the infrastructure and liquidity of the international economy, building railways and mining gold across North America, Australia, and South Africa. Chinese migration was particularly extensive, constituting a transnational network with "nodes" in Hong Kong, San Francisco, and Singapore, and financed through highly sophisticated systems of credit. The United States had encouraged its development in the 1860s, pressuring China's imperial government to drop exit barriers based on the "inherent and inalienable right of man to change his home and allegiance." To be sure, Washington was also acting on behalf of employers eager to find low-wage workers, especially in the booming state of California.[38]

American labor leaders were trying to organize for what they considered a living wage. They claimed that they did not oppose immigrants as such, but only those who were ensnared and exploited. It is not clear whether any significant number of the Chinese in California were actually "coolies," or whether workers of European descent would have accepted them under any circumstances. When Chinese attempted to strike or start their own businesses, it only brought more ferocious and broad-based attacks. The real issue was whether people, no less than capital, goods, and ideas, could move across the planet and compete on equal terms. Ironically, settlement colonies, built on the graveyards of their original inhabitants, would give rise to movements that demanded an end to unrestricted migration.[39]

The turning point came in California in July 1877, just after the Besant-Bradlaugh trial. Here too the El Niño effect had caused a devastating drought, the worst in a quarter of a century. By this point both the Democratic and Republican parties had come out against further Chinese immigration. But the migratory workers and unemployed of San Francisco, aroused by the oratory of Denis Kearney, instead chose direct action and a new Workingmen's Party under the slogan "The Chinese Must Go." For three nights they rampaged through the Chinese quarter, and from this point forward the cause of labor unity was tied to a relentless campaign against Asian immigration.[40]

Many on the East Coast were alarmed at the unrest. Lurid reports of the famine in Shanxi Province helped sway the debate. One writer projected that as many as a hundred million might seek escape. In fact, most Chinese emigrants to the Americas came from areas unaffected by the famine, particularly the southern provinces of Guangdong and Fujian. They tended to stick to the paths established by forerunners from the same family or village, sometimes decades earlier. The great majority of Asian migrants, whether from India or China or elsewhere, tended to remain within Asia.[41]

But the *perception* that Asians constituted a limitless pool of hungry people who would pour through every opening dictated the response. Each border closing intensified political pressure for closures worldwide. Rather than more piecemeal responses by individual provinces or states, national or even international policies seemed imperative. One after another, the United States, Australia, South Africa, and New Zealand moved to restrict immigration from Asia. And once the numbers of Chinese immigrants declined, labor leaders found new menaces to replace them.[42]

Social Darwinists like Herbert Spencer, who first coined the term *survival of the fittest,* opposed immigration restrictions. With Europeans not only surviving but spreading across the globe, why should they fear the Chinese? The pressure of population growth and economic competition could only spur further improvement. Advocates replied that, in Darwinian terms, "the fittest" only meant those who could subsist on less and reproduce more. In 1877, a special U.S. House-Senate committee asserted that, although the Chinese lacked sufficient "brain capacity" to sustain self-government, they could survive in conditions that would starve other men. To compete head-to-head, "the American must come down to their level, or below them." If they were not stopped in California, the legislators warned, the Chinese would eventually have to be fought on the banks of the Mississippi, or even the Hudson. Writing for the *North American Review* the following year, M. J. Dee contended that older American stocks in New England were being replaced by recent arrivals from Europe, who were accustomed to raising larger families at a lower standard of living. But even the Irish would suffer in direct competition with the Chinese. "Constant over-population" had taught them "to live in swarms," Dee explained, to "drive the vulture from his prey, or devour the unclean bird itself." Dire consequences would therefore ensue, should they "withdraw the intelligence of artificial selection from the environment, and leave the battle to the chances of natural selection alone."[43]

A decade later this argument for artificial selection would issue from the tribunes of the most eminent scientific authorities. But that same year it began to be echoed in a series of novels and short stories proclaiming an imminent Chinese "invasion" of Europe and the United States. Chinese were depicted as lacking individual initiative or intelligence, instead forming "hordes" or "floods." The same images sometimes featured in European journalistic and fictional accounts of migration. Kaiser Wilhelm II became so obsessed that he sketched his own version, in which the Archangel Michael rallied the nations of Europe against the Buddha riding a fierce dragon—an allegorical reference to the potential power of Japan united

Hermann Knackfuss's rendering of "The Yellow Peril" depicted European cities aflame and Christian nations rallying to fight the Chinese dragon and Japanese Buddha. Bildarchiv Preussischer Kulturbesitz / Art Resource, New York.

with China. It was redrawn by the court artist, engraved, and then dispatched to the crowned heads of Europe. Yet the phrase *yellow peril* just as often designated East Asia as an *economic* threat that could imperil European standards of living, even if Asians did not actually invade. Whether they were called "hordes" or "coolies," both terms treated them as a population rather than as individual people, a population that, by its very nature, was said to imperil a certain "quality of life"—whether that quality was eugenic, economic, or both at the same time.[44]

Ironically, after Congress acted to exclude Chinese immigration in 1882, many thousands continued to secure admission by compelling courts to recognize their individual rights to due process. These rulings only further infuriated American mobs. In 1885, anti-Chinese violence culminated in a series of mass expulsions up and down the West Coast, with massacres even farther afield. Apologists described this violence as a means by which workers everywhere expressed their citizenship. In a theme that would resonate through the history of population control, they insisted that freedom of

movement—the first and simplest way people express control of their own
bodies—was limited by obligations to society, beginning with a duty to belong. As W. C. Owen explained the following year:

> The workingman may not be able to explain his motives with scien-
> tific precision, but . . . a great part of the repugnance felt to [the Chi-
> nese] upon this coast is that they do not act as citizens, that they
> have no concern in the solidarity of the nation. A precisely similar
> sentiment has dictated the persecutions of the Jews in Germany,
> Austria and Russia, persecutions which have been justified precisely
> on this ground. . . . It is, in short, but the public method of voicing
> the sentiment, 'no rights without duties,' or, as Comte puts it, 'Man
> has no rights except to fulfill his duties.'
>
> In a word, I believe that we ought to welcome every opportunity
> which presents itself to the proletariat of developing itself in the only
> way, as evolution proves conclusively, in which anything ever did
> develop itself since the beginning of the world—that is to say, by
> struggle.[45]

Whether the sentiments that impelled anti-Jewish pogroms really were
"precisely similar" is less important than the fact that even socialists like
Owen perceived them to be. Both the Chinese and the Jews were depicted
as disease-carrying cosmopolitans who excelled in economic competition
and conspired to rule the world. And both now figured in political projects
intended to define nationalism and delimit citizenship through both state
policies and popular violence. When Owen wrote, Prussia was in the midst
of driving out some thirty thousand Russian émigrés; this action was a culmination of years of Jewish expulsions, though it was also spurred by anxieties about Polish nationalism. This period also marked the peak of both
right-wing anti-Semitism and labor attacks on foreign workers in France,
where anxieties about "race" or "blood" had earlier been nearly absent in
discussions of immigration. And beginning in 1892, Great Britain's Joseph
Chamberlain showed how immigration restrictions might form part of a
nationalist program with appeal to labor. Opposition to Jewish immigration would finally drive Parliament to pass restrictive legislation in 1905.
The difference between Poles and Chinese was considered "but a question
of degree." Two years later, the socialist Sidney Webb was still worried that
Britain was "gradually falling to the Irish and the Jews . . . there are signs
that even these races are becoming influenced. The ultimate future of these

islands may be to the Chinese." However fantastic, such fears reflected an increasingly common belief that global economic competition was a biological process, in which a race to the bottom might ultimately lead to racial degeneration or even extinction.[46]

Working-class solidarity largely withstood the movement against immigrants in the United Kingdom and France as well. But in other cases, including the United States and Germany, labor leaders accepted that maintaining high wages and the welfare state required rejecting whole classes of potential citizens. Ironically, the need to treat aspiring immigrants as populations—rather than individuals with inalienable rights—entailed creating faceless bureaucracies that were not even accountable to the federal courts. The American Federation of Labor's Samuel Gompers claimed that he opposed the Chinese only because their history revealed them to be a people who accepted exploitation. But many more nationalities would now be subject to the same exclusionary logic. Commentators like Richmond Mayo-Smith could discredit Italian and Slovak immigrants merely by describing their frugality and willingness to work at low wages, concluding that "it is the question of Chinese labor all over again." In Canada, Australia, and several European nations as well, Italians came to be known as "the Chinese of Europe." Even violent attacks against immigrants, such as the 1891 lynching of eleven Italians in New Orleans, were blamed on the immigrants themselves. According to Congressman Henry Cabot Lodge, soon to become senator, they were "breeding race antagonism and national hostilities which never existed before." Here again, competing with them might cause degeneration. Expelling them was a public duty.[47]

Anti-immigration spokesmen like Mayo-Smith could also cite statistics purporting to prove that newer immigrants were disproportionately represented among criminals and the mentally ill. The United States had already denied admission to convicts and the insane as well as those with various disabilities. Activists intent on controlling social spending kept pressing authorities to develop more discriminating ways to sift, sort, and exclude immigrants. But as much as opponents exploited fears of unfit or criminal immigrants, their admitted purpose was to exclude many who would otherwise become law-abiding and tax-paying citizens.[48]

In the 1890s, the census superintendent and MIT president Francis Walker found a way to combine new and old critiques of immigration, targeting both the individual degenerate and the degenerate horde. According to his infamous epithet, immigrants were "beaten men from beaten races;

representing the worst failures in the struggle for existence." As such, they would flock to countries that could still afford to be charitable. Once there, these "vast hordes of ignorant and brutalized peasantry" would depress wages and discourage "native stocks" from forming new families. Echoing M. J. Dee's argument against the Chinese, he predicted that in time descendants of those able to live on less and reproduce more would supplant everyone else.[49]

Walker's argument was taken up by Karl Pearson, Galton's chief disciple in the newly christened science of eugenics, and spread to both sides of the Atlantic. It resonated with the social Darwinist idea that public charity impeded progress, which was gaining influence in Great Britain and Germany. Demonstrating differential fertility and projecting population trends appeared to prove that society was already degenerating. In different contexts this approach might focus on the growth of disease, criminality, or degraded classes rather than ethnic minorities, but the figure of the immigrant often served to embody these diverse concerns. Immigrants represented the immanence—and imminence—of this apparent threat in a way more easily and vividly conjured than dry tables detailing differential fertility. In time, even scientists would begin to refer to "old stocks" they considered unfit as "alien," as Charles Davenport described the notorious Jukes.[50]

As fears grew about the vitality of European peoples, there was increasing skepticism about their prospects for populating new territories, especially outside temperate zones. When, in 1890, the British geographer Ernest George Ravenstein was asked to lecture on "Lands of the Globe Still Available for European Settlement," he concluded that there were very few left. European peoples were too ill-suited to nontemperate environments to permanently inhabit them. He was instead inspired to predict that world population would peak at some six billion, the first recognizably modern, quantitative estimate. In reviewing Charles Pearson's *National Life and Character,* Theodore Roosevelt agreed that the British would eventually be "absorbed" in India and Africans would always threaten to "swamp" white settlements. "It is impossible," he concluded, "for the dominant races of the temperate zones ever bodily to displace the peoples of the tropics." Even Benjamin Kidd, still supremely confident that whites would rule over the remaining tropical lands, admitted in 1898 that not many of them could actually live there. The man who, as much as anyone, made such settlement possible by identifying mosquitoes as carriers of malaria, Sir Ronald Ross, considered "the whole of the tropics comparatively unsuitable for the full development of civilization." By 1905, U.S. Army doctor Charles E. Woodruff argued, based on his experience in the Philippines, that blond Teutons might not survive even in warmer parts of the United States.[51]

The colonization of areas like coastal Australia, the Argentine pampas, and the North American plains therefore seemed to signal the end of an era, marked in the United States by Frederick Jackson Turner's famous address to the Chicago World's Fair of 1893, "The Significance of the Frontier in American History." In depicting frontier settlement as a process of turning Europeans into Americans, Turner was speaking to concerns about whether and how newer arrivals from an ever-growing array of nations would acquire this identity in increasingly crowded cities. Many of his contemporaries feared that, rather than being an assimilating mechanism, taking and defending lands from lesser breeds turned Europeans into savages. This way of thinking can be inferred from fictional depictions of frontier types, ranging from Kurtz of Joseph Conrad's *Heart of Darkness,* traumatized by King Leopold's Congo, to Bram Stoker's *Dracula.* It was also embodied in the figure of the métis, a growing obsession for colonial authorities in such places as Indochina and the Dutch East Indies. Whether abandoned to squalor or aspiring to equality, the mixed-race offspring of colonial liaisons both exposed and challenged the arbitrariness of racial and national classifications. Mixed unions and their multiplying offspring created havoc along the "interior frontiers" of the nation.[52]

The fear of degeneration increased demand for scientific methods to measure the level and progress of civilization. Cesare Lombroso's physical measurements of criminals, which first became widely known in 1878, purportedly proved that they were evolutionary throwbacks. For Gustave Le Bon, by contrast, the very act of joining a crowd caused an apparently cultivated person to become prey to the basest emotions. In addition to mob rule, he thought civilization was also endangered by the intermixing of distinct peoples. Henry Cabot Lodge borrowed liberally from Le Bon in arguing for legislation to close borders to immigrants from southern and eastern Europe, who threatened "the very fabric of our race."[53]

There were many differences among the aforementioned authors, reflecting the great diversity of ideas inspired by population trends. Some, such as Le Bon, doubted that an improved environment could improve the quality of peoples. Lombroso, at least initially, insisted that most criminals responded to treatment. Similarly, social Darwinists differed on who represented the fittest in any particular society, whether hereditary aristocracy, professionals, or the working class. And in their skepticism toward permanent white settlement in tropical zones, authors like Roosevelt and Ravenstein were not predicting the decline of European peoples, only recognizing their limitations.

Yet together their writings contributed to the sense that both spatial boundaries and social hierarchies all over the world were becoming increasingly unstable. There was not yet very much concern about what would now be called the earth's "carrying capacity," and even Ravenstein thought the world would not reach its maximum population until 2072. But throughout the 1890s a geopolitical vocabulary preoccupied with space and its relative scarcity entered common usage. German geographer Friedrich Ratzel, perhaps the first European to draw attention to California's "Chinese question," popularized the notion of *lebensraum*. The term was exceedingly vague, and purposely so, merely suggesting that nations were like bodies, and that political borders had better accommodate biological processes of growth and movement. To this way of thinking, trends in fertility, mortality, and migration provided the best measures of improvement or degeneration. While racial differences were depicted as more real and more determinative than political structures, races might rise, fall, or even disappear through miscegenation or misplaced charity. Frontiers were key sites for reproducing civilization, but they could also shift into reverse. And although cities might showcase the highest cultural achievements, they were also the first place to look for signs of decline in social pathologies such as sterility and suicide.[54]

Scientific advances that made it possible to plumb the origin of species and prolong life also provided a language for—and lent prestige to—those who would depict social, economic, and political inequality as manifesting biological differences. The germ theory of disease improved public health and opened up theretofore fatal shores and jungles to exploration and exploitation. At the same time, it provided powerful metaphors to represent the dangers of degeneration, both from within and from without. Similarly, the idea of the "survival of the fittest" both flattered and excused those who were prevailing in a competition that now encompassed the world. But it also signaled that no level of success, however great, could be secure, especially as industrialization, migration, and social welfare measures rapidly changed the terms of this competition—a competition that was increasingly tracked through censuses and vital statistics. For those inclined to look for possibilities of improvement or signs of decline, social science also offered figures on crime and illness. Broken down by race, class, or ethnicity, correlations could be construed to prove that people fit into biological categories differentiated by both instincts and abilities—even by the size and shape of skulls.

In a world that seemed out of control, scientists and their popularizers responded by systematically counting and categorizing people, assigning each one an exclusive ethnic identity. This was already an international project in the 1870s and 1880s. In South and Southeast Asia, colonial authorities created classifications and hierarchies that in some cases were figments of their own imaginations but soon had very real consequences. In one instance, British commentators on India perceived in caste the aftermath of a failed effort to preserve the biological purity of white Aryan invaders. Their efforts to systematize and order caste differences through decennial censuses mobilized groups who demanded recognition, protection, and promotion, unleashing a political revolution that would have a long-term impact on the history of population control. All over the world, intellectuals would appropriate and adapt social Darwinist ideas, ideas that would seem even more salient to Asians and Africans confronting what some began to call "the white peril."[55]

National societies were beginning to be viewed as populations that shared a common ancestry, whereas "uncontrolled" immigration was coming to seem unnatural. Although annual migration rates continued to increase until World War I, the United States, in restricting Asians, had developed the legal rationales and administrative mechanisms for a systematic shift in how states controlled their borders, and thus controlled their populations. By 1908 Roosevelt was calling for a concerted policy to exclude Asian immigrants from all English-speaking countries, an idea with high-level support in Canada and Australia. He said that he had dispatched the White Fleet to the Pacific, ready to go to war with Japan, in defense of "white civilization." This marked the first official attempt to establish an international population control policy. It failed because British authorities feared containing Asia would provoke a backlash in India and other colonies. Even the minimal duties that came with formal rule over subject peoples seemed to preclude policies explicitly intended to reduce their number—a problem that would later present an opportunity for organizations prepared to assume the civilizing mission.[56]

Though states could not formally coordinate exclusionary policies, by World War I some who were outside of government had begun to see them as the starting point for a comprehensive system of de facto population control. For a generation already, regulating the rate and composition of immigration had been recognized as a means of safeguarding the fertility and thus the supremacy of "native stocks." Roosevelt popularized the idea, borrowing the concept of "race suicide" from the sociologist Edward Alsworth Ross. In 1912 Ross developed it further. Rather than accept destitute immigrants, he argued, Western nations had to hold fast to all settlement colonies and fill them with their own offspring, or else see them "filled with the children of the brown and the yellow races." He predicted that "the world will be cut up with immigration barriers which will never be leveled until the intelligent accommodation of numbers to resources has greatly equalized population pressure all over the globe." Two years later, Ross argued that shirking this responsibility would invite foreign immigrants to take the place of future offspring of native stocks, and thus constitute a policy that would "pity the living but not the unborn."[57]

Because migration was barred precisely from those areas with the highest rates of mortality, in time exclusion would be recognized as a way to control the population of other countries too. People were already comparing the claims and merits of immigrants and "the unborn" in terms of who should make up the future population of the nation and even the world. Through the regulation of migration, they had developed both the inclination and the capacity to assess the value and control the number of prospective citizens. Just as parents were calculating the cost of children, and generally electing to have fewer, states and the experts who sought to advise them began to look at their own people in a new, harsher light.

It was decades yet before a family planning movement emerged with the slogan "Every child a wanted child." But social planning had already begun to reflect the goal of making every citizen a wanted citizen. By 1903, virtually every government had dropped the requirement that individuals receive permission to emigrate, and some, such as Germany and Great Britain, encouraged it. The fact that free emigration became the policy of every state, from Swiss cantons to Meiji Japan, and in no particular order, suggests that authorities were moved not by a desire to expand the scope of freedom, but rather to free themselves of "surplus" population.[58]

Controlling the birth of new citizens was a much more difficult proposition. In countries with sizable Catholic communities, the Church would brook no interference in the moral instruction of families. Yet practice in the confessional varied widely, and priests hesitated to broach the subject of birth control before mixed congregations. When ecclesiastical authorities finally took action, they identified private morality with the national interest. In 1908 the Lambeth Conference of Anglican bishops condemned the increasing practice of contraception as "demoralizing to character and hostile to national welfare." The next year the Catholic hierarchy of Belgium issued an equally strong pastoral letter and stricter instructions to confessors. Rome remained wary of the growing government role in family life, but stricter bans on birth control and prosecution of abortionists provided a basis for cooperation—even if that cooperation was to be organized along national lines. As the German bishops declared in 1913, offering a small but crucial amendment to the 350-year-old Roman catechism, the main point of marriage was procreation to ensure "the continuation of the Church *and the state.*"[59]

Whether it was proper, on the other hand, to impose sterility on those deemed unfit was more problematic. Among Malthusians there were always some, including Besant, who suggested that reducing their fertility should be the focus. Beginning in the 1890s, proposals to use eugenics to eliminate social problems like crime and alcoholism became commonplace. But most proponents were at least equally committed to maintaining the fertility of the fittest. Darwin himself had warned Besant that her work would hinder natural selection. Catholic theologians debated whether Rome might deem sterilization permissible as part of a larger program that could isolate those who would indiscriminately promote contraception among Christian peoples. New national organizations emerged to unite eugenicists, beginning with Germany's Society for Race Hygiene in 1905. They attacked Malthusians for "endangering the future of our race" in its "great combat for lasting supremacy," and rapidly overtook them in influence and effectiveness. The first international meeting of eugenicists in 1912 included among its speakers Arthur Balfour, two French senators, and a host of leading scientists and university presidents.[60]

But in the United States and the UK, eugenicists often focused on the relative growth of Catholic communities. Madison Grant exemplified how opposing immigration and cutting the social costs of deviance could form part of a program with far less appeal to the Catholic hierarchy. A New York lawyer and conservationist who would serve as treasurer to both the second and third international eugenics congresses, Grant wrote a bestseller in 1916 called *The Passing of the Great Race.* He accused Catholic, Slavic, and Jewish immigrants of crowding jails, hospitals, and insane asylums. But sterilization offered "a practical, merciful, and inevitable solution of the whole problem, and can be applied to an ever-widening circle of social discards, beginning always with the criminal, the diseased and the insane, and extending gradually to types which may be called weaklings rather than defectives, and perhaps ultimately to worthless race types."[61]

However changeable in its program and indefinite in its ultimate scope, eugenics zeroed in on human weaknesses, especially the craving for distinction and the fear of seeming different. Malthusianism, on the other hand, always offered the same diagnosis and the same bitter medicine. It did not speak to the trends in migration and trade that were linking distant societies together while driving living standards apart. By contrast, the social problems posed by immigration provided "a golden opportunity to get people in general to talk eugenics," as one of its proponents noted. And once they began talking, they finally produced a synthesis that appeared to offer a global solution to population problems. As the veteran leader of the Immigration Restriction League, Prescott F. Hall, explained:

> The moral seems to be this: Eugenics among individuals is encouraging the propagation of the fit, and limiting or preventing the multiplication of the unfit. World-eugenics is doing precisely the same thing as to races considered as wholes. Immigration restriction is a species of segregation on a large scale, by which inferior stocks can be prevented from both diluting and supplanting good stocks. Just as we isolate bacterial invasions, and starve out the bacteria by limiting the area and amount of their food supply, so we can compel an inferior race to remain in its native habitat, where its own multiplication in a limited area will, as with all organisms, eventually limit its numbers and therefore its influence. On the other hand, the superior races, more self-limiting than the others, with the benefits of more space and nourishment will tend to still higher levels.[62]

Who would lead this movement for "world eugenics"? One might assume that governments would be at the forefront, given their growing capacity to monitor and control populations. This process has come to appear so natural that scholars often describe the state as "penetrating" society, and depict official documentation of individual identity—whether through fingerprinting or passports—as an "embrace." Yet even though governments were developing an interest in controlling populations, it should not be supposed that state officials were the first to recognize and act on this interest. Besant and Bradlaugh had to pester British officials before they agreed to prosecute the two of them for publishing instructions in how to avoid pregnancy. Similarly, prosecutors and judges across the United States showed great reluctance in enforcing the Comstock Law against trade in contraceptives. Proponents of alien registration, improved population statistics, immigration restrictions, and sterilization were continually frustrated with government officials' inability or unwillingness to implement even those laws and regulations that were already on the books. Labor movements were usually the driving force behind stricter regulation of immigration, and eugenics societies proliferated precisely because of the perceived failure of governments to consider the effects of their policies on the quality of population. Although states had the authority to enforce definitions of "normalcy" and "deviance," the process typically depended on people's eagerness to identify themselves through these categories. Thousands of Londoners lined up outside Galton's anthropometric laboratory in the 1880s to have their heads examined even though no official ever asked them to do so. This included Prime Minister William Gladstone, who was certain his cranium was larger than it turned out to be.[63]

Population control was truly a *mutual* embrace, in which states, churches, and social movements would grow closer because of the stress of war, and not always at the initiative of governments. Nowhere is this more apparent than in the regulation of sexuality, where nongovernmental organizations made the first moves. Indeed, as far as sex is concerned, the state did not penetrate society, nor even accept an embrace, so much as blush at what was asked of it.[64]

2

TO INHERIT THE EARTH

In the fall of 1917, waves of troops from across the British empire advanced on German trenches near Passchendaele, American reinforcements poured into French seaports, and repeated attacks by Italian forces pushed Austria-Hungary to the brink of collapse. General Erich Ludendorff worried about an even graver threat. He urged the imperial chancellor, Theobald von Bethmann Hollweg, to heed a warning from the director general of the medical services, who predicted that the war would not be decided on the day peace was signed: "Germany's future is dependent to a far greater degree on the question within what time, and to what extent, she will have repaired her losses, especially *in men*. Worse than the losses through the war is the decline in the figure of our population owing to the falling birthrate." Ludendorff agreed that Germany must undertake a crash program to restore fertility rates: "the decisive issue for our whole future is that we should remain a healthy, growing nation."[1]

Though we remember the trenches, what made World War I frighteningly new was not just what happened to soldiers, but the systematic violence inflicted on civilians. Enemy nations were subject to aerial bombardment, blockade, deportation, forced labor, and mass executions. As great as were the military casualties—about eight and a half million dead—the number of civilians killed may have been even greater. Afterward, all the combatants worried about whether and how they would recover. "In the event of a war similar to that which we have just experienced," Britain's National Birth Rate Commission asked in its 1920 report, "what would happen to us with a greatly reduced birth-rate? Surely all we have would be taken, and we must become slaves." No power had mobilized so large a proportion of its population as France, and almost three-quarters were dead, missing, wounded, or captive at war's end. Prime Minister Georges Clemenceau suggested that the Versailles Treaty should have required his country to increase its birth rate. "For if France turns her back on large families," he warned the Senate, "one can put all the clauses one wants in a treaty, one can take all the guns of Germany, one can do whatever one likes, France will be lost because there will be no more Frenchmen." Russia's losses continued with the civil war, and by 1920 less than a third of Petrograd's population remained. In Moscow half had died or fled.[2]

Catholic authorities in many countries joined in warning against "race suicide," enlivening the old theological argument that contraception was contrary to nature with a new demographic imperative. "The theories and practices which teach or encourage the restriction of birth are as disastrous as they are criminal," the French bishops insisted. "It is necessary to fill the spaces made by death, if we want France to belong to Frenchmen and to be strong enough to defend herself and prosper." Similarly, Cardinal James Gibbons of Baltimore spoke for all his American brethren when he called contraception "the crime of individuals for which, eventually, the nation must suffer."[3]

But it was not just the numbers that had been killed that aroused concerns about the population of the European peoples. It was also what mobilization revealed about their condition even before they started fighting, and what the losses meant for the character of future generations. Even in the United States, which had armed a relatively small proportion of its people, many of them volunteers, new IQ tests of nearly two million recruits purported to show that their average mental age was thirteen. Britons worried more about those who were rejected from the military and had thus survived to sire a new "C-3" generation. French conscripts had almost to be lame, blind, or both to escape service. Eugenicists shuddered to think about having to repopulate France with "hare-lips, the rachitic, tubercular, the insane, idiots, the feebleminded and the impotent."[4]

After the Great War, the sense that the flower of a generation had been sacrificed for mere yards of shell-blasted soil imparted an urgent lesson: the power and pride of nations were vested in their population, not territory, and governments must strive to increase their numbers or at least their "quality." The response varied widely between different countries, but it was often dramatic. In addition to rapidly stepping up maternal and child welfare work, Britain planned to create a new, hardier generation of pioneers to populate the open, vulnerable spaces of its dominions, aiming to assist in the emigration of sixty to eighty thousand people a year. The Weimar Constitution explicitly favored the "child-rich," and authorities forbade contraceptive advertising and elective abortions. In France, even to advocate limiting population growth became a crime. Italy joined its northern neighbor in offering subsidies for large families while also taxing celibate men.[5]

Religious authorities backed many of these measures, and Catholic spokesmen were often leading proponents. But government's growing role in marriage, child welfare, health care, and migration affected Church interests across the board, pushing bishops in France, Germany, Switzerland, England, Ireland, Poland, and Yugoslavia to organize into national conferences. The Americans were particularly active, creating a National Catholic Welfare Council (NCWC) with legislative and press offices and affiliated organizations of laymen and -women. In 1919 a top Vatican official assured Gibbons that Rome "looks to America to be the leader in all things Catholic, and to set an example to other nations." When Pius XI assumed the papacy in 1922—and thereby became a prisoner of the Vatican, since popes refused to recognize Italian sovereignty—he decided to rein them in. National conferences subverted the Church hierarchy, in which all authority emanated from Rome. The curia applauded efforts to keep birth control and abortion illegal, but worried that a spirit of independence was developing not just in the United States, but in France and Germany as well. The American bishops only managed to save the NCWC by arguing that competing with the enemies of the Church in national politics required a national organization. If the bishops did not lead, lay organizations—even more prone to "Americanism"—would fill the vacuum.[6]

The stakes in these struggles were growing ever higher. Eugenicists were able to win passage of compulsory sterilization laws across the United States. California alone sterilized at least seventy-five hundred people by 1931. And the 1924 National Origins Act led to not just a drastic reduction in immigration from southern and eastern Europe, but the expulsion of immigrants that courts did not consider to be white and the forced sale or seizure of their property. But the bishops focused on a different danger: "The feminist movement has grown strong here as elsewhere, and the granting of the franchise to women has made many of them aggressive, and their new power has excited many and led them to adopt the most radical ideas." Catholic leaders insisted that birth control remain illegal and that abortionists belonged in jail.[7]

Women who organized internationally to win suffrage, including birth control stalwarts like Aletta Jacobs, followed up with a broad feminist agenda, such as the right to employment and equal pay. Their adversaries framed all of this as threatening "race suicide"—that is, as a threat not just to the survival of individual nations, but to the preeminence of "the West" relative to the rest of the world. Fratricidal bloodletting among Europeans raised doubts about whether they remained sufficiently robust to rise above ever-greater numbers of Asians, Africans, and Amerindians. Many called for a concerted response.

But those endangered by European imperialism—now at its peak with the scramble for former Ottoman territories—instead warned of a "white peril." Concerns about population "quality" were also prevalent and perhaps more understandable in poorer countries. It was estimated that in the 1920s the annual toll of premature deaths in China far exceeded the number of people killed each year in the course of World War I. Roughly a million lepers, half of the world's total, lived in China, which at that time probably had a population smaller than Europe's. In India around twenty million people died in the influenza epidemic of 1918–19. During the following decade most infants did not live to see their first birthday in places such as Madras City. Infant mortality rates in Egypt were even higher than in India. Japan was making measurable progress in health and sanitation, but tuberculosis mortality was still two to three times greater than in western Europe. Similarly, despite vigorous public health campaigns, more than 90 percent of Brazilian army recruits were found to be afflicted with parasites after the first nationwide draft in 1916.[8]

Eugenicists and birth controllers in Latin America and Asia insisted on participating in what was becoming a global debate about population trends. They were sometimes ignored or rebuffed, as their counterparts in Europe and the United States concentrated on reducing the fertility of ethnic minorities, the poor, or the otherwise "unfit" of their own societies. But when American and European scientists and activists became more concerned about the future of world population, they would realize that they needed allies abroad.

In the realm of population policies, like so many others, the war made international cooperation seem both more difficult and more imperative. It spurred nationalism and an inclination toward isolation, most dramatically shown by the rapid falloff of migration. In the famous 1927 *Buck v. Bell* decision that upheld forced sterilization, Oliver Wendell Holmes likened it to the sacrifice of America's fallen soldiers. The chairman of the International Commission of Eugenics, Leonard Darwin, advised Americans that personal ambitions would have to be "sacrificed on the altar of family life" to ensure racial progress. Fitter people should choose superior spouses and have more children, not merely to serve the nation, according to Charles Davenport's "creed for the religion of eugenics," but for "love of the race." Men's and women's bodies did not belong to themselves or even to God. They were elements of social bodies that could be enlisted in collective political projects.[9]

Many reacted with revulsion to a summons to "breed before you die," as one pacifist opponent put it, especially because it followed on the deaths of so many millions. "Public magistrates have no direct power over the bodies of their subjects," Pope Pius XI would insist in his 1930 encyclical, *Casti Connubii*. He insisted that procreation remained the primary aim of any Christian marriage, but warned of divine punishment for any who forgot that "the family is more sacred than the State and that men are begotten not for the earth and for time, but for Heaven and eternity." Feminists asserted the rights of individuals and especially women to control their own bodies as the very foundation of peace and freedom. "A free race cannot be born of slave mothers . . . ," Margaret Sanger argued. "No woman can call herself free who does not own and control her body." She hoped that a global movement to secure this individual freedom could vanquish militarists in every country.[10]

Politicizing reproduction—whether to enlist individuals, empower them, or insist that God ruled over them all—was always potentially destabilizing, since it opened the way to alternative social formations, from white solidarity to international sisterhood to a worldwide community of the faithful. Concerns about population growth and movement inspired doubts as to whether the church or the state was meeting the challenge. Even those who favored a "league of low birth-rate nations" or "Great Barrier of the peoples of Europe, the Americas, and Australasia against those of Africa and Asia" were implicitly conceding that any adequate response had to be international or even transnational in nature. Thus, while population trends encouraged people to think of the world as a whole, it also provided new reasons and new ways to divide it.[11]

If there was one person who represented all the contradictions, the possibilities as well as the perils of population control, it was Margaret Sanger. It was she, after all, who first coined the term *birth control,* and she did mean *control.* As she related years later, she was with friends in her New York apartment one evening in 1914 when she realized that a "new movement was starting, and the baby had to have a name." As a visiting nurse in Manhattan's impoverished Lower East Side, she had seen women on the brink of death from septic abortions and was now determined to help. They considered "neo-Malthusianism," "voluntary parenthood," "voluntary motherhood," but none sounded right. "Then we got a little nearer when 'family control' and 'race control' and 'birth-rate control' were suggested. Finally it came to me out of the blue—'Birth Control!'" The original keynote for Sanger's efforts was not volition, then, but control. Who exercised that control, by what means, and for what purposes, were questions that persisted throughout a public career that spanned almost half a century.[12]

Like Annie Besant, Sanger first gained the public's attention and kept it by challenging state authority. She started by publishing contraceptive information in her newspaper, *The Woman Rebel,* and defied the Comstock Law by sending it through the mail. She fled the country, then returned to conduct her own defense. Sanger provoked reluctant prosecutors once again by opening a birth control clinic in Brownsville, Brooklyn. After being arrested, she refused to be fingerprinted, threatened a hunger strike, and finally served thirty days in jail. It was all wonderful publicity. Sanger soon discovered the same phenomenon at work abroad. When she traveled to Japan in 1922, authorities initially decided to deny her entry as an agitator. But Sanger observed that the "fact that the government was against the idea of birth control threw the sympathy of the people everywhere decidedly for it and for me." This Asian tour also reinforced in her mind the idea that birth control was a universal value, transcending history and geography. To Sanger, a Shanghai sex worker's lament, "Me no want baby," seemed to have "a thousand years of tragic sorrow behind it."[13]

Yet Sanger's experiences abroad also helped her realize that the success of birth control did not just depend on its recognition as a human right. In the Netherlands she discovered an extensive network of clinics staffed by doctors and nurses trained to dispense diaphragms. She credited them with improving maternal health, reducing the birth rate, and eliminating differential fertility. But she recognized that governments were in the best position to deliver reproductive health services to broad segments of the population, or at least permit doctors to do so. To advance her cause, Sanger sometimes used state opposition to obtain free publicity, at other times tacking to the right to convince officials that distributing birth control to the poor was in the public interest.[14]

Sanger felt she "never had a country," as she confided in her journal on her first day at sea in November 1914, fleeing from prosecution. This reflected her conviction that "any country where one can pay [the] rent is yours." She came to enjoy the itinerant lifestyle and continued globetrotting until she was almost 80, at significant risk to her health. Travel created opportunities both to raise money and to spend it among a growing circle of affluent friends, since it was invariably accompanied by fundraisers, receptions, and celebratory dinners. But even when Sanger became personally secure through her writings and a second marriage to Noah Slee, a millionaire twenty years her senior, she understood that money went farther when combined with adroit networking. She pursued relationships, and sometimes affairs, with literary figures, society doctors, scientists, and politicians. As she traveled across the world creating webs of influence and nodes of communications, she could draw on a variety of political currencies based on social connections, professional credentials, and public notoriety.[15]

Yet as much as Sanger was sought after, she was sometimes resented. National differences were often transcended but never irrelevant. Professional qualifications, or the lack thereof, could also constitute barriers that blunted her influence—as when Dr. Aletta Jacobs, who had founded the first birth control clinic in Amsterdam, refused to meet Sanger because she was a mere nurse, or when Japanese women doctors found her unable to answer their questions. And even when scientists accepted Sanger's money, attended her conferences, and joined her organizational efforts, that did not stop them from marginalizing or even excluding her. While they liked to pretend they were making the study of population a true science, for them, too, it was all about control.[16]

The great question of how this nascent movement might reconcile individual rights and social responsibility was often posed during the interwar period in terms of the relationship between birth control and eugenics. After the long decline of orthodox Malthusianism, Sanger and other activists—including Marie Stopes in Britain, Shidzue Ishimoto in Japan, and Elise Ottesen-Jensen in Norway and Sweden—revitalized the cause by appealing to individuals' rights to health and happiness while organizing services to meet their needs. In a sense they were preaching to the converted, as birth rates were declining in all these countries. But they too realized that reaching everyone required promoting birth control not merely as a personal choice, but as a public good.[17]

Eugenicists had long criticized the indiscriminate promotion of birth control for reducing the fertility only of those who were educated enough to use it. The women who would rival Sanger in claiming international leadership tried to win them over. Ottesen-Jensen argued for the sterilization or isolation of the unfit so that society would not be unduly burdened. The first editorial of Marie Stopes's *Birth Control News* emphasized that her organization did "NOT desire to see the numbers of the English Peoples reduced (and must explicitly and emphatically separate ourselves from many cranks with whom the public may at first confuse us)." Instead, it wanted to "recruit" from "healthy, well-conditioned individuals only. Constructive Birth Control will fill the comfortable cradles, and empty the gutters." At this point Sanger agreed that it was important to try to increase births among the better-off. But she resented Stopes's emphasis of the constructive side of birth control, "as if it were something new." She declared, "There has never been any birth control movement that did not lay stress on the eugenic side of it." The question was whether to advocate contraception as a tool to advance eugenics or insist that eugenics was only one reason, among many, for giving more people the means to control their fertility.[18]

Even some Catholics appealed to eugenics in opposing birth control. Before *Casti Connubii,* theologians continued to debate whether eugenic sterilization might be permissible. But most Catholic spokesmen emphasized "positive" eugenics: preventing "race suicide," opposing immorality, and improving the conditions for raising children. Debating Sanger, one bishop warned that "the races from northern Europe," which he deemed the "finest type of people," were "doomed to extinction, unless each family produces at least four children."[19]

Eugenicists were inclined to reach out to Catholics rather than accommodate birth controllers. They prominently reviewed the works of Catholic thinkers in their journals, included them in conferences, and asked priests to serve on consultative committees. On the other hand, the Eugenics Education Society of Britain rejected a proposed lecture by Stopes, and their American counterparts refused Sanger's invitation to a national conference on birth control in New York. Archbishop Patrick Hayes dispatched the police to break it up, which backfired when Sanger declared that Americans had to choose between "Church Control or Birth Control."[20]

The scene could scarcely have been more different when eugenicists opened their Second International Congress in the new Hall of the Age of Man at the American Museum of Natural History. Amid displays celebrating the feats of Nordic man in his prehistoric struggles with nature, notables including Herbert Hoover and Alexander Graham Bell together with more than three hundred foreign delegates heard reports of eugenic research and organization in eighteen countries. The *New York Times* covered the Congress daily under headlines warning of "Tainted Aliens" and "Deterioration of Race." Eugenicists used their influence to demand immigration reform. Some of their exhibits—which included one warning of the "Approaching Extinction of the 'Mayflower' Descendants"—were reinstalled in the U.S. Capitol Building. Under the guidance of race theorist Harry Laughlin, the chairman of the House Immigration Committee, Albert Johnson, came to view their work developing the National Origins Act of 1924 as "more and more like a biological problem." The Department of Labor commissioned Laughlin as a special immigration agent "to find out whether it is possible to go into the matter of individual reputations and family pedigrees in selecting immigrants."[21]

Considering that this legislation would drastically reduce immigration from Catholic countries in southern and eastern Europe—a fact duly noted in Rome—American bishops offered surprisingly little opposition. But most of the bishops were of German or Irish descent. None were Italian. Whereas NCWC lobbyists sought exceptions for members of religious orders and other minor amendments to immigration restrictions, they were told to make "every effort" to defeat "iniquitous legislation" that would repeal bans on birth control. As for sterilization statutes, Catholic opposition was initially ad hoc and local, depending on the initiative of individual bishops. "Mainline" eugenicists like Davenport and Paul Popenoe opposed any alliance with the birth control movement because such an alliance would "take on a large quantity of ready-made enemies which it has accumulated"—that is, Catholics. Popenoe still considered the Catholic hierarchy neutral, if not potentially friendly, to their cause.[22]

But other American eugenicists believed that their very success in immigration and sterilization made it apparent that little else could be achieved without acquiring more allies at home and abroad, even if that also entailed taking on more enemies. If eugenicists could not find some way to reverse differential fertility, the proportion of pure-bred Nordics or Anglo-Saxons would continue to shrink. For more than twenty years, warnings of race suicide—even when they came from the White House—had had no discernible effect on the fertility of the fitter classes. Madison Grant himself died a bachelor. As the Harvard geneticist Edward M. East complained in one of the first academic studies of global population growth, one that had considerable influence on Sanger, "The eugenists have seldom gone farther than to make the pathetic suggestion that the whole current of society can be changed by interfering with the two little rills which flow from either side"—that is, by targeting the offspring of the most and least fit.[23]

East helped convince other scientists that those who seemed "fit" carried recessive genes, and "feeble-mindedness" could not be the expression of only one of them. Blunt instruments like forced sterilization or marriage restriction would therefore do little to improve the quality of populations. And even if it were possible to identify desirable genes, and even if premarital screening, prizes for "fitter families," and mass propaganda encouraged enough Americans to propagate them, this would not begin to address growing concerns about the proliferation of the unfit worldwide.[24]

That was the view of Lothrop Stoddard, a disciple of Grant who had a Ph.D. in history from Harvard. In 1920 he wrote a book called *The Rising Tide of Color,* which popularized the arguments for "World Eugenics" first put forth by Prescott Hall and E. A. Ross over the preceding decade. He even provided a pull-out, color-coded map indicating the "overcrowded colored homelands" from whence would come the "outward thrust of surplus colored men." Moreover, Stoddard claimed that white men were responsible for their own demise, because they had reduced famine and disease and had thus removed checks to population increase among other races. And because racial identity was more fundamental than national differences, World War I was really a civil war, one that left European peoples even more vulnerable to invasion.[25]

Stoddard therefore backed Hall's argument for isolating inferior races like "bacterial invasions . . . limiting the area and amount of their food supply." He also quoted extensively from Ross, including his forecast of a system of "dams against the color races" to "enclose the white man's world." He communicated this view together with his own experiences in characteristically vivid fashion before the House Immigration Committee. After traveling overseas, he concluded that foreigners generally disliked America but that almost the entire population of countries like Syria would immigrate if given the chance. If Congress did not stop them, tens of millions were ready to move, though most would be of the "organ-grinder and cart-pushing variety."[26]

The Rising Tide of Color was reissued in British and Australian editions and translated into German, French, Japanese, and Hindi. Though Stoddard was often dismissed as an alarmist and a crank even in his own day, key elements of his argument continually recurred in what was becoming a worldwide debate about population trends. The earliest exchanges occurred in Britain and the United States, the two countries that provided most of the leadership of population movements for decades to come. Some in America, like Edward East, shared Stoddard's vision of international relations as race relations even while dismissing the idea that white people could be overwhelmed. "Both heredity and environment," East insisted, "combine to render black-brown threats powerless." Only East Asians possessed the requisite intelligence and initiative to lead them, and they were too chauvinistic to unite with one another, much less with all the non-European peoples of the world. As long as barriers to immigration and miscegenation held, the real danger was that areas inhabited by Europeans would one day become as overpopulated and squalid as India and China.[27]

Other observers thought that population growth among European peoples was already well in hand, but that they could not be content with controlling their own numbers. Though he did not indulge in lurid race war imagery, Cambridge economist Harold Wright was skeptical about whether it would be possible to contain the growing population of China and Japan as they acquired modern armaments. He argued for "a world policy," but worried that national rivalries were leading in precisely the opposite direction. The influential demographer A. M. Carr-Saunders thought that war would inevitably result from differential growth rates among nations and races unless declining populations were provided "some form of international guarantee." Similarly, the former MP and editor of the *Edinburgh Review,* Harold Cox, thought that low-fertility nations needed to band together to defend themselves "against any race that by its too great fecundity is threatening the peace of the world." The psychologist and future leader of International Planned Parenthood, C. P. Blacker, worried that Asia and Russia might become a "solid block, determined to shake off the yoke of the Western Powers and of America." Birth control offered the only way to avoid a world war between East and West, but that required that every culture accept it.[28]

Even in France, Nobel laureate Charles Richet conceded that fertility decline was a parochial concern: "There will always be enough human beings on the face of the earth. In the near future, it is a plethora, and not a scarcity of men, that is to be feared." Reacting to an influx of immigrants, in 1926 high officials in the Interior Ministry warned that "the first waves of Orientals and Slavs that are breaking on France presage the invading flood that threatens to submerge what remains of our civilization."[29]

"Yellow peril" themes and imagery were now perennial features of discussions about world population trends. More remarkable is that almost all of these authors suggested that the populations of Asia and Africa had been growing relative to those of the European peoples. In fact, the situation was just the opposite. By 1930 their share of world population had declined to barely 60 percent, its lowest level in history. Many people seemed to *want* to believe that the populations of Asia and Africa were growing rapidly. India was already the most often cited example. Examining more closely what was happening there can reveal the agenda behind such misrepresentations.[30]

In 1924 Ross travelled to India vowing to Sanger that he would write a book that would entirely change the terms of debate. "I want to make my fight solely on the necessity of limiting births to balance the great reductions in mortality and to avoid entirely the moral question. Whatever is entirely necessary will in time have to be declared to be moral." Ross's book, *Standing Room Only*, included chapters titled "The Muzzling of Famine," "The Conquest of Disease," and "The Throttling of Pestilence." Most of his evidence came from the United States and Europe, but he did not hesitate to write that his argument applied to all of Asia as well, and especially India. Thus, he described how "holy places like Hardwar and Allahabad, where half a million pilgrims may throng, are carefully guarded against outbreaks of cholera." He praised provincial health departments and their publicity campaigns. "Crown of it all" was the Calcutta School of Tropical Medicine with its fifty researchers. He quoted approvingly the 1921 census report's contention that "systematized attack is being made on mortality at every point."[31]

It is understandable that British officials would insist that they were trying to reduce mortality, because all the census data indicated otherwise. Be-

tween 1911 and 1921, India's population grew from 315 to 319 million. So
discouraged were officials at the results that, for the first time in fifty years
of census taking, they decided not to make a life table, the standard means
of estimating mortality rates and life expectancy. When, years later, demographers decided to fill this gap and reexamined the 1921 census results,
they were amazed to discover that Indians had been dying at the annual
rate of almost 50 per 1,000, meaning that every two years the population
of India was literally decimated. This implied a life expectancy of only
twenty years. As for guarding against cholera, the festivals at Hardwar
and Allahabad regularly coincided with new outbreaks in northern India:
120,000 deaths in 1918; 150,000 in 1921; 67,000 in 1924. The Calcutta
school, which had only opened in 1922, was a step forward. But outside of
Calcutta and its sister city of Howrah, there were fewer than twenty health
officers for all of Bengal, which had a population of over 47 million people.
Nevertheless, East agreed that "English brains have made a new India in
50 years. Famine-stricken, pestilence-smitten, cobra-bitten India has been
given a new lease on life." Wright simply insisted that, although growing at
less than a quarter of the rate of England's population, Indians were "increasing with disquieting rapidity owing to the removal by British rule of
many of the checks to population."[32]

When East, Wright, and Ross were celebrating Britain's achievements in
bringing health and welfare to India, a scant decade and a half had elapsed
since famine struck the United Provinces and carried off some two to three
million souls—the fourth major famine in thirty-two years. It was only a
few years after the great influenza epidemic of 1918–19, when mortality
rates were as high as one in five among low-caste Hindus. There was some
improvement in life expectancy in the 1920s, but that would not become
apparent until the next census, in 1931. Until then, the Public Health
Commissioner did not announce any general decline in mortality, estimated little if any increase in population, and did not once mention any
prospect for rapid growth.[33]

Not long after Ross returned from India, another American arrived
there and decided to take a different, though complementary, approach.
Katherine Mayo was worried about Asian immigration and wanted Americans to understand the good work of the Raj. But British officials gently
dissuaded her from inquiring into public health programs. Instead, an
agent of Indian Political Intelligence suggested she should write a book
about the oversexed and unsanitary native, and the threat independent In-

dia might pose to the public health of the world. Mayo agreed that a report on prepubescent marriage, venereal disease, and bride-burning would be explosive. Her book, *Mother India,* ignited international debate about whether Indians were too backward to merit self-rule, especially because of their treatment of women.[34]

All of these books portrayed Indians as apathetic about public health, despite much evidence to the contrary. For Ross it was "humane administrators of Western training who are able to narrow the exits of Indian life *but not the entrances.*" In this way, they had upset "the age-old equilibrium of the Orient." Was it not therefore up to more hardheaded Westerners to right the balance between fertility and mortality, especially after what Mayo revealed about the plight of Indian womanhood? Even without making any claims about a "rising tide of color," which many did not find credible, Westerners could in this way be made to feel it was their duty to reduce birth rates abroad.[35]

For eugenicists more generally, it was necessary "to build the higher ideal of eugenic duty on foundations already laid for us by the Social Hygiene movement," as the 1927 program of the American Eugenics Society (AES) stated. That same year Holmes asserted in the *Buck v. Bell* decision that "the principle that sustains compulsory vaccination is broad enough to cover cutting the Fallopian tubes." Of course, the movement eugenicists intended to build would go in a dramatically different direction. Cooperation with physicians, public health officers, social workers, and reformers "should have the effect of converting much hygienic effort now distinctly dysgenic into effective eugenic influences."[36]

For those who were contemplating global population trends, like Edward East, the question was nothing less than, Who shall inherit the earth? His friend the biologist Raymond Pearl, who thought that a law of nature governed the rate and extent of any population's growth before "saturation," drew the following implication for the 1921 International Eugenics Congress:

Projecting our thought ahead for a moment to that time, *at most a few centuries ahead,* we perceive that the important question will then be: what kind of people are they to be who will then inherit the earth? Here enters the eugenic phase of the problem. Man, in theory at least, has it now completely in his power to determine what kind of people will make up the earth's population of saturation.

Pearl believed that politicians and the public should give scientists "a chance at directing the course of human evolution," as he told the 1912 Eugenics Congress. And now, as one of the highest paid professors at Johns Hopkins, a member of the governing councils of the National Academy of Sciences and the National Research Council, he had begun to insist. It was pointless, he said, for eugenicists to go on urging the fitter classes to become more fertile as a "transcendental social duty," or pretend that it was enough merely to target the obviously unfit. Real eugenics required much more:

> It is not only desirable in the eugenic interest of the race to cut down, indeed completely extinguish, the high birth rate of the unfit and defective portions of mankind, but it is also equally desirable because of the menacing pressure of world population, to reduce the birth rate of the poor, even though that unfortunate moiety of humanity be in every way biologically sound and fit.[37]

Like the anti-immigration campaigns of a generation before, the scope and ambition of calls for controlling fertility grew to become truly global. Beginning with popular attacks against the patently unfit, progressing with expert opinion that abjured cruder kinds of racialism or class prejudice while excluding whole nations, it had finished by targeting "that unfortunate moiety of humanity." But blaming the "pressure of world population" still served to mask the more personal antagonisms of insecure elites. Receiving a letter from a young Harvard biologist asking what works he ought to be reading, Pearl asked East, "Who is this Jew of yours named Gregory Pincus[?] . . . I like his nerve and think he ought to go far"—Pincus would invent the birth control pill—"but just how did he ever get the notion that I have no other amusements in life except making bibliographies for lazy Jews." Pearl worried that they would inherit the world and later argued in print that blacks had smaller brains.[38]

In 1927 Pearl broke with mainline eugenicists like Laughlin and Charles Davenport of the Eugenics Record Office, ridiculing both their scientific pretensions and the overt class prejudice that had limited their political effectiveness. Others, including East, Blacker, and Ross, along with NYU sociologist Henry Pratt Fairchild and C. C. Little, president of the University of Michigan, chose to retain leadership posts in eugenic organizations even while working with Sanger to create new institutions to control *populations,* not just "problem families," an approach that could define whole nations and races as problematic.[39]

For East, "The really useful eugenics is properly directed birth-control. . . . The remedy proposed is to promote birth control at the lower end of the social scale." Though he worried about birth rates among the poor within the West, Ross and Pearl were thinking of "the social scale" as extending to the rest of the world. Domestic and international concerns were linked not just conceptually, but concretely, such as in the development of new contraceptives. The first Rockefeller-funded study in 1927 aimed at "some simple measure which will be available for the wife of the slum-dweller, the peasant, or the coolie, though dull of mind."[40]

This alliance between eugenicists and birth controllers was not the only way a transnational population control movement could have developed. Eugenicists in Europe and the United States might instead have linked up with their counterparts in Asia and Latin America. Some were willing to act as patrons and cultivate protégés. Charles Davenport taught Guangdan Pan, who founded the Chinese Eugenics Association in 1924, and mentored Domingo F. Ramos, a Cuban physician who organized Pan-American eugenics conferences. Similarly, the Japanese journalist Shigenori Ikeda became an admirer of Popenoe when both resided in Germany in the early 1920s, and he helped found the Japanese Eugenics Movement Association upon his return. Both Davenport and Laughlin assisted in the development of the principal eugenics journal in Japan. And the NYU sociologist Henry Pratt Fairchild, who compared immigrants to inassimilable bacteria, was tolerant enough to serve as dissertation advisor to Sripati Chandrasekhar, who would one day become India's minister of health and family planning.[41]

But eugenicists in the United States and Europe had difficulty deciding "the desirability of inviting other races" into their organizations. After the International Commission on Eugenics was founded in 1921, it turned down an application from the Indian Eugenics Society in Lahore out of deference to the British. And while Rockefeller money flowed freely to eugenic research institutes in Europe, eugenicists in India, Japan, and China received virtually no monetary assistance.[42]

The most ambitious effort to create an international eugenics regime outside Europe and North America was the First Pan-American Conference on Eugenics and Homiculture held in Havana in 1927. The conference revealed the problems any such effort would entail. Working with Ramos as secretary-general, Davenport tried to persuade the delegates to adopt a "Code of Eugenics and Homiculture"—a kind of eugenic NAFTA. It called

for empowering authorities in each country to identify individuals who
were eugenically "bad" or "doubtful," isolate or sterilize them, and annul
their marriages. Free migration in the Western Hemisphere would be limited to nationals of countries that conformed to these principles. In presenting the code Ramos delivered a eulogy to "the superior white type" and
a warning against racial crossing, doubtless insulting or at least discomfiting
several members of the audience. Many delegates were shocked by his proposals. Though parts were accepted in principle, the preceding discussion
made it clear that the whole code would remain a dead letter.[43]

Rather than accept U.S. immigration laws as a model, some of the Latin
American representatives in Havana seized the opportunity to attack them.
These laws had alienated eugenicists in other regions as well. Ikeda, for instance, raged against "Yankee-Monkeys" for abrogating the diplomatic ac-
cords that had permitted Japanese immigration. Italian eugenicists also criticized the unilateral nature of the U.S. legislation and called instead for an
international convention. But American eugenicists considered the National Origins Act to be their proudest achievement. Indeed, following the
failure of the Pan-American initiative, the AES launched a campaign to
broaden the law to exclude Latin American immigrants.[44]

American and European eugenicists valued international networking as
a way to establish the credibility of eugenics as a real science. But many
viewed their ultimate goal as preserving and promoting the Nordic race,
not aiding all races equally. Such a project would obviously not appeal to
those they believed to be their biological inferiors. Some eugenicists in Japan, China, Mexico, and Brazil actually considered ways to "whiten" their
populations as a way to improve them, or at least defended racial intermixing as having potentially eugenic results. Increasingly disaffected with his
Anglo-Saxon counterparts, Corrado Gini of Italy called for more research
to determine which kinds of racial intermixing produced favorable results.
But for people like Davenport, Ross, and Leonard Darwin, preventing further miscegenation was a top priority.[45]

The most important reason that birth controllers succeeded where
eugenicists failed in organizing a worldwide movement is simply that they
took the initiative. With few accomplishments, less public credibility, and
little access to policymakers, they agreed on the need to ally with eugenicists
in every country. The eugenicists, for their part, were divided, especially
over contraception, and could respond only individually to these initiatives.
Many opted out. Davenport, for instance, who had the best connections in

Asia and Latin America, would never go along with colleagues who wanted to work with birth controllers. Pan, the leading eugenicist in China, also worried that widening access to birth control would reinforce reverse selection. Japanese eugenicists subjected it to even more severe attacks as "racial (or national) suicide."[46]

Such commonalities encouraged orthodox eugenicists to continue exchanging ideas and comparing research, but they did not add up to a positive program that could inspire a broader movement. Upon assuming the presidency of the International Federation of Eugenic Organizations in 1928, Davenport still thought that the greatest potential for growth would come "where whites are in contact with other races," proposing a meeting in South Africa. Only seventy-three people could be induced to attend the Third International Congress when it was held in New York in 1932, the last such meeting. The International Federation never linked up with new and influential eugenics organizations that were springing up across Latin America. By that point Sanger had helped to launch population organizations with global ambitions, and birth controllers had begun to mount campaigns across Asia.[47]

Of course, eugenicists failed to organize a vibrant international movement of their own not merely because of what they did wrong but because of what Sanger did right. If Pearl, East, and Little were confident that providing the poor and ethnic minorities with birth control would reduce differential fertility, it was because Sanger collected socioeconomic data at her clinic to prove it to them. She invited them onto her board, deferred to their judgment, and rewarded them with research money. She was also adroit at co-opting eugenics arguments and drawing analogies between controlling immigration and controlling births, suggesting that parents should be required to "apply" for babies as immigrants applied for visas. The backing of scientists like Pearl, in turn, helped Sanger network with another key constituency, medical doctors, including the leaders of the Committee on Maternal Health and the American Medical Association (AMA).[48]

The increasing caliber, scope, and synergies of Sanger's network became apparent as she began organizing an international conference to take place in New York in 1925. Consider, for instance, her itinerary during a preparatory visit to London. In a thirty-six-hour period she dined with her friend and lover Havelock Ellis, the world's leading sex researcher, saw John Maynard Keynes the next day for lunch, had tea with the author and MP

Harold Cox, shared another meal with reigning Malthusians Alice and Charles Vickery Drysdale, and met the next morning with H. G. Wells (another lover). Among others whom she saw on this trip, often multiple times, were Lord Buckmaster—the former lord chancellor—George Bernard Shaw, and two physicians to the royal family.[49]

Even so, when Sanger soon afterward heard that professors in India had been fired because they advocated birth control, she did not check with her friends in London before offering financial assistance and requesting a plan of action. Unlike the eugenicists, with their clumsy national committee structure and cumbersome procedures, Sanger acted decisively in the international sphere when a danger or opportunity arose. Thus, when she heard that Gandhi had condemned artificial contraception, Sanger immediately solicited an endorsement from the Nobel Prize winner Rabindranath Tagore.[50]

Thanks to Sanger's leadership, the Sixth "International Neo-Malthusian and Birth Control Conference," held at the Hotel McAlpin in New York City, was the most important population meeting to date. What was most impressive was not so much the extensive press coverage, or the sixteen countries represented, or the notables in attendance—such as Keynes, Lytton Strachey, and Norman Thomas—or the many more who sent supportive messages, including W. E. B. Du Bois, Bertrand Russell, and Upton Sinclair. In these ways it merely equaled earlier meetings of eugenicists. Rather, it was the fact that Sanger had finally convinced the most influential eugenicists to attend *her* meeting and join with every other constituency to discuss their common interests. These eugenicists included the president of the AES, Irving Fisher, the two presidents who would succeed him, and nearly a dozen members of its Advisory Council. Together they met with veteran Malthusians like the Drysdales and Binnie Dunlop, feminist birth controllers including Ishimoto, Jacobs, and Thit Jensen, race theorists Georges Vacher de Lapouge and Ellsworth Huntington, eminent geneticists such as Julian Huxley and Norway's Otto Lous Mohr, and leading demographers like Gini and Louis Dublin.[51]

So broad and distinguished was the assemblage that Pearl peevishly complained, "The Birth Control movement after a long and bitter struggle has attained a certain academic sort of respectability. Harmless people are at any rate allowed to meet together and discuss it." He insisted to fellow delegates that "the time is hard upon us when a certain militancy in its advocacy, even at some possible expense of respectability, is called for. The highest interests of humanity demand that the menace of population growth and of the differential birth rate shall be effectively met if our civilization is to persist in anything like its present form." Others used the conference to call for "an international movement for the restriction of population as a necessary step to the avoidance of warfare," as Cox put it. Will Durant was more pointed. "To offset the so-called 'yellow peril,'" he urged the conference "to spread Birth Control knowledge abroad so as to decrease the quantity of peoples whose unchecked reproduction threatens international peace."[52]

This can be considered the first meeting of a population control movement that would one day span the globe. Yet major differences remained unresolved, and some would never go along with the idea that birth control was a means to control populations rather than liberate individuals. The population "menace" was defined in multiple, incommensurable ways. Any campaign against the "yellow peril" would alienate potential allies abroad even before it started. An Indian nationalist in attendance, Taraknath Das, countered with evidence demonstrating the "white peril." He endorsed the dissemination of birth control, but only as part of a program that aimed at poverty, poor health, and ignorance. Indeed, he thought that its use would increase the size of population by reducing mortality. It would also "afford greater freedom to women and greater opportunity for real education," inculcating pacifist ideals. But championing birth control as a human right also aroused dissent. The demographer Louis Dublin argued that the "very life of a state is involved as soon as we begin to tamper with who shall and who shall not be born." He also reminded them that there had been little or no testing of contraceptive safety and reliability.[53]

The conference did agree to resolutions calling for improved contraceptives as a way to reduce war, poverty, infant mortality, and immorality. But attendees could not entirely paper over their differences. Sanger felt compelled to repudiate another resolution, which had encouraged larger families among "persons whose progeny give promise of being of decided value to the community." Drafting an editorial to quiet the ensuing controversy, she argued that the resolution was irrelevant because "the progeny of all parents will give unsuspected promise" if they were provided with the means to plan their families rationally. Pearl and East warned her that this faith was unwarranted. She was reduced to suggesting weakly that "at least all normal children . . . give promise," and agreed that the principal task was to remove any obstacles that prevented "diseased, overburdened and poverty-stricken mothers" from controlling their procreation. "No matter what you say Birth Control is part of a eugenical program," East insisted. "It is an aspect of a larger whole, but it is the key."[54]

The most important questions remained unresolved: how would fertility be controlled, by whom, and to what end? Broad agreement could be found between birth controllers and eugenicists to concentrate on "those who need contraceptives most," finessing the question of whether that need was felt by the parents themselves. But they still needed to decide how any international campaign would be organized. This was a question not only of leadership, as between Sanger and her new allies, but also of structure. There was no consensus, or even focused discussion, on the respective roles of nongovernmental organizations, state agencies, or international institutions.

After the conference Sanger advanced on all fronts. She encouraged the development of national voluntary organizations both abroad, as in India, and at home, through the American Birth Control League (ABCL). She wanted not merely to overturn the Comstock Law, but to involve the U.S. government in regulating reproductive behavior—even to the point of calling for "Civil Service examinations for parenthood and a 'parenthood license.'" The Catholic bishops thwarted this campaign. But, realizing that having Sanger arrested or engaging her in debate only brought more publicity to her cause, they began working behind the scenes. To counter her effort to enlist doctors in demanding the right to prescribe contraceptives, the NCWC persuaded allies in the AMA and the American Gynecological Association to pass resolutions refusing support for legalization. Sanger decided to redirect her energies abroad. In June 1926 she announced that she was taking a leave of absence from the ABCL to plan an international conference in Geneva.[55]

The fight over birth control in the United States redounded in the halls of the Vatican, where Cardinal Raffaele Merry del Val, secretary of the Holy Office, demanded reports on the spread of this "plague." The apostolic delegate in Washington assigned the task to Rev. John Montgomery Cooper, an anthropologist at Catholic University. Cooper was a member of the AES Committee on Cooperation with Clergymen. He could therefore pass along the latest eugenics research indicating that wealthier, better-educated, and urban people were more likely to use contraception. He suggested the Church carry out a census in some dioceses, especially if it could be done without parishioners' knowledge. Even without hard data on contraceptive use, he predicted that the decline in immigration would make the reproductive behavior of American Catholics more like that of the rest of the nation: "We are destined almost inevitably to witness a great increase in the prevalence of the practice among Catholics in the United States within the next generation or two."[56]

The unilateral U.S. decision to isolate itself from global migration had the paradoxical effect of making population even more of an international concern. In Japan it stoked renewed interest in Manchuria as an outlet for surplus population. In Italy it renewed calls for the colonization of Africa. It also contributed to Gini's determination to unite pro-natalists in Catholic and "Latin" countries. He had a powerful backer in Mussolini, who appointed him to lead a new agency charged with centralizing all statistical work in the Fascist state. The normalization of Italian-Vatican relations between 1926 and 1929 created a favorable climate to promote areas of common interest. Publishing contraceptive information became illegal, and doctors had to report all suspected abortions. In his famous May 1927 Ascension Day speech, Mussolini decried the diseased state of the Italian population, launching a "demographic battle" to increase population by 50 percent before 1950. He vowed to govern according to the dictates of statistical research. Gini became Il Duce's chief demographic advisor.[57]

In Sweden, on the other hand, U.S. immigration restriction had "caused an almost complete change in public opinion on these matters," as G. Westin Silverstolpe, a young socialist and lecturer in political economy at the University of Göteborg, remarked. "Nowadays, very few people look upon birth control as an evil." Conversely, many in France feared that America's new policy put them "in danger of becoming a receptacle for the outcasts," as the chair of hygiene at the Sorbonne, Léon Bernard, put it. Like Congressman Johnson and his eugenicist advisors, high officials of the French Interior Ministry now considered immigration "a biological problem," and proposed rejecting not only the sick but also those judged more susceptible of becoming sick based on their race or religion. The majority of political elites still favored immigration, though they too viewed it as a biological process: a means of repopulating France. In August 1927 they succeeded in passing a new law that accelerated the rate of naturalization, only to provoke vicious criticism in the press of how the true French were being replaced by the "vermin of the world." France's naturalization law was also denounced in Fascist Italy. Mussolini sought to halt emigration, even going so far as to authorize border guards to shoot unauthorized migrants on sight. He also promoted nationalist organizations among emigrants abroad and paid the return fare of pregnant women so that their babies could be born in Italy.[58]

Immigration was also a growing concern in Latin America. In Brazil as in France, U.S. immigration restrictions were invoked when congressional deputies introduced legislation that would bar blacks, Asians, or both. And like Henry Cabot Lodge a generation before, Brazilian legislators cited the French race theorist Gustave Le Bon about the dangers of racial intermixture. Although there was resentment in revolutionary Mexico about the treatment America meted out to migrant workers, there was also an increasingly violent campaign against Chinese immigrants. Such moves, in turn, encouraged the United States to regulate migration even more strictly and deport legal immigrants now judged to be undesirable. As Laughlin explained, "We can expect little or no further superior human seed stock through exiling or 'dumping' by foreign countries. Countries are continually increasing their appreciation of good human stock."[59]

Bitterly nationalist debates about migration were a particular concern of the International Labour Office (ILO) and its French director, Albert Thomas. This bearded, bespectacled socialist had tremendous ambitions for the new organization, founded along with the rest of the League of Nations by the Versailles Treaty in 1919. For years he had worked for international conventions requiring equal treatment of foreign workers. He saw them as a way "to encourage the working class to become interested in the ILO, and more generally in international questions." Beginning in 1921, meetings of the ILO had occasioned calls for a more balanced distribution of world population. By 1927 Thomas had a truly ambitious idea for how such questions might be arbitrated. But he needed an opportunity, and an audience, to present it.[60]

Thomas therefore encouraged Sanger in her efforts to hold an august gathering in Geneva that could seize the attention of the international community. He directed the chief of his Migration Service, Louis Varlez, to provide researchers and translators. Sanger opened a local office, secured the use of a friend's château, and hired a full-time organizer, Edith How-Martyn. A suffragette educated at the London School of Economics, How-Martyn was willing to do the "donkey work" for a nominal salary.[61]

At first Sanger had planned another conference that would promote birth control as a panacea for the world's problems. But officials from the

ILO along with Pearl and others persuaded her to avoid propaganda that would repel the influential people she most needed to reach. Even so, while the announcement insisted the proceedings would be "strictly scientific," it also suggested that population problems merited "concerted international action," and that "from such a conference will come an international movement." When Pearl discovered that people he considered cranks were invited, he demanded that their invitations be withdrawn. Otherwise he not only would withhold conference funds he controlled through the National Research Council but also would use them to sabotage the entire effort. Sanger went along, even though it was she who had led the fund-raising. She also ceded the presidency of the conference to Sir Bernard Mallet, apparently because, in addition to being president of the Royal Statistical Society and former registrar general, he was a friend of League of Nations secretary-general Sir Eric Drummond.[62]

Nevertheless, Drummond refused to attend the conference and actively discouraged Secretariat officials from going, explaining that it would raise questions "which arouse the strongest national feelings." A devout convert to Catholicism, Drummond seems to have also found his own feelings aroused. Officials from the Health Section warned that the League would eventually have to confront population problems "whether it desires to do so or not." Some were sympathetic to Sanger's cause, but even they ruled out official participation. Albert Thomas was the only one to accept an invitation, and he addressed the conference as a "citizen of the world."[63]

The meeting would therefore be dominated by academics, especially biologists. Three days before it was to begin, Mallet struck the names of Sanger and the women helping her from the program. Sanger prevailed on them to continue working, even while bitterly complaining that "the subjugation of woman is inherent in the European male." In a draft speech she noted that "the question of population has been considered by specialists. . . . Few however have viewed the question from the viewpoint of those who are producing the populations, the parents—and especially the mothers." She was not permitted to deliver the speech.[64]

How-Martyn tried to buck up Sanger's spirits. Unable to sleep, she stayed up late one night and wrote her boss with this advice:

Keep your name off every bit of printed trash but keep yourself in everything—wear all your prettiest frocks—let all your brilliant organizing ideas have free play behave as though Pearl & co were kissing your hands instead of stabbing you in the back, smile and charm everyone as though you and Sir B. were joint presidents and by God you surely shall not wait for heaven for your reward. . . . We will be even with the "distinguished scientists" yet and send them back to their flies and mice having been taught how unwise it is to deal unjustly or to scorn a beloved woman.[65]

The World Population Conference opened on August 29, 1927, in the Salle Centrale of Geneva. Over the following days Pearl presented his theory of how populations of yeast, fruit flies, and people all grew on an S-shaped logistic curve. Gini argued that the fertility rate increased over the life of races, nations, and "all other animal and vegetable species" before "demographic decadence." And Henry Pratt Fairchild suggested that, at 104 million, the United States had exceeded "optimum" population. All of these papers were illustrated with mathematical formulas, the insignia of their scientific rigor. Papers showing, on the one hand, that low fertility correlated with high intelligence and, on the other, that fertility was entirely dependent on environmental factors, were equally replete with exhaustive tables of data. And if there was one thing upon which all the assembled scientists agreed, it was the need for governments to provide more statistical data. Otherwise, Sanger's hope that they would agree to take action to control world population was bitterly disappointed.[66]

On the fifth day, near the very end of the conference, Albert Thomas finally took the podium to address the assembled delegates on "International Migration and Its Control." He knew that presenting his ideas might cause a "scandal," but judged it essential nevertheless, "for the future of the world." He began by noting that migration, fertility, and mortality were the three determinants of population, and that migration was "the most susceptible to direct intervention and control." It was less and less a function of individual free will and increasingly an instrument of state policy—whether the "protectionist policy" of the United States, or the recruitment practiced by the British empire and France, or the nationalist agenda of Italy, all of which had aroused international tensions.[67]

"Has the moment yet arrived," Thomas asked, "for considering the possibility of establishing some sort of supreme supernational authority which would regulate the distribution of population on rational and impartial lines, by controlling and directing migration movements and deciding on the opening-up or closing of countries to particular streams of immigration?" He quickly admitted that the obstacles seemed insurmountable, but insisted, "If it be no more than an ideal, it is at least an ideal which can and should serve to direct our deliberations and our efforts."[68]

Thomas advanced a set of principles to resolve migration questions: individuals had no absolute right to migrate as they pleased, but could migrate only on certain conditions; these conditions could be set by an international authority with the agreement of the concerned parties; and receiving communities, subject to some oversight, had the right to select immigrants. Such rules, he concluded, "might, failing a policy of compulsory birth control, form a protection against excessive growth of certain sections of the world population where such growth may represent a danger for neighboring communities." In the meantime, he called for an international scientific institute to both study population problems and "make constant and untiring efforts to acquire the necessary authority" to address them.[69]

Thomas's speech was followed by a respectful discussion, but nothing else. The leaders of the conference ignored his proposals, just as they ignored the woman who had invited him. Fairchild, a member of the conference's Advisory Council, was on record as supporting "the complete elimination from international law and international thought of any recognition of migration as a potential remedy for population evils." It was only at the concluding dinner later that night, during Mallet's farewell address, that Sanger's presence was finally acknowledged. Mallet said that he could not sit down without first giving "an expression, perhaps too long delayed," of appreciation. Clapping began at one table, almost certainly that of her long-suffering staff, and slowly spread, building momentum. Finally, at least a third of those present were on their feet, and the British led a round of "For She's a Jolly Good Fellow." For Sanger, who now had the job of editing all of the proceedings, it must have been bittersweet.[70]

Afterward, in a meeting with League officials, the conference leadership explained what they were really after: an international union of population scientists. Pearl would head an organizing committee dominated by like-minded British and American scientists, including F. A. E. Crew, Mallet, East, and William Welch, a colleague of Pearl's who was also chair of the Milbank Memorial Fund's advisory board. The International Union for the Scientific Investigation of Population Problems (IUSIPP) would merely be a federation of national committees and would have no formal connection with any international organizations, further reducing the possibility that it

Margaret Sanger's absence from this commemorative caricature of the 1927 World
Population Conference shows how marginalized she and other women were from the
proceedings. Margaret Sanger Papers, Sophia Smith Collection, Smith College.

would spur an international movement, or even an internationalist spirit.
Thomas thought the meeting was "ridiculous" and left early. "Is the idea of
an institute completely abandoned . . ." he asked Varlez. "Has Miss Sanger
completely disappeared?" In fact, Sanger had left town complaining of ill
health, and canceled a trip she had planned to India. She wrote Pearl asking
about a rumor that the money he controlled through the National Research
Council was obtained only on condition that she be kept off the program.
He pretended not to understand. With the IUSIPP safely launched and the
proceedings of the conference completed, he had no further need of Sanger.
Shortly afterward, he resigned from her board.[71]

The opportunity to push for an international organization to promote
solutions to population problems was now lost. As Thomas and others
noted, of all the factors that made up population, migration was the easiest
to regulate. Unilateral moves by individual states had created tensions that
made international coordination seem desirable, even imperative. If it was
impossible to have a proposal along these lines seriously considered at an

international conference with no official government representation, how much more difficult would it be to advance international policies in the field of birth control?

Though ineffectual in promoting birth control, the Geneva conference provoked organizations favoring large families to mobilize against the "blind, fanatical, or criminal proponents of doctrines of death." Largely at the inspiration of the NCWC and its lay affiliates, they came together in the Ligue Internationale pour la Vie et la Famille. The test came five years later, when the League of Nations Health Organization published a report suggesting that, in cases where ill health precluded bearing children, providing instruction in how to avoid pregnancy was preferable to abortion. The Ligue Internationale, the NCWC, and the Vatican diplomatic corps launched a coordinated attack, working through the Irish, Polish, and Italian delegations in Geneva. The report was first amended to acknowledge religious objections and then suppressed in its entirety. The incident contributed to the ouster of the director of the League Health Section.[72]

The IUSIPP, the only tangible achievement of the Geneva conference, was hobbled from the start by academic rivalries. Pearl had been allowed to set the agenda because he came to Geneva in control of sizable foundation grants and the promise of obtaining more—enough to support an international research program. Together with East and Welch he succeeded in obtaining $10,000 a year (about $115,000 in today's dollars) for three years from the Milbank Memorial Fund, even though the head of its Technical Board was "a little disappointed" that the Union would not research birth control. They sought backing with a sense of entitlement and scant regard for what anyone but biologists thought of population. This prejudice was only confirmed when the Laura Spelman Rockefeller Foundation rejected a much larger grant application. Blaming the rejection on Edmund E. Day, an economist on the foundation staff who would go on to become president of Cornell, East wrote that sociologists and economists "exhibit the lowest order of human intelligence in their researches, and this results in an inferiority feeling about biologists." East added, "Day is out to knife us," and Pearl agreed.[73]

Though Pearl did not at first realize it, he had acquired an even more dangerous adversary. In the spring of 1929 he was nominated to join Harvard's faculty and reorganize its biological research program. But an influential professor at the School of Public Health, Edwin B. Wilson, held him in contempt. As president of the Social Science Research Council (SSRC), he blocked grants to the IUSIPP. Wilson even persuaded the Harvard Board of Overseers to reject Pearl's appointment, forcing him to return to the president of Johns Hopkins hat in hand to ask for his job back.[74]

Wilson then had the grim satisfaction of responding to an inquiry from Gini about whether the United States would be represented at the next meeting of the Union he was organizing in Rome. Wilson described the whole debacle, culminating in the joint decision of the SSRC and National Research Council—really, his own decision—to dissolve the U.S. committee of the IUSIPP. Any future grant application would likely meet "a reaction of fatigue or disappointment." Two months later, apparently having no reply from Gini, he wrote again to make it painfully clear: "Just one word more. . . . What Pearl led you to believe might be forthcoming I do not know. He had no business to promise anything which had not been agreed to." Pearl's work was widely viewed as unsound, Wilson wrote, noting for good measure that Gini himself was seen as so close to the Fascist movement as to make a new grant all the more unlikely. "I do not know whether there is any such thing as the scientific study of the population," Wilson concluded, ". . . it begins to look as though the organization of the Union had been a mistake, that things had not been thought out well enough in advance."[75]

This feud had far-reaching consequences. A new U.S. committee was formed, but under Gini's friend Louis Dublin, who was vehemently opposed to birth control. Without adequate funds, the IUSIPP's research program languished. Most serious of all, Pearl's failure together with Wilson's insulting rebuff both enabled and encouraged Gini to assert leadership of pro-natalists worldwide dedicated to a more "positive" vision of eugenics. The schism would hobble the new Union for a decade. And, with the onset of the Depression, fund-raising of every kind became far more difficult.[76]

The World Population Conference of 1927 was therefore a massive setback, not only because of the wrongs done to Sanger, but also because she had not yet come up with the right strategy to mobilize an international movement. In the months leading up to the conference, even sympathetic observers at the ILO found her uncertain and even confused about what she meant to achieve. She surrendered control over the proceedings to Mallet and Pearl without securing guarantees for what they had promised in exchange. And she let them set the agenda without letting her most important ally, Albert Thomas, know what he was walking into. Sanger seems to have assumed that science would provide an apolitical and respectable foundation for an international movement, not realizing that it was an arena with personality clashes, patriarchal attitudes, and bureaucratic politics as vicious as any other.[77]

Yet the conference did show how people had begun to reimagine international relations in terms of the relations between different populations as well as the institutions—actual or prospective—that might regulate them. The "scientific" papers so valued by Pearl and his colleagues now seem less significant than the more free-ranging discussions about the political implications of population trends. Speakers drew analogies between free migration and free trade, between population control and arms control, and between human and social bodies. They were grasping for a new analytical and legal framework. Was migration akin to trade or invasion? Were migrants like commodities or like weapons? Could a nation really be construed as a body, with its own biological processes and pathologies—including parasitism, overfeeding, and old age?

One of the common themes in these discussions was that population growth together with improvements in transport and communications had made the world seem small. "Our hands, our voices, stretch from continent to continent. We have become neighbors, whether we cared to be neighborly or not," as East put it. "This problem of an isolated population does not exist," another conference participant agreed. "In reality, the whole area within which migration is possible is a unit."[78]

Even some of those who bitterly opposed birth control, such as the powerful Catholic layman Constantine McGuire, agreed that population problems could not be addressed or even contained within one country. An economic consultant who had already been knighted by the Vatican, McGuire was asked to undertake a major study on fertility declines for the apostolic delegate and Cardinal Patrick Hayes of New York. McGuire noted that this "pathological condition" was spreading to all the more important nations of the world, and could not be understood in "geographically water-tight compartments." The whole world was headed toward "a catastrophic decline in population."[79]

Most of the leading demographers disagreed. Pointing to what they viewed as the danger of population growth in resource-poor areas, they suggested that states might have to cede territory, dismantle tariffs, and drop immigration restrictions. What all had in common was a belief that narrow definitions of national interest and a strict defense of state sovereignty would stand in the way of addressing population problems—indeed, might even bring on a new world war. This internationalist sentiment was not necessarily benign. A Finnish diplomat—soon to become foreign minister—argued at the conference, "The worst enemies of the great statesmen are not the nations living on the other side of the political frontiers, but are all the anti- and a-social forces keeping even the highest of nations internally weak and socially sick."[80]

During the following decade, more states followed Italy's example in launching pro-natalist campaigns while demanding new territories to settle. But the tendency to treat population trends as matters of international concern opened for consideration a whole range of potential responses, not just war and empire. Albert Thomas suggested that the right of overpopulated nations to occupy underpopulated territories came only on condition that they first tried to solve their own population problems. As Julian Huxley observed in a report for the *New York Times,* "All present realized that the United States or Australia has precisely the same right to demand that the population-exporting countries should take steps to lower their birth rate as each has to demand an open-door from the population-importing nations." Suggesting that nonracist immigration policies and population stabilization were international obligations was a potentially world-changing idea, one that could also link population control to a range of other new or emerging international norms, including trade liberalization and aid for economic development. Its full potential would become apparent once there were international institutions that could issue such demands in the name of all humanity.[81]

Equally revolutionary was a "little germ of an idea" that came to Edith How-Martyn when she spent a day with Sanger after the end of the conference: "that the message of sex freedom [as] the basis of all freedoms for women shall be carried to the women who need it most all the world over." The idea of birth control as a progressive cause would have to be carried forward at the grassroots level. Meanwhile, state population policies became increasingly nationalistic and even belligerent. They would continue to rest on the backs of the most disenfranchised members of society, women and children first.[82]

3

POPULATIONS AT WAR

In September 1931, four years after Margaret Sanger had proclaimed the gathering of a peaceful movement to shape the demographic destiny of all humankind, population activists and scientists the world over were arrayed in hostile camps. Beginning a year earlier, four different international conferences had convened to claim leadership. The first, organized in Zurich by Sanger herself, was limited to discussions about contraceptive technique among doctors and clinicians. "All theories, all propaganda, all moral and ethical aspects of the subject," Sanger explained, "were left in abeyance—practically forgotten—in the unanimity of cool scientific conviction that today contraception as an instrument in racial progress is on the way to be reliable and efficient and may in the very near future be perfected."[1]

In fact, Sanger's conference was sharply divided over the very definition of birth control—specifically, over whether birth control included abortion. A week later a congress of the World League for Sex Reform convened in Vienna to take on a broader, and braver, agenda. Even though roughly a thousand came to Vienna—ten times as many as were in Zurich—they managed to agree on a common set of policies, including birth control, sex education, and "the liberation of marriage (and especially divorce) from the present Church and State tyranny." They called homosexuality abnormal, but not sinful, and insisted that "sexual acts between responsible adults, undertaken by mutual consent, [were] to be regarded as the private concern of those adults." But Sanger was not there, having withdrawn for fear that the cause of reforming sexual mores would discredit her efforts to bring birth control into the mainstream.[2]

Sanger's scientific conviction that racial progress through birth control would soon be perfected was not shared by scientists themselves. By this point, nearly all the leading geneticists had disowned eugenics as a practical program. Raymond Pearl had described a field in disarray in his parting address as president of the International Union for the Scientific Investigation of Population Problems at its London assembly in June 1931: "It requires no expert to perceive that the growing hordes of people on the face of the earth are constantly and increasingly adding to the economic and social difficulties of an already sufficiently harassed world." But while for some the falling rates of fertility in Europe and the United States promised greater prosperity, others saw "a danger to Western civilization in an inevitable struggle with teeming yellow and black populations."[3]

Corrado Gini allowed for no doubts when he opened a World Population Congress in Rome three months later in defiance of the Union's governing council. "For too long," he declared, "students of population problems have based their discussions on Malthusian premises." Many more eugenicists assembled before him, such as Herman Muckermann of Germany and Lucien March of France, were equally convinced that population "quality" depended above all on maintaining quantitative growth. With the backing of Mussolini himself as honorary president, Gini persuaded the Congress to adopt a resolution warning all the advanced nations that they were facing decline.[4]

Over the following decade several attempts were made to reconcile competing activist groups and professional associations. Cooperation seemed all the more imperative as fertility decline now affected "all of modern civilization," as Fernand Boverat declared in the Salone Giulio Cesare at the Rome Congress, "as much in Moscow as in Tokyo." This inspired an outpouring of books with such titles as *The Twilight of Parenthood, Le Destin des Races Blanches,* and *Sterben die Weissen Völker.* Some leading researchers still assumed, based on the European experience, that population growth was slowing worldwide. But during the 1930s all the most important imperial possessions, including Egypt, India, Indochina, Korea, North Africa, the Philippines, and Java—the core of the Dutch East Indies—reported steady and sometimes rapid growth. Even Gini's conference, intended as a concerted attack on Malthusianism, heard a Japanese official endorse birth control because Korea and Formosa were already too crowded to permit further settlement. The chief of statistics in the Dutch East Indies agreed, voicing an argument that would soon be heard across the colonial world:

"For decades to come the problem of native prosperity in Java will mean the problem of Java's demography."[5]

Nevertheless, negotiations to forge an alliance of population scientists and activists repeatedly failed. Indeed, these efforts are all but invisible alongside the genocidal programs of the period. One of the hallmarks of the more extreme population politics was a common rhetoric that construed political problems as biological problems and human history as natural history. Class, ethnic, or racial enemies were rendered as existential threats, variously semihuman, subhuman, vermin, or worse, requiring a purge of the body politic. Women who failed in their duty to produce healthy workers, soldiers, and settlers were also targeted, as the pro-natalist policies of France and Italy spread across Europe and beyond. The home front of the coming world war witnessed hostilities—and casualties—even before armies clashed, as belligerent nations drafted women to fight their "demographic battles."[6]

A week after the conclusion of the Rome congress, the Japanese Army inaugurated this new era when it invaded Manchuria. Propagandists argued that settling its supposedly open spaces would purify a rural society grown diseased and restive because of overpopulation. During the long struggle with China, commentators on both sides portrayed it as a war of populations. By the end, the imperial Japanese government had not only imprisoned birth control advocates and abortion providers, but also set demographic targets to make a purified "Yamato race" overlords of Asia, including the overseas settlement of twelve million people.[7]

Fascist Italy had already adopted some of the harshest penalties for any practice that interfered with people's social duty to contribute to the national "stock." Squadrons of destroyers escorted convoys of settlers to dramatize the need for colonies in Libya and beyond. Mussolini demanded figures on births and deaths month by month, province by province, pressing Gini for explanations when they did not add up. Excited when Italy's population appeared to be overtaking that of Britain and France, he grew dispirited when Italy started to fall farther behind the USSR and Germany. Gini was finally removed from his post when he suggested that Il Duce's "demographic battle" was unwinnable.[8]

Soviet officials also tracked fertility and mortality closely, and therefore knew that, rather than prevailing in the struggle over population, collectivization and the ensuing famine had brought an astonishingly rapid decline in births and ten times as many deaths in some districts of the Ukraine. After 1936, they withdrew contraceptives from the market, made abortion illegal, and offered mothers cash incentives to bear large families. More prosperous peasants continued to be described as "anti-Soviet elements" who required "extraction" in order to "cleanse" the workers' paradise. By the end of the decade deportations increasingly targeted whole ethnic groups for the unacknowledged purpose of consolidating control over border regions.[9]

Even in this appalling era, Nazi population control stood out, and not just because of its single-minded pursuit of the Jews. Before Hitler came to power, popular and professional movements—many considered progressive—had pushed for pro-natalism, improved public health, eugenics, and sex reform. The Nazis plundered them for ideas and put them all in the service of racial purity and anti-Semitism. In part because of the equally fractious nature of Hitler's state, contending groups were empowered to implement increasingly radical measures, including eugenic matchmaking to breed a new generation of colonists, massive sterilization programs of the unfit, and secret medical killing of the disabled. Hitler would explain the conquest, depopulation, and resettlement of Eastern Europe as "the planned control of population movements" to restore the numbers and quality of the Aryan race.[10]

If the Nazi program was unparalleled, the idea that the state had an interest in ensuring a racially sound citizenry cut across the political spectrum and circled the world. While never approaching the pace set by Nazi Germany, programs to sterilize people in and out of institutions—whether for "feeblemindedness" or just antisocial behavior—accelerated across the United States. At the same time, eugenic sterilization laws were passed in Denmark, Norway, Finland, and Sweden with hardly a word of protest. This trend was apparent even in those areas where the more "positive" version of eugenics had once prevailed. In the late 1930s Mexico experimented with eugenic sterilization, Brazil adopted racial quotas in its immigration law, and both countries together with Argentina passed prenuptial laws intended to prevent marriage among the unfit. There was a similar hardening of attitudes and proposals for harsher measures in France, targeting both prospective immigrants and citizens who interfered with the state's biological goals. The Vichy regime sentenced some women to a lifetime of hard labor or the guillotine for abortion, at the same time pressuring working women to stay home and raise more children.[11]

While such moves were polarizing, they also encouraged those with

The Nazi Volkswohlfahrt, or People's Welfare Organization, was one of many non-governmental organizations all over the world that promoted "Healthy Parents—Healthy Children" in the 1930s. The Nazis differed more in their extreme methods and Aryan mythology than in their eugenic goals. Deutsches Historisches Museum.

more positive visions of population control to find common ground and make themselves heard. Birth controllers, pro-natalists, and eugenicists in democratic countries strived to devise a common platform that could win popular support. Among the most promising approaches was to make contraception part of a broader program to help parents raise healthy children. Promoting maternal and infant health also rallied some of the most influential proponents of the welfare state, including Gunnar and Alva Myrdal and Sir William Beveridge. All shared a concern for both the quantity and the quality of population along with an aversion for the antidemocratic and pseudoscientific measures imposed by dictators.[12]

The lengthy negotiations among population activists and organizations did not achieve unanimity, but they did lay the ideological foundations of a postwar movement for "family planning." The phrase first emerged in this period, and was meant to represent fertility regulation as both "family friendly" and essential to social planning. Even the Vatican came to accept a version of family planning, albeit under the same conditions: individuals could not plan their families without professional guidance to guarantee the greater good.[13]

Birth controllers and eugenicists also pursued practical experiments in international coordination and assistance. Activists from Europe and the United States conducted publicity tours across Asia, and dozens of clinics and contraception information centers were founded in India and China. While the Great Depression drastically increased the difficulty of organizing international meetings, the new Birth Control International Information Center (BCIIC) in London brought activists around the world into closer contact than ever before. By sharing experiences, they became all the more aware of the need to develop inexpensive, reliable, and easy-to-use contraceptives, and this period also witnessed the beginning of international research and field trials. No less important, it also saw the first foundation-backed studies of the use and effectiveness of existing methods. Some of the organizations that pioneered these efforts—such as the BCIIC and the Birth Control Investigation Committee—did not survive the war, but their failures as much as their successes prepared the way for the international campaigns of the 1950s and 1960s.

The aftermath of World War II would parallel the period following World War I, in that both witnessed ambitious designs for truly global population programs under the auspices of new international organizations, once again culminating in a clash with the Catholic Church. But these in-

tervening years, marked by doubt and pessimism, were no less important in
the practical and ideological development of population control as an international movement. Reduced resources forced birth controllers and more
liberal eugenicists to rein in their ambitions and focus on experimentation
and organizational work. And a more hostile political environment, including a growing reaction against both Malthusianism and racism, compelled
them to consider the common interests—or at least common enemies—
that could unite them in a long-term campaign that would eventually span
the globe.

Shaping every discussion of population problems throughout the 1930s
was a preoccupation with the international economic crisis. As it intensified
and spread with the financial panics of 1931, and millions of unemployed
parents struggled to feed their families, it began to appear ludicrous to deny
them the means to prevent unwanted births. Birth control was one of the
few American industries to prosper, serving a $250 million market by
1938. Not only was there growing acceptance of contraception in many
countries, there was increasingly an *expectation* that parents should have no
more children than they could manage. For instance, that same year a Gal-
lup poll found that 76 percent of American women thought that family income was the most important consideration in reproductive decisions, and
almost 80 percent favored the use of contraceptives. With the state assuming unprecedented importance in providing for people's everyday needs
even in liberal democracies, it became less shocking to assert that it also had
a role in helping them to regulate their reproduction.[14]

Yet it was not at all clear what that role would be, and whether states
could or should further reduce birth rates by making contraceptives more
widely available. Using new—but faulty—techniques to project population
trends, demographers had begun to forecast absolute declines within the
decade for France, Britain, and Germany. In 1933, U.S. fertility fell below
the estimated replacement level for the first time in history. No less disturb-
ing, international labor migration not only had come to a halt, but had
shifted into reverse. In 1931 more Europeans returned from the United
States than emigrated there, and that same year some two hundred thousand disappointed workers also left France. Eminent authorities, including
John Maynard Keynes and Louis Dublin, suggested that the state might
have an interest in stimulating the reproduction of workers and consumers,
just as it was attempting to jump-start economic production and consumption.[15]

Population scientists and activists also disagreed about how to influence policy. Even among those committed to widening access to contraceptives, there were bitter personal and organizational disputes—especially in Britain and the United States. These two countries had the largest and richest constituencies for birth control, especially after their counterparts in France, Germany, and Japan were silenced. But Sanger had resigned from the ABCL in 1928 when she found she could no longer control it. The dispute began to play out across the pages of the *Birth Control Review,* which Sanger now derided as no different from "all the little Catholic papers." Her long-running feud with Marie Stopes also continued unabated, as both competed to win the favor of eugenicists.[16]

At the end of 1930 Sanger tried once again to create an organization to encompass all those concerned with population trends, raising the necessary money from the Milbank Memorial Fund. And once more she allowed someone else to take the credit: the sociologist and anti-immigration activist Henry Pratt Fairchild. Fairchild invited a dozen of the most prominent birth controllers, eugenicists, and population researchers to his office at New York University; these included Laughlin, Dublin, and ABCL president Eleanor Jones. Now that the time had come when population could be "rationally manipulated," he called for an association "to present a united front to students and the world at large." The Milbank representative, John Kingsbury, was emphatically in favor of more activism, proposing as its name the "National Association for the Scientific Study and Control of Population." They opted for a more neutral moniker, the Population Association of America. No one disagreed with the idea that the PAA should study eugenics. Its very first research project investigated the fertility of "the socially inadequate classes." But advocating birth control was still too controversial. Fairchild pointed out that the association would receive all kinds of inquiries. "Shall we reply, 'We have been studying these problems for ten years and have nothing to say'? 'Yes, yes,' came from all over the room."[17]

Just as in Geneva, Sanger was persuaded to recede into the background. The organizers created a governing body limited to scientists, the College of Fellows. And as a further safeguard, Dublin continued to lead the American National Committee of the IUSIPP as an independent entity. The leadership of these organizations was virtually identical, and they went to great lengths "to keep out all but the purest of the academically pure," as one participant, Frank Notestein, later recalled.[18]

Once again, disappointment in effecting change in the United States led Sanger to redirect her efforts abroad. Thanks to Edith How-Martyn's tight management of the Geneva conference, there had been enough money left over to launch the Birth Control Information Center in London, with Sanger as its president. In 1930, the Lambeth Conference of Anglican bishops accepted the practice of contraception within marriage when there was "a clearly felt moral obligation to limit or avoid parenthood." The Center was encouraged to canvass members of Parliament as part of a successful campaign to secure government authorization for the distribution of birth control at health centers. After the Zurich conference, it was rechristened as the Birth Control International Information Centre (BCIIC). London was the seat of the largest empire, with local clinics now available for training doctors from around the world. It was ideally situated to become headquarters of a birth control movement with global aspirations.[19]

The defection of the Anglican bishops together with the spread of contraception and marriage restrictions even in the most Catholic countries finally provoked a public response from Rome. On New Year's Eve 1930, Pius XI declared that marriage was divinely instituted, not only to ensure that the number of worshippers "should daily increase," but also to encourage the "blending of life as a whole," aiming at the mutual improvement of husband and wife. This may have been an oblique response to governments inclined to keep apart those judged unfit for parenthood. But *Casti Connubii* was above all an uncompromising defense of patriarchy as well as the pope's authority—as Vicar of Christ—to act as father to believers everywhere: "The man is the ruler of the family, and the head of the woman." As the "heart" of the marriage, "the chief in love," the wife deserved her dignity. But this required "the ready subjection of the wife and her willing obedience." The pope insisted that states had no right to deny marriage to their citizens or sterilize them, any more than people had a right to sterilize themselves.[20]

The Church hierarchy closed ranks behind *Casti Connubii*. Bishops disciplined dissenters and instructed organizations of lay men and women to help defend the faith. A crucial test came early with the sterilization campaign launched in Germany in 1932. The Vatican bitterly protested, especially when Catholic institutions and personnel were required to participate. Together with threats to the independence of its schools, newspapers, and youth organizations, Vatican secretary of state Eugenio Pacelli—later Pope Pius XII—condemned it as an extreme form of the hypernationalism that everywhere imperiled the Holy See's "supernational mission." But this mission did not include protecting Jewish communities. Pacelli offered hardly a word of protest about the mounting threat to their survival. The Vatican's official newspaper, *Osservatore Romano,* at other times condemned eugenics as an "aberrational type of internationalism," and suggested that Nazi racism, like birth control, was really a manifestation of the global problem of infertility.[21]

Instead of focusing on murderous regimes, or even eugenic sterilization, the Vatican fought with equal fervor any attempt to legalize contraception—indeed, any legislation that "does not give to heads of family the place that normally belongs to them!"—as the Vatican's secretary for extraordinary ecclesiastical affairs, Cardinal Giuseppe Pizzardo, put it. National hierarchies in Europe cheered pro-natalist policies, including harsh penalties for violators. At the same time, lay organizations dedicated to promoting the Christian family proliferated, coming together for the first time in an international congress in Paris as part of the Universal Exposition of 1937. In the United States as well, the bishops' fight against "modern paganism" and the "menacing decline in the birth rate" reached a fever pitch.[22]

By this point, Malthusians and eugenicists also had a quasi-religious faith in the possibility of a worldwide population control crusade. As the Malthusian League's official organ, *The New Generation,* put it: "The next great task for birth controllers is to convert the backward countries of the world—China, India, and Japan and the coloured peoples of Africa and South America." Lord Horder, private physician to the Prince of Wales, endorsed its work as "missionary . . . making known as widely as possible, and therefore in as many countries and languages as possible, the gospel of birth control." Similarly, a public appeal issued by Keynes, J. A. Hobson, and Julian Huxley, among others, asserted that a shared need for birth control knowledge and a common recognition of the "interdependence of classes and nations" might form the basis for "the ultimate union of self-interest and religion."[23]

The BCIIC was a thin reed upon which to place such hopes. Initially, it had correspondents in just six countries, and the addresses give the impression of an exclusive boutique rather than an evangelical religion: Geneva, Cannes, Paris, New York, Pasadena, Sydney, Tokyo, and Shanghai. The problems Sanger had with her representative in China showed how difficult real missionary work could be, and how even handpicked emissaries might escape her control. In 1929 she began sending fifty dollars a month to Agnes Smedley, a friend who was working as a journalist there, on the understanding that she would use it to start birth control clinics. Though she was a communist revolutionary and was already under surveillance by British intelligence, Smedley seemed devoted to the cause. While living in Germany she had used Sanger's money to establish several clinics on the American model, offering contraceptives exclusively rather than a broader program in maternal health and sexual counseling preferred by local advocates. When Smedley witnessed profound poverty in the streets of Shanghai, and met some of the local supporters of birth control among wealthy Chinese and Christian missionaries, she developed an entirely different attitude. "I am more and more convinced," she told Sanger, "that no b.c. work is possible until there is a national revolution that will wipe out the whole capitalist class, the land-owner class, and the foreign imperialists." Sanger considered the venture an abject failure. Shanghai was "the weakest spot in the BC movement," she wrote How-Martyn, "far more difficult than where there is no movement at all."[24]

Most of Sanger's missionaries would remain loyal to her doctrine of birth control, which sought to absorb and reconcile other approaches to reproduction, including Malthusianism, eugenics, and feminism. While Sanger herself was devoted to eugenics, associates like How-Martyn and Marjorie Martin discussed it as one argument among many in attracting new followers. But whatever the rationale, all of them insisted that widening access to contraceptives was appropriate in every circumstance, and was quite possibly the solution to social, economic, and political problems. As such, this broad church of birth control came into conflict with other faiths, secular as well as religious, including Marxism and revolutionary nationalism. And the foreign or elite identity of its evangelists and early converts—not to mention their quasi-religious rhetoric—inevitably raised doubts as to whether they were any different from the missionaries who had preceded them. In 1935, when How-Martyn met and debated Gandhi, the Mahatma told her not to leave India "until you have converted me or converted yourself."[25]

The cause of population control could never be entirely foreign in places like China and India, which had rich intellectual traditions concerning its quantitative and qualitative aspects as well as hundreds of years of practical experience in regulating fertility. Even so, these countries had long served as negative models in both Malthusian theory and such pivotal mo-

ments as the Besant-Bradlaugh trial, when Annie Besant had blamed their
famines on overpopulation. Though neither Gandhi nor How-Martyn mentioned Besant in their 1935 meeting, she could not have been far from their
thoughts when they debated who would convert whom. After all, Besant
had renounced the cause of birth control to convert to Theosophy and,
through it, embrace Hindu spiritualism. She had finally passed away in
Madras only two years earlier.[26]

The politics of population in the West and the larger nations of Asia did
not just intersect; they sometimes ran parallel. The similarities presaged
both the problems any global movement would encounter and some possible solutions. Eugenics organizations active in China, India, and Japan
shared a concern that birth control endangered the survival of the fittest.
Much of the history of population control in Asia, as elsewhere, revolved
around efforts to reconcile widening access to contraception with programs
aiming at national or racial renewal. And here too, common ground could
often be found by focusing on the welfare of mothers and their children.

By the time Sanger and others began sustained campaigns in the 1930s
in China and India, these were hardly "backward countries" as far as birth
control was concerned. In some respects, they were at the forefront. For instance, concerns about population growth had been voiced in the Assembly
of the princely state of Mysore since 1881, and in 1920 the Maharajah had
donated a tidy sum to Britain's Eugenics Education Society. A decade later
he took up birth control. At the request of the state's senior medical officer,
How-Martyn supplied a list of eminent persons who were only, first, or second children, apparently satisfying the Maharajah that "it is not necessary
to have big families in order to get people of genius." When two state hospitals were authorized to provide contraceptives as part of maternal health
care in June 1930, Mysore achieved the distinction of having the world's
first government-sponsored birth control clinics outside the USSR. Meanwhile, the founder of the new National Midwifery School in Beijing,
Marion Yang, included contraception in the curriculum and required all
students to apply their knowledge at a local clinic established that same
year. Graduates were then dispatched to the provinces with instructions to
train new midwives. Yang would go on to establish fifty-four regional midwifery schools. While all this was happening, the U.S. government would
scarcely even discuss birth control. Organizers of a White House conference on child health and protection in 1930 refused to give it any place on
the agenda and confiscated all records that broached the subject.[27]

The fight for birth control in Asia occasioned alliances no less complex than in the UK or the United States. Yang worked hand in hand with nationalist population experts like Guangdan Pan and Da Chen, who were mainly concerned with eugenics. In Japan as well, some advocates of birth control allied with eugenicists—whether out of sincere conviction, as in the case of Raichō Hiratsuka, who was inspired by much the same maternalist ideology as Alva Myrdal, or merely political expediency, as with Shidzue Ishimoto. All could agree on the ideal of improving maternal health care even while differing on organizational tactics.[28]

But it was India, not China or Japan, that became the launching pad for a movement that would make population control the overriding priority. Like China, India had long served to exemplify the problem of overpopulation. But whereas China—beset by banditry and warlordism, communist revolution and Japanese invasion—approximated the Malthusian nightmare all too well, India remained relatively accessible and accommodating. There were regular sailings from England and extensive rail and telegraph networks. Although Sanger was now accustomed to traveling first class, even she was impressed when a "lovely car and spiffy driver" awaited her at the end of a red carpet before a Maharajah's palace. "Life is so easy and charming and warm and bright for those who have money." Sanger had also received the royal treatment in Japan—even Marie Stopes was treated like a star. In both countries demographic growth seemed to threaten international repercussions. But unlike Japan, where fertility soon became the object of a pro-natalist policy by a state increasingly unwilling to brook foreign interference, India under the Raj remained open to inspection and instruction.[29]

Innumerable Americans and Europeans therefore traveled to India, witnessed "overpopulation" firsthand, and returned ashen-faced, suitably appalled, to tell others of their experience. "But how humanity breeds here," How-Martyn exclaimed upon her arrival. She reported to Sanger that Indians, however attractive, "have the animal's unquestioning acceptance of life as it is and its surroundings." The layout of Indian cities, with narrow paths following irregular patterns, gave the impression that the whole country was impossibly overcrowded—more than a century earlier British visitors had come to the same conclusion. This outsider's view of Indians as being one with nature, and a force of nature, remained common long after independence. Such figures as Julian Huxley, Claude Lévi-Strauss, Dwight Eisenhower, and Paul Ehrlich—whose *Population Bomb* provided the most famous account—all reported a feeling of being overwhelmed by the corporeality of Indian crowds at close quarters, and a new commitment to the cause of population control.[30]

As in Japan and China, considerable numbers of Indian academics, doctors, businessmen, and officials were already aware of the possibility of shaping population for political purposes. But in this case nearly all spoke and wrote in English, and could therefore more easily contribute to the formation of an expanding but still predominantly Anglo-American network of birth controllers. Sanger personally received about a thousand letters from India between 1922 and 1935. Authors like Pyare Kishan Wattal, Bhalchandra Trimbak Ranadive, and Radhakamal Mukherjee were published in London and New York. And for a time the Bombay-based *Marriage Hygiene,* under the direction of Aliyappin Padmanabha Pillay, was considered one of the world's leading journals of sexual research. Indians regularly attended international population conferences and would soon begin to host their own. While official bodies in China and Japan also promulgated positions on population by the early 1940s, a decade later it was India that finally launched the world's first national policy to limit growth, a policy that was designed in consultation with international organizations and funded from abroad. Narrating this earlier history, and placing it within the larger context of a crisis of colonial rule, can help explain why.[31]

Before the 1931 census, colonial administrators described the state of India's public health with weary resignation. "The people multiply like rabbits and die like flies," complained Sir John Megaw, soon to become director-general of the Indian Medical Service (IMS). "Until they can be induced to restrict their rate of reproduction there is no hope of doing much good by medical relief and sanitation, as the population is very nearly up to the possible limit." The discovery that Indians had somehow managed to surpass this limit and were living a bit longer created a degree of alarm that bordered on panic: "Nature now threatens to take her revenge for our interference with her destructive powers," Megaw warned. "The country is in a state of emergency which is passing rapidly toward one of crisis."[32]

Over the preceding decade, death rates had declined by about a quarter. Those who survived their first five years could now expect to live until the ripe age of 38. The resulting rate of population increase was still not particularly high: about 1 percent a year. Nevertheless, it was enough to increase India's already sizable population by almost thirty-three million in a decade. It is still not clear how much of this can be attributed to improved public health and famine relief efforts rather than the development of natural immunity, the emergence of less virulent strains of microbes, or more stable agricultural productivity due to more consistent rainfall. In 1935 there were still no qualified health officers in most of the municipalities in India, malaria continued to carry off a million or more every year, malnutrition remained common, and tuberculosis may have actually increased.[33]

Yet, for all these deficiencies, health budgets were finally under the control of Indian assemblies and ministers. They spent more money on health than ever before, both in absolute terms and per capita, with programs that increasingly reached beyond favored enclaves. Career British officials fiercely resisted what they considered political interference, fighting to defend the independence of the IMS and retain control over budgets. Coming in the midst of an economic crisis and pressures for austerity, the possibility that spending on public health had created new problems further complicated this continuing struggle.[34]

Population growth also put into sharper relief the continuing controversy over the status of Indian women, a controversy that had become international in scope after Katherine Mayo published *Mother India*. Indian nationalists had taken up the challenge by demanding that the government prohibit child marriage, exposing as a sham its claim that the Raj stood for progress. The British actually depended on the support of the most conservative Hindu and Muslim leaders. Women's groups, organized into the All India Women's Conference (AIWC), successfully argued that fulfilling their duty as mothers to the nation required having rights and dignity. They were critical of the passage in 1929 of the Sarda Act, the first legislation prohibiting child marriage, though they would have to keep fighting to see it enforced.[35]

In this charged environment, concerns about population growth would therefore inspire assemblies of Indian women, international conferences, and a prolonged debate in India's own Council of State. Running through it all was the now old question of control: Who would control fertility, and for whom? But in India, the ideological, political, and practical considerations assumed unprecedented complexity, as a gathering movement with global aspirations finally came to ground.

When the representative of Bihar and Orissa, Imam Hossain, rose in India's Council of State to introduce a resolution calling on the government to check population growth, he immediately captured the paradox of any such proposition: "It is a matter," he explained, "which primarily concerns the masses, the dumb millions, whose voice is never heard in the councils of the Government." When challenged as to whether he was speaking for these masses, he did not hesitate: "It is not what they want, but what is good for them." Hossain went on to betray that his own concern was above all the threat that "class warfare" and "violent revolution" posed to both the wealthy elites represented in the Council and their imperial overlords. "It is for the safety of the British Government itself that they ought to take steps now, otherwise," he warned, "they will be swamped by the coming hordes."[36]

British officials were no less worried. Megaw, now president of the India Office's Medical Board, had gone public in a special meeting of the East India Association in London. But what was to be done? The renowned feminist Eleanor Rathbone suggested that government doctors were duty-bound to provide contraceptive advice. Megaw refused to accept this rhetorical "stick" and "proceed [ed] to beat the government with it." He warned that any birth control legislation would only become another Sarda Act.[37]

Just as Megaw predicted, the Council of State debate put government ministers in the most awkward position—especially when members quoted Megaw in urging official action. But others accused Hossain and his allies of advocating infanticide, or worse. One Bengali suggested that he would urge contraceptives on Muslims only so that his co-religionists could out-reproduce them. Another representative held up an illustrated birth control manual, outraged that anyone would suggest distributing it to their wives and daughters. It was impossible, he concluded, "to convert east into west as the twain can never meet!" ("They will, Sir," a colleague replied, "and they have!")

When at last the home secretary, Maurice Hallett, responded for the government, he suggested that he was given the assignment only because his ministry was already so unpopular. Along with the new director-general of the IMS—claiming to cover "the scientific aspect of this question"—he resorted to Pearl's now-discredited logistic curve theory to suggest that, as with laboratory mice, India's fertility rate would naturally and inevitably decline. But the crux of the matter was political: considering the controversial nature of the question, and the position the British held in India, it was "very difficult and dangerous for them to step in and take an active part in measures of this kind."

Before the resolution was finally defeated—just as similar resolutions had been defeated in the Delhi and Bombay municipal assemblies—some representatives suggested that it would be better for voluntary associations to advance the cause. In fact, the AIWC had already endorsed the dissemination of birth control. But as a group of elite women, it too seemed to lack the standing to speak for "the population of India." Some members of the AIWC claimed only to speak *to* "the ignorant and the poor . . . who need our guidance and advice," as a Kashmiri Brahman put it. Other AIWC leaders would back sterilization of the "unfit" and broader dissemination of birth control as a means to reduce differential fertility. In the Council of State debate, one representative wryly observed that he had "never known them to carry propaganda to the outlying villages, to these dumb millions with whom we are very much concerned."[38]

These problems of class, religion, nationalism, and gender did not exhaust all the possible complications of advocating population control in India. The most vocal proponents were upper-caste Hindus upset that popular movements were challenging their privileges. This included Ragunath Karve and Narian Sitaram Phadke, both Brahmans from Maharashtra, whom Sanger had aided when they lost their jobs because of birth control advocacy. Pillay, a Tamil Brahman, complained that charity was permitting the unfit to survive and reproduce, and called for either voluntary or coercive sterilization. Wattal, one of the most widely cited authorities, was a Kashmiri pandit who asserted that lower castes and Muslims had higher rates of fertility. Lacking actual data, he inferred a general relationship between superior and inferior social strata from the 1911 census of Scotland. But Radhakamal Mukherjee, a Bengali Brahman, claimed to have statistical proof when he convened the first Indian Population Conference in Lucknow in 1936. Mukherjee later headed the subcommittee on population of the Congress Party's National Planning Committee.[39]

It would be too simple to equate the cause of population control in India with the interests of upper-caste and upper-class Hindus. The sponsor of the Council of State measure was a Muslim, and the leader of the Dalits—or "untouchables"—of Maharashtra, B. R. Ambedkar, would introduce another resolution backing birth control in the Bombay Legislative Assembly in 1938.[40] Spokesmen for many different communities used notions of both population quantity and quality to express their anxieties and aspirations because they seemed to make political, cultural, and social changes precisely measurable and amenable to scientific explanation. But terms like *biological scale, fitness,* and *overpopulation* remained vague enough to be translatable into countless local contexts, and would have to be if they were to define policies and programs acceptable to each one. This process of translation produced new ideas with potential applications that went far beyond the particular community in question, and far beyond India. Because, despite their many differences, all the participants—whether colonial officials, or high-caste Hindus, or American feminists—had to consider ways to convince or coerce people unlike themselves to accept new standards of reproductive behavior.

India proved, for example, that even the very poorest people could not be relied on to want fewer children. Those who shared Sanger's faith that the value of birth control was self-evident would often ignore this lesson. Others approached contraception as just a tool to fix political and economic problems. Public health officials, in particular, had more experience in the challenges of large-scale social engineering projects, and they were already proposing what would later be called "information, education, and communication" programs aimed at "demand creation." Megaw, for instance, suggested devoting fully a third of India's education budget to "well-organized propaganda" in film and radio to promote what he called life planning. "The whole outlook on life of the people of India," he claimed, "might well be revolutionized within a few years." Similarly, Wendell Cleland, a professor at American University in Cairo, argued that Egypt's overpopulation problem required national reeducation under international supervision: "By arousing the people's latent desires for better health, more creature comforts, intellectual growth, and the spiritual satisfactions which flow from these, the matter of population increase would probably take care of itself." Imam Hossain was more pragmatic when he called for including in secondary school curricula "hints" about contraceptive technique and discouragement of large families. To reach rural areas he suggested mobile birth control clinics, which would become a fixture in postwar campaigns. So too was visual propaganda, which his supporters in the Council of State already recognized as key to swaying illiterate audiences.[41]

There were also, already, serious discussions about what message the new media might convey, and the impressions they must absolutely avoid—especially at the first Birth Control in Asia conference, convened at the London School of Hygiene and Tropical Medicine in November 1933. It was organized by How-Martyn and the BCIIC, and included such eminent speakers as Harold Laski, Eleanor Rathbone, and Krishna Menon, indicating the kind of attention such questions were beginning to attract. Wattal warned against appearing to advocate population control in places like India in terms of the menace it posed to the West, perhaps alluding to *Mother India*. This would only discredit their local allies. Indeed, opponents cited pro-natalist policies in Europe when questioning why Westerners wanted to limit the population of India. Another speaker, Helena Wright, concluded that "the control of the movement ought to be in the country in which the work is taking place." She would go on to train generations of foreign doctors and nurses in contraceptive technique in her London clinic. And this idea of protecting and promoting local leaders—or at least avoiding the impression of imposing a Western agenda—would recur again and again in years to come.[42]

Scientists and activists still confronted the same question of who would control birth control, and to what end. As Robert Kuczynski argued at the Birth Control in Asia conference: "Birth-control will be a very good thing for millions of individual women in India and China, but this does not imply that it will necessarily be the best solution for the community of India and China as a whole." These potentially conflicting interests could be reconciled through the formula "fewer births and better in quality," as a Bombay public health official, Nasarvanji Choksy, put it before the Council of State. Borrowing another idea from eugenicists, it could be easily and vividly represented by contrasting a small, happy family with the miserable fate of an unplanned family of fourteen—or twelve, or twenty. This was already the theme of a film Choksy had promoted in Bombay, and it would recur in virtually all the family planning campaigns of the 1950s and 1960s. By such means, he explained, these campaigns could be justified "not only for the sake of reducing the population but for the sake of humanity, for the sake of the women, the mothers of our children."[43]

Some went even further—not just recognizing women's role in reproducing society, but promoting their personal autonomy. The vice chancellor of the University of Madras, Ramunni Menon, told the Council of State, "It is now fairly well accepted that the progress of education will almost automatically bring about a reduction in the birth rate . . . particularly female education." Taraknath Das had said much the same thing before Sanger's 1926 conference in New York. But while birth control proponents were quite diverse and usually divided, none took up the cause of women's education. That would have undermined efforts to forge an alliance with eugenicists, because it would only remind them of how contraception helped educated women avoid contributing to the gene pool. Instead, they could agree that the solution was to find a simpler, cheaper contraceptive that could be used by uneducated people.

In the half-century since the first rubber diaphragms, there had been no major advance in contraceptives. The prospect of increasing demand only made the shortcomings of this method all the more apparent. It required individual fitting by trained personnel and even then failed many who used it. The only practical results of the London conference were to endorse the plans for the BCIIC to send an organizer to India and to train Asian medical students at local birth control clinics. Everyone realized that only a tiny minority of Indian women would ever see a Western-trained doctor, and many would refuse examination by a male physician. There were only about four hundred women doctors in the whole country.[44]

But the search for "methods adapted to the wives of dull-minded natives"—as the secretary of the Rockefeller-supported Committee on Maternal Health (CMH) put it—had already begun. In 1927 a CMH initiative led to the creation of the Birth Control Investigation Committee, which grew into an international network of researchers run out of London. Many of those intent on developing simpler contraceptives were primarily concerned with influencing the reproductive behavior of simple people. "The future of Birth Control necessitates the discovery of a method which is simple and effective and which does not require the cooperation of the individual," as a 1934 paper presenting research on injectable contraceptives put it. In the 1920s and 1930s there were at least twelve such studies. Contraceptives had to be made foolproof, in other words, if fools were to use them. As C. P. Blacker explained in a successful 1935 appeal to Britain's Eugenics Society to back the Birth Control Investigation Committee:

> Even the most sanguine supporter of sterilization must expect a considerable amount of time to elapse before dysgenic persons are sterilized in sufficient numbers to produce racial effects. And even after the lapse of years, it is possible that not more than a few hundred persons will be sterilized annually. Such results, excellent in themselves, would not have effects in any way comparable to those which would follow the discovery of a simple, reliable, and fool-proof contraceptive. As an achievement of negative eugenics, such a discovery would have racial consequences thousands of times more important.

Simpler contraceptives promised "racial consequences" not just locally, but globally. The principal contraceptive evaluator for the Birth Control Investigation Committee, Cecil Voge, aimed at a contraceptive that would be "so easy to use that the most ignorant woman in the Orient, the tropics, the rural outposts or the city slums might be protected."[45]

Sanger was also impatient for a technological fix. Unlike many eugenicists, she had faith that individual women would make good use of simple contraceptives under their own control. As early as 1932 she spoke of "the 'pill,' the magic 'pill'" as the solution. "Until that is found we will have to fight on and on." She was also meeting disappointment in organizational efforts. In November 1933, leaders of the American Birth Control League and the American Eugenics Society agreed to ask the Population Association of America to try to bring the two organizations together in a federation. An ABCL representative explained that the falling fertility rate required them to "promote births among the more intelligent." Fairchild invited Sanger together with the leaders of virtually all of the other major U.S. organizations to a February 1934 meeting. They discussed collaborating in publications and fund-raising, finally agreeing to form a consultative "Council on Population Policy." But nothing concrete ever came of it. The aims of these groups were still too divergent, and the PAA delegates were not even allowed to speak on behalf of what the leadership insisted was an apolitical association of scientists.[46]

Whatever their differences, Sanger insisted that all of these groups shared "one common enemy, one group of opposition objecting to everything we do or what we say—the Catholic Church." For five years she had been working to repeal the Comstock Law, and came tantalizingly close. But every time the Catholic Church had defeated her. When a Senate bill was close to winning a majority, the NCWC solicited support from some three thousand organizations and individuals. It was therefore able to testify in the name of sixteen national, nine state, and seventeen hundred local bodies. "The main burden of making articulate the opposition of the various religions, social, and labor groups has fallen to the National Catholic Welfare Conference," as one official reported to the apostolic delegate, Giovanni Cicognani. Many had no presence in Washington. Since it was important that the cause not be completely identified with the Catholic Church, the official reported, "we arrange whereby the opposition hearing is technically at least held under the supervision of a 'neutral' organization."[47]

The NCWC also had access to top officials in Roosevelt's administration, which had benefited from strong Catholic support in the 1932 election. When Cicognani came to present his credentials together with the general secretary of the NCWC, Father John J. Burke, they were ushered from the executive office—walking past those who actually had ambassadorial status—into the White House itself. They congratulated Roosevelt for appointing Catholics to govern the Philippines and Puerto Rico, America's two largest colonial possessions, but warned about the spread of the birth control movement abroad. Smiling, FDR replied that he "had given orders to Governor Gore [in San Juan] not to make any speech in favor of birth control." Roosevelt bid his eminence to return so that he could introduce his wife. Sanger, for her part, could not even get a meeting with Eleanor, whom she had once counted among her allies.[48]

While Sanger directed her American organization to try to overturn the Comstock Law in the courts, she decided to redirect her own energies back to international work—India in particular. The Catholic hierarchy would have a harder time stopping her in countries where Catholics were a tiny minority. As the NCWC grudgingly acknowledged, "Mrs. Sanger is an indefatigable propagandist, and seems never to tire or be discouraged." With the BCIIC she had a base upon which to build. Its budget was small, but the staff volunteered under How-Martyn's capable management. A tight-knit board of English patrons, especially Harry and Gerda Guy and Maurice Newfield, were devoted to the cause and deferred to Sanger's leadership. Official correspondents now reported from twenty-five countries, the newsletter enjoyed a growing circulation, and visitors from abroad regularly dropped by for weekly talks—among them, Jawaharlal Nehru. How-Martyn had already made an exploratory visit to Egypt, Palestine, and Syria, and had just returned from a tour of India. During her three months on the subcontinent, she addressed seventy-nine meetings.[49]

With her own much higher profile, and without a Briton's colonial baggage, Sanger thought she could bring enough publicity to the cause of birth control in India to make it an example for the world. It would also appeal to wealthy donors. The BCIIC had already had two successful fund-raisers in London to send How-Martyn abroad: a "Malthusian Ball" under the patronage of Princess Alice, and a dinner given by Lady Dhanvanthi Rama Rau, wife of a prominent Indian official. Writing to Albert Milbank, Sanger claimed that her visit "may well affect our future civilization." Critics of the birth control movement had charged it with reducing the population of Western countries without doing anything in the East. "Here now is our opportunity to balance the populations of the world."[50]

Sanger appealed to the British Eugenics Society through C. P. Blacker, explaining that she had two aims and both merited their support: "first, to bring to the poorer and biologically worse-endowed stocks the knowledge of birth control that is already prevalent among those who are both genetically and economically better favored; and secondly, to bring the birth rates of the East more in line with those of England and the civilizations of the West." She mentioned only the first when she stopped in London and described her plans in a BBC radio address. She was going to India, she announced, not because it suffered from any absolute overpopulation, but rather because birth control could bring happiness to individual families. But in India, as in the West, it was unevenly distributed, leading to "dysgenic" differential fertility between the "well endowed" and the "not so well endowed." She would therefore encourage its dissemination among "the social, economic and biological classes in which it is most urgently needed."[51]

Sanger received a tremendous send-off in several London fund-raisers, the most glittering of which took place in the onetime operating theater of the Barber-Surgeons' Hall. "When the history of our civilization is written it will be a biological history," H. G. Wells declared, "and Margaret Sanger will be its heroine." Julian Huxley thought she had affected the structure of the world more profoundly than Franklin Roosevelt. Perhaps after one too many toasts from Henry VIII's Royal Grace Cup, Wells insisted her historical importance would surpass that of Alexander and Napoleon.[52]

On her arrival in Bombay, Sanger disembarked from the *Viceroy of India* and was met by a delegation of almost fifty along with a personal invitation from Gandhi. Though she would address sixty-four meetings over the following nine weeks and travel ten thousand miles, she knew the journey to the Mahatma's ashram in Wardha would attract the most attention. She was assisted by a publicist who wrote daily press releases, and some 377 American newspapers in forty-three states reported her travels. But Sanger's encounter with Gandhi was the one that made headlines.[53]

Sanger had been warned of what to expect. When Gandhi had argued with How-Martyn earlier that year, he had complained that, rather than seeking to convert him through their correspondence, Sanger had "cursed" him in the newspapers. Gandhi, for his part, had condemned contraceptives as a curse of modernity, making possible the celebration of sensual pleasure as an end in itself. This only exhausted body and mind. Gandhi echoed many of his contemporaries, like Oswald Spengler, who linked the rational control of reproduction to the decline of spiritualism and predicted the ultimate demise of modern civilization. But whereas Spengler equated fertility with the vital force of a people, Gandhi thought spiritual life required mastering "animal passions." In fact, he thought Indians should have smaller families, which would also be healthier families. But the only acceptable means was abstinence. When How-Martyn had pointed out that some women might be at the mercy of their husbands, he insisted that no woman could ever be raped if she were prepared to die fighting. Margaret Cousins considered Gandhi and his "medieval views" on women to be "the greatest stumbling block to the B.C. movement in India."[54]

Sanger found Gandhi cordial but unyielding. A decade earlier, when they had debated in print, Sanger appeared to claim greater spiritual enlightenment than the Mahatma. Arguing that sex was the most spiritual of all experiences, she presented birth control as a moral instrument of self-development. Perhaps recalling this gambit, Gandhi now claimed to be the greater feminist, because he would leave childbearing decisions entirely up to women. But abstinence was still the only acceptable means, even if wives must resist their husbands. He did not seem to consider the possibility that women might also have some interest in sex—despite having claimed to know "tens of thousands of women," so completely identifying with them as to be "half a woman" himself. He insisted that he was better qualified to speak for his "unsophisticated sisters" than Sanger's allies in the AIWC. Sanger, for her part, asserted universal sisterhood, having worked with tens of thousands of women around the world. "I believe firmly that the heart of the Indian woman is not different from the heart of the American, Chinese, Italian, or European woman where love is concerned."[55]

The duel continued for a day and a half. Gandhi claimed to represent the common people, whereas Sanger appealed to universal sisterhood. Both invoked the power of large numbers. Moreover, they agreed that uncontrolled reproduction was a social problem, making India's population sickly and weak. And both saw a solution in making people—and especially women—understand that they had a duty to plan their families. If anything, Gandhi was even more emphatic about individuals' need to take responsibility for the size, health, and even aesthetics of India's population. "Is it right," he asked, "for us who know the situation to bring forth children?"

We only multiply slaves and weaklings if we continue the process of procreation whilst we feel and remain helpless. . . . Not till India has

become a free nation . . . have we the right to bring forth progeny. . . . I have not a shadow of doubt that married people, if they
wish well to the country and want to see India become a nation of
strong and handsome, well-formed men and women, would practice
self-restraint and cease to procreate for the time being.[56]

Sanger felt she won a concession when Gandhi said that he did not necessarily favor lifelong celibacy, and that the "safe period" might be the answer subject to further study. Researchers had recently confirmed when
ovulation occurred in the menstrual cycle. Sanger did not publicize Gandhi's openness to the rhythm method at the time, perhaps because it was
endorsed by growing numbers of Catholics. *Casti Connubii* was silent on
the practice, and a veritable cottage industry had sprung up to offer instruction. The problem was—and is—that the menstrual cycle is variable,
making even advanced techniques unreliable.[57]

But with no "magic pill," and no other way to reach Indian women
lacking access to a doctor, Sanger herself could only offer a new and untested method: a foaming powder that, when applied to a square sponge inserted into the vagina, was meant to form a spermicidal barrier. How-
Martyn thought it "would revolutionise the whole B.C. propaganda as it is
so simple clinics and doctors are not necessary." But there had been little or
no investigation of its safety or effectiveness. It was only after Sanger had
crisscrossed India promoting foam powder that she decided to send it to
the USSR and China for clinical trials. She was probably reacting to the
news that another foam powder had caused painful irritation when tested
on dogs. Lydia DeVilbiss, a notorious racist, had been promoting it to
black Floridians and had tried to pass it off for use in India and China as
well. It was only later still that Sanger had animal tests performed on her
own formula. By that point she had helped establish some twenty clinics
and forty contraceptive information centers in India, everywhere promoting foam powder as the method of choice.[58]

The problems with foam powder would eventually undermine all of
Sanger's work in India. But while she was still there she was preoccupied
with an unfolding crisis within the BCIIC. The immediate cause was resentment toward How-Martyn for operating independently of the Center's
board even while using its funds to pay for her overseas travel. In reply to
concerns about where the money would come from, How-Martyn had
joked to Newfield that she would borrow it if need be—she was living off a
small inheritance. "It is thrilling to live dangerously sometimes, don't you

agree?" A major Canadian donor, Alvin Kaufman, demanded that she return to London rather than accompany Sanger to Southeast Asia and China. Sanger tried to smooth over these differences but resented the board's efforts to exert remote control.[59]

As in the failed effort to coordinate American population organizations in 1934, these difficulties reflected disagreement over basic strategy. Kaufman, a businessman worried about world revolution, insisted on reducing the number of poor people at the least possible expense, an increasingly common attitude among donors. Birth control volunteers also found it difficult to contend with questions about what, exactly, all their globe-trotting accomplished. "A report of contacts in many foreign countries may sound inspiring . . . ," Kaufman observed, "but is not so inspiring to me when I hear nothing further in regard to actual results." There had been no follow-up, for instance, to How-Martyn's tour through the Middle East. He complained that "too much money is spent on traveling," and suggested giving greater responsibility to local organizers able to subsist on lower wages—like the two he already employed in Korea. Sanger countered that, in India, with its population of 350 million, "we must have sufficient propaganda and publicity to elicit native forces—favorable and adverse so that we may foster and guide indigenous effort." Just as Sanger used one illegal pamphlet and clinic in Brooklyn to spark a debate across America, she appeared to hope that a publicity tour and a few dozen clinics could begin to move India.[60]

Rather than turn over control to her counterparts in India, Sanger tried to raise funds to hire How-Martyn as a full-time organizer. "What I want most to do," she explained to Harry Guy, " . . . is to avoid a personal conflict and a growing resentment which will most certainly result in an opposition movement." She was still feuding with both the ABCL and Marie Stopes, British eugenicists were bucking Blacker's efforts to push a reform program more palatable to public opinion, and Nazi race theory as well as Gini's splinter group posed a growing threat to the unity of the IUSIPP. With her international work barely getting off the ground, Sanger wanted desperately to prevent any further fragmentation.[61]

Nevertheless, an embittered How-Martyn resigned from the BCIIC, and illness forced Sanger to cut short her trip before touring China. But at the same time, in Britain there emerged signs of a potential break-through—not in contraceptive technology, but in ideology and organizational politics, a new consensus that came to be known as "family planning." The problems with Kaufman were mere "trivialities," as Newfield put it, compared with the "opportunities that are being created by the co-operation of the [National Birth Control Association] and the Eugenics Society." Both organizations needed a more positive vision than "family limitation" on the one hand and "sterilization of the unfit" on the other. Birth controllers feared that a declining population would make fund-raising impossible for cash-strapped clinics. British eugenicists, on the other hand, were financially secure but frustrated at their inability to translate propaganda campaigns into actual practice. They had long been interested in ascertaining "how poor and incompetent a section of the community it may be hoped that Birth Control would penetrate if it were introduced into Welfare centers."[62]

A merger would give birth controllers the resources to secure and expand their network of clinics. Eugenicists, on the other hand, would have the assurance that these clinics provided contraceptives to the type of people who "needed them most," at the same time offering infertility treatment and other encouragement to parents with better prospects. In this way birth control could be "the basis of eugenics," as the leaders of both organizations concluded in a joint memorandum: "We mean a selective control by which some births are restricted and others encouraged." And the birth control clinic, as "the organization which can best provide guidance and instruction," would be "the operative unit of all eugenic policy."[63]

The concept of family planning had broader origins than the immediate needs of Britain's two leading population organizations, and it would ultimately have a global impact. In different places it took on different inflections, but Gunnar and Alva Myrdal offered the most comprehensive and influential articulation. Like their counterparts across Europe and the United States, they were concerned that Sweden was headed for a decline in population, which the ever wider dissemination of contraceptives would only accelerate. Notwithstanding Gunnar's later contributions to the understanding of American racism, he and Alva did not consider immigration to be an acceptable solution. They worried that it would likely originate in southern and eastern Europe or even Africa and Asia, and thus threaten the Swedish *volk*. Instead, they called for a comprehensive program of providing contraceptives to everyone to ensure that every child would be a wanted child. At the same time, they sought to improve the conditions for child-rearing so that good parents *would want more children*. Everyone, even bachelors, had an interest in the next generation, so all of society should share the costs of properly caring for them. In 1937–38, Sweden's "Mothers and Babies" parliament accepted these proposals, legalizing birth control and abortion while providing maternity relief and subsidized housing.[64]

Yet improving the social conditions for child-rearing presented a problem. As Alva Myrdal noted, it could lead to "increased fertility in some groups hereditarily defective" at a time in which mortality was already declining among the "deficient." She concluded that the situation "demands some corresponding corrective." In their writings as well as their participation on government commissions, the Myrdals therefore urged "quite ruthless" policies for sterilizing people deemed seriously defective, including the use of force against "those incapable of rational decisions." In 1941 Sweden greatly broadened grounds for sterilization, and by the end of the decade authorities had cited eugenic indications in sterilizing more than eight thousand people. Hundreds more received the same treatment simply because of "antisocial" behavior.[65]

In the Myrdals' model of the welfare state, giving everyone access to birth control did not preclude controlling populations. It was a necessary precondition. The genius of family planning was to imply that parents would do the planning, whereas the Myrdals expected social engineers to create the conditions that would shape parents' preferences (and in some cases compel more rational choices). Ideas like free school lunches provided an appealing package for this program, and won the Myrdals an international audience. In Great Britain, this is what most impressed Sir William Beveridge when he issued his famous report in 1942, a blueprint for the Labour Party's postwar social program. Yet very specific concerns about population quality and quantity continued to recur even in the most general discussions of the welfare state. In an influential critique of protectionist agricultural policies that inflated food prices, the Australian economist F. L. McDougall underlined the importance of improving nutrition for what he called "the young human breeding stock of the advanced countries" as a way to reduce the number of defectives. By reducing infant mortality rates, it was also a more effective way to reverse quantitative declines than paying parents to have children.[66]

As the debate about population quality broadened to include the whole range of environmental factors, and included many more people as well, subjects like nutrition and disease displaced old-fashioned negative eugenics. McDougall's ideas inspired the League of Nations to establish an expert panel to investigate the prevalence of malnutrition and come up with universally applicable standards. The panel's report was the League's all-time best-seller, and helped persuade some twenty countries to establish national nutrition committees.[67]

Some of the worst conditions were in the colonies. British officials sought to shield India from any inquiry. But the Colonial Office commissioned its own report covering forty-eight territories, and concluded that widespread malnutrition "must result not only in the prevalence of specific deficiency diseases but in a great deal of ill-health, lowered resistance to other diseases, and a general impairment of well-being and efficiency." As they focused on malnutrition-related diseases like pellagra and the debilitating effects of parasites such as hookworm and schistosomiasis, public health officials came to appreciate how poverty and poor health created a vicious cycle. Even more important, they realized that it could be reversed. This had a long-term impact on population policies worldwide, demonstrating how problems of "feeblemindedness" or the "lazy native" could be caused—and cured—by environmental factors. At the same time, this realization helped discredit the racial and class prejudices that had inspired a whole generation of eugenicists but also curtailed their political appeal.[68]

When Nazi Germany began to implement a comprehensive "race hygiene" program, some eugenicists in the United States and the UK admired and defended particular aspects, notably the sterilization tribunals. Hitler had praised U.S. immigration restrictions in *Mein Kampf,* and the authors of Nazi sterilization law gave fulsome and public praise to American pioneers like Paul Popenoe. But the number who returned such praise and repeated Aryan race ideology, such as George Pitt-Rivers, general secretary of the IUSIPP, and Clarence Campbell of the Eugenics Research Association, steadily dwindled. Most American and British eugenicists considered events in Germany to be a growing embarrassment. Campbell's counterpart in the Eugenics Society, C. P. Blacker, warned that his Nazi sympathies made him "one of the most dangerous enemies of the eugenics movement!"[69]

Blacker led efforts to rally British eugenicists and birth controllers to a new family planning agenda, one that would reject overt prejudice and work to improve the conditions for bearing and raising children. A Freudian psychiatrist, Blacker had personally witnessed the demise of many of Britain's fittest during World War I. But he was more open-minded than his older colleagues and did not share their class bugbears and scientific blind spots. Like Raymond Pearl a decade earlier, he believed that eugenicists would have to win popular support and shape reproductive behavior across the whole social spectrum. As he explained to one of the Birth Control Investigation Committee's researchers, "Since our eugenic proposals are all voluntary, it seems to me to be in the highest degree necessary to enlist the cooperation and support of dysgenic people. . . . You are not likely to enlist their sympathy if you speak about them disparagingly as dregs and scum." Blacker had better political skills and a more advantageous position than Pearl to advance this agenda: as general secretary, he represented the Eugenics Society in merger negotiations with the National Birth Control Association (NBCA) and damage control operations meant to distance them from the Nazis. He was also secretary of the Birth Control Investigation Committee and, after the war, drafted the constitution of the International Planned Parenthood Federation (IPPF).[70]

In the United States the job of reconciling birth control and eugenics through the concept of family planning fell to the wealthy and well-connected Frederick Osborn, who was a friend of Blacker's ever since the two met at the 1927 Geneva conference. Osborn had devoted himself to years of independent study of the nature-nurture question in the office of his uncle, Henry Fairfield Osborn, president of the American Museum of Natural History. Whereas the elder Osborn had promoted compulsory sterilization and rebuffed Sanger, his nephew concluded it was a mistake—as a matter both of science and of policy—to focus only on the heredity of the most or least fit. After chairing the short-lived Council on Population Policy, he recognized in the Swedish model a way to win broad public support. "Greater freedom of choice as to size of family," he declared in 1937, "should be regarded as a major aim of eugenics." Over the following two years, with Osborn as its secretary, a rejuvenated American Eugenics Society doubled in membership.[71]

Neither Osborn nor Blacker was a scientist, but both appreciated the usefulness of promoting the scientific study of population. They were in a better position than Sanger to manage male scientists' egos and publicize new research. It was Osborn who had dissuaded Sanger from running for office in the PAA, which was now a home to researchers of every political persuasion. He had also financed the Eugenics Research Association. And in 1936 he induced the Milbank Memorial Fund and Princeton University to create an Office of Population Research (OPR). Under Frank Notestein, the OPR became a leader in policy-oriented demographic studies and a model for similar centers in the United States and abroad. Blacker, for his part, used the British Eugenics Society's considerable financial resources to promote a broad range of research, from genetics to social planning. He also reached out to the most eminent scientists, like Huxley, even when they were the most critical of older, more dogmatic members. In this way, the scientists helped Osborn and Blacker to marginalize activists with controversial political agendas, whether white supremacy or sexual liberation. Together they began to build an intellectual foundation for policies to shape reproductive behavior that, in their own way, would be no less revolutionary and far-reaching.[72]

What has been called "reform eugenics" did not reject the mainline idea that more privileged socioeconomic and racial groups tended to display more desirable characteristics. It simply did not emphasize it, stressing instead the potential for improved conditions to nurture talent and ability at every social level. Though Osborn acknowledged that every group had worth, under his watch the American Eugenics Society continued to campaign against Mexican immigration and never showed a concern for fertility declines among African Americans. Even more than by their understanding of the role of environment in how genetic inheritance was expressed, leaders like Blacker and Osborn were distinguished by a more subtle appreciation of the *political* environment. Consider how Blacker anticipated voluntary sterilization would work in practice: "Defectives being, for the most part, readily suggestible and open to the influence of the people around them, [they] should in most cases be easily persuaded." When deemed incapable of giving their consent, Blacker pointed out, others could do it for them. Similarly, Gunnar Myrdal thought that "a border-line group" ineligible for mandatory sterilization could be influenced toward "severe family limitation by direct propaganda and instruction in contraceptive methods." While Blacker and the other reform eugenicists refrained from calling "dysgenic" people "dregs and scum," that did not mean that they accorded them any more respect. In fact, reform eugenicists believed that the whole future of family planning depended on their ability to manipulate a large segment of society they considered unfit for parenthood.[73]

Even those who did not accept birth control eventually endorsed the idea of family planning, especially if it remained under male control. The Catholic hierarchy was initially disconcerted by the spread of the rhythm method across northern Europe and the United States. The publicity given to it by Catholic periodicals and supply houses constituted "a real scandal," according to the bishop of Seattle. In 1935 he urged the NCWC to hire a

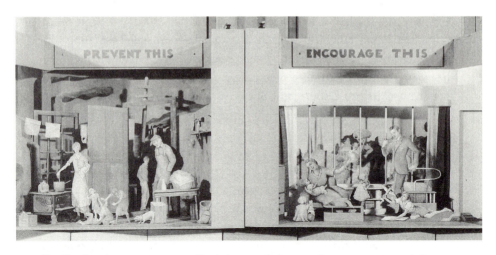

Family planning proponents realized they needed strong images to convey their message. Like countless more such depictions in the decades to come, this 1936 ABCL diorama attributed poverty, violence, and poor health to unplanned families. Conversely, the prosperous couple is encouraged to have four children. IPPF Archives.

private detective to intercept one wayward monsignor before he showed his "Wheel of Life" to FDR advisor Harry Hopkins. This prompted Cicognani to write to Rome for instructions.[74]

After more than a year, the Holy Office finally replied that the rhythm method might be "tolerated as an extreme remedy and a means to turn the faithful away from sin." But they must not "give the impression that the way to the limitation of births is left completely open." There was to be no further publicity. Instead, bishops would be quietly told to convey the Church's teaching on the rhythm method to priests. In their capacity "as physicians and directors of souls," priests would determine who among their penitents might be permitted to use it. The bishop who presided over the NCWC's laymen's and laywomen's organizations argued that it proved that "virtue and self-discipline and real manliness are on the side of the Catholic Church, and self-indulgence and weakness and a surrender to the lower instincts are on the other side."[75]

When U.S. federal courts overturned the Comstock Law that same year, they too gave professionals—in this case, medical doctors—the power to determine what constituted legitimate use. Justice Augustus Hand explained that he had new information "as to the evils resulting in many cases from conception." Doctors would be able to prescribe contraceptives for the purpose of "promoting the well being of their patients" without in any way sanctioning their use for "illicit sexual intercourse." In 1937 family planning under a doctor's supervision received the seal of approval of the American Medical Association.[76]

Removing the stigma of illegality from contraception—even while embedding it within institutions of social control—had required a joint effort by all of the American birth control organizations. In its aftermath, the ABCL and Sanger's Clinical Research Bureau began negotiating a merger with the ultimate goal of persuading governments to offer contraceptives as part of federal and state health programs. An outside consultant suggested that the new organization promote family planning, because birth control appealed only to women and it was the support of influential men they needed most. Indeed, he advised that a man lead the new group.[77]

In 1939 the ABCL and Sanger's Clinical Research Bureau united as the Birth Control Federation of America. Sanger was made honorary chair, but Kenneth Rose of the same consulting firm was put in charge as national director. The new organization encouraged couples sound in body and mind to have children and instructed clinics to disseminate contraceptives only to married women who met a set of medical criteria. The next year Fairchild declared at their annual meeting that "these two great movements [eugenics and birth control] have now come to such a thorough understanding and have drawn so close together as to be almost indistinguishable." In 1942 Rose arranged a national referendum, which ratified the new approach. Henceforth the largest and most powerful family planning organization in the world, with over eight hundred clinics nationwide, would be known as the Planned Parenthood Federation of America.[78]

In Britain the NBCA had already changed its name to the Family Planning Association (FPA) in a general meeting under the chairmanship of Lord Horder. It could not ultimately close the deal with the eugenicists because the clinics would not go along—birth control stalwarts thought it was the Eugenics Society that needed a name change. But the headquarters staff moved in with the eugenicists at 69 Eccleston Square, and promised their new patrons that they would instruct clinic workers in eugenics. The FPA also announced its intent to establish women's health centers that would provide assistance for infertility, gynecological ailments, and marriage problems. The new name was meant to "emphasize the essentially constructive nature of its work."[79]

Yet even while American and British population activists began to construct a broad coalition, one that could accommodate most pro-natalists, eugenicists, and birth controllers—and marginalize everyone else—the international effort began to fall apart. Soon after How-Martyn's resignation from the BCIIC in 1936, the other four members of the staff left as well, marking the start of "a completely wasted year," as Harry Guy admitted. Nearly half of the Centre's affiliates abroad also chose to stick with How-Martyn when she set up a short-lived rival organization. How-Martyn's replacement at the BCIIC began with ambitious plans but soon feuded with Sanger, who was incensed by the discovery that she was warning correspondents in India about the problems with foam powder. Sanger remained reluctant to side against How-Martyn, and the new director resigned. The Centre's board finally decided to amalgamate with the FPA without even consulting Sanger. Most of its contacts and institutional memory were already lost. A former worker visited the old premises just as the center's files were being carried out in bags for the trash.[80]

The FPA's new International Subcommittee seldom met, but even if it had shown more commitment and ended the feuding, the international climate was rapidly turning hostile. Commenting in 1937 on "the militarist psychology" sweeping Europe, Sanger thought it a "waste of time and money" to attempt anything there. She decided to devote herself to Asia but found it impossible to raise funds for a return trip to India and dipped into her own savings to tour East Asia. While she was en route the Japanese bombed Shanghai, almost killing Sanger's local coordinator, and Sanger decided to return home after just a week in Japan. Three months later Sanger's host in Tokyo, Shidzue Ishimoto, was arrested and interrogated about her American friends and work for birth control. A clinic Sanger had helped her open was closed for the duration, and Japan would subsequently make five-child families the national standard.[81]

Even birth control "missionaries" who were able to work in Asia, like How-Martyn and her deputy, Eileen Palmer, had little to offer. In India Sanger's onetime ally, Pillay, had publicly repudiated foam powder. The FPA, for its part, could only suggest local manufacture of cervical caps—which were useless without trained personnel to offer instruction. In 1939 Blacker's Birth Control Investigation Committee dissolved itself with only a marginally improved spermicide, Volpar—standing for "voluntary parenthood"—to its credit. The FPA did, in the end, hire an Indian organizer with the help of the Eugenics Society, but by that point they could offer little support. When they sent a shipment of Volpar in 1941, it took seven months to arrive.[82]

Everywhere the coming of war disrupted already faltering efforts to maintain the international networks of population scientists and activists that had been so painstakingly assembled over the preceding decades. A preview was offered by the progressive disintegration of the International Union for the Scientific Investigation of Population Problems, which peaked at just thirteen national affiliates. In 1935 the U.S. and British committees boycotted an IUSIPP congress held in Berlin because of Nazi abuses of academic freedom. That same year, Gini declared his independence from the Union, accusing it of being dominated by Malthusians. He established a federation for "Latin" eugenics and established relations with like-minded groups in India, Mexico, Brazil, and Spain. At the 1937 congress of the IUSIPP in Paris, Pitt-Rivers argued for revoking membership of Czechoslovakia and barring entry to the USSR, considering their respective treatment of Sudeten academics and Mendelian geneticists. German delegates, for their part, used the tribune to make the case for "eliminatory eugenics." Georges Mauco of France, who replaced Pitt-Rivers as general secretary, later claimed to have opposed the Nazi occupiers when they tried to co-opt the Union and its journal for their propaganda. But he too was a strident anti-Semite and would be charged with collaboration. Conveniently for Mauco, most of the Union's archives and library in Paris were destroyed during the war.[83]

In London Blacker soldiered on through the Blitz, sleeping in the offices of the Eugenics Society rather than seeking shelter. He avoided injury, but the Luftwaffe scored a direct hit on the offices of *The New Generation,* which was all that remained of the old Malthusian League. How-Martyn safely escaped to Australia, but a good deal of her correspondence sunk en route. Years earlier she and her husband had decided to spend their fortune traveling the world, then commit suicide when the money ran out. But they could not bring themselves to do it, and eked out the rest of their days in a decrepit apartment in Sydney.[84]

Birth control activists who came under Nazi rule risked a far worse fate. In Germany, the Netherlands, and Scandinavia—much more so than in the United States and UK—they tended to be outspoken advocates of legalizing abortion and tolerating homosexuality, and were thus anathema to the new regime. Indeed, the abortionist and sexual deviant were favored themes in anti-Semitic and anticommunist propaganda. Some German proponents of eugenic sterilization attained prominence and power, but only those who had endorsed it as an instrument of Aryan race hygiene and were prepared to move on to other instruments, such as medical killing. The many more who saw sterilization as integral to the welfare state, like birth control, day care, and progressive sex education, were silenced or arrested, driven to exile or suicide. And in countries considered to be Aryan, like Denmark, Norway, and the Netherlands, German occupation brought the immediate closure of birth control clinics and imprisonment of their directors. Among Eastern Europeans, by contrast, the Nazis threw up obstacles to marriage, kidnapped children considered worthy of the master race, and encouraged and even compelled abortions. And by this point, what began with compulsory sterilization and medical killing to purify Aryan bloodlines had progressed to mass murder of Jews and Gypsies.[85]

Far more was at stake in this struggle for blood and soil than the fate of the birth control movement. But it was especially heartbreaking to the activists who had worked so hard in the interwar years to make friends overseas and who were now the recipients of their plaintive letters. Sanger in the United States, Ottesen-Jensen in Sweden, and Helena Wright in the United Kingdom tried to use these same international networks to arrange safe passage. It was all they were now good for, and often they were not good enough.[86]

In the aftermath, one might have expected the whole idea of shaping populations for political purposes to be discredited, considering the ways in which Nazis tried to control reproduction. Instead, the cause of increasing access to birth control was about to enjoy a remarkable revival. In the years immediately following World War II it won outspoken converts among the leaders of new United Nations agencies. Tentatively at first, but with increasing largesse, it gained the support of the world's richest foundations. And it would become the official policy of the largest nations.

This revival could not have been predicted at the time—indeed, proponents of population control could hardly believe their luck. But it was simply a matter of opportunity meeting preparation. Accelerating population growth in the poorest parts of the world, first noted in British India, was now occurring in many more countries. Outside observers found that the Japanese "deplored and feared" the relative increase of their colonial subjects in Korea, more and more of them crowding into Seoul or migrating to the Home Islands, even if nationalist ideology insisted that they were one people and welcomed population growth. A 1936 census of the French empire revealed that, rather than slowing as was previously thought, population growth in Morocco, Algeria, and Tunisia had begun to pick up. One public health official in French Indochina remarked that "they are born too much and they don't die enough." The governor-general appointed a council to advise how continued growth could be accommodated. Concern about population increase in the British empire had now spread far beyond India. In 1939, a Royal Commission insisted that reducing birth rates was "the most pressing need of the West Indian colonies."[87]

The director of the U.S. Division of Territories and Island Possessions, Ernest Gruening, agreed that "Birth Control is the only hope" for Puerto Rico. He worked with the American Birth Control League to make it available in government clinics, and defied repeated protests from the National Catholic Welfare Council to secure passage of a law that legalized sterilization. It was a rare defeat for the Catholic hierarchy, and perhaps merely payback for the bishops' failure to rein in the rabidly anti-Roosevelt radio personality Father Charles Coughlin. But along with a similar public-private program in Jamaica, this marked the first time colonial powers had assisted fertility limitation.[88]

The West India Royal Commission observed that high fertility was a phenomenon "throughout the whole tropical and sub-tropical world," and that its "sharp contrast" to the trend among Europeans and their progeny was "of the most profound importance, with far-reaching implications." In fact, birth rates were relatively low in many parts of the tropical world—most of Africa, for instance. And even where they were high, and mortality had begun to fall, the resulting rate of increase was still not extraordinary by historical standards. But areas like Korea, Indochina, and India had "appallingly high" infant mortality—as a British report admitted about the West Indies. One in five infants died in many parts of the empire. In some places it was a third or even more. Infant and child mortality was three to four times higher in French Africa and Indochina than in the metropole. The life expectancy of Korean women in 1935 was 38.5 years, almost 10 years less than their Japanese counterparts. All of this meant that there was the potential for much more rapid growth. Even in Algeria, which had achieved a truly impressive rate of population increase of over 2 percent a year—enough to cause a doubling in thirty-five years—the governor-general admitted that it was still "very backward" in public health.[89]

Once large American and European conscript armies began to march into places like North Africa, the Middle East, and South and East Asia and were exposed to the same diseases that had long plagued inhabitants there, more effective public health measures suddenly seemed like an urgent necessity. Colonial powers felt pressured to act or else answer for their failures to the United Nations. At that point population growth in poor countries truly took off. The Algerian rate of increase would be replicated—even exceeded—in all of these areas. It was already enough for a participant in the 1937 IUSIPP session to ask whether France "will be conquered—demographically speaking—by her conquest."[90]

To meet this threat—and opportunity—advocates of population control had prepared a set of ideas and strategies with tremendous potential. They had already seized on improving public health as both a justification—"death control" had to be balanced by birth control—and a model, because both were meant to minister to society's ills. And just as indigenous peoples had sometimes opposed public health measures, it was argued, one could not assume they would accept contraceptives with alacrity. Advocates had therefore begun to think about "demand creation," including the use of new media and even some of the specific images—such as the "unplanned family" and fewer, better births—that might attract them. In countries like India, China, and Japan they had also discovered allies and learned they needed to let them take the lead. Moreover, Nazi population policies had shattered the international sexual reform movement while discrediting eugenicists with an explicitly racist agenda. That cleared the way for those who wanted to promote more "family friendly" policies to improve reproductive choices and child-rearing while regulating such choices through institutions of social control.

Family planning could win the allegiance of so many and different kinds of population activists, including Catholics and pro-natalists, because it simply meant encouraging what certain individuals with science and power on their side considered a more rational approach to reproduction. It required calculating the value of human beings and implied that society must have a say—locally, nationally, and globally. Applying this concept in particular cases would occasion fierce struggles, but it could easily accommodate the idea that poor nations, like poor people, should plan to have fewer children, while rich nations and rich people should have more. The concept of family planning brought human reproduction squarely within the realm of public policy. Population control was now recognized as a tool of social engineering. It could also be a weapon.

4

BIRTH OF THE THIRD WORLD

War has sometimes been called population control by other means. But the worst war in history did not stop the growth of world population. The best contemporary estimates suggested that even as warfare laid waste to great cities and ravaged the countryside, the earth gained an average of fifteen million inhabitants a year. One of the most battered nations, Japan, grew by two and a half million people between 1940 and 1945. Like many of the belligerents, it continued trying to persuade couples to have more babies even as the bombs fell. Birth rates increased even more in the United States, the UK, the British dominions, and Scandinavia. Full employment and hasty marriages, rather than pro-natalist propaganda, were what helped renew population growth. Another important reason for that growth was that so many countries—Canada, Denmark, the Netherlands, Sweden, the United States, the United Kingdom, as well as Japan—cut infant mortality in half in the 1930s and 1940s.[1]

The war accelerated trends that were combining to create a baby boom that would be heard around the world. Mothers and infants living behind blockades sometimes ate better because a rationed diet was more nourishing than any they had known before. The leading British authority, the future director-general of the UN's Food and Agriculture Organization (FAO), Sir John Boyd Orr, remarked in 1943 that gains in nutrition could add more "man years" in one lifetime than had been lost in all the wars of the modern era. The eminent French demographer Alfred Sauvy would later argue from such evidence that this was the first war to increase world population. Against total population losses of some sixty million—including births that did not take place—one had also to tally all the advances the conflict had occasioned in antibiotic drugs, vaccines, and pesticides, which by the 1950s were adding at least five million people annually who would otherwise have died.[2]

Consider the impact of just one such wartime innovation: DDT spraying against malaria-carrying mosquitoes. In the South Pacific, malaria caused five times as many allied casualties as enemy fire. Military transports and bombers therefore flew in front of advancing marines in Iwo Jima, Saipan, and Okinawa to wipe out mosquitoes. The campaign continued after the war, as hope swelled of eliminating one of the world's greatest killers. The most famous case was Ceylon, where the parasite afflicted more than half the population with anemia and chronic fatigue. In November 1945 public health authorities dispatched trucks, jeeps, and men on foot to begin spraying the walls of more than half a million homes. Within two years the total number of malaria cases had been cut by three-quarters, and six years later life expectancy had increased from 46 years to 60, largely because of the decline in infant mortality. For that brief period the population of Ceylon experienced something that had never happened before in all of recorded history: the statistical equivalent of everlasting life. With every passing year the average lifespan increased by a year or more. This feat would be repeated in such places as Mauritius, Costa Rica, Mexico, and Barbados.[3]

But, of course, individuals do not experience greater life expectancy as equal to and opposite from a life cut short. Improving public health is something that happens to a population, but seemingly no one in particular. Even in Ceylon, all most people would ever see were a few men with spray canisters who came and went every six weeks, leaving behind nothing more than an invisible film and the faint smell of chlorine. A great war like the one that had just ended, on the other hand, pulls people in or drives them out, forcing tens of millions to fight or flee.

That helps explain why, in the years that immediately followed World War II, even demographers tended to assess population growth as important in direct ratio to the increased danger of new wars and future famines. Sauvy himself noted that World War II would also be the *last* war that would add to world population, considering the probable outcome of a clash between the superpowers. "The population bomb" was routinely compared to nuclear bombs in its potential to make the world unlivable.

Fears of a population "explosion" could now be aired without appearing to credit Axis propaganda about the need for lebensraum. Moreover, commentators could invoke the authority of new international institutions in citing seemingly implausible projections of future growth. When the rate of growth proved even more rapid, the need for accurate population projections seemed all the more pressing, attracting still greater resources and attention to the burgeoning field of demography. The researchers responded with "demographic transition theory," which made reducing fertility integral to the "modernization" process. At the same time conservationists offered a very different argument for population control: the economic development necessary to support growing populations was already destroying the environment. They could point to millions of "displaced persons" crowding refugee camps and living off UN rations. Harry S. Truman, like many leaders of the era, concluded that "overpopulation" in Western Europe was "one of the gravest problems arising from the present world crisis."[4]

Yet older fears, especially about the quality of the lives being saved through improved public health, persisted in the postwar period. The defeat of the Nazis and the revelation of all their efforts to elevate a master race did not suddenly and irremediably discredit eugenics. The process had started earlier and continued episodically. In most places where compulsory eugenic sterilizations were carried out before and during the war they continued into the 1950s or even later. There was little popular reaction against the practice even in countries that suffered Nazi occupation, much less in Georgia, Nebraska, North Carolina, and Iowa, where the pace of eugenic sterilizations accelerated after 1948. That same year a new eugenic law was passed in Japan, and authorities carried out far more compulsory sterilizations than ever before. In Germany itself, even doctors and professors who had killed "incurables" for the Third Reich escaped prosecution, and many resumed their posts.[5]

Governments that had never adopted negative eugenics as an explicit policy, including Great Britain, France, India, and Nationalist China, gave high-level attention to differential fertility after the war. It also found a place on the agendas of United Nations bodies like the Population Commission and the Educational, Scientific, and Cultural Organization (UNESCO). Prestigious new demographic research institutions, such as the Population Council in New York and the Institut National d'Etudes Démographiques in Paris, as well as a reformed IUSIPP—now called the International Union for the Scientific Study of Population (IUSSP)—included issues of quality in their charters. Some of the same individuals and institutions that identified with the cause of eugenics would also champion international aid for family planning. Yet because eugenics could signify so many things—for instance, investing more in "human capital" rather than fear of degeneration—the nature of its influence has to be investigated, and not simply assumed.[6]

It might be supposed that what we now call the Holocaust should have made any new experiments to improve population "quality" unthinkable. But most contemporaries did not consider the Nazis' attempted genocide of the Jews to have been a defining feature of their era, or even a defining feature of the Nazis. Dissident socialists, nationalists, and a host of others were usually grouped together as victims of a persecution that was thought to be primarily political—not racial—in its nature and purpose. The defeat of the Nazis did not, therefore, silence those in other countries who still worried about biological degeneration, especially when they were careful to specify that every ethnic group included people of quality. Indeed, in 1950 Margaret Sanger pointed to the death camps as conclusive proof of the "widespread devaluation of human lives" and the urgent need for policies to improve them, beginning with the sterilization of those with "dysgenic qualities of body and mind."[7]

Where eugenics did begin to give way, it was not without a struggle. In 1953, for instance, Alva Myrdal carefully selected a committee of scientists to consider whether differential fertility might cause a decline in the quality of population. She hoped they would issue a joint statement for UNESCO analogous to its famous declaration denying the scientific validity of racial hierarchies. But after four days of debate, only a bare majority could agree that it was "not proven," and most urged additional research on negative eugenics. By that point, Myrdal herself had become convinced that differential fertility was harming the quality of population, even while remaining hopeful that it was a passing phase as "lower social groups" reduced their birth rate in line with the rest of society.[8]

The most noteworthy immediate consequence of Nazi population programs was to provoke retaliatory ethnic cleansing across Central and Eastern Europe through the expulsion or flight of roughly thirteen million German-speakers. In Asia, three and a half million Japanese civilians were made to return to the home islands—even those settled for decades in such colonies as Taiwan and Korea. France adopted an entirely different approach. Foreign ministry officials called for reducing the population of Germany by thirteen million people, with different countries assigned a quota of emigrants—even Mexico was expected to absorb two hundred thousand. Otherwise, they warned that a Germany swollen with refugees from the East would launch new wars of aggression. Charles de Gaulle himself approved a policy in which France set out to recruit some four million immigrants to replenish its population losses. They were to be selected according to racial categories, reflecting its authors' concern that the nation had become too "Latinized" and needed an infusion of Nordic blood— even to the point of encouraging German POWs to remain and marry French women. Just as a genocidal war had the paradoxical result of accelerating world population growth, it encouraged increasingly bold policies to shape demographic trends, including trends in population quality.[9]

Yet the most ambitious of the postwar policies—and the ones that were most fully implemented—aimed not merely to prevent the reproduction of the unfit, or push and pull people across international borders, but to effect a qualitative transformation so sweeping that it would end racial hierarchies once and for all. This was the import of the new international initiatives to improve public health. "The undernourished eastern peasant, afflicted with chronic malaria, and host to a rich assortment of internal and external parasites," is generally "a weak and lethargic worker," the State Department observed in introducing the Point IV program, the first American foreign aid initiative for "underdeveloped areas." Improving public health, especially through new techniques promising quick results, would not merely make them more productive, but bring "intangible changes in outlook on life."

> Certainly some of the fatalism and lethargy of the peoples in some of the less developed countries derives from nothing more mystical than malnutrition or the debilitating diseases that sap the physical vitality of the people. A more vigorous physique and a more satisfying knowledge of their power to control diseases which have subjugated them will contribute to the initiative and the receptiveness of these people to new methods.[10]

This faith in the transformative power of public health had deep roots, and the Point IV planners themselves cited the international campaigns of the Rockefeller Foundation dating back to 1913. Then too proponents claimed that a dramatic demonstration could have a "regenerating influence" on "whole families and even communities." All along, the ultimate goal was to change the lethargic and fatalistic sharecropper or peasant into a modern worker and consumer who could participate in global markets. It would guide not only America's own efforts, but also UNICEF's campaign against tuberculosis, and WHO's effort to eradicate malaria. In this way what came to be known as "development" was conceived of as the triumph not merely of economics over politics, and of man over nature, but of man over himself, in an evolutionary process that was at least partly biological in nature.[11]

From the beginning, some worried that nature would have its revenge. Improved public health, the Point IV planners acknowledged, "will at the same time intensify one of the great problems in the success of the program—increases in the population of areas already overpopulated under present economic conditions." The belief that communism was a "malignant parasite which feeds only on diseased tissue," as George Kennan put it, gave Western policymakers a new reason to worry about the prospect of rapid population growth in poor countries. But the fear of becoming a small, persecuted minority in a world overrun by nonwhite hordes was so deeply felt that it required little rationalization, sometimes overwhelming more conventional geopolitical calculations. Japan's early victories and plans to unite Asia had reignited these anxieties. Winston Churchill was haunted by the fear that the war would arouse Asia and eclipse Europe. Even Hitler grew worried upon hearing of the fall of Singapore, reportedly suggesting that "he would gladly send the English twenty divisions to help throw back the yellow men." Roosevelt, on the other hand, wanted to accelerate decolonization, which he saw as a way to avoid making "1,100,000,000 potential enemies."[12]

More precise figures became available on population growth in Asia, Africa, and Latin America at the same time that imperial powers began to lose control of these regions. Differential fertility between North and South came to be seen as part of a crisis in the colonial world. Yet it continued to elicit varying responses, and promoting birth control more broadly was not the obvious choice. For the head of the National Catholic Welfare Conference's Family Life Bureau, birth control remained "a decided menace to the future leadership of the white race," and Japanese birth control devices were more destructive than Japanese bombers. Pearl Buck argued for the repeal of discriminatory immigration policies against America's Chinese allies. Otherwise, she warned,

> we shall have to make up for our inferiority in numbers by military preparations of the most barbarous and savage kind. We must prepare superweapons, we must not shrink from chemical warfare on a

mass scale, we must be willing to destroy all civilization, even our own, in order to keep down the colored peoples. Is this a future which any human being wants to face?

Similarly, the novelist Aldous Huxley, Julian's brother, warned that "military leaders of the countries with low birth-rates will come to believe that their only chance of survival consists in using, before it is too late, their technical superiority in atomic and biological weapons, in order to offset the effect of the big battalions."[13]

Westerners directly responsible for maintaining control of overseas possessions increasingly viewed population growth as their nemesis. It was a source of anxiety for American officials in occupied Japan. In the European empires, high fertility was thought to drive poverty and political unrest. "If there is any single cause to which the difficulties of the African people can be ascribed," the governor of Kenya insisted, " . . . it is their astonishing rate of increase." Another British official claimed in 1944 that in the West Indies, Mauritius, Cyprus, and Malta population growth might pose an insuperable obstacle to increasing living standards. That same year a French commission in Algiers was told that "the decisive problem" for the future of the North African *départements* was "a demographic problem." By 1947 Sauvy had concluded that the demographic vitality of the colonies was driving their independence movements: "The relation of cause and effect is no more in doubt. It is the demographic factor that commands political expansion."[14]

Algeria posed a particular problem. Once Muslims were made citizens of the republic, authorities could find no legal basis for denying them entry at French ports.[15] With as many as forty thousand reported to be disembarking in a single month, the prime minister himself demanded a solution. How could countries like Britain and France encourage population growth at home while insisting it was "pathological" in their overseas possessions—as the eminent political theorist André Siegfried insisted—especially if their colonial subjects were free to move into the metropole? "If one has to consider France and Algeria as communicating vessels," Fernand Boverat, now head of a national population commission, warned in 1952, "there will no longer be any possibility of planning a pro-family and pronatalist policy in France: in the future millions of Muslims would come to fill our empty spaces." Great Britain faced a similar problem with the start of large-scale immigration from the West Indies.[16]

At times colonial officials could be no less lurid and imaginative than novelists like Buck and Huxley. T. H. Davey, a member of the Colonial Advisory Medical Committee, warned in 1948 that if new public health techniques spread throughout the empire, Britain might soon confront hopelessly overpopulated and impoverished nations, and find itself "dragged into a war for survival, using against them the most terrible of the weapons which science had produced." But whereas Davey concluded that there was still time to avoid placing "undue emphasis on medical advance" without first planning social and economic development, other members of the committee warned against discouraging long-delayed public health programs. Along with the obvious humanitarian considerations, the gathering anticolonial movement compelled both British and French officials to prove that they were improving the lot of their charges.[17]

Though Davey himself did not actually say it—few did—the idea was in the air: Rather than resort to weapons of mass destruction to cut down growing populations, Westerners could simply refrain from using pesticides and vaccines to reduce mortality. For Frank Notestein, who directed Princeton's OPR before heading the UN's Population Division, such a policy only made sense as a means to preemptively weaken those who might one day threaten the West. He agreed that some overpopulated regions would imperil the peace, while others would simply be "increasingly expensive and troublesome to administer, and unsatisfactory to do business with." But the European powers had proved either unwilling or unable to modernize their colonies. The solution was therefore to end outside rule and assist social and economic development, even if it meant that "the now dominant powers would in effect be creating a future world in which their own peoples would become progressively smaller minorities, and possess a progressively smaller proportion of the world's wealth and power."[18]

In one of the first general formulations of what came to be known as demographic transition theory, another Princeton demographer, Dudley Kirk, predicted in 1944 that the Western pattern of mortality declines and rapid population growth would spread across Asia. Any attempt to preserve "white supremacy" would bring "an intercontinental conflict that might well dwarf the present war in ferocity." There was no alternative but to assist development in order to reduce birth rates. While miracle cures and DDT might rapidly cut mortality, fertility would not begin to fall until peasants moved to cities, earned paychecks, and enrolled their children in school. Until then, distributing contraceptives would achieve nothing. From a historical and descriptive model, one that had initially gained little notice even among demographers, "the demographic transition" was transformed into a policy prescription.[19]

Kingsley Davis, who is credited along with Kirk and Notestein for winning acceptance of this theory, acknowledged that the "possibility that Asia's teeming millions will double or even triple within the next few decades, acquiring Western instrumentalities at the same time, appears as a Frankenstein appalling to many observers." But he argued that "the Asiatic *races* [are not] going to cause the whole world to 'sink' to the level of present-day Oriental civilization." Instead, the "Asiatic peoples" would become "Westernized." The prospect of Western populations shrinking relative to the rest of world appeared less threatening to the extent that the rest would become more like the West.[20]

American demographers thus rose to prominence by promising to slay monsters, or at least domesticate them. This required more than just reducing mortality and morbidity, which could add to both the poverty and the latent power of nonwhite peoples rather than transform them into ersatz Westerners. As Notestein put it, one had "to apply in synchronized fashion every device for the creation of a social setting favorable to reduced fertility." Women's access to education—which Notestein's Princeton colleague Irene Taeuber was proving had a direct and proportionate relationship with fertility—was given no particular importance. Instead, modernity was understood as an integrated whole, and achieving it required "a great project of social engineering," as Notestein described it to FAO director-general Orr. Similarly, UNESCO director-general Julian Huxley called for nothing less than "the formulation of a world population policy." Like ILO director Albert Thomas a generation earlier, these men were attracted to the sheer scale, complexity, and controversy of population problems as an opportunity to coordinate efforts of individual states—even to go where colonial administrations feared to tread.[21]

Some UN officials began to see in population policy a step toward global governance. New international institutions could transform the old civilizing mission into a modernizing mission, taking up unfinished work in public health and education in the name of global norms. As Orr argued, politicians were hung up on adjusting borders, even though advances in communications and technology made absolute sovereignty impossible. But UN agencies, by focusing on improving "borderline" populations, especially their health and nutrition, could make a much greater contribution to reducing international tensions. For Orr, concrete action in areas of common interest was a way of moving toward world government. Similarly, Julian Huxley pointed out that the population problem—including both uncontrolled growth and qualitative decline—"affects the future of the human species as a whole, and not merely the separate nations into which the human species now happens to be divided." If the population explosion posed a threat akin to nuclear war, then population control was no less urgent than arms control. The UN seemed to provide the appropriate forum and agency. Here too, Orr insisted, the choice was "one world or none."[22]

The UN's work in population picked up where the League of Nations left off. By the late 1930s it had finally begun to dawn on people that population trends could provide an international organization's *raison d'être*. Many League members feared demographic decline, others claimed to be running out of room, and migration, by definition, defied the conventional separation of domestic from international affairs. Pro- or anti-natalist measures remained controversial, but since the League was trying to improve nutrition and control disease, it could not be indifferent to the consequences. Moreover, almost everyone agreed that these were matters of fundamental importance that remained poorly understood. In September 1937 the League Assembly had therefore called on the secretariat to begin investigating population problems.[23]

At the League's last assembly before war broke out, representatives agreed that problems of population had been neglected and their work had suffered because of it. But at this late date it proved impossible even to agree on how to define population problems, much less make policy recommendations. A senior League official, Alexander Loveday, resolved to carry on when the war forced him to decamp to Princeton together with a skeleton staff. A grant from the Carnegie Corporation enabled Loveday to commission Notestein and the OPR to undertake demographic studies. He assured Sean Lester, the League's last secretary-general, that he retained total editorial control: "anything explosive I shall sit on." As war raged, officials of even defunct international organizations still feared the mere mention of possible links between population trends and international conflict.[24]

On the face of it, the OPR studies for the League were innocuous. Notestein was hardly one to speak up for programs to change reproductive behavior, considering his skepticism that they could have any impact. But he understood the "importance of the unimportant," as he later put it. Defining population problems, much less doing anything to solve them,

first required better data. In 1946 some 44 percent of Asians and 60 percent of Africans were living in areas where no census had been held since
1919. Obviously, the actual number of those uncounted could only be a
matter of conjecture. Estimates of China's population, for instance, ranged
from 325 million to 582 million. The situation was little better in the European empires. There had been no recent census for more than half of
the territories administered by Britain's Colonial Office. Both British and
French authorities knew that official estimates were less accurate than publicly admitted. Recording every birth and death could produce more "real
time" data. But it posed an even greater administrative challenge. In British
India, which had the longest and most impressive tradition of tracking its
subjects, experts found that reported figures on fertility and mortality were
largely "worthless." Even if vital statistics were collected and considered accurate, there were no qualified personnel to analyze them in such places as
Nigeria, the Cameroons, French West Africa, or the Belgian Congo.[25]

An international organization could therefore lay the groundwork for a
world population policy merely by reporting on official data and publishing some of its own. To begin with, it would draw together demographers
from different countries and provide them with a neutral setting to develop
common practices and common positions vis-à-vis their own governments.
By establishing standards for vital statistics and censuses, they could more
readily compare and aggregate the results. A fuller accounting made the
gaps easier to identify, and marked disparities begged questions that could
be answered only by further investigation. Projections of growth or decline,
even when properly interpreted as an extrapolation of present trends, implicitly stood as a judgment on past and current policies. When read as predictions about the future, they were the stuff of headlines. And the very
categories used—especially "world population" when issued under the imprimatur of the United Nations—had political implications, because they
suggested collective interests and a common destiny.[26]

As the UN took over the League's work in population, a tension persisted between those, like Notestein, who thought the UN should ascertain
facts and let the facts speak for themselves, and others who continued to
demand that it not only define problems like overpopulation but advocate
means to address them. The first approach found acceptance in the annual
meetings of the UN's Population Commission together with its permanent
staff in the UN's Department of Social Affairs, the Population Division. In
principle, the Commission always represented the interests of the govern-

ments that appointed its members. But American demographers were well placed to push an agenda that emphasized improving official statistics and projecting future growth: Notestein was the first director of the Population Division, succeeded by Pascal Whelpton of the Scripps Foundation, and then John Durand, who had completed his Princeton Ph.D. under Notestein's direction.[27]

The UN's new specialized agencies, such as the World Health Organization (WHO) in Geneva and the FAO in Rome, exemplified a more action-oriented approach. Officials who planned programs to eradicate malaria and boost global food production were well aware that, if successful, their efforts would rapidly reduce mortality. As "executing agencies" they were more inclined to contemplate programs to deal with the consequences. All of these organizations ultimately answered to governing bodies made up of government delegates. But the FAO and WHO provided more of a bully pulpit for activist leaders to propose new initiatives.

No one demonstrated both the potential and the limitations that came with leadership of a UN agency better than Julian Huxley, the first director-general of UNESCO. In 1939 Huxley had helped formulate a statement by Britain's leading biologists denouncing Nazi racism while upholding the long-range potential of eugenics to improve the human species. He was now in a position to forcefully advance this agenda. Shortly after assuming his post in Paris in 1946, and without consulting any of the member nations, he issued a statement titled *UNESCO, Its Purposes and Its Philosophy.* One of its key themes was the danger that "The Age of the Common Man" might lead to "discouragement of high and unusual quality." He wanted UNESCO to take on the most pressing world problems, beginning with population control, if "man's blind reproductive urges are not to wreck his ideals and his plans for material and spiritual betterment." Government delegates insisted that the statement be published as his personal opinion, not UNESCO policy. Huxley was undeterred, and would continue pushing population problems onto the UN's agenda.[28]

Orr attracted even more attention to the need for a concerted response to population growth. He and his successors at the FAO worried that a revival of Malthusianism might discourage efforts to boost food production. But they could not help using the specter of global famine in appealing for support. This redoubtable Scotsman insisted that the only solution was a world food policy together with an organization empowered to enforce it. When member states seemed ready to limit the FAO to an information-gathering role, Orr stood up at its first conference in Quebec to protest: "The hungry people of the world [want] bread, and they [are] to be given statistics." After delegates first considered seventeen other candidates for director-general, they finally offered him the job. Orr reluctantly accepted. The result was an awkward match between an institution constitutionally indisposed to challenge the status quo and an idealistic leader intent on feeding the world.[29]

Orr's dream was a "World Food Board." It would manage a global reserve that could be added to or drawn down to stabilize agricultural prices year-to-year. It would also make concessionary sales to help poor countries meet basic nutritional requirements and extend long-term credits to increase productivity. In this way farmers could become prosperous by maximizing production, and thus provide an adequate diet for a growing world population. At a time in which hunger persisted even in the heart of Europe, governments destroyed surplus foodstuffs, and farmers were paid to leave their fields fallow, this was a tremendously appealing idea. But Britain's Foreign Office and the U.S. State Department eviscerated his proposal. The United Kingdom worried about the start-up cost and the prospect of paying more for food imports. The United States insisted that the proposed International Trade Organization and the goal of deregulation should govern all commodity exchanges. Utterly disgusted, Orr embarked for home. "I took out my handkerchief," he later recalled, "wiped the dust of America from the soles of my shoes with it and threw it into the harbour."[30]

The failure of the World Food Board demonstrated that ambitious new UN initiatives would get nowhere without the support of the most powerful member nations. Nevertheless, in 1948 Huxley chose to tackle the even more controversial demand side of the resources-vs.-population equation. In a message to UN secretary-general Trygve Lie he warned that wherever WHO was successful in reducing mortality, a rapid increase of population would follow, and that the globe was already facing a Malthusian crisis. He considered the "qualitative aspects" to be equally important. Blacks in the United States and parts of Africa were multiplying more quickly than their white counterparts, and Russians were growing faster than any other white nation. Research showing an inverse correlation between fertility and intelligence portended an imminent decline in innate ability, one that might be only masked—temporarily—by better nutrition and education: "the cumulative effect on the human species and its capacities would be disastrous." Huxley therefore proposed a world population conference to prepare governments and public opinion for "the formulation of a world population policy."[31]

The proposal rang alarm bells at the temporary UN headquarters in New York, already preoccupied with the recent communist takeover of Czechoslovakia and the intractable problem of Palestine. UNESCO's representative reported that Lie's top advisors were "perturbed." There was "already too much high tension obtaining here at Lake Success to warrant another provocation." They had therefore refrained from circulating Huxley's memorandum to the Population Commission. Assistant Secretary-General Henri Laugier explained that a population conference would merely provoke an ideological debate pitting Malthus against Marx. "We covet one or two quiet years in which the technicians of the population commission can probe to discover their areas of agreement and disagreement." UNESCO dropped the proposal when Huxley left office.[32]

By the end of 1948 the cause of a comprehensive international response to population growth had lost its two leading advocates. But Orr and Huxley did help provoke a dramatic increase in press coverage, much of it on how resource competition could threaten new wars. So too did the publication of two books that offered a different argument for population control, one that could help keep the cause alive even without high-level support. Rather than warning of future calamities, Fairfield Osborn's *Our Plundered Planet* and William Vogt's *Road to Survival* pointed to tangible signs that time was already running out, including deforestation, desertification, and extinction of wildlife. Even without nuclear war, man had become a "geological force," as Osborn put it, and had begun to ruin his only home. Both authors also warned about a decline in the "quality" of population. With the survival of the species at stake, they stressed the essential kinship of humankind.[33]

The residual differences between these books—and the even greater difference between their authors—reveal which arguments would strike a chord in the postwar period. "Fair" Osborn had both the pedigree and the social position to give environmentalism a sharper eugenic edge, if he had wanted to. He presided over the New York Zoological Society and had just established the Conservation Foundation. His father, Henry Fairfield Osborn, had made New York's Museum of Natural History a shrine of "Nordic Man" and co-founded the American Eugenics Society. But when his son sat down to write his first book, he—much like his cousin, Frederick Osborn—rejected the idea of a rigid racial hierarchy. To illustrate the power of environment and not just heredity in determining performance, Osborn related the story of a stable of thoroughbred horses. In 1933 its fortunes had begun to decline, and by 1941 it did not win a single race. Even worse, it started to sire stillborn or deformed colts. It turned out that the problem was not in the blood, but in the soil. The grounds had grown barren from overgrazing and no longer supplied essential nutrients. A program of restoration was begun at once, and at last a new generation had begun to win again. The story supposedly illustrated the importance of preserving the "precious sensitive earth" upon which all life depended. But it also paralleled the history of Osborn's own family: In 1933 his father had been forced to resign from the museum, his work largely discredited. And in 1942 an internal review determined that he had misallocated funds.[34]

Vogt, by contrast, was a self-made man. An ornithologist by training—in the field, rather than through advanced degrees—he had become head of the conservation section of the Pan-American Union. But he seemed most obsessed with the undeserving poor—not just in the United States, but all over the world. He insisted that the FAO "should not ship food to keep alive ten million Indians and Chinese this year, so that fifty million may die five years hence." Absent "a rational population policy these nations . . . have no right to expect aid from the rest of the world." Even unregulated trade might "subsidize the unchecked spawning of India, China, and other countries." Echoing the nineteenth-century warnings of M. J. Dee and Francis Walker, he claimed that they would supplant better-paid workers. Vogt was particularly critical of the medical profession, which "continues to believe it has a duty to keep alive as many people as possible." Indeed, he described diseases like sleeping sickness as positive "advantages," because they "prevented the development of overgrazing and overpopulation." Vogt also backed sterilization bonuses:

> Since such a bonus would appeal primarily to the world's shiftless, it would probably have a favorable selective influence. From the point of view of society, it would certainly be preferable to pay permanently indigent individuals, many of whom would be physically and psychologically marginal, $50 or $100 rather than support their hordes of offspring that, by both genetic and social inheritance, would tend to perpetuate the fecklessness.[35]

Despite Osborn's prominence and social connections, it was not his version of environmentalism—"precious" and "sensitive"—but Vogt's diatribe that became an international best-seller, available in nine languages. Largely because of this success, three years later he was selected to be national director of the Planned Parenthood Federation of America (PPFA). Vogt's chief concerns—cheaper contraceptives, education and incentives to increase demand, and linking food aid to population control—helped set an agenda that would persist for thirty years. He showed how, even while agreeing that "we be of one blood" and that everyone shared the same "road to survival," environmentalists could insist that many poor people would not make it and must be left to die.[36]

These arguments would gradually rouse an enormous new constituency for population control, especially after Rachel Carson took up another theme shared by both of the books: condemnation of DDT spraying. But before *Silent Spring*, environmentalists did not have confidence they could stand alone. Osborn and Vogt warned that population trends would lead to degeneration and war, not just environmental degradation. Similarly, later that same year Aldous Huxley argued that the "demographic and ecological crisis" was part of a "double crisis," because it caused international political tensions. In a UNESCO-sponsored study, he warned that differential fertility posed a danger to Western Europe and North America from both within and without. Their future population "will be constituted, in the main, by the descendents of the least intelligent persons now living in those areas." At the same time, the West's quantitative decline would invite aggression from the East.[37]

Less skillful authors stretched metaphors to link environmental and eugenic concerns. In 1951 Robert C. Cook, the new director of the Population Reference Bureau—a key source for journalists writing on these subjects—argued that "the scramble for bare subsistence by hordes of hungry people is tearing the fertile earth from the hillsides." At the same time, in the West "misplaced and badly distributed human fertility is leaching away the inborn qualities of tomorrow's children," in a process he called "biological 'erosion.'"[38]

All of these writers dismissed assisted economic development as a solution to population problems, in terms of both numerical growth and qualitative decline. For Vogt, the "capitalistic system" was "ruinous" and industrial development was a "parasite" that depended on the exploitation of new lands to survive. Similarly, Huxley described industrialism as "the systematic exploitation of wasting assets." Reproducing the Western pattern of fertility reduction would merely reproduce the problem of differential fertility. "When the nature of the human universe is such as to discourage the more sensitive, intelligent and prudent individuals from reproducing their kind," Huxley argued, "the deterioration of entire societies comes about with explosive rapidity." Eugenicists and conservationists would therefore have trouble working with demographers who saw urbanization and industrialization as the only means to equalize fertility rates worldwide.[39]

Yet another constituency—potentially larger than all the rest put together—considered birth control as a way to liberate individuals and especially women, not save "Western Civilization" or save the earth. The Swedish Association for Sex Education, for instance, had grown stronger during the war under Elise Ottesen-Jensen's leadership, and now counted one hundred thousand members. But the sex reform movement on the Continent was still reeling from persecution by pro-natalists. Vichy, Fascist, and Nazi laws against abortion and birth control were maintained by postwar governments. Even Heinrich Himmler had tolerated condoms to prevent the spread of venereal disease. Catholic youth organizations assaulted the kiosks where they were sold.[40]

When Ottesen-Jensen called a "Conference on Sex Education, Family Planning and Marriage Counseling" in 1946, some sixty participants—mostly from Scandinavia—agreed on the need for a new international association. They affirmed that "parents have the right to decide upon the number of children they shall bring into the world." Margaret Sanger came away determined to organize a much larger meeting to launch a new organization under her own leadership. She wanted to bring together all those concerned about eugenic degeneration or environmental scarcity and also attract the interest of new international institutions with a modernizing agenda.[41]

In August 1948 about 250 participants from twenty-two countries convened in Cheltenham, England, for a conference titled "Population and World Resources in Relation to the Family." Both Orr in his keynote address along with the conference president, veteran eugenicist Lord Horder, hit on a theme that could appeal to the different constituencies for family planning: the idea of a "global family." Horder described the interrelationship between food resources, standards of living, and population trends as the "tripod of international citizenship." Orr confided that he, like many others, was still getting used to the idea that all nations and all races were equals. Only recently, a British minister had greeted an FAO report with incredulity: "'That means the people of India would have the same diet that we have.' Why not?" Orr responded, "We have to think of the world as a whole."[42]

Orr had tremendous credibility in making this argument, and progres-

Elise Ottesen-Jensen leads 8-year-old Swedish schoolchildren in a sex-education class. She was an indefatigable proponent of birth control as essential to emotional well-being and maternal health, not population control. IPPF Archives.

sive internationalists like Ottesen-Jensen could hardly disagree. But while Orr stressed equality and common rights, the idea of a "human family" emboldened others to argue that world citizenship entailed common duties that might limit individual freedoms for the good of all. "The only possible ultimate solution—apart from a disaster of one sort or another—is the establishment of a world population policy," Julian Huxley explained in his agenda for a UN conference, "including the means of its implementation, whether by means of incentives and penalties, by persuasion or by compulsion, by direct methods or as the indirect result of social changes." Huxley was unable to get to Cheltenham, but the head of UNESCO's science department, Joseph Needham, gave the concluding address. "Conscious world control of population in relation to natural resources is not an impossible dream," he argued, "but a certain development having all the authority of social and biological evolution behind it."[43]

Needham spoke vaguely of the need for "some kind of representative body, which will have to take administrative responsibility for introducing the various controls which may be necessary." Clearly, the time was not yet ripe. But conference participants did agree to establish a provisional international committee made up of leaders of the American, British, Swedish, and Dutch associations. It was to support research and education, exchange information, and "promote the supply and exchange of goods needed by any country." Nothing was left out, including sex education. But the very first resolution made clear that the organization would be oriented toward "control of population increases."[44]

Margaret Pyke and Helena Wright of the British Family Planning Association set up a small office in London with space and furniture provided by the Eugenics Society. But much time was wasted debating how to proceed. It took more than a year just to agree on the name: "International Committee on Planned Parenthood." Sanger and the demographers who were advising her wanted a large, well-funded organization that would support contraceptive research and publicize the problem of population growth and resource scarcity, ultimately aiming at government programs rather than private clinics. This was what prospective American donors were demanding, but it was problematic for the Dutch and the Swedes, who still emphasized sex education and marriage guidance. From the perspective of the committee secretary in London, Vera Houghton, it was enough simply to respond to all the requests that had started to pour in. By 1950, activists were writing or visiting from over twenty countries. They "don't want a mass of facts and figures—they want to know how to start clinics, what methods to use, what other countries are doing in the same field and, above all, they want a simple contraceptive."[45]

Complicating matters was a marked decline in Sanger's health following a second heart attack, which for a time left her addicted to Demerol. She was short-tempered and impatient, and sometimes seemed incapable of understanding what her British correspondents, ever so politely, were asking of her. As ever, they resented Sanger but also needed her, even to the point of needing her to send them care packages (Britons were still living on strict rations). Houghton was told that "one of her main tasks was 'to keep the Americans happy because they had all the money'—collected by Margaret Sanger."[46]

There was just one way in which everyone—demographers, eugenicists, conservationists, birth controllers—could be kept happy. All might be united under the banner of family planning, provided that most of the "global family" actually wanted help in planning fewer children. If they did, then "control of population increases" would not require penalties and compulsion, or even broad-based development, just adequate supply and distribution of contraceptives. Authors like Vogt straddled the issue, pointing out that birth control education had hardly been tried, whether for individual welfare or the world's survival, while at the same time suggesting that some people would have to be bribed. Orr and some other speakers at the Cheltenham conference, including T. H. Davey, the Colonial Office consultant, argued from Western Europe's experience that standards of living first had to improve before people would want smaller families. A representative from the Gold Coast—latter-day Ghana—disagreed. Industrialization, urbanization, and trade were already bringing rapid changes, Dr. Okli Ampofo noted. "Nowhere are more tablets and pills of all kinds for contraceptive purposes sold than in West Africa," he reported, "and I learn this morning that these things are more or less useless." The conference treasurer, Gerda Guy, knew from her experience with the Birth Control International Information Center that available methods were inadequate even for existing demand, and suggested that they should try harder to meet this demand without waiting for social and economic development. Davey blithely replied "that that had been tried, and that these primitive populations would not have anything to do with family limitation in any form, whether by contraceptives or other means; they believed in large families."[47]

The question of whether poor people wanted to plan smaller families could be answered only by empirical research. Frank Notestein shared Davey's skepticism, but by November 1947 he was also becoming worried about the speed of mortality declines. "We need to know how to reduce birth rates in an agrarian society," he argued. "The problem is too urgent to permit us to await the results of gradual processes of urbanization." In September 1948 he stepped down as director of the UN's Population Division and joined a Rockefeller Foundation mission to Japan, China, Taiwan, Korea, Indonesia, and the Philippines. John D. Rockefeller III had returned from a visit to the region concerned that the foundation had not considered how its work eradicating disease might exacerbate overpopulation. The mission he inspired was the first of its kind, and would influence all of those that followed.[48]

Of all the countries visited, Japan seemed to offer both the most pressing reasons and the best prospects for "concrete experiments" in fertility reduction. After losing most of its merchant fleet and industry in a vain attempt to acquire settlement colonies and raw materials, Japan saw its population growing again at the fastest rate in its recorded history. The victorious allies had decided to repatriate all 6.6 million Japanese soldiers and settlers scattered from Micronesia to Manchuria. In the postwar chaos, hundreds of thousands would die from depredations, starvation, and exposure, or forever disappear into Soviet POW camps. But by the end of 1947, roughly 5 million had made it back to the home islands. Upon arrival, American occupation authorities dusted everyone with DDT to kill the body lice that spread typhus. Tens of millions of Japanese would be vaccinated against smallpox, cholera, and tuberculosis. Together with emergency food shipments, the public health program brought mortality rates down below prewar levels. As families reunited, fertility rates also recovered. So in addition to all the returned soldiers and settlers, many roaming the countryside or subsisting in shanty towns, twelve million more Japanese were born between 1945 and 1950, about double the number who died in the same period.[49]

U.S officials closely monitored these trends, issuing monthly reports on repatriation, births, and deaths and holding no fewer than five censuses in five years. The head of the Public Health and Welfare Section, General Crawford F. Sams, was a student of demographic transition theory and worried about what would happen if Japanese cities and industry failed to recover quickly. Douglas MacArthur's initial directive as Supreme Commander was to assume no responsibility for restoring Japanese living standards. But when it became clear that this growing population might remain dependent on U.S. aid, Sams helped to convince him and the allied Far Eastern Commission of the necessity for industrial reconstruction—in part because it could reduce fertility rates. Sams was pleased that Japan's parliament, the Diet, had passed legislation legalizing birth control and abortion, albeit under strict controls, and with the express purpose of preventing "the increase of the inferior descendants." But he considered it "ridiculous" to think that a country's demographic trends could be altered by directives from on high. Just before the arrival of the Rockefeller delegation, he pronounced himself confident that "they will agree that all measures which are practical and sound have already been undertaken."[50]

In Japan there had been a strong tradition of research and activism in birth control before it was driven underground during the war. Kyusaku Ogino was the first to accurately chart ovulation, and Tenrei Ota was one of the inventors of the intrauterine device. The landmark ruling that

overturned America's Comstock Law had been occasioned by the import of
Japanese pessaries. Now that militarist elites together with the pro-natalism
they espoused were discredited, Ota together with Shidzue Katō—the former Mrs. Ishimoto was recently remarried to a Socialist minister—won
seats in the Diet. It was they who had introduced the bill that eventually
became the Eugenic Protection Law, the first significant legislative initiative
to originate outside a government ministry.[51]

In its original version, the bill appropriated funds to supply contraceptives to government clinics. But it was finally passed with the sponsorship
of a new, more conservative government and what Katō considered crippling modifications. Instead of a network of clinics offering birth control
services, it created Eugenic Protection Commissions empowered to order
compulsory sterilizations and rule on women's petitions for abortion. They
would be granted only on grounds of hereditary disease or to protect the
health of the mother. While some occupation officials were troubled that
the law offered "abundant opportunity for abuse," they decided that criticism "would be interpreted as an unwise imposition of Western ethical, religious, or social ideas upon an essentially different Oriental civilization."
Sams shepherded the bill through the approval process in the belief that,
despite its flaws, it would be "useful in control of population," and as such
had to be seen as a purely Japanese initiative. Ironically, when it was debated in the Diet, sponsors noted that financial distress had been rejected as
grounds for abortion in Norway and Sweden. As ever, proponents of population policies looked to precedents and parallels abroad. At this point, a
woman's plea to end an unwanted pregnancy was deemed a less compelling
reason for abortion than preventing the birth of unfit offspring.[52]

Katō was embarrassed by the outcome, and finally had a chance to vent
her frustrations when U.S. officials organized a meeting with the members
of the Rockefeller mission. One can only imagine their reaction when Katō
delivered a bitter indictment of government inaction and American indifference. "We have to face this problem, but unfortunately most people
from the Prime Minister to the masses of people do not think seriously
about this matter; they do not think about it at all." She had warned the
last two prime ministers, but they did not have any plans or ideas, apparently waiting in vain for U.S. guidance. How could occupation officials
pretend to be neutral about birth control, she asked, when they pushed
public health campaigns and thus population growth that would outstrip
Japan's natural resources? "How," she asked, "can you be thinking of human welfare?"[53]

When the Americans could offer no answer, a colleague reminded Katō of the argument that population growth would decrease in time, apparently referring to demographic transition theory. She would have none of it, protesting that such a fundamental problem should be addressed by elected representatives, not academics and bureaucrats. She may have been thinking about how, earlier that year, the sponsor of the Eugenic Protection Law had proceeded under the mistaken impression that it provided for birth control clinics. Afterwards no one could explain why the authorizing clause had been omitted, but it bore the fingerprints of bureaucrats in the Health and Welfare Ministry. A year and a half later, they had still not approved a single contraceptive. This made abortion and sterilization a lucrative monopoly for doctors. For Katō the solution was clear: the government had to provide free clinics for birth control as well as abortion. Encouraged by her example, the other Japanese present spoke up for liberalizing access to abortion and reeducating doctors.[54]

The two demographers with the Rockefeller mission, Notestein and his Princeton colleague Irene Taeuber, started to rethink transition theory. Taeuber pointed out that Japan's death and birth rates both fell in the 1920s and 1930s, as would be expected in a society with improving living standards. Yet there had been no such improvement. Gains from increased productivity had been reinvested in industrial expansion. Relatively low fertility persisted nonetheless even among peasants during the chaotic postwar period. Taeuber worried that it might not be sustainable, but speculated that birth control education could help maintain the evident preference for small families. Notestein became "more and more convinced" that "the unprecedented process of population control prior to material well-being is more likely to occur here than anywhere on earth . . . it might well set the precedent for the whole of East Asia."[55]

The mission's final report offered an ingenious adaptation of demographic transition theory that could both explain why some Asians wanted smaller families and justify a policy to help them. Borrowing from the ideas of sociologist Talcott Parsons, it argued that reproductive behavior could be changed without trying to change everything in a culture. "No social system, however coercive, maintains absolute homogeneity of behavior. All systems have their dissident extremes open to innovative suggestion: all have those who conform only because of the absence of alternatives." Accordingly, a farmer's wife who told Taeuber that her husband was happy with one child was not just a statistical outlier, but a leading indicator of a broader social change that could be assisted and accelerated. And whereas previously Notestein had treated the fertility rate as a dependent variable, reflecting a culture's social and economic development, now he suggested that reducing fertility might be a necessary condition for such development. This idea of "dissident extremes" as the vanguard for modernization would eventually lead to alliances between aid agencies and birth control activists all over the world.[56]

The report warned that rapid population growth was imminent in Indonesia, China, and Korea. But the authors admitted that there was no simple, cheap, and effective contraceptive suited to people who were mostly illiterate and never saw a doctor. When pressed for recommendations in a rural Chinese community, they felt embarrassed and "spoke lamely of folk methods and of what might be done under different economic circumstances." Notestein later confided that this predicament made him feel more helpless than at any other time in his life. But it also appears to have provoked him and his colleagues to rethink the whole issue. Rather than considering contraceptives as adequate, or not, and motivation as sufficient, or not, they concluded that the two were dynamically interrelated: "Without the motives for family limitation, no means will suffice; and with weak motives efficient means are required. . . . Both of these problems are critically important, both are amenable to attack and neither has received the attention it deserves." This insight had the potential to shift control of population increase from individual parents and their preferences to policymakers determined to make contraception easier than any conceivable alternative, or even to ensure "the absence of alternatives."[57]

The Rockefeller report was sent to a host of influential figures in Washington, perhaps the first time they saw birth control presented as a U.S. foreign policy interest. It merely recommended further research, but all of the findings pointed to occupied Japan as the country most "amenable to attack." There was sufficient literacy and health infrastructure to provide contraceptives without waiting for the invention of simpler methods. Moreover, occupation authorities could protect and promote the "dissident extremes" who were already demanding such an initiative. Indeed, Notestein and Taeuber may have considered the whole of Japanese society a dissident extreme, one that could potentially lead the rest of Asia to accept population control.

Daniel Luten, a chemist for Shell Oil and advisor in the occupation's Natural Resources Section, had accompanied the Rockefeller mission and immediately started lobbying for a birth control education campaign. He thought it was "the only hope for East Asia," and Japan was "the best place in the world to demonstrate its feasibility." Trying to increase production without restraining fertility was like pumping water into a paddy field with broken dikes. "The other carrier of this burden is not doing his share," he wrote, apparently referring to the Public Health and Welfare Section under Crawford Sams, "and seems unlikely to do it unless aggressively driven to action."[58]

Help was on the way in the person of Warren Thompson. He was known both for publishing one of the earliest statements of demographic transition theory and for his sympathetic appreciation of how mortality declines would pose a particular problem for Japan. Even during the Pacific war he had suggested that areas of Australia would eventually have to be opened to settlement. He was therefore ideally placed to lead a publicity campaign to convince Japanese leaders of the need for population control.[59]

After Thompson's arrival in 1949 there was a fourfold increase in the number of articles on population in the Japanese press. To Luten's relief, he no longer considered emigration to be a feasible solution and urged fertility reduction through birth control. The fact that Thompson was an advisor to the occupation suggested this was official policy. Even Sams was now convinced that government clinics should provide contraceptive information. On April 14, shortly after Prime Minister Shigeru Yoshida met with Thompson, his chief cabinet secretary announced that he agreed that birth control was "a fundamental solution to Japan's population problem."[60]

Over the following month the cabinet established a Council on Population Problems, reportedly to develop a population control program. The Health and Welfare Ministry removed onerous regulations of birth control clinics and approved twenty-seven different contraceptives. And the Diet moved to revise the eugenic protection law to make economic hardship grounds for abortion and provide contraceptives at government clinics. On May 10 the House of Representatives passed a resolution declaring that "Japan is extremely overpopulated" and that the public should support a reduction in the birth rate.[61]

Unfortunately for population control advocates, MacArthur himself had not changed his mind. During an hour-long interview with Thompson, he agreed that the home islands could not support even their present population, emigration was no solution, and such powerful incentives for smaller families would make it impossible to stop the nascent birth control movement. But he was skeptical that the process could be accelerated, and therefore concluded that there was "no need to take active measures." Ironically, this was pure demographic transition theory, much as Thompson had first described it twenty years earlier.[62]

As a man with presidential ambitions, MacArthur had powerful incentives to steer clear of the whole issue. When the Allied Catholic Women's Club of Tokyo protested some of Thompson's public statements, MacArthur claimed that his office was "not engaged in any study or consideration of the problem of Japanese population control. Such matter does not fall within the prescribed scope of the Occupation and decisions thereon rest entirely with the Japanese themselves." Media coverage of population issues dropped precipitously, the Health and Welfare Ministry reversed approval for contraceptives, and police even arrested someone for showing a birth control film. Occupation officials, for their part, censored publications with passages that might imply a policy favoring population control and canceled visits by other demographers. When Luten's boss continued to press, he was reprimanded and made to sign a letter reassuring the Catholic Women's Club.[63]

Authorities went on to deny Sanger permission to visit Japan, which finally provoked a reaction back home—including a critical newspaper column by Eleanor Roosevelt. MacArthur received an almost unprecedented number of letters, and had a three-page, single-spaced response prepared. He explained that the Japanese were already working on the population problem in a reasonable and dispassionate fashion—indeed, they had adopted a position more advanced than certain American states that still outlawed birth control. If Sanger came it would lead to propaganda charging that the occupation "had imposed measures upon the conquered Japanese people leading to genocide."[64]

The UN genocide convention did include the deliberate suppression of fertility of a subject people. But MacArthur's concern is difficult to credit. Sanger had already visited Japan twice before the war, and had been invited back by a Japanese newspaper chain. MacArthur was obviously most concerned about his political viability at home. That is precisely what makes his response more than merely disingenuous. It gave population control proponents an invaluable lesson: it was better to talk about the potential sensitivity of Asians to outside interference than to acknowledge the real and obvious opposition of the Catholic hierarchy to any move anywhere that liberalized access to birth control. Even the Rockefeller Foundation declined to accept its mission's recommendation to pursue the population issue at least partly because Cardinal Francis Spellman of New York told them not to. But if Asians took responsibility for controlling their own numbers, even someone as imperious as MacArthur could not oppose them. The development of "local capacity" was already a concern of activists since the 1930s. After this setback, Notestein, Thompson, and Whelpton would elevate it to a cardinal principle.[65]

Japan's birth rate did begin to decline, but it was almost entirely through the use of condoms and abortion. The Health and Welfare Ministry refused approval for new contraceptives women could use on their own. On the other hand, obstetricians and gynecologists won authorization to perform abortions at their own discretion. By 1955 it was estimated that there were 30 to 50 percent more abortions than live births. Rather than setting an example for the rest of Asia, family planning in Japan was unlike that in any other country. After a complicated dance among birth control activists, Occupation authorities, foundation consultants, government bureaucrats, and Catholic opponents, the outcome reflected the triumph of special interests rather than a national policy to limit population growth.[66]

More generally, the population policies of the biggest countries in Asia were developing in isolation according to local circumstances. There was not yet any institutional network connecting proponents to one another or to external sources of support. But in one respect Japan's experience reflected a broader trend: India and China also displayed an abiding interest in eugenics. In India, the Congress Party's National Planning Committee (NPC) ordered a report with recommendations on population problems even before it assumed power. Under Radhakamal Mukherjee, a subcommittee that included Nehru's own sister, Vijaya Lakshmi Pandit, warned in May 1940 that "disparity in the natural increase of different social strata shows a distinct trend of mispopulation." Upper classes and advanced castes were more likely to practice contraception, leading to the "gradual predominance of the inferior social strata." The subcommittee urged removing barriers to intermarriage among upper castes as well as birth control propaganda among the masses to prevent "deterioration of the racial makeup." But any population policy also had to target the estimated eight million insane and feebleminded Indians, who were "at large and producing abnormals and subnormals"—indeed, reproducing more rapidly than normal parents. While citing precedents from the United States and Europe—including Nazi eugenic tribunals—it noted that "caste has created the outcastes and contributes to make the problems of eradication of the defective types probably easier than in the West." Mukherjee's subcommittee made a contradictory but nonetheless chilling recommendation: "*selectively* sterilising the *entire* group of hereditary defectives."[67]

The NPC, which was chaired by Nehru, discussed the report over two days and finally passed a set of recommendations that emphasized broad-based economic progress as "the basic solution." But it also acknowledged that "measures for the improvement of the quality of the population and limiting excessive population pressure are necessary." It backed fertility limitation, cheaper contraceptives, and, as part of a "eugenic programme," removal of barriers to intermarriage along with sterilization of epileptics and the insane. The war temporarily suspended operations. But in 1947 the NPC published Mukherjee's report in full, together with an introduction by its general secretary, K. T. Shah. He complained that "attention to eugenics or race-culture are matters hardly yet in the public consciousness of this country." A national policy—still less "scientific breeding of the human race"—was not yet in sight.[68]

In 1945 Chinese nationalists under Chiang Kai-shek also adopted a population policy premised on eugenic concerns, in this case promoting internal migration and ethnic intermarriage while prohibiting abortion. But the civil war prevented implementation. By the time the Kuomintang accepted family planning as a method to improve population quality in February 1949, it had already lost the decisive battles and was soon driven off the mainland. The State Department's famous White Paper defending its China policy cited the nationalists' failure to feed a growing population as one of the major reasons for their defeat. This elicited an immediate rejoinder from Mao Zedong, who insisted that "of all things in the world, people are the most precious." Communist revolution and increased production would create "a new China with a big population and a great wealth of products." This polemic made it difficult for any Chinese to advocate a program to reduce fertility.[69]

In India and China, as in Japan, proof of rapid growth would spur governments to take action. By the end of the decade, the UN's work in this area began to pay off. UN training programs in Mexico City, Cairo, Paris, and New Delhi had graduated two hundred technicians from forty countries. At the same time, both the Population Commission and the Statistical Commission encouraged every country to hold a census around 1950. An unprecedented number of censuses were held between 1945 and 1954, covering more than 150 out of 214 countries and territories and about two billion people. And because most adopted UN-approved procedures, the results were increasingly comparable, so that Ceylonese and Swedes, Argentines and Australians, could be combined in a single number representing all of humanity.[70]

Producing a census that could stand up to international scrutiny became a matter of national pride. India's home minister sought to inspire the "six hundred thousand patriots" who would conduct their first census as a free people. It was imperative, he said, that they set about their task in perfect synchronization. "Like a swarm of bees that build a beautiful hive according to the laws of geometry, each doing its part in obedience to a mystic urge, you should do your part according to conscience and the sense of truth inherent in us all." It was the largest census in history, but it was only part of the UN's plan for the world as a whole.[71]

In 1951 the Population Division published its first high, medium, and low projections of growth in world population, then estimated at 2.4 billion. As recently as 1945, informed observers had not expected the 3 billion mark to be reached before the millennium. Now it was the low projection for 1980. The medium estimate, 3.3 billion, was based on the assumption that Africa, the Middle East, and Asia (excluding Japan and the USSR) would not further reduce mortality rates, only managing to avert major calamities. In fact, even the maximum expected growth—to 3.6 billion—would fall far short of the actual increase in the intervening years.[72]

By any estimate, population growth appeared likely to outpace expected increases in food production, and per capita consumption was still below prewar levels. Whereas about 60 percent of the world's population was thought to consume more than 2,200 calories per day before the war, in 1952 it was estimated that 50 percent or fewer had an adequate diet. As the economies of Western Europe began to recover, attention increasingly focused on the disparity between haves and the growing number of have-nots, and how it could possibly continue.[73]

Even before the UN began to issue world population projections, Sauvy questioned the very idea of a "world population"—both the agenda it served and the invidious comparisons it permitted. In 1949 he insisted that the term made no sense, because there was no world government or even the requisite sense of solidarity that might support it. In his view, the Japanese population or the French population actually existed because within their borders people and goods moved without impediment. While he allowed that there was at least the intention of permitting capital and trade to cross international borders more freely, "national sovereignty in the matter of immigration, more than ever, rules supreme."[74]

For Sauvy "a strengthening of world solidarity" was positively dangerous, because it "would lead to a closer examination of the problem of the distribution of land among peoples." This would "favor the birth of a new international legal order; less aggressive, no doubt, than the Hitlerite doctrine of *Lebensraum,* yet one dangerously breaking with the past." It would be better if people were shielded from accounts and images of misery in overpopulated countries. "The resulting reflex is one of anxiety and withdrawal," Sauvy argued, partly because of "the fear, more or less admitted, that some day some redistribution will become a duty."[75]

Sauvy claimed to have received death threats from proponents of population control. He took his fight to the Population Commission in its May 1950 meeting. He accused the staff of a Malthusian bias, receiving strong backing from the Yugoslav representative (the leading communist spokesman until the Soviets ended their boycott on behalf of mainland China). In fact, the most any Population Division report said about limiting growth was that "some governments may consider it advantageous," and even then improvements in education, public health, and the status of women—not access to birth control—were proffered as the likely means to that end. "To show partiality to one political philosophy or another," an indignant director, John Durand, insisted, "I think would be a grave offense, if consciously committed by a servant of the United Nations. It might be called the international equivalent of treason."[76]

Such open conflict was rare. Clever UN officials preferred indirection when taking initiatives that exceeded their mandate—initiatives that reflected their ambitions for the organization as much as any particular political philosophy. Thus, the next year Durand's boss, Henri Laugier, went outside of the UN to revive Huxley's idea of a world conference on population. He quietly suggested that the IUSSP propose joint sponsorship to the Population Commission. The IUSSP quickly acted on this suggestion, no longer representing a hodgepodge of national committees but a transnational community of population experts.[77]

While still divided over whether birth control provided a solution, the Population Commission increasingly took it for granted that population growth created problems that merited further study. In fact, Sauvy's own government was pressing hard for international action on "the problem of population surpluses" in Europe, though solely through assisted migration—especially to France, preferably with international funding. Together with Italy, it held that migratory movements merited no less attention than cross-border trade and capital exchanges, and all were necessary for European reconstruction.[78]

The IUSSP's proposal of a world conference—which had seemed so divisive three years before—easily won UN approval. Catholic and communist representatives no longer tried to censor mention of fertility limitation in official documents. Indeed, in its first field study, the Commission set out to assist India in ascertaining whether people were already planning their families, and "whether fertility differentials exist between different social and economic groups." The results would demonstrate that there was untapped demand for contraception even in poor, rural communities, suggesting that a program to lower fertility could be both feasible and ethical.[79]

In May 1951, Alva Myrdal—now director of social sciences at UNESCO—told Notestein that she was encouraged by the prospects for controlling population growth. States were now in the business of planning social change, and could therefore plan the reduction of fertility. They could change the "psychological background" through "high-pressure propaganda" for emancipation of women, education, and human rights. Medical services could introduce family spacing as a means to protect the health of mothers and children. Finally, there was "the possibility of introducing differential economic measures . . . placing particular advantages on the first children but not all the later ones in a family." Notestein basically agreed: "To me it seems important that we constantly stress the need for speeding the process of transition." He was discouraged by the lack of leadership in China, Japan, Indonesia, and the Philippines, but was "enormously impressed" with the news that India was finally at the point of announcing a "forthright policy."[80]

The previous month Prime Minister Nehru had been given advance notice of the census results. India's population now stood at 362 million, a gain of 43 million people in ten years. Despite all the calamities of the preceding decade—World War II, the Bengal famine, the partition massacres—population growth had scarcely slowed and was now almost certainly accelerating. Nehru was again chair of the renamed National Planning Commission, and once again created a subcommittee on population and family planning. After three days of sharp debate, it recommended fertility limitation both for the sake of mother's and children's health and to stabilize population "consistent with the requirements of national econ-

omy." The report called for free sterilization and contraception when recommended on medical grounds, and suggested that where feasible it should
be given for social and economic reasons as well.[81]

Among the members, Lady Dhanvanthi Rama Rau, president of the
newly formed Family Planning Association of India, would have liked to
have seen free family planning services required as part of every health program. Sushila Nayar, an ally of Health Minister Rajkumari Amrit Kaur, argued instead that the only proper state policy on population was to promote education and improve living standards. Both Nayar and Amrit Kaur
had been disciples of Gandhi, and for the next fifteen years they would
wage a rearguard action against birth control. The Planning Commission,
on the other hand, was a powerful and persistent advocate. In this first
round it backed the subcommittee's recommendations, including state-
funded research centers to develop "birth control suitable for all classes of
people." But Amrit Kaur continued to insist that the rhythm method was
the only one that was acceptable.[82]

Kaur finally asked the World Health Organization to help design a
study to test the rhythm method. WHO's first director-general, Canadian
Brock Chisholm, wanted to do much more. He favored a comprehensive
UN program to control population growth. WHO's Expert Committee on
Maternity Care agreed that advice on fertility limitation should be provided in support of national population policies. Chisholm began to push
the UN Economic and Social Council to make population a priority and
not rest content with sponsoring a conference. He was said to have twenty
pages of recommendations and the strong backing of UNESCO's new director-general, Jaime Torres Bodet.[83]

The Vatican's watchdogs already had Chisholm in their sights and concluded that he was pursuing an "anti-Catholic" agenda. The Information
Center of the International Catholic Organizations had already set up a liaison committee of medical associations to bolster their presence within
WHO. Pope Pius XII would likely have been informed about what was
coming, which may help explain why, in October and November 1951, he
spoke out on the subject. For the first time, in a speech to the Italian Catholic Society of Midwives, he said parents could avoid childbirth if they had
"serious motives," whether economic, social, medical, or eugenic. But he
also affirmed that the primary aim of marriage was procreation. "The individual and society, the people and the State, the Church itself depend for
their existence, in the order established by God, on fruitful marriage." A

month later, he told an association for large families that he hoped medical science could make the safe period "sufficiently certain" to permit "regulation" of births. The ban on contraception, on the other hand, was divinely inspired and could never change. Pius XII also opposed abortion even to save the life of the mother.[84]

Chisholm's push for a broader birth control program would therefore constitute a crucial test. The issue was particularly fraught for an institution that, from its first days, had aspired to be apolitical and inclusive, based on the principle that diseases know no boundaries and public health is indivisible. The very name—*World* Health Organization—was meant to indicate that it would be a global, and not merely an international, endeavor. But communist states had already walked out over the issue of Chinese membership. The organization could ill afford another rift.[85]

The struggle began at the executive board meeting in Geneva in January 1952. Chisholm brought a representative from the UN Population Division, Halvor Gille. Gille offered to support demographic research and data collection as part of a joint project in which WHO would test contraceptives and manage distribution. Chisholm also circulated a supportive memorandum from the WHO regional committee for Southeast Asia, which quoted extensively from the Rockefeller mission report. It suggested that WHO might begin by creating an expert committee to formulate advice for all member states.[86]

Opponents responded with a frontal attack on the very idea of birth control. It was "contrary to the laws of nature," according to the Lebanese representative. For Greece, giving contraceptives to people in developing countries was "tantamount to giving them weapons with which to commit suicide." The British representative pointed out that WHO did not have to rule on moral or religious questions. The board finally accepted with near unanimity a resolution that "welcomed *exclusively technical* collaboration with the United Nations [Population Division] in demographic problems." Saying that he had explored "every possibility I could think of," Chisholm concluded that he could go no further. Even so, he told supporters that the India study was "the most important thing that WHO has done up to this time, and it would be very difficult indeed to stop now."[87]

Difficult, but not impossible. Catholic medical organizations urged affiliates to send members as part of national delegations to the World Health Assembly, WHO's governing body. When they arrived at the Palais des Nations in Geneva in May, they found that supporters of birth control were

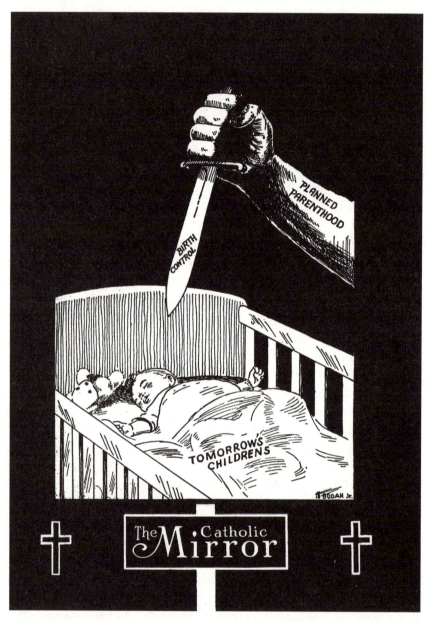

Catholic propaganda, such as this successful 1948 campaign against birth control legislation in Massachusetts, made little distinction among contraception, abortion, and euthanasia. After Pius XII endorsed "Natural Family Planning," the Catholic position became even more difficult to explain. *The Catholic Mirror*, Springfield, MA.

ready for a fight. The Indian delegate, Sir Arcot Mudaliar, insisted that his country had every right to request assistance from WHO in addressing the "great menace" of population growth. Ceylon and Thailand supported his stance. The charismatic Karl Evang, one of the architects of Norway's welfare state, called on the program and budget committee to approve an expert committee on health aspects of "the population problem." The Swedish representative declared that this would be the first "real step" following years of delay, "for reasons which all knew and very seldom talked about officially."[88]

This was "rather dangerous ground," the American delegate objected. "The problem was a burning and controversial one, one which made headlines." Any report by a WHO committee would be construed as an official position and might give ammunition to those who opposed funding the organization's work—the United States was far and away the largest contributor. Besides, how were they to define "the" population problem when the world had so many population problems? Ceylon's representative considered it obvious enough: it was overpopulation in particular countries. But Mudaliar thought infertility should also be considered. Interestingly, Evang was a longtime advocate of eugenic sterilization. When the delegates reassembled two days later, he listed hereditary disease among the "health aspects" he had in mind. But while Evang took pains not to mention birth control specifically—even while confessing to surprise that it should embarrass an audience of doctors—he suggested that the committee could focus on rapid population growth.[89]

Belgium, Italy, and Lebanon responded with a joint resolution insisting that "from the purely medical standpoint, population problems do not require any particular action on the part of WHO at the present time." No delegate denied that overpopulation could be a problem. Countries so afflicted were assured that "they had Ireland's deepest sympathy." Nevertheless, the Irish delegate was also the first to warn that his country would withdraw from WHO if it developed a birth control program. The representative from the Philippines appreciated the sympathy, but hoped that other countries would give assistance. Mexico and Yugoslavia also spoke up for the Norwegian resolution. But the threat of a boycott hung heavy in the air as the meeting broke for the weekend.[90]

At the third and final session, Mudaliar said that in calling on WHO for advice, India had not intended to create a rift, and suggested a compromise resolution that would merely have Chisholm and regional committees consider the matter and report back. Evang offered to amend his own resolution, specifying that the population committee would include all schools of thought. Nevertheless, he refused to withdraw it, because population problems were "of the greatest urgency"—indeed, "might turn out to be a decisive factor underlying a third world war." The Italian representative now hinted that his country might also be prepared to leave WHO. Austria, Costa Rica, Panama, Portugal, and even Japan all spoke in support of the joint resolution denying the need for any WHO action on population. If it was approved, which appeared increasingly likely, it would make impossible any future WHO work in the field—indeed, might even require canceling the India study. But it was thought that if the Norwegian resolution passed, as much as a third of the assembly might get up and walk out.[91]

Italy and Belgium called for a vote, sensing their advantage. They repeated their demand even when Ceylon proposed that all of the resolutions be withdrawn. In the end, at the urging of the U.S. representative, they finally agreed to this face-saving compromise. They did not wish to "jeopardize the existence of the organization," but insisted that the record show that it had acceded to their objections. The next day the *New York Herald Tribune* quoted an anonymous WHO official suggesting that they were not bound by a discussion that did not lead to a formal decision. Pressed by Paul Van de Calseyde, the Belgian delegate, Chisholm then had to make a statement that "the job of the Secretariat is clearly to carry out the wishes of the national delegations . . . whether in formal resolution or in expressed wish . . . the job of the Secretariat is to carry out wishes." It was everything Van de Calseyde could have wished for. He then read into the record the following statement, insisting the stenographer take down every word:

> It must be made clear beyond any possibility of doubt that the outcome of the debate which has taken place is that we, in this committee, are agreed that although the problem of over-population in certain regions may call for WHO's advice and assistance for the protection of the health of these peoples—advice and assistance already being given—there can be no question of our organization, which is universal and neutral, becoming the advocate, still less the means of implementing, certain economic and social theories which are far from receiving universal approval.[92]

The outcome was disastrous for those who were looking to WHO as the natural home for a UN program in family planning. After Chisholm left in

1953, governments seeking assistance were told that it was not within the organization's mandate. The WHO debate had a chilling effect on other UN agencies as well. When preparations for the World Population Conference started soon thereafter, such topics as the effects of "cross-breeding," differential fertility, and the reproductive behavior of "special problem groups" met no objection. But it was decided that future population growth would be subject to a "pure-statistical demographic analysis," as Myrdal put it, ". . . leaving aside the social and economic consequences in order not to invite the prophets and propagandists." Ironically, the opponents of WHO involvement had insisted that overpopulation was "dominated by preponderant economic and social factors." The medical aspects, and thus a specialized health agency like WHO, could come in only if the UN decided that they offered some solution to these social and economic problems. Now the UN would not even provide a forum for outside experts to discuss them.[93]

Myrdal would insist to Population Division officials—"old colleagues and friends" like Whelpton and Durand—that they could hardly avoid the subject of population policies in what, after all, was an academic gathering that would pass no binding resolutions. She argued "with more force than I have ever had to," and only prevailed after threatening that UNESCO would withdraw its support. Soon the Population Commission and UNESCO were hardly communicating, and the Division was struggling to survive a 60 percent reduction in staff.[94]

The opponents of population control seemed to have had the last word. As might have been expected, the UN population conference in Rome in 1954 was "massive, but non-conclusive," as Myrdal put it. Appropriately enough, Pope Pius XII greeted the delegates at the Vatican. Their deliberations attracted so little public interest that American press coverage of population issues that year was less than half as extensive as in 1948, the year Huxley had first proposed the conference.[95]

Advocates had placed their hopes in the United Nations, a body that could claim to speak and act for all humanity. For a time its highest offices seemed to be filled with allies and sympathizers. But now it was clear that neither the UN secretariat nor any of the specialized agencies were willing to lead a campaign for population control, discuss whether it might be necessary, or even coordinate the few, limited initiatives they allowed themselves to undertake. Far from providing a means of advancing toward world government, the struggle over population control proved so contentious that it impaired the functioning of international agencies. Indeed, the very attempt made them appear irrelevant.

The similar fate of nuclear arms control in the same period is usually construed as a Cold War story. But in this case the Soviets and their allies were not even represented on the governing bodies when the critical decisions were made constraining the FAO, UNESCO, and WHO. Their primary concern in the Population Commission was the issue of Chinese representation. They did not begin to take a more active role until 1955, when they aligned with the Catholic countries—ironically, the same year abortion became legal again in the USSR. On the other side there was a disproportionate number of Britons and Americans, including scientists and civil servants, activists and philanthropists. But their governments stuck to a neutral stance both in international forums and the territories they administered, even when—as in India and Japan—there was local support for population control. The U.S. government role in international debates about population was often merely to stop ugly spats among allies like Belgium, Italy, and Lebanon, on the one hand, and Ceylon, Norway, and Mexico, on the other. Cold War neutrals like Sweden, India, and Ireland were among the most vigorous and persistent combatants in the continuing struggle over international family planning assistance.

The international politics of population control did not fit into Cold War categories. But in truth, any geopolitical analysis would be equally unhelpful. As amusing as it might be to imagine a world divided between a Scandinavian–South Asian alliance on the one hand and a Gallic–Latin American–communist axis on the other, the very idea only reveals the need for a deeper analysis. One cannot view populations and those who would seek to control them in just two dimensions, color-coded on a political map of the world. For John Boyd Orr, and many like him, in every society "borderline populations" exemplified divisions more profound than national boundaries. Scientists and activists all over the world were trying to renegotiate these borderlines, seeing parallels in the global growth of urban conglomerations, positing causal relationships between apparent overpopulation in poor countries and pressures for immigration, and seeking simpler contraceptives for the "unfit" both at home and abroad. Transnational connections were easiest to discern when territorial boundaries were so fluid that they created "communicating vessels"—as Fernand Boverat put it—whether between Algiers and Marseille, San Juan and New York, or

Kingston and London. Soon such connections would proliferate, as growing rural unemployment in the Global South and booming industrial expansion in the affluent North brought massive new migration flows.

No matter where they resided, advocates of population control shared both a sense of belonging to a "world population"—albeit the small, conscious part of it—and a determination to remake their own societies, whether by eliminating "social problem groups," or merely preventing unwanted children. They were fired by a quasi-religious fervor, but they confronted an organization that was already global in scope, had immeasurable influence with dozens of governments, and answered only to God. The Church needed only to defend the status quo, since pro-natalism was embedded in most political systems. Promoting birth control—or even legalizing it—was a more difficult proposition.

Viewed from this perspective, one can begin to see the full implications of a famous article Alfred Sauvy wrote for a French weekly in August 1952, shortly after population control proponents were defeated at the World Health Assembly. Six years earlier Orr had argued that population growth posed the choice of "one world or none." Sauvy now proposed a new way of envisaging the globe. Up to that point, statesmen had been paying more attention to population pressure in Europe than anywhere else. Overpopulation was something that could happen anywhere and—with the idea of "world population"—everywhere. But Sauvy argued that, rather than one world, or even two, there were really three: the communist bloc, the capitalist West, and what he called the "Third World."

For Sauvy, ideological rivals East and West actually depended on each other because the conflict defined their identity. But as they continued along their two paths to modernity, the distinction between them would inevitably disappear. The differences with the South were far more profound—to Sauvy, they inhabited an alternative universe. It was not the case, as the American demographers optimistically assumed, that this Third World merely had to advance along some imagined continuum from tradition to modernity. According to Sauvy, "these countries have our mortality of 1914 and our natality of the 18th century." Saving lives was cheap, but giving people something to live for was expensive. They were therefore caught in a "cycle of misery." This Third World—like the Third Estate of revolutionary France—was desperately poor and increasingly overcrowded. In Egypt, in Tunisia, rebellion was in the air. "Do you not hear, on the

Cote-d'Azur"—many of his readers were vacationing there—"the cries of misery coming to us from the other side of the Mediterranean . . . ?" Sauvy asked. "The pressure is growing constantly in the human furnace."[96]

Equally significant was the metaphor that Sauvy chose in describing how this Third World was emerging. He likened it to a "slow and irresistible push, humble and ferocious, towards life." The new babe could not speak for itself. For Sauvy, it was born not of political protest against imperialism, but through a biological process not of its own making. In effect, it needed nothing so much as care and feeding until it was mature enough to choose between the two paths to modernization. "Because," Sauvy concluded, "in the end, this third world, ignored, exploited, and misunderstood like the third estate—it also wants to be something."[97]

Sauvy's idea of three worlds—seemingly progressive in its criticism of the Cold War and advocacy of development aid—was, in fact, deeply conservative. The whole point was to banish forever the thought that there might be only one world, in which all humanity shared mutual obligations. As he knew only too well, proponents of population control were pushing an idea more radical than national self-determination and assisted development. They insisted that everyone had the same right and the same duty to plan their families. Starting with India, soon to be followed by Pakistan, South Korea, and a host of others, they persuaded leaders of newly independent countries to not just choose between capitalism or communism, in which reproductive behavior was a byproduct of modernization or a matter of indifference. They presented population control as a means to jump-start the process. By rationalizing and redirecting reproduction, they could make their people modern in a single generation. But for the leaders of a new, transnational population control establishment, controlling the birth of this new Third World was just part of a larger plan to remake humanity.

5

THE POPULATION ESTABLISHMENT

For a movement that claimed to speak for all humanity, even future generations, losing the battles for UN recognition was like losing the commanding heights in the middle of a long war. Proponents of population control craved the visibility and legitimacy it would have conferred. The United Nations and its specialized agencies might also have provided places to marshal and direct their efforts. In the aftermath of the 1952 World Health Assembly, their movement became even more diffuse and diverse, and it was ever more difficult to dispel the suspicion that adherents from rich and poor countries really did live in different worlds.

Yet considering all of the ways in which population control proponents were divided, they had no choice but to advance on multiple fronts, and even underground. While they sometimes worked at cross-purposes, they were better able to survive attacks as well as public apathy. Over time, philanthropists and the researchers they supported, a federation of voluntary associations, and a growing number of government programs would form a dense network of relationships, one that remained flexible in its operations but became increasingly resistant to challenge. This could not have been planned in a smoke-filled room. But that is not to say that some people did not try. A close look at a secretive gathering in June 1952, just after the debacle at the World Health Assembly, reveals what was driving some of the most important leaders as well as the direction in which they intended to take this movement.

The idea for an invitation-only "Conference on Population Problems" was first conceived in a chance meeting between John D. Rockefeller 3rd

and Lewis Strauss in the men's room on the 56th floor of Rockefeller Center. Strauss, shortly to become chairman of the U.S. Atomic Energy Commission, pointed out that most scientists had given little thought to population problems and might benefit from a meeting with experts. Rockefeller
himself had been keenly interested in birth control for more than twenty
years and was resolved to make it a major focus of his career as a philanthropist. He enlisted the support of the head of the National Academy of
Sciences (NAS), Detlev Bronk, soon to become president of Rockefeller
University, and paid all the expenses. For the venue he chose Colonial
Williamsburg, a pet project of his elderly father. The third Rockefeller was
already worth over one hundred million dollars—more than three-quarters
of a billion dollars in today's money—and was giving it away at the rate of a
million a year. Not surprisingly, the NAS invitation attracted some of the
most eminent scientists and administrators of the era, representing fields
ranging from botany to physics, embryology to economics, including several past and future recipients of the Nobel Prize, a former surgeon general,
and the president of MIT. The only woman, Irene Taeuber, was one of the
four demographers Notestein chose to guide and inform the discussion.
They also included Davis, Whelpton, and Thompson. And among population activists there was Vogt, director of the Planned Parenthood Federation of America (PPFA), along with Fairfield Osborn and Frederick
Osborn—the last a friend of Rockefeller's.[1]

As the heir to a famous name who had yet to make a name for himself,
Rockefeller sometimes felt insecure at such gatherings. Several officials
from the foundation were also present, including Warren Weaver, the head
of the natural sciences division. He had helped defeat Rockefeller's effort to
follow up the foundation's mission to East Asia, concerned that it would divert resources from his growing program in agriculture. The setting must
have added to Rockefeller's discomfort, even apart from the heat of Virginia
in June. By the time the conference convened, he was on a leave of absence
as chairman of Colonial Williamsburg, Inc., because his father no longer
trusted him with its management.[2]

Yet rather than sinking into depression, as he had in the past, the son
was now determined to strike out on his own. It was actually Rockefeller
who managed to catalyze the discussion on that first morning, showing
himself to be more informed and thoughtful than some of the heavy hitters
in the room. MIT's Karl Compton, for instance, had started proceedings
by asserting that indigenous chiefs in New Guinea killed someone with

every new addition to the tribe—quickly adding that he did not recommend it as a policy. Rockefeller then asked a question they could actually discuss intelligently: "whether or not there is a limit to the number of people that wisely can be permitted in the world, even assuming there is enough food to supply them." Beyond food production and population levels, Rockefeller pointed out that "there is a third element that is equally important, the humanity side." Fairfield Osborn suggested using the term *optimum population* as opposed to "the mere physical supporting of the population as, let us say, living animals." But this immediately started a debate about whether such an idea was culture-bound, with some insisting that the United States also had a population problem.[3]

The verbatim transcript suggests that the most sensitive and contentious debates—with participants accusing each other of "being provocative" before going off the record—concerned whether "industrial development should be withheld" from poor, agrarian countries like India. Vogt and Fairfield Osborn appear to have been in the minority in making this argument, to the point that Osborn admitted to "wondering whether we should have come to Williamsburg." Warren Weaver considered the environmentalists too pessimistic, but he openly worried that development aid would only make Indians "nigger rich." Weaver translated: "a man who finds out that he has a little income.—And what does he do? Well, at that moment he just stops working four days or a week, and he just sits there. I do not think that is what we want to bring to India." "I hesitate to use this language," Weaver had said, "but I guess it's all right at the moment." The Williamsburg Inn admitted only white patrons.[4]

Indians were represented at this meeting, but they did not represent themselves. Instead, participants projected their prejudices onto the subcontinent as they speculated about its future. The only one who had actually published research on India, Kingsley Davis, had visited the country for the first time six months earlier. Consequently, Weaver was not the only one who fell back on his experience of divisions in American society to understand relations between rich and poor countries, particularly regarding "the potential degradation of the genetic quality of the human race"—as Detlev Bronk put it. When the conference reconvened the next morning, Bronk worried about the interaction of diverging fertility and improving public health, "making it possible for individuals to survive, who would not under natural conditions be able to survive." Summarizing the first day's discussion, he said that "there was the recognition of the fact that a very great obstacle to the achievement of much that was defined as being desirable is the level of intelligence in those areas of the world where these controls and these developments are most needed." Frederick Osborn painted a truly apocalyptic picture: "This little group of three or four hundred people, who produce most of the freedom of the human mind . . . may be engulfed by a great mass of people to whom these conceptions are largely alien."

Some of those in the room, such as Irene Taeuber, considered that throughout the Middle East and Asia "the political survival of westernized groups is at stake." In January 1954 she observed that they understood the population problem "not as a theory but as a nightmare." But at Williamsburg some may have felt the danger was that elite *societies*—distinguished by conditions of low fertility and low mortality permitting "freedom of the human mind"—would be engulfed by those peoples with a lower "level of intelligence" and no elite to speak of. Even discussions about the quality of America's population kept coming back to India. On the afternoon of the second day, one of the participants, the economist Isador Lubin, tried to explain to himself and the others why that should be:

> At luncheon today I raised the question as to why it was that almost everybody who spoke this morning talked about India. What is there about India that makes this situation so acute? And I think unconsciously we are scared, and I think we have a right to be. In other words, that is where the ferment is taking place. That is where the pressure is the greatest.

Communists were filtering in, he said, promising solutions that did not depend on technological advance. "If that part of the world accepts another political philosophy of life then the pressure on us will be such that we will have less time and less men and less interest—I am talking about the Western civilization—to do these things that we are talking about." Similarly, Davis warned that "the advanced countries, the places where the scientific developments are being made, are beginning to be leveled down by the tremendous demands of the rest of the world for sheer subsistence, at low levels of living." Thus, "Western Civilization" along with its technocratic elites would be dragged down through the diversion of energies to emergency aid, or even to self-defense, before most of the world's population could be raised to the point where they could stand on their own.[5]

For forty years, population researchers had been asking "who shall inherit the earth." But it was only now that controlling the growth of poor countries took clear priority over "race suicide" and differential fertility within Western societies. One reason was that birth rates were rising again in Europe and North America, including among better-educated and affluent people. Some of those at Williamsburg thought that the United States might already be overpopulated, but no one disagreed with the idea that other societies had far greater problems, problems of global significance. Moreover, no one dissented from Notestein's proposal that they should help these societies understand their population problems and act on them.

Reflecting the sharp lesson recently delivered in Japan, the demographers and the activists agreed that Asians had to want population control for themselves. Imposing it would only provoke them (not to mention—they did not—Catholic opposition.) Even so, Notestein thought that "there is considerable opportunity to influence opinion and policy, perhaps directly, to channel such influence through international agencies." He therefore urged training "local scholars" and setting up research centers, while admitting that "some of the research, of course, would be pretty bad."

On the last day, the meeting accorded a "fuzzy mandate" to a committee consisting of Rockefeller, Notestein, and Frederick Osborn—"a license for JDR to do pretty much as he liked," as his biographers put it. What he liked, and what he paid for, came to be known as the Population Council. Rockefeller was the first president, though Frederick Osborn handled day-to-day management as executive vice president. It was also Osborn who hosted a series of lunches that brought together representatives from the International Planned Parenthood Federation (IPPF), the United Nations, the Ford and Rockefeller foundations, as well as major pharmaceutical firms. In this way the Population Council became not only the world's preeminent institute for policy-oriented research in demography and contraception, but also a nexus for all the other major players in the field.[6]

The discussions at Williamsburg foreshadowed their agenda over the next several years. The Council's very first major program was to provide fellowships, most of which went to Indians and Americans, with about half training under Notestein at Princeton's Office of Population Research. And the first time the Population Council received money from anyone but Rockefeller—a Ford Foundation grant in February 1954—it was used to create UN research centers in India and Chile. As Notestein had anticipated, these centers served a political as much as a scientific function. Directors were to "combine the qualities of scientist, pioneer, diplomat, and salesman," as a Ford-sponsored meeting agreed. They were not expected to contribute to the understanding of population problems outside their particular regions, much less in Europe or North America.[7]

The first draft of the Population Council's mission statement specified that it would also seek to create conditions such that "parents who are above the average in intelligence, quality of personality and affection will tend to have larger than average families." But one of the trustees, the former surgeon general Thomas Parran, objected on both political and intellectual grounds. The final version simply backed research "in both the quantitative and qualitative aspects of population in the United States." In its first three years, the Council provided grants for studies of twins and of differential fertility among social classes, and also supported experimentation with sterilizing drugs on "women with bad hereditary history." Direct contributions to the American Eugenics Society continued for more than two decades. As secretary of the AES, Osborn tried to promote higher fertility among Americans with more intelligence and character while "diminishing the burden of hereditary disabilities." In 1959, shortly before he succeeded Osborn to the presidency of the Council, Notestein wrote, "All of us were convinced that, so far as the western world was concerned, the important issues were likely to be qualitative rather than quantitative."[8]

Why were qualitative issues in poor countries considered relatively unimportant? A number of demographers, including Taeuber, were interested in how women's access to education and paid work was correlated with lower fertility and might therefore bring down birth rates in poor countries. If Taeuber had not been the only woman at the Williamsburg Inn, women's education might have been on the agenda. She tended to hold back in such meetings for fear of crossing male demographers, including Conrad Taeuber. "The basic problem," she explained to Osborn, "has always been to keep out of the way of my husband and Frank Notestein."[9]

The separation of men and women, the qualitative and quantitative, and the First and Third worlds, meant that this crucial insight about the relationship between women's education and fertility was overlooked when it came to designing policies and programs. Another reason is that, in contrast to the United States, there was no lessening of concern about differential fertility in India. India's first official research program included studies comparing birth rates among caste, class, and religious groups as well as the development of intelligence tests appropriate for each one. One of the "main goals," a Council representative learned, was to determine if a birth control program "will reduce family size in much greater degree among the more desirable than among the less desirable groups of the population." They might not, therefore, have welcomed a suggestion to promote education as a means to reduce fertility.[10]

If American population control proponents ever considered "qualitative" issues in countries like India, they appear to have concluded that reducing quantitative growth by promoting contraception was the best way to address them. The danger of differential fertility was unproven, whereas high birth rates seemed to drive infant and maternal mortality. Reducing fertility was thought to be integral to modernization, enhancing the health and productivity of both poor people and poor countries. Moreover, with the onset of the Cold War, the communist victory in China, and the United States and its allies on the defensive in Korea, defusing the population bomb was becoming the overriding priority.

Chinese "human wave" tactics revived visions of a racial apocalypse. The fear of being "engulfed" voiced at Williamsburg, however inchoate, was not uncommon. It found echoes in both the popular press and academic studies. "Few generations in the long history of Christendom have faced a darker prospect than our own," wrote the lay Catholic editors of *Commonweal,* recalling the days of Augustine when "the Roman empire fell all about him before the onslaught of the barbarians." In 1953 a massive study of population and resources by W. S. and E. S. Woytinsky pointed to the growth of "the redoubtable colossus in the East," observing that "the decline in northern and western Europe seems to foreshadow the engulfment of Western civilization by the peoples of Russia and Asia." The Soviets seemed ideally positioned to lead "Eastern hordes" in a march on the West. As Pearl Buck and Aldous Huxley had predicted, American chemical warfare specialists argued that only superior technology could defeat such "great hordes of military manpower." When he assumed the presidency in 1953, Eisenhower was determined to end the attrition in Korea, even if it required brandishing nuclear weapons at the outset of any new war.[11]

The population of poor countries was now growing much faster than either Notestein or Davis had anticipated. Because of gains in longevity, the rate of increase was accelerating. Rather than waiting for "modernizing" societies to begin using birth control, both now called for experiments to reduce fertility. If they were still sanguine about the rise of Asia and the relative decline of the West, they never said so. Indeed, Notestein considered economic policies that met minimal needs as "worse than useless," since they were "expanding the base populations," and the situation was already causing "political explosions." Davis warned that expanding food aid would have the effect of "building up ever larger populations on the basis of charity. As time went on, it would become ever more difficult to remove the prop." Leaders of impoverished, overpopulated countries would resort to blackmail, especially if some industrial power supplied these "youthful hordes" with weapons of war. Nevertheless, both men professed optimism that Asians could be encouraged to see their self-interest in smaller families, just as Westerners did. After all, they were "not stupid." But the demographers did not, on the other hand, assign any particular importance to expanding access to education, which had been a crucial factor in most Western fertility declines.[12]

For others, population control was simply a means to cut poor countries down to size. The most outspoken advocate of this approach was Hugh Moore, who had made a fortune in founding the Dixie Cup Corporation. Reading Vogt's *Road to Survival,* he experienced "a religious revelation." He resolved to devote the rest of his life, not to paper recycling, but to publicizing the menace posed by growing numbers of poor people. He began by recruiting establishment figures, starting with Will Clayton, former undersecretary of state, and Ellsworth Bunker, soon to become ambassador to India. In 1954 he began mass mailings of a pamphlet, *The Population Bomb,* with a circulation that started with ten thousand notables listed in *Who's Who* and eventually totaled over 1.5 million. "We are not primarily interested in the sociological or humanitarian aspects of birth control," they told Rockefeller. "We *are* interested in the use which Communists make of hungry people in their drive to conquer the earth."[13]

This version of population control updated the "world eugenics" Prescott Hall first proposed in 1919. Dudley Kirk, the man who headed the Population Council's demographic division, and who recruited and selected its first fellows, had been concerned about how population trends were driving "the eastward movement of power" since 1944. In a 1989 interview he was still worried:

> I hate to go on the record saying this, but I think there's a real problem in the Western Civilization in that we are approaching a stationary population and the rest of the world, the less developed world, is rapidly becoming an increasing proportion of the total population. . . . The supremacy of Western civilization, it was a rapidly ex-

panding civilization in numbers, in population, as well as in technology. . . . this is a difficult thing to express really—numbers are really going to count.

While he was still with the Population Council, Kirk emphasized that they "should advocate birth control as a humanitarian gesture and not because there are too many Asians, too many Arabs."[14]

It is difficult to know who else among the leaders of the population control movement were making these kinds of calculations. But their attitudes toward immigration are revealing. Kingsley Davis wrote that the United States was helping Mexico deal with its overpopulation "by acquiring each year tens of thousands of impoverished, illiterate, superstitious, non-English-speaking, and in many cases diseased new citizens." He warned of "a gradual Mexicanization of the southwestern border states." Similarly, C. P. Blacker privately acknowledged a concern that the increasing population in tropical and subtropical areas posed a direct threat to Great Britain: "Emigrants from this country [are] being quickly replaced by coloured people accustomed to lower standards than ours," he wrote. "Instead of a quantitative reduction of our population there will take place a qualitative change which (despite dogmatic pronouncements to the contrary by Unesco) will almost certainly be for the worse."[15]

But Blacker believed that they had to pursue a strategy he called "crypto-eugenics." In essence, "You seek to fulfill the aims of eugenics without disclosing what you are really aiming at and without mentioning the word." This is how the Eugenics Society conceived of its funding for the IPPF. Blacker credited Frederick Osborn with the idea. In public Osborn insisted that the genetic potential for intelligence did not vary much between rich and poor or among different races. Even if the average was higher in certain groups, individual differences mattered more: "We need the greatest possible number of births among genetically superior individuals, whether they are among the able individuals who are the majority in one group or among the able individuals who are the minority in another group."[16]

The pope's position was not so different. Pius XII had accepted eugenics as a valid reason for some parents to forgo childbearing, and Catholic marriage manuals now counseled parents to plan larger or smaller families depending on their means. Osborn tried to forge a consensus. He agreed with Catholic scholars that their disagreement was over means, not ends. For Osborn, family planning had to be made available to everyone:

Family planning propaganda, such as the ubiquitous planned/unplanned family posters, continued to reflect the influence of eugenics in the 1950s and 1960s. This example from the IPPF was designed for distribution across all of Southeast Asia and
Oceania. IPPF Archives.

If they are unable to feed the children they have, if they are afraid of responsibility, if their affectionate responses are weak, people don't want many children. If they have effective means of family planning, they won't have many. Our studies have shown this to be true all over the world.

Conversely, Osborn considered it crucial to use "psychological conditioning" to encourage parents with higher intelligence and income to want children—indeed, to want large families, even families of fourteen.[17]

Even "mainline" eugenicists like Paul Popenoe now saw marriage counseling as a better way to advance eugenics than compulsory sterilization. Long friendly with like-minded Catholics like John Montgomery Cooper, Popenoe became a TV celebrity and wrote the "Can This Marriage Be Saved?" column for the *Ladies' Home Journal.* The answer, invariably, was yes, provided wives and husbands accepted the biological basis for their assigned roles. Authors of college textbooks, such as Garrett Hardin, taught students they should have more children than those with lower IQs. Eugenicists cheered as middle-class birth rates continued to rise in the 1940s and 1950s, especially among Catholics. Even so, when not addressing the public, Osborn still worried that in the United States "the socially handicapped are contributing more than their share of children," and asked whether they should consider more direct action to "reduce or reverse present socio-economic differentials in fertility." In poor countries, on the other hand, the Population Council sought to reduce birth rates indiscriminately.[18]

On one point Osborn was emphatically clear: eugenicists would get nowhere continuing to "humiliate one half of the individuals who comprise the human race by telling them that they are not as fit as the other half to procreate the next generation." This approach "all but killed the eugenic movement," and he was not going to see the same error repeated with family planning. Rockefeller and Notestein, for their part, insisted that population control could not be overtly linked with a Cold War agenda, much less the supremacy of Western Civilization. Even Moore and company, after hearing Rockefeller's critical reaction to the *Population Bomb,* insisted "We *are* interested in the humanitarian aspects of birth control" when they wrote to the World Bank. After just a year as director of the PPFA, Vogt also started to choose his words more carefully. "It is commonly said in the Orient that we want to cut their population because we are afraid of them . . . ," he pointed out at Williamsburg. "But the program can be sold on the basis of the mother's health and the health of the other children, . . . There will be no trouble getting into foreign countries on that basis." Nevertheless, Vogt doubted population control could be a populist movement and endorsed a focus on foreign elites. For the next decade or more, he judged that "we are going to get much further, working from people to people, than trying to work through governments, because it is a hot potato to most governments, that they don't want to pick up."[19]

By and large, Vogt was right. For the next ten years family planning advanced most rapidly wherever people of money and leisure took an interest and formed voluntary associations. These activists also played key roles in the few cases in which governments backed population limitation. They forged international alliances on the basis of maternal and child welfare as well as economic development. They generally did not discuss the eugenic implications of their work, whether locally or globally. Instead, "population quality" provided a convenient catchall, because it could encompass concerns about both poor environment and differential fertility. And proponents could also agree on providing birth control to those "who needed it most," which usually meant the very poorest people, especially when they were so benighted as not to realize that they should want fewer children.

For activists, the most important country to "get into" was India. It had to be "continually prodded into a national consciousness daily, hourly, for at least five years," as Sanger put it. But the International Committee on Planned Parenthood established at the Cheltenham conference turned out to be ill-suited for the task. Sanger complained that its mandate—"furthering human welfare through planned parenthood and progressive sex education"—did not inspire potential donors. Sex reformers, particularly the Dutch, considered the Americans to be "obsessed," as the IPPF official history puts it, with "attacking population problems, and especially those of coloured people." Sanger had tried and failed to keep the head of the Dutch association, Conrad van Emde Boas, away from the Cheltenham conference. In 1951 the plucky doctor was pressing to play host to the next international conference.[20]

At this critical juncture, Sanger decided that Bombay would be the ideal site for such a meeting, even though the newly established Family Planning Association of India (FPAI) was not even a member of the International Committee. Rama Rau accepted, albeit with some trepidation. From that moment, she and the rest of the FPAI leaders would depend on Sanger's contacts and money to make the conference a success. Sanger, for her part, could count on her hosts to pack the meeting, aside from those few foreign participants to whom she had granted travel funds. Van Emde Boas and the rest of his organization boycotted the conference. Most important of all, those who disagreed with Sanger's priorities but came anyway would hear from a procession of Indian leaders pleading the case for population control.[21]

On the first day of the conference, November 24, 1952, almost five hundred delegates from fourteen countries gathered in Sir Cowasji Jehangir Hall, the public galleries packed to overflowing. Messages of goodwill were read out, beginning with a note from Nehru. The prime minister wrote that India should limit population growth, but that this would not solve economic and social problems by itself—indeed, it might divert attention from them. This concern was not shared by everyone in his government or his party. The message from the minister for planning declared without equivocation that population growth was "our most crucial problem." According to Kanialal Mareklal Munshi, governor of Uttar Pradesh—the most populous state—it was the cause of most of India's difficulties. Similarly, India's chief justice insisted that its importance "cannot be exaggerated." The commander in chief of the army, General Kodendera Madappa Cariappa, was even more emphatic, calling for "an all out drive." In his written message of support he argued that it had to be treated as a military operation. Cariappa had already ordered the army to provide soldiers with contraceptives and urged them not to father a nation of unplanned and sickly children. Governor Munshi agreed, deeming it "essential that decrepit, diseased, infirm and incurable adults should be prevented, by enforced surgical treatment, from adding an unhealthy and infirm element in our national composition."[22]

All of this belied the impression that population control was just a way for wealthy, insecure Americans to keep down poor, dark-skinned people. But Elise Ottesen-Jensen was never particularly impressed by famous names listed in conference programs—not even if they now included Eleanor Roosevelt and Albert Einstein. For her, the most persuasive message was the one delivered in person by Dr. Sarvepalli Radhakrishnan, a noted philosopher, who also happened to be the vice president and future president of India. For someone who had spent forty years insisting, before audiences of Lutheran farmers, on the importance of frank discussion of sex—and was sometimes spat on for her trouble—it was deeply affecting to hear the same message applauded in Bombay. Radhakrishnan patiently and thoughtfully demonstrated how the cause of planned parenthood was a crucial theater in the struggle for human rights, a safeguard for women's and children's health, a cornerstone of the welfare state, consistent with Gandhi's teaching on self-control, and a fulfillment of God's wish that people use their intelligence to alleviate suffering. New contraceptives could be misused, of course, but "all knowledge is a double-edged sword, whether it is atomic energy or whether it is contraceptives." Most important, Radhakrishnan said that India could not wait. "Our need is desperate, the claims of humanity appeal to us, and it is essential that we should do something for regulating population."[23]

Now, with Ottesen-Jensen's acquiescence, Sanger could proceed with plans for an international federation drawn up according to her own specifications. C. P. Blacker was just the man to do it. When he stepped down as secretary of the Eugenics Society, Sanger invited him to serve as director of the new organization. He agreed, provided he did not have to deal with sex educators, and the two collaborated in marginalizing the Dutch. Ottesen-Jensen worried that Blacker was a racist but finally went along. While the conference continued in the main hall, the three of them met privately in Sanger's suite at the Taj Mahal Hotel together with Rama Rau, Vogt, and Helena Wright. On the last day, November 29, Blacker presented a series of resolutions that created the International Planned Parenthood Federation, constituted by the four original associations now joined by India, Hong Kong, Singapore, and West Germany. Together they would form regional offices and committees empowered to accept new members. The Federation would have two honorary—that is, voluntary—presidents: Sanger and Rama Rau.[24]

Even after the incorporation of the Population Council in November 1952, even with the founding of the IPPF three weeks later, the population control movement could not really get moving without the backing of at least one government. Field-workers and funds for field experiments were useless without an actual field of operations, preferably one made free and accessible through the backing of responsible officials. On December 7, 1952, this final element fell into place. On that day Nehru presented to India's parliament the world's first explicit policy of population limitation, part of India's first five-year plan of economic and social development.

The plan did not specify targets, but only called for reducing birth rates to "a level consistent with the requirements of national economy." At the same time, it acknowledged that family planning's "main appeal" was the improvement of individual welfare, and therefore recommended that it be part of the public health program. Amrit Kaur, the health minister, insisted on collecting "reliable data" before they "launched out on an uncharted sea." The government allocated just 6.5 million rupees for the program's first three years, or $480,000 a year—an annual budget of $3,645,000 in today's money. Nevertheless, the long-anticipated news "profoundly influenced" John D. Rockefeller 3rd, helping to convince him to fund the Population Council with $100,000, and to pledge another $1.3 million within a year. Up to this point, the five U.S. centers of demographic research had subsisted on a total of only $160,000 a year.[25]

Officials from the Ford Foundation, now even wealthier than the Rockefeller Foundation, were also closely monitoring developments in New Delhi. Waldemar Nielsen, just hired from the Marshall Plan's information office, approached demographers and activists to ask how Ford could help. They must have felt that their ship had finally come in. Kingsley Davis, for instance, advised that Ford provide funds on a "liberal grants-in-aid basis to currently productive scholars, simply to advance the useful work they are already conducting." But they also suggested training more demographers from countries like India so the research produced would have "the flavor of a domestic investigation of a domestic problem."[26]

In a June 1953 meeting of the Ford Foundation board of trustees, Nielsen together with Frank Abrams of the Standard Oil Company of New Jersey presented a plan to promote public awareness along with demographic research and training. Nielsen emphasized the danger that population growth would contribute to international tensions by causing unrest, migration, and aggression. It was "the basic problem of the world," according to John Cowles, publisher of Minneapolis's *Star* and *Tribune*. Abrams and Cowles were married to "Planned Parenthood nuts"—as a Ford official put it—who insisted their husbands do something to help. But Robert Hutchins, president of the University of Chicago, said that demographers were "charlatans" with a record of faulty predictions. The board decided to go forward, with no great enthusiasm. It was not until February 1954 that they authorized the first major grant, $600,000 for the Population Council ($141,000 of which established the UN demographic training center in Bombay). By that point the Rockefeller Foundation had also gotten involved, setting aside almost a quarter of a million dollars between 1953 and 1956 for a single population control project in Punjab, the Khanna study.

And in 1956 the Rockefeller Brothers Fund gave the Council $540,000 to inaugurate a program in biomedical research.[27]

When the demographers advised the Ford Foundation to work in India, they argued that foundations could play a "pump–priming role," encouraging governments to take a greater interest. Ironically, in this case it seems that India had primed the pumps. By 1956 it had spent only a fraction of the small sum allocated: 1.5 million rupees, or about $110,000 a year. In India's federal system, officials at the Ministry of Health decided whether and how states would receive funds for family planning. They made the approval process cumbersome and would not even pay for contraceptive supplies. Much of the money therefore remained unspent. International financial assistance for population control exceeded monies expended by India's own government, though much of it was spent in Princeton and New York.[28]

The demographers hyped India's program in order to prime foundation pumps. But however slow-moving and underfunded, it did prove historic, because it helped the population control movement get off the ground. "Following the money" will take us only so far in explaining how that happened, because this was one of many instances in which a relatively small sum had a major impact. Moreover, money was only one of the more tangible signs of the growing influence of an informal network that outsiders would one day come to know, and fear, as "the population establishment." It already included academic empire-builders, foundation grant-slingers, and "Planned Parenthood nuts" around the world—this last group unpaid.[29]

Funding fellowships and training centers drew more people into this network and created new places to make connections. In 1953 it was estimated that there were fewer than one hundred trained demographers outside the United States, the great majority in Europe. There was not one, for instance, in all of Indonesia. By 1957 the Council had spent about $100,000 on 32 fellows, 25 from abroad, in a program that continued to expand and diversify. In 1960 alone there were 29 new fellows from fifteen countries. By that point the UN centers in Bombay and Santiago were graduating more than thirty people a year. This greatly enlarged a pool of "population experts" who shared the same theoretical assumptions and methodological approaches—especially the need to hurry along "the demographic transition."[30]

But most Population Council fellows just took one or two years of classes. There were few professors offering courses in demography, much less family planning, and they were increasingly in demand. When a foundation or the United Nations instead hired staff in places where population control programs were actually being carried out, they would often end up doing much the same work they might have done anyway—only now as highly paid consultants. The lead researcher of a UN fertility study in Mysore, Chidambara Chandrasekaran, was reputed to earn more than Prime Minister Nehru. On the other hand, this doubtless made population studies more attractive as a career, and recruiting a better class of students was something that demographers repeatedly emphasized in their appeals to foundations—with no apparent embarrassment as to what that implied about their own abilities.[31]

In other cases large grants only demonstrated how difficult it would be to reshape reproductive behavior. The Khanna study, which eventually cost roughly a million dollars, remains a notorious example. When the Rockefeller Foundation agreed to support the Harvard School of Public Health project, it was hoping to learn "how to implant this small family system in a peasant population." No one had ever undertaken such a study with a control group, without which it was impossible to prove that providing contraceptives had really made a difference, or whether instead people would have had smaller families without outside assistance. Obtaining statistically significant findings required intensive surveillance of some eight thousand people in seven villages in the Punjab. Researchers visited every household monthly to ask whether people were using the contraceptive foam tablets they offered, and if not, why not. They tracked eight thousand more villagers as a control group. It was difficult, painstaking work, but the team agreed it was worth the effort, and that there was no better place to do it. "India," they decided, "is the cauldron in which mankind will be tested."[32]

This was American social science at its most hubristic. From their headquarters in the market town of Khanna on the historic Grand Trunk Road, the Harvard researchers could look at their maps, point to one family's dwelling, and pull out a file. There before them were answers to questions few people would have been so bold as to ask. They would know the mother's menstrual cycle, how often she had sex with her husband, whether she had suffered any miscarriages, and whether she was trying to become pregnant. The data were transferred to keypunch cards and processed with IBM computers. "The approach is biological," the Rockefeller board was told, "concerned with the physiology of reproduction but with the population as the unit of observation, as distinguished from the individual." As the principal investigator, John Gordon, put it, they were exploring "the natural history of birth and conception in an Indian rural community."[33]

But Gordon and his collaborators soon realized that there was no such thing as a typical Indian village, or a "natural history" of human reproduction. In one village the headman told them "that they didn't need American help since every person in the village was a Communist." The very factors that made this particular area of the Punjab seem ideal for such a study—its relative isolation from the refugee flows following on the 1947 partition—also made it atypical. So too did an extraordinarily high ratio of males to females, beginning in infancy, such that polyandry was common and one in five men over 25 were unmarried, three times the national rate. Moreover, large numbers of their subjects moved away—a sensible strategy for sons of landless farmers, though one that made them more difficult to study. Similarly, husbands and wives gave contradictory answers—about frequency of sex, for instance. In sum, they simply did not stay put and behave like bacteria in a petri dish, or allow themselves to be studied as such. Most worrisome, there were already indications during the pilot study that some people were accepting the tablets but not using them.[34]

Gordon claimed that "the aim of population control is more than mere control of numbers," and he designated "improved health and social status" as the second goal of the project. But as soon as the researchers realized the difficulty they would have even to reduce birth rates—and had secured their funding—they were inclined to put measurements of health and welfare "on the shelf." Consequently, although they noticed that death rates were almost 50 percent higher for infant girls than boys, they made little effort to find out why, merely noting that males received better medical care. The study staff, for its part, made it a policy not to hold clinics, provide medicines, or otherwise take responsibility for sick villagers who came to them for assistance, lest it affect their findings. By contrast, they became increasingly, obsessively concerned with why so few people wanted help planning smaller families. After five years, the birth rate of those provided with contraceptives was higher than that of the control group. After a follow-up study featuring even more intensive efforts, it was still higher.[35]

The researchers never understood why people would not accept their help, or even accept that they were trying to help. Though at first villagers suspected that the project aimed at pulling down houses to straighten the roads, they correctly apprehended that these outsiders were trying to make their community more "legible." Only gradually did they realize, to their amazement, that it was to control their fertility. Whereas the researchers concluded that the villagers were ignorant and did not perceive the true value and potential of human life, the villagers did not understand why the researchers did not see the value of sons to parents with no other form of social security. Some grew deeply resentful, noting that those who found it hard to conceive children received no help with *their* "family planning." Rockefeller Foundation president Dean Rusk grew worried when he heard that Health Minister Kaur had visited the study site and gained the mistaken impression that it was a success. He instructed officials there, "in self-defense," to caution the minister.[36]

Whereas Rockefeller officials managed to spend hundreds of thousands of dollars on a study only to distance themselves from it, others gave comparatively small sums that had a far greater impact. Clarence Gamble, for instance, a grandson of the soap manufacturer of Procter & Gamble fame, saw himself as a provider of "risk capital" for start-up projects. It was actually Gamble who had provided the funds necessary to begin the Khanna study, though he managed to get out before it crashed. He preferred to write checks for no more than $500 or $1,500—subsidizing a journal here, a conference there, all the time demanding results. His most successful investment, which started in 1950 and eventually totaled about $38,000, allowed the director of Japan's National Institute of Public Health, Yoshio Koya, to introduce family planning to small communities of farmers, coal miners, and fishermen. There was no effort to determine if these three villages were representative, nor was there any control group. But in just two years Koya was able to show a marked decline in birth rates (never mind that birth rates were declining all over Japan). Amrit Kaur visited these villages as well, and was no less impressed. Koya delivered the message to family planning conferences that they could succeed anywhere.[37]

The Three Villages study worked in part because it featured the "cafeteria approach," offering people a choice between foam tablets, diaphragms, spermicidal jelly, the rhythm method, sterilization, and a sponge dipped in salt solution. Hardly anyone chose this last method, and all who did quickly dropped it. It was included at Gamble's insistence, part of his quixotic search for a contraceptive that would cost nothing and could be used by the most impoverished people. Simpler methods were essential in any effort to "control the dangerously expanding population of an unambitious and unintelligent group," as he explained in advocating another study in Puerto Rico:

> It has been said that birth control has been injurious to the race since it has been used by the intelligent and foresighted. It seems to me that only by some . . . demonstration can this accusation be refuted and our nation protected from an undue expansion of the unintelligent groups.

As Gamble traveled through Asia promoting simple methods, even enlisting his family and hiring "missionaries," people sensed his hidden agenda and resented him for it. It did not help that he insisted on using words like *coolie* and *native* with the government officials he was constantly importuning. "He just will not learn anything about the people he is dealing with," one volunteer in Ceylon complained. "They are all natives and sex to him." Rama Rau accused him of pawning off untested methods that no American would willingly choose.[38]

Yet by this point everyone was growing desperate for simpler and more reliable contraceptives. Not only was there still no "foolproof" method, it seemed impossible even to interest researchers in studying the problem. The most important single piece of philanthropy, once again from outside the world of major foundations, was therefore Katherine McCormick's bankrolling of the first oral contraceptive. A smart, physically imposing former suffragette, McCormick also commanded great wealth. For many years she had used it to search for a cure for her husband, who developed schizophrenia soon after their marriage. At the same time she developed a keen interest in birth control, possibly because of her fear that any offspring might also be afflicted. When she inherited his stake in the International Harvester Company, she wrote to Sanger asking how she might help. The reply came quickly:

> I consider that the world and almost our civilization for the next twenty-five years, is going to depend upon a simple, cheap, safe contraceptive to be used in poverty stricken slums, jungles and among the most ignorant people. I believe that now, immediately there should be national sterilization for certain dysgenic types of our population who are being encouraged to breed and would die out were the government not feeding them.[39]

This could not have come at a more critical time for Gregory Pincus, director of the Worcester Foundation. After he was denied tenure at Harvard, his independent laboratory had survived hand-to-mouth. When he suggested to the Searle Company that they salvage a failed venture in synthesizing cortisone by exploring whether other steroids had contraceptive properties, he received only a rebuke. Even the PPFA declined to renew funding, despite early success in demonstrating that oral administration of progesterone prevented rabbits from ovulating. Vogt apparently did not understand the significance of this finding.[40]

Luckily for Pincus, McCormick had majored in biology at MIT and was already familiar with the Worcester Foundation's work. At Sanger's urging, the two women traveled there to ask whether, and when, Pincus could develop a hormonal contraceptive. Pincus promised to start immediately, once he had adequate backing. McCormick then pulled out her checkbook. She even moved to Boston to look in on John Rock, a respected Catholic obstetrician, who tested an early version on women being treated for infertility. When Pincus and Rock were ready to test its effectiveness as a contraceptive, they elected to conduct the first large-scale studies in Puerto Rico.[41]

American researchers had for many years been using Puerto Rico to try out all kinds of theories and techniques intended both to explain and change poor people in foreign places. This included one of the only government-sponsored programs to distribute contraceptives and perform sterilization in a colonial dependency. The study participants wanted a better alternative, but they were not warned of potential side effects. Quite apart from the high incidence of nausea, dizziness, and headaches—Pincus and Rock would later realize that they were administering many times the required dose—the researchers did not know whether subjects might become infertile or develop cancer. They were relieved when test subjects who went off the pill had children of both sexes. "There was this terror," one lab worker admitted "—are you going to get all girls, or all boys, you know, what's going to happen?"[42]

When Pincus announced the coming of the pill at the IPPF Conference in Tokyo in 1955, a British official relayed an "authoritative American view" that "the 'pill' would, in fact, be launched prematurely, before there was full knowledge of its 'side-effects.'" Subsequently, advocates would argue that risky contraceptives were better than none in places like Puerto Rico, considering the high toll of unsafe deliveries and botched abortions. This left unexamined—and unchallenged—the low priority given to medical care for women. In fact, new contraceptives would generate demands for infrastructure, personnel, and "incentive" payments that diverted scarce resources from primary health care. By the time the FDA gave Searle approval to go to market in May 1960, it was already clear that this was not the "magic pill" Sanger had been waiting for all these years. The problem was not merely questions about its safety and expense, though the FDA had only approved a high-dose—and high-cost—formulation. Clinical trials had shown high dropout rates even in Los Angeles, so it seemed unlikely to work in places like the Punjab. Subsequently, researchers focused on developing cheaper methods that did not require so much motivation, or any at all.[43]

Yet over time the pill did have a global impact, primarily because of how it changed the most affluent societies, beginning with the United States. Placing a contraceptive that was 99 percent reliable in the hands of women helped them to take charge of their lives. This was part of Sanger's original vision, though a part she had soft-pedaled during the intervening years. After the FDA decision, Mike Wallace asked Sanger in a television interview whether her work had already made women "too independent." By that point she was in no shape to debate, but countless women would answer for her. Even her critics implicitly conceded that preventing unwanted childbirth was crucial to gender equality. Doctors found that they could not deny women access to the pill without losing much of their business, and priests discovered that parishioners would stay away if they continued to insist it was a sin. In the spring of 1963, Dr. Rock began arguing in the pages of *Good Housekeeping, Redbook, Life,* and the *Saturday Evening Post* that even the pope could not, must not, object to the pill. At the same time his friend Cardinal Richard Cushing of Boston said that the Church would not campaign to uphold bans on birth control. Finally, in the Supreme Court's landmark 1965 ruling *Griswold vs. Connecticut,* it overturned an 1879 ban on birth control and recognized a constitutional right to privacy, which eventually was stretched to include access to abortion. Altogether, this made many millions of women feel they had a stake in defending these rights as human rights wherever they were threatened.[44]

While undermining the power of doctors and priests to control family planning, women's demand for the pill strengthened suppliers. Searle reaped a windfall, with the company's revenue more than doubling in the first five years. In 1960, British Drug Houses defeated a takeover bid based on the news that it was preparing to market its own contraceptive pill. Shortly afterward it began making donations to the IPPF, as did several other pharmaceutical companies. Britain's FPA solicited contributions from all of those with products on its "Approved List," apparently unconcerned about the conflict of interest. This was another Sanger legacy. At the Bombay conference she had blocked Ottesen-Jensen's proposal that the IPPF develop its own capacity to manufacture and distribute contraceptives at cost, realizing that it would not pay to alienate private industry.[45]

Yet the specific ways in which population control had begun to gain momentum could not have been planned in advance or coordinated by one person. Instead, it resulted from very different people and institutions working independently while at the same time drawing strength from their combined efforts. The IPPF exemplified this tendency to combine formal decentralization with informal coordination. Its constitution vested power in a Governing Body composed largely of representatives appointed by each national affiliate. In addition, there were representatives from four regions, each of which had its own governing body, budget, and staff: Western Hemisphere, East Asia, South Asia, and Europe–Near East–Africa (a division reflecting the priority accorded these different regions). Because of overlapping memberships—leaders of national associations were also leaders of regions—the same person would sometimes occupy more than one voting position. This system stacked the Governing Body with like-minded people who could define "family planning" differently depending on whether they were planning for their country, for their region—including the poorer, "overpopulated" countries of that region—or for the world. As Blacker wrote in *The Eugenics Review* in 1956, "the balance between the need for compulsions and the claims of freedom is one which every country and every generation must interpret in its own manner."[46]

There was a time when people could simply appoint themselves birth control missionaries to the world and offer poor people untested methods. The consolidation of a diversified but single-minded population establishment now required putting such people in their place. It began with Clarence Gamble. It was not just that he pushed dubious techniques—many shared his interest in what Blacker called an "interim method" for "backward" countries. Nor was it his continued use of Western "missionaries." Vera Houghton, secretary of the IPPF, admitted that they too lacked true field-workers and suggested recruiting some of Gamble's staff. But Gamble insisted on choosing these people himself without consulting the national affiliates they were meant to assist, provoking objections from Burma, Egypt, and Iran. The IPPF's central office gathered up these complaints and—increasingly resentful about their dependence on American money—

added even harsher criticisms, describing Gamble as "a tribulation." But while they played up his miscues, implying that he was unacceptable to their increasingly diverse membership, this was not really the core issue. He actually diversified the IPPF by helping establish many new family planning associations. As Frances Ferguson of the PPFA noted, they could not easily get rid of Gamble, because "when it came to a vote, all the Asian countries wanted to keep him because of his money."[47]

The real problem was that Gamble's money and free time gave him the ability to act as both a philanthropist and an organizer, hiring field-workers while acting as one himself. Moreover, as a medical doctor—albeit one without a practice—he fancied himself a researcher. He combined in one person—and a very annoying person, by all accounts—roles that needed to be kept separate, undermining a strategy that depended on a division of labor and local autonomy. Thus, when the Rockefeller Foundation supported a Harvard study with the cooperation of the government and the Family Planning Association of India, failure could be blamed on the Rockefeller Foundation or Harvard or India or the FPAI. None of their bona fides were in dispute. But when Gamble hired someone in the name of the IPPF with no experience or training to study what happened when women agreed to put "salt rice jelly"—something he had cooked up in his own laboratory—in their vaginas, it threatened to discredit the whole idea of international support for planning "the global family."[48]

As the IPPF grew, internal divisions grew with it. Ceylon's FPA objected to Gamble, but it was no less concerned about the Indian association's pretensions in speaking for all South Asia. Some of the bitterest disputes occurred within national associations—Japan had to form a "federation" of its own because its many independent groups were so fractious. These were growing pains for an organization that was inspiring passionate commitment from people all over the world. But the most important leaders thought it would be fatal—not just to the IPPF, but to the movement more generally—if family planning came to be seen as a centrally directed campaign to control poor countries. That is why it was so important that its different components—including contraceptive research, grassroots organizing, government lobbying, and mass propaganda—not be represented by any one institution, much less one individual.[49]

The showdown with Gamble came at the IPPF conference in Tokyo in 1955. Sanger tried to stand up for him in the meeting of the Governing Body, but explained afterward that she "had a heart attack the night before and was under the influence of sedatives." Gamble was told that his field-workers and studies did not represent the IPPF, and he should not pretend otherwise. The Federation would accept his money, but they would decide how to use it. Gamble decided to go his own way and, in 1957, created a new organization: the Pathfinder Fund.[50]

Even Sanger could no longer do as she liked, as shown by the reaction to her proposal to host an IPPF conference in Washington that same year. Osborn worried that it would backfire. "IPPF is concerned not with the United States birth rate," he argued, "but with the birth rate in the high fertility areas of the world." In fact, America's baby boom had been growing every year since 1940. Betty Friedan would complain that the U.S. birth rate—by 1957, it was close to four children—was overtaking India's. This may be another reason why Osborn did not want to draw attention to the United States. He anticipated speeches in Congress proposing that population control be a condition for foreign aid: "Anything coming out of Washington proposing to reduce the birth rates of other countries would be dynamite in the hands of nationalist groups and might immeasurably set back the important movements already under way with government support in high fertility countries." After making certain he had support from London, Osborn persuaded her that a Washington conference would jeopardize the "remarkable" progress population control had been making in such countries as Japan, India, Egypt, and China. "As a result it is the United States which is a backward country in respect to the control of population growth, and not the Asian countries."[51]

Osborn was not exaggerating. Visitors like Rama Rau and Ottesen-Jensen perceived the United States as a primitive place where natives could not even speak publicly about birth control. In other countries the trend toward population policies was picking up momentum. In 1957, when it was still illegal for doctors in Massachusetts and Connecticut to prescribe contraceptives for their married patients, China removed all restrictions on sterilization and abortion. The 1953 Chinese census, the first in more than a century, had reported the population to be 583 million, 100 million more than expected. The State Council was increasingly concerned about the "chaotic condition" created by the "blind influx" of peasants into the cities. It issued directives to discourage urban migration and ordered the Ministry of Health to help people control their fertility.[52]

Mao himself complained to a group of cadres about how overcrowded schools were turning away students and adding to the unemployed:

[We] need planned births. I think humanity is most inept at managing itself. It has plans for industrial production, the production of
textiles, the production of household goods, the production of steel;
[but] it does not have plans for the production of humans. This is
anarchism, no government, no organization, no rules.

When he went on to suggest establishing a "planned birth department,"
there was nervous laughter. His audience could not believe what they were
hearing. As Mao "let a hundred flowers bloom," some writers began to suggest that Malthus had been right all along.[53]

China's first national debate on population control had developed independently of the international movement. But invoking Malthus was apos-
tasy in a communist country. An incident three months later involving the
former director of the FAO, Josué de Castro, was no less problematic. He
was lecturing members of China's Academy of Sciences on fertility reduction. When asked if it was all right to adopt birth control if it proved impossible to boost agricultural production, he replied, "If the food situation
cannot be changed, then the government ought to be changed," provoking
more laughter. Shortly afterward, those who had pushed for population
control came under attack. Contraceptives and abortions were still available
in theory, but with no promotion of smaller families. China's leadership instead decided to attempt a "great leap forward" to achieve communism in
one bound, organizing peasants into large communes and small-scale industry. The resulting chaos brought on a famine that left as many as twenty
million dead.[54]

Beginning in 1953, Egypt also seemed on the point of implementing a
population control policy. Nasser, like Mao, made a point of publicizing
population figures as a fundamental concern for the nation. He established
a National Committee for Population Affairs with plans for both public
and private clinics. "If we continue to reproduce with the maximum biological impetus, without regard to the capacity of society to provide for the
basic needs of its members," the minister of social affairs warned, "we shall
have more weaklings, vagrants and beggars."[55]

Advice was solicited from both the Ford Foundation and the UN (and
received unsolicited from one of Gamble's missionaries). The Commission
took two years to report and opened just eight clinics, in part because of resistance from the rector of El Azhar University. While some religious leaders were supportive, Nasser feared the Muslim Brotherhood. He backed

away from a national policy and turned the clinics over to a private association. A more appealing alternative was at hand in plans for the Aswan High
Dam. It was conceived as a way to solve the population problem by expanding agricultural output and increasing electric power for industrialization.[56]

Demographic changes appeared to demand a dramatic response, even if
it was not always, immediately, to instruct everyone to start using birth
control. Colonial officials were even more hesitant. By 1956 local governments in several small, densely populated British territories, such as Singapore, Hong Kong, and Barbados, had started to fund clinics run by private
family planning associations. The population of Hong Kong, for instance,
had swelled with a million migrants both legal and illegal between 1946
and 1958. But the colonial ministry concluded that an official campaign
would be "socially injurious and politically disastrous," implying "a desire
by a white government to restrict the growth of the non-white population."[57]

The French faced a still more acute dilemma. After losing Indochina,
they were forced in 1956 to cede independence to the protectorates of Morocco and Tunisia as well. At the same time, they escalated a counterinsurgency campaign in their Algerian *départements*, which they insisted
were not colonies but an integral part of the Republic. Rapid population
growth—not gross political and economic inequality—was deemed the
fundamental cause of unrest across North Africa. Like the British in the
West Indies, only on a much larger scale, they feared increasing migration
to the metropole. *Bidonvilles*, or shantytowns, started to sprout, not just in
the cities of North Africa, but in France itself. As soon as authorities bull-
dozed them into the ground, they arose again in another neighborhood.
Projections indicated that by 1982 the annual number of Muslim births in
the Algerian *départements* would exceed the number of French births in the
metropole. Muslims already outnumbered European settlers ten to one.[58]

Yet any campaign to encourage Muslims to have smaller families would
violate France's 1920 law prohibiting advocacy of birth control. When the
government solicited advice from population experts, some, like Sauvy,
along with several other members of its Haut Comité de la Population, suggested that it need not enforce the law in Algeria. This provoked sharp debate, as did various proposals for rigging child benefits to encourage European but not Muslim fertility. Salary-earning Muslims in Algeria, like their
European counterparts, were entitled to cash payments for every child. In-

stead, the government sought to enroll more girls in school and prohibit early marriages.[59]

India was still the only country with a formal policy and a program to control population growth. Here as in China, the West Indies, and North Africa, leaders worried not merely about the rate of growth but about the way it seemed to give rise to uncontrollable migration. Nehru was concerned that "people continue to come to Delhi from outside and put up shacks without any authority. . . . Some slum owners are ejecting people on the plea that they want the houses for themselves." The government still assigned priority to rural development and rapid industrialization, and Nehru professed optimism that food production could keep pace with population growth, no matter how rapid. But officials in the Planning Commission increasingly viewed population limitation as essential for raising standards of living.[60]

In making this argument Indian officials had the support of Western consultants like Notestein and Davis. Ironically, the two researchers most closely identified with demographic transition theory were now among the most critical of its capacity to predict a decline in fertility. On the one hand, they now cited historical cases like France and Bulgaria, which demonstrated that agrarian societies could adopt the "small family norm." On the other hand, the latest data suggested that "urbanization provides no mystical means for the reduction of fertility," as Notestein put it. Economists had begun to theorize that population growth might actually prevent the capital accumulation necessary to fund industrial development, because savings would be diverted to providing basic services—especially if people kept moving into cities. India was already considered a case study for what came to be known as the "low-level equilibrium trap."[61]

With its second five-year plan in 1956 the government of India therefore decided to give population control a higher profile. It established a Central Family Planning Board presided over by the minister of health, and sometimes by Nehru himself. A new director of family planning, Lieutenant Colonel B. L. Raina of the Army Medical Corps, would take charge of the program. While population control was still a tiny part of the plan budget, an annual allocation of ten million rupees represented an almost fivefold increase—twenty times the amount that was actually spent in the first plan. This one called for establishing 2,500 clinics nationwide to provide free contraceptives for low-income clients. By 1959 Raina had a staff of twenty and was subsidizing family planning boards and full-time directors in most of India's states. Together they had established 473 rural and

Throughout her career, Sanger promoted birth control as a panacea for social problems. Nehru was more skeptical, but gave way to Indian proponents—like Lady Dhanvanti Rama Rau, behind him—who shared Sanger's outlook. Margaret Sanger Papers, Sophia Smith Collection, Smith College.

202 urban clinics. At the same time, they launched a nationwide publicity campaign, printing almost half a million posters and broadcasting hundreds of radio programs a year in multiple languages.[62]

All this seemed impressive on paper, but what was happening on the ground was another story. In rural areas—where four-fifths of India's population lived—opening a clinic usually meant that just one additional worker was hired at an already overburdened primary health center. These workers were responsible for servicing a population averaging sixty-six thousand people. In 1958 they supplied an average of only eighteen condoms per clinic per day. Officials had at least some notion of what they were asking of them. With no more than two months of training—and sometimes none at all—they were expected to do everything, from motivation to education, screening their clients while also supplying them. Since it proved impossible to recruit enough individuals with degrees in health care or social work willing to serve in rural areas, officials stressed personal qualities, including "infinite patience."[63]

Officials themselves began to lose patience, and some concluded that sterilization provided the only long-term solution. This was an even greater logistical challenge, because New Delhi insisted that both husband and wife agree and that the operation be performed only in a medical facility. But the number of sterilizations doubled and doubled again, totaling at least twenty-five thousand by 1958. R. A. Gopalaswami, formerly India's registrar-general and now chief secretary of Madras, increased the pace by paying people thirty rupees ($6.30) to undergo sterilization, and fifteen more for each person they persuaded to do the same. These were significant sums, considering that per capita GNP was less than seventy dollars a year. In February 1959 the Central Family Planning Board decided to follow his lead, strengthening the staff at three thousand hospitals and maternity homes to enable them to conduct more sterilization operations free of charge while compensating low-income patients for travel expenses and lost wages. Public-sector employees were offered a week's vacation.[64]

India's Family Planning Association was well positioned to keep pushing the program forward. Rama Rau was married to the governor of the Bank of India, had been in on the early planning, and was now a member of the Central Family Planning Board. She ensured that the FPAI received a portion of the board's growing budget, with annual grants amounting to over one hundred thousand rupees by 1958. National and private programs were thus mutually reinforcing.[65]

A voluntary Family Planning Association—founded by the wife of the foreign minister—also played a key role in Pakistan. The government started supporting FPA clinics in 1957, and the following year appointed its leaders to a National Family Planning Board with authority to approve additional grants. President Ayub Khan declared that "a big concentrated drive is necessary to educate the people about the evils of overpopulation." In 1960, $7 million was set aside for family planning over the next five years, as Pakistan became the second country with both a policy and a program to control population growth.[66]

The IPPF's central office in London provided "a gigantic switchboard" for activists in different countries. By 1958 Houghton had a staff of six, and the Governing Body allocated grants totaling almost $100,000 annually, quadrupling the amount given during the first five years. Yet the IPPF's principal asset remained its capacity to facilitate international networking, especially between activists in rich and poor countries. Thus, it helped Ottesen-Jensen develop close relations with her counterpart in Ceylon, Sylvia Fernando. Beginning in 1954 the two women lobbied their governments to undertake a joint project in family planning, finding an ally in Sweden's new ambassador to India and Ceylon, Alva Myrdal. This effort paid off in May 1958 when Sweden agreed to provide expert advice and $80,000 (about $558,000 in today's money) to distribute contraceptives and measure the results in two Ceylonese communities. The project had many of the same flaws as the Khanna and Koya studies, and could not, after seven years, prove its effectiveness. But it marked the first time that any government included family planning in its foreign aid program. Sweden did not wait for the results before following up with a much larger grant to Pakistan in 1961, appropriating almost $400,000 to send three medical teams.[67]

The biggest prize in this field was the U.S. foreign aid program. Here activists had a "Planned Parenthood nut" in a very high place. Meeting with his National Security Council in December 1958, President Dwight Eisenhower professed frustration about how little United States aid actually bought:

In all our discussions of the problem of underdeveloped countries and the kind of assistance which we could effectively provide them, we had not faced up to what was really the most serious problem, namely, that of exploding population growths. As far as he could see, continued the President, the only solution to this problem throughout the world was finding an effective two cent contraceptive.

Eisenhower thought that "something drastic had to be done to solve this problem," though "he certainly did not know how to get started on this solution and he furthermore could not himself get it started."[68]

The president said he was looking for ideas. He may have been looking to General William Draper, who was sitting in because of his work leading a presidential commission on U.S. military assistance. An investment banker for Dillon, Read, and Company, Draper was a personal friend of Eisenhower's, raised money for his presidential campaigns, and had held a string of high-level appointments. It was Draper who as undersecretary of the army had granted permission to the Rockefeller mission to visit Japan and Korea. A decade later he followed in its footsteps and was struck by Japan's progress in reducing population growth while boosting productivity. With Eisenhower's words still ringing in his ears, Draper managed to expand his commission's mandate to include economic aid. He passed around Hugh Moore's *Population Bomb* pamphlet and invited experts to brief his staff. One internal report warned that population trends might lead to "international class war." The commission members were all heavy hitters from the world of national security and international finance, including Joseph M. Dodge, an architect of occupation policies in Germany and Japan, General Al Gruenther, former NATO supreme commander, Arthur Radford, who had just stepped down as chairman of the joint chiefs, and John J. McCloy, a legendary "wise man" who had led the World Bank, the Ford Foundation, and Chase Manhattan Bank.[69]

It was therefore all the more striking when—charged with examining military aid—this group came back with a recommendation that the United States meet requests for assistance in "maternal and child welfare." The phrase was a fig leaf for members who still could not countenance explicit mention of contraception. Nevertheless, with the Draper report the American foreign policy establishment had begun to view population growth as a national security issue and see birth control as critical to the nation's defense.[70]

The Catholic hierarchy then rolled out their big guns, including Cardinals Spellman of New York, Cushing of Boston, and O'Hara of Chicago, and more than a dozen other bishops and archbishops. "For the past several years," they noted in a joint statement, "a campaign of propaganda has been gaining momentum to influence international, national and personal opinion in favor of birth prevention programs." This approach to the population problem was "morally, humanly, psychologically and politically disastrous." The real solution was "human solidarity," including increasing food production and international migration. John F. Kennedy, now running for president, was asked if he agreed. He claimed that he had long believed that "it would be the greatest psychological mistake for us to appear to advocate limitation of the black or brown or yellow peoples whose population is increasing no faster than in the United States." Once again, the alleged danger of a backlash abroad provided cover for those concerned about Catholic opposition at home.[71]

Eisenhower himself was hardly a profile in courage. In a press conference held almost a year to the day after he called for "something drastic," he categorically rejected the Draper committee recommendation for aid to family planning: "I cannot imagine anything more emphatically a subject that is not a proper political or governmental activity or function or responsibility." If other countries wanted to do something about this "explosive question" of population growth, "that is their business." But Eisenhower had not lost interest in the subject. He confided to the NSC that it was "a constant worry to him and from time to time reduced him to despair." Shortly after the Draper report was released, he complained that American aid had focused excessively on the communist threat: "We have had a narrower view than we should have. The real menace here was the one and a half billion hungry people in the world."[72]

The Draper report and the ensuing campaign controversy was a godsend for the publicity and fund-raising efforts of family planning organizations. There was a spike in coverage by the *New York Times* and American magazines, including articles in *Time, Newsweek,* and *U.S. News and World Report* and a cover story in *Life.* An hour-long CBS TV documentary focusing on India, "The Population Explosion," was seen by nine million people, with another nine and a half million viewing a rerun two months later. In a series of columns in the *New York Times,* Arthur Krock painted a vivid portrait of "billions of half-alive, starving peasants, condemned to short, miserable lives of hatred and hunger." The danger was not just that they might succumb to communism, but that there "will be forcible invasions of national borders by hordes of human beings who cannot subsist within their own."[73]

To the bishops all of this seemed like a concerted campaign. In fact, it was quite decentralized, even contradictory. Hugh Moore wanted to capitalize on the Draper report by launching a fund-raising campaign to support the IPPF and pressure policymakers. He considered it essential to play up population growth as a threat. Blacker, for his part, was helping to increase Eugenics Society support for the IPPF as part of its continuing strategy of "crypto-eugenics." But he insisted that the Federation itself, to remain effective, had to stand for universal principles such as "family welfare, community well-being, and *international goodwill.*" They were in the middle of delicate negotiations to bring in new affiliates from Eastern Europe. If the issue could not be settled, Blacker warned, "the IPPF itself may be threatened at best by the severest crisis it has yet encountered and at worst by a complete split." Notestein, for his part, testified *against* U.S. support for population control programs in a Senate committee hearing. It would provoke accusations that they were "fearing the 'rising tide of color.'"[74]

Moore finally agreed to set up a new organization, the World Population Emergency Campaign (WPEC), which would raise money for the IPPF and, at the same time, organize public pressure to change U.S. policy. Draper joined the organizing committee and became an indefatigable fund-raiser. The only expert included was Harrison Brown, a professor of chemistry at Caltech. He had written a popular book proposing that industrialized countries form a world federation empowered to control access to artificial insemination and abortion and pursue "a broad eugenics program." With the first $100,000 raised from wealthy donors, the WPEC launched direct-mail campaigns that targeted as many as six hundred thousand people at a time. A typical brochure featured the headline "The Population Explosion Can Shatter *Your* World" over a photo of Africans with grasping hands, adding, "It is a fact of life that people will not passively starve. They will fight to live." By the end of 1961 the WPEC made $170,000 in grants for population control and held another $100,000 in reserve.[75]

The floodgates had also begun to open at the Ford Foundation, in part because it wanted to promote Notestein's low-key approach. Since 1952, Ford had contributed a little over $2 million for work in population research and education, with the Population Council receiving the lion's share. At the March 1959 meeting the board approved another $1.6 million for the Council, including the first direct support for contraceptive research. At the same time, it signaled that it would also approve support for family planning overseas. Population grants jumped to $2.8 million in 1960, then to $3.4 million in the following year. A State Department official reported that "money was not a serious limitation upon what it could consider doing. Expenditures of up to one hundred million dollars would present no great problem." The Ford Foundation's total outlay added up to more than just the grants, since the foundation had started to hire staff to administer its own programs. But the Population Council still received almost half of its $3.1 million annual budget from the foundation in 1962, enough to employ almost forty people. It even had enough money lying about to start passing some along to the IPPF.[76]

As American money poured in, the IPPF performed a difficult balancing act. With twenty-seven national associations, now presided over by Ottesen-Jensen—affectionately known as Ottar—the ideal of representing a worldwide movement acting without fear or favor in the service of all humanity never seemed closer to realization. Yet an increasing proportion of its budget came from the United States for the express purpose of controlling the population of poor countries. Refusing the money would reduce the IPPF's capacity to contend with wealthy U.S.-based institutions like the Population Council or smaller, more entrepreneurial groups like Gamble's Pathfinder Fund. The obvious solution was to diversify the IPPF's sources of support and launch more affiliates. Yet six months after the British FPA announced a major appeal in 1960, it had raised only a tenth of its $250,000 goal. When the Swedish association called on the public to contribute to an "Ottar Fund" in honor of its founder, four months of fundraising netted an embarrassing $3,000. Some national associations did not even send the IPPF the nominal five or ten pounds sterling in annual dues. Even so, they were still entitled to a vote in the increasingly unwieldy Governing Body.[77]

In May 1960 a Management and Planning Committee was created in which the Americans secured a commanding voice in the key decisions. Even as the IPPF grew to include more than eighty national associations in the decade to come, Americans would continue to hold a quarter or even a third of the committee's twelve to seventeen seats. One of its first decisions was to accept that the IPPF Western Hemisphere Region—essentially run by U.S. representatives—would decide how WPEC grants were spent. They could therefore buy influence among the national affiliates who received this money—which in some cases was critical to their formation—including the first members from Central and Latin America. The IPPF's American treasurer, Rufus Day, advised that the WPEC should also dole out money to Western European associations, considering it "important to acceptance of our objectives on a worldwide basis." In 1961 the PPFA and WPEC would join to form Planned Parenthood–World Population.[78]

The London central office accepted the American money, and the tactics necessary to obtain it, with more or less good grace. Blacker abruptly announced his departure as administrative chairman at the first meeting of the Management and Planning Committee. He resented its decision to hire George Cadbury, a veteran UN official, as "special representative and field director." Blacker could not have blocked the Americans and their emergency approach even if he had wanted to. But in his last months, reviewing WPEC publicity, he sometimes asked that they "make sure that this document does not get out of North America."[79]

Extreme rhetoric was not, however, the preserve of American publicists. In November 1960 eminent citizens from nineteen countries, including more than one hundred scientists and thirty-nine Nobel laureates, signed a public "statement of conviction" urging the UN to take action. "Unless a favorable balance of population and resources is achieved with a minimum of delay," they warned, "there is in prospect a Dark Age of human misery, famine, under-education and unrest which could generate growing panic, exploding into wars fought to appropriate the dwindling means of survival." Even legislators in countries officially hostile to Malthusianism, such as France, shared this vision of a "demographic deadline" when "misery and hunger will become the lot of all humanity." The same rhetoric of how barricades would be "overrun by advancing population" came from diplomats representing countries like India, the alleged source of the danger. Hamid Nawaz Khan of the All Pakistan Women's Association agreed that "states ought to adopt vast programs of controlled reproduction if they don't want to remain powerless before a human tidal wave which will certainly bring about an immense decline of civilization."[80]

The least one can say about these apocalyptic visions is that contemporaries had alternative ways of looking at population growth. It is curious, for instance, that Africans were already appearing as poster children, considering that the continent was a net food exporter and featured some of the lowest rates of growth in the world. It is also odd that even sophisticated analysts continued to assume that starvation would present the first constraint on continued population increase. A broader view of the earth's "carrying capacity" might have noted declines in commodity prices, as well as the fact that poor countries consumed relatively little and their share had actually been shrinking. Moreover, the 1950s had witnessed increasing food production per capita. And even if more people went hungry, it did not necessarily follow that they would overrun the wealthier parts of the world.[81]

But people had a new way of thinking about how population growth and poverty could lead to political unrest. It was summarized in a catch-phrase, "the revolution of rising expectations," which was first coined by Harlan Cleveland in 1950 but only began to catch on at the end of the decade. As Davis explained in *Foreign Affairs,* "Envy and revolt are nurtured not by absolute but by relative poverty." Radio, film, and rising rates of literacy were making poor people more aware of how much better rich people were living, and they would not stand for it. Observers might have welcomed increasing enrollments in primary and higher education as a long-term solution to poverty and unsustainable population growth. Instead, more literate people only seemed more threatening.[82]

There is no doubt that people in rich countries found the proliferation of images of the poor—in TV documentaries, in *Life* magazine, in their mailboxes—difficult to tolerate. The authors of WPEC direct-mail appeals offered reassurance, pausing to let the ideas sink in: "These millions who can turn to revolution . . . or who may explode out of national boundaries to find food, status and shelter are not evil . . . not cruel . . . not truly enemies. They are just hungry. They will fight to live." Population control presented itself as a charity like any other, helping less fortunate people. But it was the only one that promised to *make them go away.*[83]

There was really little need to exaggerate the rate of world population growth. It was and is astonishing. Beginning with the Chinese census in 1953, the UN had continually raised its projections. In 1950 the medium figure for world population in 1980 was 3.28 billion; in 1954 it was expected to be 3.63 billion; in 1958, 4.22 billion. It had taken almost all of human history to reach 1 billion people around 1800, and 130 years to grow from 1 to 2 billion; from 1930 to about 1960 world population grew by another billion. This was the first time anyone really noticed, because previously there was no United Nations office to mark the moment, much less offer projections for the future. In this new age, however, obscure bureaucrats could miscalculate the impending arrival of another billion human beings as if it were an accounting error.[84]

The new projections caught even the Soviet representative to the 1959 Population Commission off guard. They were "wrong," he insisted, "too high." The Belgian delegate objected to "sensational figures" that were "unnecessarily alarming the public." In fact, with the completion of the next round of censuses in 1961, they turned out to be too conservative. Asian countries were growing at least 50 percent faster than in the previous decade. Indian officials were shocked to discover that their five-year plan was premised on a population estimate at least ten million short of the actual figure. Other countries, like Pakistan, Thailand, and Korea, were growing even faster. But some of the most rapid increases were in Latin America. Costa Rica was growing at the unprecedented rate of 3.9 percent a year, fast enough to double in two decades. "What disconcerted demographers," one of them recalls, ". . . was not so much how high the projections were but how rapidly population growth was outstripping our ability to project it."[85]

Once again, India led the way. Even before the census results appeared, it was ramping up its population program and, for the first time, specifying a target. The third five-year plan, announced in 1960, placed family planning "at the very centre of planned development." It provided for a sixfold increase in funding and projected a fivefold increase in the number of clinics. But there was also a shift to an "extension approach," like that of the Khanna study, based on the idea that family planners cannot wait for people to come to clinics but have to find them where they live and breed. Raina defined it as a strategy "whereby the forces of group pressure can be mobilized." Thus, every village and town was directed to form a family planning committee, and "natural group leaders" were paid an "honorarium" of four thousand rupees ($800) to develop the "small family norm among their group."[86]

The most dramatic example of the new approach was the mobile vasectomy camp, which first appeared in the state of Maharashtra, formerly Bombay. During a five-week "intensive Family Planning campaign" in 1960, almost fifteen thousand people were sterilized—more than the nationwide total for 1957. This was held up as a model for other states. Sterilizing men was preferred because a competent surgeon could perform the operation in ten or fifteen minutes under local anesthetic. If all went well, the patient could return to work the next day, as opposed to the week of hospitalization typically required after women underwent sterilization. But the drive to rapidly reduce fertility at minimal cost in mobile camps would make it difficult to maintain basic standards, including medical screening and sterile instruments. In 1962, 158,000 Indians went under the knife, as the Ministry of Health began to encourage the use of mobile units to sterilize people institutionalized for tuberculosis, leprosy, or mental illness.[87]

India was now committed to the goal of reducing the birth rate by 40 percent by 1972. Not since Mussolini's Italy and Imperial Japan engaged in "demographic battles" had a government pursued a population program with specific fertility goals, and this was the first in history aimed at *reducing* population growth. Many more would follow, but it was already becoming difficult to pretend that the phrase "family planning" was not just a euphemism for population control. "Birth control, prevention of population explosion and family planning; in essence they mean the same thing," Ambassador Mahommedali Chagla declared in May 1961. "We use them according as to whether we are courageous or timid advocates of the same cause, whether we like to speak sotto voce or shout our faith and our creed from the house tops."[88]

Chagla really was shouting from the housetops—the legendary Starlight Roof at the Waldorf Astoria on Park Avenue. He was addressing three hundred dinner guests who had come to honor Margaret Sanger and inaugurate a "Conference on the World Population Crisis." Sanger herself had barely made it to what would turn out to be her last public appearance. But she was rewarded for her exertions. Forty-five years after she had opened an illegal clinic across the East River in Brooklyn, she could behold at last the gathering of something more than a movement. It was a veritable establishment, with representatives of government and industry, science and medicine, academia and philanthropy, and an increasing number of full-time professionals all working for the cause.

Among the audience were six ambassadors—from Lebanon, India, Liberia, Israel, Ceylon, and Nigeria. There were some of the old warriors, like Julian Huxley, who had masterminded the Nobel petition. But there were many new faces, too, like Marriner Eccles, former chairman of the Federal Reserve, warning that capitalism could not "survive as an island of abundance in a sea of poverty." And there was Katherine McCormick, "rich as Croesus," as Dr. Rock described her, who there and then donated another one hundred thousand dollars to the IPPF to honor her friend, and perhaps to encourage the other millionaires in the room, like Moore and Lammot du Pont Copeland. Sitting with them were media moguls like Cass Canfield, chairman of Harper & Bros. and president of the PPFA, as well as media darlings like Rock.[89]

Those distracted by the skyline view might have felt the presence of still more powerful forces now supporting the cause—whether the offices of the Rockefeller Foundation, facing the Waldorf on Fifth Avenue, at the top of a seventy-story skyscraper; or the Ford Foundation, just around the corner on Madison, which was planning a landmark headquarters across from the UN Secretariat building; or the Population Council, housed in the famed New York Central Building, the crown of Park Avenue. George Cadbury, the IPPF's new special representative, proclaimed that advocates of family planning now spoke for the majority of mankind. The number of their opponents, he said, was "so small in fact that their attempts to impose their views on the rest of the world are not only improper, but also impertinent."[90]

Sanger struggled to the podium, assisted by her son, and managed to express a few words of gratitude for a "very rich life." After returning to her seat, she nodded off and was taken back to her suite. But even if she had remained, she would not have heard the most telling tribute of all. It could be found only in confidential correspondence among members of the Catholic hierarchy. They had grown increasingly alarmed at what a National Catholic Welfare Conference memorandum described as a "systematic concerted effort" to convert U.S. policymakers as well as international bodies to the cause of population control. But they also warned against any temptation to deny that population growth created real problems. The Vatican's representative in the United States urged the NCWC to "propagandize by every legitimate means the Catholic view on this matter."[91]

Proponents of population control were now ready to storm back to the commanding heights of world politics. Yet this movement now marched under a new banner. Contrary to the Indian ambassador, "birth control, prevention of population explosion and family planning" had meant many, many things over Sanger's long career, and the struggle to redeem the liberating potential of her message continued long after she died. But this new population establishment was now united in a drive to control the population of the earth.

6

CONTROLLING NATIONS

In the life of a social movement, supporters must prove that their cause is not only just but important and worthy of the world's attention. In the case of population control, official recognition came in a complicated and roundabout fashion, revealing how even discussing contraception was still difficult for most governments. It started in August 1961, as the latest census results surprised observers, and Sweden and Denmark insisted that population growth be added to the UN General Assembly's agenda. This was the work of Ulla Lindström, Sweden's minister for foreign aid and a longtime ally of Elise Ottesen-Jensen. They had some hope for success. Earlier that year, the Vatican and its allies had only just managed to prevent the governing bodies of WHO and the ILO from discussing family planning on procedural grounds. No delegate objected when the Population Commission heard that more governments would request advice for national programs, in that way appearing to accept the technical aid already given to countries like India and Egypt. And at the Economic and Social Council, the president of the World Bank, Eugene Black, argued that population growth imperiled economic progress. Finally, Lindström lined up support from countries like Pakistan and Tunisia—deflecting, if not entirely defusing, the charge that only white, wealthy countries cared about the issue.[1]

The item was assigned to the Second Committee of the General Assembly, responsible for economic and social matters, but there was not time enough for a debate before the end of the session. The sponsors asked that their resolution be given priority in 1962 because of its importance—a standard motion for closure. Instead, representatives of Argentina and

Spain, deprived of the opportunity for a fight, demanded that the words "given priority" and "important" be put to a vote. They had the satisfaction of seeing them stripped from the motion. It was a petty gesture, and one that would only call attention to the weakening of opposition to population control when the debate was rejoined a year later.[2]

Lindström had called for "a debate, realistic and tolerant, of what active role the United Nations could play." When the Second Committee took up the matter once again, realism and tolerance were largely absent. The Argentine delegate claimed that the UN had not conducted enough studies, underdeveloped countries did not have adequate statistics, and scientists knew nothing about either fertility or sterility—as if more studies, statistics, and biomedical science would convince him to accept something he considered "repugnant to the moral conscience." Supporters of the resolution retorted that, when not actually "absurd," their opponents' arguments were "moral or religious," and thus "quite extraneous." They claimed that the resolution did not concern "policies, propaganda or financing of birth control"—which, if it were true, meant that it concerned very little.[3]

When the debate resumed, delegates began to get at the heart of the matter. The French representative, Maurice Viaud, admitted that the resolution itself did not actually call for reducing births, but only for UN assistance to states seeking to develop programs to deal with population problems. But how, he asked, could governments define population growth as a problem and create programs to address it without telling citizens to use birth control? And if people could not be persuaded, might that not lead to "more brutal measures"?[4]

Defining population as a problem had certainly limited people's reproductive rights in France, which for decades had made it a crime even to advocate use of contraception. Yet it was increasingly implausible to dismiss the cause of reducing population growth as just a means to prevent a "sharing of riches between the starving and the more fortunate," as Viaud put it, at least in these simplistic terms. After all, supporters of the resolution included countries like Algeria, which had gained impeccable revolutionary credentials by throwing off French rule. Its representative insisted that population growth was "one of the most important problems confronting the under-developed countries." The Pakistani delegate was particularly passionate, declaring his country was "fighting for its life." It was "utopian," he said, to suggest that sharing the wealth could feed everyone. This would make overpopulated countries dependent on handouts.[5]

The debate extended over four meetings and sometimes continued late into the night. Opponents tried, and failed, to remove a passage calling for UN assistance before the resolution went to the plenary session. With the Vatican representative coordinating their defenses, they invoked the decline of the Roman Empire and warned that tinkering with "genetic processes" had brought "the collapse of civilizations." If the resolution passed here, under the gold leaf emblem and dome of the General Assembly Hall, it would give the stamp of UN approval to policies to control population growth.

Opponents' only recourse was a parliamentary maneuver. Resolutions emanating from the Second Committee were typically subject to a simple majority vote. If a majority agreed, the matter could be explicitly designated as "important," and therefore require two-thirds of the Assembly to approve passage. A year earlier these same delegates had insisted it was not important. But how could supporters disagree? The crucial language on UN aid won only half the votes. The rest of the resolution passed unanimously.[6]

The fact that opponents had to recognize the importance of population problems in order to defeat an endorsement of UN action marked the end of one era and the beginning of another. After this setback, the forces backing population control began to sweep aside opposition and advance from strength to strength. Within a year Pope John XXIII convened a commission to prepare a statement on population. In the heady atmosphere of the Second Vatican Council, as laypeople and demographers joined the deliberations, most observers assumed that the Church was preparing to retreat from *Casti Connubii*. The budgets of national population programs and private associations began to double and double again. New techniques and new technologies were deployed in campaigns across Asia, Africa, and Latin America. Hesitantly at first, but with increasing determination, the United States began to use public exhortations and quiet coercion to induce other countries to embrace the cause. It would finally culminate with preparations for an official United Nations campaign to convince poor people all over the world to plan smaller families—and the beginning, too, of a groundswell of dissent that finally threw the whole field into crisis.

At the time, what seemed most important about the General Assembly debate was the qualified backing of the United States for a UN role in population control, the first public indication of a significant policy shift. Since Kennedy took office, senior U.S. officials had been debating whether to reverse Eisenhower's position. Some White House staffers assumed that the president's Catholicism made this impossible. The new U.S. Agency for International Development (USAID) did not want to risk venturing into such treacherous terrain. The State Department's Policy Planning Council under George McGhee—a veteran of the Draper Commission and also married to a Planned Parenthood volunteer—recommended a more active role. George Ball, undersecretary for economic affairs, quipped in a memo to Dean Rusk that "the people who get so excited about population control are often 'garden club' types whose only concern about the underdeveloped areas is that the brown and black-skinned population seems to be growing so rapidly."[7]

But the secretary of state himself had long been concerned about population control, and he decided to approve several of McGhee's recommendations. Thereafter, AID could give information "on the various aspects of human reproduction." Because Rusk presumably did not intend instruction in the birds and the bees, this could only mean advice about organizing family planning programs. Moreover, a State Department official, Robert Barnett, would be assigned to the issue full-time. And henceforth the United States would give "maximum support consistent with avoiding undue publicity to work in the population field by the UN." After a representative of the National Catholic Welfare Conference threatened a backlash against the United Nations, U.S. spokesman Richard N. Gardner abstained on the operative passage, asserting that it would not give any authority the UN did not already possess. He described the unanimous backing of the rest of the resolution "the clearest kind of mandate . . . to proceed with vigor to a further exploration of population growth and its relation to economic and social development."[8]

President Kennedy was still unwilling to take on the Church. This became clear when Draper briefed him about population growth and poverty in South America. "Why doesn't the Ford Foundation," Kennedy asked, "concentrate all of its resources on the population problem around the world?" When Draper related that the Foundation felt there were other problems too, Kennedy again professed not to understand why they did not use all their assets—worth over $20 billion in today's money—on population programs.[9]

During Kennedy's administration, the United States continued to encourage the UN, foundations, and private associations to take the lead. Ball considered it "a field which is pre-eminently suited to private activity." Yet it was increasingly difficult to draw a line between public and private activity. Barnett owed his State Department job to the Ford Foundation and Planned Parenthood staff who had helped McGhee draft his recommendations. Gardner's UN statement was based on a form letter Barnett wrote to people provoked by Hugh Moore's full-page newspaper advertisements. Barnett circulated the statement to AID missions and embassies all over the world to encourage requests for U.S. help. Foundations and family planning associations then set about drumming up such requests to prove the demand for more such assistance.[10]

The president himself finally endorsed aid for family planning in April 1963 in response to a reporter's question planted by Barnett and Planned Parenthood staff. That summer, at Draper's urging, Senator William Fulbright added an amendment to the foreign aid bill that explicitly authorized "skittish" bureaucrats to support research on population problems. Planned Parenthood was assured that "population was now AID's Number One problem." Reallocating budgets and hiring personnel took much longer. But all along, encouraging the United Nations, foundations, and family planning associations to take the lead led government officials to follow up with more overt U.S. assistance. Their contacts became more frequent, the connections proliferated, until the collective weight of this burgeoning establishment began to move bureaucracies.[11]

The interpenetration of public and private agencies went even deeper in India. As a matter of policy, the Family Planning Association, like the rest of the IPPF, was increasingly looking to work with government health departments, reserving for itself the role of publicizing these programs and developing new service models. What made the case of India even more complex was the "crowd of Americans here," as Nehru put it, "who nose about far too much." He was upset to hear that senior officials were soliciting travel grants from U.S. foundations. In 1958 he upbraided a minister who accepted that the government would merely have a veto over Rockefeller Foundation nominees, rather than designate the recipients itself. He finally acceded after Rusk—then president of the Foundation—refused to budge, reluctant to risk losing hard-currency development aid.[12]

The Ford Foundation had a much bigger presence in India, with commensurate influence. The number of Ford personnel in Delhi rivaled the American embassy staff. A single foundation grant in 1962 would give the government $5 million—more than $33 million in today's money—to pursue a more intensive population control program in selected districts.

The head of Ford's office in India, Douglas Ensminger, did not trust the minister of health, Sushila Nayar. So the grant would support a new Central Family Planning Institute to train the personnel and a National Institute for Health Administration and Education (NIHAE) to run the program. Ensminger even agreed to pay for half the cost of constructing the headquarters.[13]

Even more Ford Foundation money went to hiring consultants to work side by side with Indian officials, typically for five years or more. By 1966 there were seventeen on long-term contracts, most working in the Health Ministry's Department of Family Planning. They monitored Ford projects and identified new funding opportunities, giving them leverage among their Indian colleagues while also reinforcing ties to their paymaster. Even before matters reached this point, a senior Ministry of External Affairs official noted that they were "watching with anxiety the increasing penetration and power of Foundations like the Ford, Rockefeller, and Nuffield in governmental spheres."[14]

Yet, from the foundations' point of view, the penetration sometimes seemed to be coming from the other direction. The Population Council, for instance, found that Indian officials were able to override its recommendations and see to it that fellowships were awarded according to seniority. Nayar retained control of the budget and staffing of the new institutions that Ford helped to establish, and a protracted bureaucratic struggle ensued. Ensminger had an ally in the minister of finance, T. T. Krishnamachari. But Krishnamachari so hated Nayar that he refused to approve the new funding needed to launch the intensified districts program. The Ford consultants therefore "floated around creating more problems than the rest of us could cope with." Ensminger was particularly incensed when some of them began to defend Nayar's position upholding health priorities against their counterparts embedded within the NIHAE. Rather than just penetrating the Indian government and propelling a more intensive population control program, Ford consultants had parachuted into a war among Indian bureaucrats.[15]

The struggle centered on a simple question: Was contraception a part of comprehensive health care, or was health care a vehicle for population control? There were tactical advantages to integrating contraception within public health programs. One of the most influential Ford consultants, Lyle Saunders, pointed out that providing services through clinics "latches on to strong, positive health motivations." Moreover, birth control could be "introduced gradually without upsetting anything, arousing opposition." The problem was that health ministries and health professionals operated according to a medical model centered on the doctor-patient relationship. Population control required looking past the individual in pursuit of a greater good: rapidly reducing fertility before it overwhelmed economic and social development.[16]

One way to reconcile this conflict was to treat high fertility as a disease. Here again, eugenicists had shown the way. For Lord Horder "indiscriminate child-bearing" was a "disease of the body politic," analogous to tuberculosis or cancer. Julian Huxley was among the first to warn that "man will turn into the cancer of the planet" after seeing crowds gathered on the banks of the Ganges. India's own ambassador to the United States thought overpopulation had to be fought with the same zeal as malaria and cholera.[17]

But when population controllers tried to "treat" whole populations, as populations, the devil was in the details. Nowhere was this more apparent than in contraceptive research. In the early 1960s the pill was little used for population control. The program in Pakistan, one of the few exceptions, was plagued with administrative problems. Oral contraceptives were still expensive, "acceptors" had to accept using them every day, and administrators risked disaster if they ran out. Moreover, by 1962 worries about side effects like thromboembolism and even cancer had begun to mount. Some countries, including India, would not permit their use.[18]

For Alan Guttmacher, chief of obstetrics at Mount Sinai Hospital in New York and new president of Planned Parenthood–World Population, all their methods shared the same defect: they were "birth control for the individual, not birth control for a nation." Discovering some means to make people "immune" to pregnancy was the "final goal." But in the meantime, as head of the Population Council's medical committee, Guttmacher worked to rehabilitate a low-tech and heretofore discredited method: the intrauterine device (IUD). No one understood how a ring, loop, or spiral inserted into the uterus worked to prevent pregnancy, and the inventor of the first IUD believed it acted as an abortifacient. IUDs were also fraught with risks. Insertion required dilating the cervix, which entailed either significant pain or local anesthetic. Removing them, typically with a long, slender hook, could scratch or even pierce the uterine wall. On the other hand, leaving a "tail" dangling through the cervix—both to ease removal and ensure that it was still in place—could provide a pathway for bacteria, and thus pelvic inflammatory disease (PID). This painful infection of the upper genital tract sometimes required a radical hysterectomy. While some doctors still used IUDs, especially in Japan, in a quarter of a century only one of them had dared to write up his experience in a Western medical journal.[19]

In 1958 one of Guttmacher's colleagues at Mount Sinai approached him with a new kind of IUD. It was a spiral molded in plastic. As such, it could be fed into a tube narrow enough to fit through an undilated cervix, then pushed out the other end, where it would become a spiral again. Even if infection occurred, new antibiotics now made treatment easier. Guttmacher agreed to give it a trial run. The results were encouraging, so much so that in 1962 the Population Council hastily convened a conference to compare the experiences of doctors in different countries. The IUD was judged more reliable than the diaphragm and the condom, and it was preferred by many patients. But others experienced pain and prolonged bleeding, or spontaneously expelled it. Even these doctors, who had championed this controversial technique against all received wisdom, generally agreed that IUDs should be inserted only by a trained physician after a thorough examination. It is now accepted that such exams should exclude women with a previous history of PID, gynecological bleeding disorders, or a congenital uterus abnormality.[20]

This is not what Guttmacher wanted to hear. Obtaining a patient's medical history "would make a more time-consuming job out of intra-uterine contraception," he objected. "We dare not lose sight of our goal—to apply this method to large populations." In countries like India, Guttmacher wanted fewer restrictions and many more people inserting the devices, including nurses and midwives. One conference participant, J. Robert Willson, chair of Obstetrics and Gynecology at Temple University, emphatically agreed with Guttmacher's approach to "worldwide population control." "We have to stop functioning like doctors," he said, "thinking about the one patient with pelvic inflammatory disease; or the one patient who might develop this, that, or the other complication from an intra-uterine device." In fact, Willson observed, "it may well be that the incidence of infection is going to be pretty high in the patients who need the device most." Of the thirty-seven patients in Willson's own study, thirty-one were black.

Now, obviously, if we are going to use these devices, they are occasionally going to be put in the wrong patient. Again, if we look at this from an over-all, long-range view (these are the things that I have never said out loud before and I don't know how it is going to sound), perhaps the individual patient is expendable in the general scheme of things, particularly if the infection she acquires is sterilizing but not lethal.

Mary Calderone, medical director for Planned Parenthood, said that she was "thrilled" to hear a clinician like Willson framing the problem in terms of population control. "For any contraceptive method at all," she said, "we must think in terms of mass distribution."[21]

Christopher Tietze of the Population Council proposed a compromise. Proving that inserting IUDs was "not the devil's work" to skeptical doctors would be their first objective, even if their ultimate goal was "mass distribution." After the conference Tietze would take charge of a program that gathered and compared data from thirty-eight institutions and clinicians in private practice. "There was such a feeling of urgency among professional people," Tietze recalled, "not among the masses, but something had to be done. And this was something that you could do to people rather than something people could do for themselves. So it made it very attractive to the doers."[22]

In October 1964 the Population Council sponsored a second conference ten times as large to report on the first year of Tietze's Cooperative Statistical Program (CSP). It was an impressive data set, including almost seventeen thousand women. But their selection and treatment varied widely, including several different kinds of IUDs. Tietze reported that in the first year about 15 percent of the participants expelled the IUD and another 11 percent had it removed—two-thirds of them because of pain or excessive bleeding. Moreover, 1.7 out of every 100 women developed PID. Was the rate of infection the same as in the general population? Given the varying protocols of the different researchers, it was impossible to know. Guttmacher actually wanted some researchers to ignore subjects' medical history so as to determine whether they could do the same in countries like India. Moreover, none of the rates computed for expulsion, removal, or side effects included patients who did not show for follow-up examinations. Their absence may have skewed the overall conclusions.[23]

In a memorandum Tietze obviously wanted on the record, he wrote to Sheldon Segal, head of the Population Council's biomedical division, that both had agreed that the "CSP is primarily designed to furnish data on effectiveness and acceptability in terms of pregnancy rates, expulsion rates, and removal rates. The CSP, as it is now set up, should not be expected to furnish the required information on PID and exfoliative cytology"—that is, tumors. By this point, Tietze and Segal had received some disconcerting reports. A Johns Hopkins physician directing a project in Pakistan noted, "It has become well known that the IUD causes bleeding; and of course every woman who is unhappy to be pregnant is looking for some safe and easy way of starting bleeding." Tietze privately admitted that the CSP data lent qualified support to the hypothesis that IUDs caused abortions.[24]

Three sizable studies from outside the Tietze program with data on side effects were also presented at the conference. Yet the findings were so varied—including wildly different rates of side effects, expulsion, and removal—that they merely provided further proof that researchers could not know what would happen if IUDs were introduced on a mass scale. Some participants tried to warn that the speed and ease of inserting many women with IUDs was deceptive: as the number with side effects and complications mounted, the demands on medical staff would start to "snowball."[25]

Two weeks after the conference, the IPPF issued a press release announcing that "the effectiveness, acceptability and safety of the I.U.D.s have now been demonstrated" and recommending their use by member organizations. The IUDs could be inserted by "specifically trained midwives and nurses under the supervision of a doctor, if this appears appropriate in a particular country." Moreover, the press release declared, "There is no evidence that it acts by affecting the implanted ovum."[26]

Up to this point the IPPF had been conservative in evaluating new contraceptives. Under Helena Wright, a critic of population control, the IPPF Medical Committee had favored barrier methods and was slow to accept the pill. But after a 1963 reorganization, Guttmacher had been put in charge. He was therefore in a position both to oversee the IUD evaluation subcommittee and to endorse its recommendations. The evaluation was led by Howard C. Taylor of Columbia University, who was also chair of the Population Council's Advisory Committee on Intra-Uterine Contraception (which included Guttmacher as a member). Taylor was then negotiating a multimillion-dollar grant from the Ford Foundation, the Council's main backer. These individuals and institutions were so entwined and so invested in the IUD that they could not possibly conduct independent and critical evaluations.[27]

When Guttmacher concluded the conference, he said that he left it with "a great sense of optimism. I leave it with a feeling that we can now talk about controlling a population." Indeed, he found it necessary to reassure the chairman of G. D. Searle—a regular contributor to the IPPF—that IUDs would not reduce sales of the contraceptive pill. He explained that they were intended for a very different clientele. Whereas if someone like Mrs. Searle or Mrs. Guttmacher had an accidental pregnancy, it would create "quite a stink," reducing the *overall* birth rate in India or Korea would be "an accomplishment to celebrate."

> As I see it, the IUD's have special application to underdeveloped areas where two things are lacking: one, money and the other sustained motivation. No contraceptive could be cheaper, and also, once the damn thing is in the patient cannot change her mind. In fact, we can hope she will forget it's there and perhaps in several months wonder why she has not conceived.

When Guttmacher and Taylor issued their recommendations on how IPPF members should use the IUD, "no specific advice was suggested as to initial and subsequent re-examination."[28]

Rockefeller along with Notestein—who now headed the Population Council—did not even wait for the IUD evaluation program to get under way before they started promoting it abroad. It promised to greatly simplify what had been the biggest challenge so far: creating infrastructure adequate to service a nationwide program. They badly needed a success. In Pakistan, for instance, Notestein worried that the Council project might "go down the drain" if they did not find a better way to organize delivery.[29]

In a March 1963 meeting, president Ayub Khan told Rockefeller that, because there was no effective method available, it made little difference. That's when Notestein showed him the Population Council's new IUD: a plastic, loop-shaped model designed by a Buffalo obstetrician, Jack Lippes. "This intrigued him vastly." Ayub said that he would have every midwife in Pakistan inserting them within two weeks. They "had to slow him down," but the president insisted he was "in a hurry." At the end of the meeting, they could not persuade him to relinquish the device. But there would be plenty more where that came from. The Council undertook to manufacture and distribute the "Lippes Loop" worldwide, investing more than two and a half million dollars by 1968.[30]

The IUD progressed most rapidly in Taiwan, with unofficial government support, and in South Korea, where the military regime of Park Chung Hee made population control a national priority. Both of these pro-

grams were largely inspired by consultants like Bernard Berelson and Sam
Keeny of the Council and Ronald Freedman of the University of Michigan Population Studies Center, one of several new Ford-funded institutes.
Rather than take the time to hire new personnel and set up clinics, they
paid recruiters and doctors for every IUD inserted. But Notestein worried
about potential abuse. He was perplexed to discover that an experimental
area in South Korea subject to more intensive efforts and monitoring
showed no better results than the national program. As he confided in his
diary: "This being high government policy, the government being very very
strong indeed, I suspect that we may get some suppression of difficulties
that do occur."[31]

Meeting with South Korea's planned parenthood association, which
trained the doctors and motivators, Notestein warned that the Council
would "pull out immediately at any signs of graft creeping in." In Taiwan,
Freedman held the entire city of Taichung under close watch. Some thirty-
six thousand households were mapped and tracked to determine the effectiveness of more versus less intensive methods in reducing birth rates. Surveillance techniques first used in the Khanna study were now being applied
to much larger populations in urban settings. Even before the data were analyzed, Taiwan and South Korea were being held up as examples for the rest
of the world. With Ford putting up the money, the Council sending supplies and consultants, family planning associations training medical staff
and motivators, and the prime minister issuing marching orders to every
ministry, they provided a model for how to control whole nations.[32]

Proponents of population control could afford to think big. At a single
June 1962 board meeting with John J. McCloy in the chair, the Ford Foundation had approved $10.7 million for population projects. This was more
than all of its support combined over the preceding decade. A big slice of
this expanding pie continued to go to the Population Council, which had a
budget of $4.2 million and a staff of sixty by 1964. The IPPF was the poor
relation, and was still spending less than a million dollars in 1965 (not
counting the national budgets of individual members). But the resources
available to its London central office were growing at an even more rapid
rate—doubling every two years—largely because of prodigious fund-raising
in the United States. The Management and Planning Committee resolved
to hire a full-time "director-general," Colville Deverell, a thirty-one-year
veteran of the British colonial service. But the IPPF's main asset was still its
international network of associations. More than 350 delegates from forty-

two countries would come to the Seventh International Conference on Planned Parenthood in Singapore in March 1963.[33]

Population controllers began to foresee a time in which they would have programs in virtually every country, dispose of all the money they could spend, provide contraceptives to everyone who wanted them, but *still* fail to rein in population growth before it was too late. Officially, the IPPF insisted that if enough people were given access to birth control, the population problem would take care of itself. But Hugh Moore argued that it was time to reconsider, and wanted compulsory measures put on the Singapore conference agenda. The Management and Planning Committee thought that it was "quite clear that this subject is unsuitable for public discussion" and reaffirmed the principle of voluntary family planning. Even the idea of accepting voluntary sterilization as a family planning method was defeated when put to a vote.[34]

Yet the most senior IPPF leaders were moving out ahead of their membership, as shown by their participation in the First International Conference on Voluntary Sterilization in New York in April 1964. It was sponsored by an old eugenics lobby that Moore had helped to revamp and reenergize, the Association for Voluntary Sterilization (AVS). Both Rama Rau and Ottesen-Jensen, who had once been hostile to sterilization as a birth control measure, said that they had changed their minds. Draper, C. P. Blacker, and B. L. Raina, India's director of family planning, also participated. The ubiquitous Guttmacher chaired the AVS medical and scientific committee.[35]

But the real standout at the conference was an economist from the Rand Corporation named Stephen Enke, who would shortly become deputy assistant secretary of defense. He was worried about growing tension between rich and poor nations and the "danger of forced immigration." A family planning official from Madras had already described their success in performing 150,000 sterilizations in four years, which he attributed to the use of incentive payments. Enke now argued that the Madras program did not go nearly far enough. He recalled being shocked to discover the payments were only $6. Admitting that he was "one of the nasty economists," Enke calculated that preventing births could increase India's per capita GNP, which was then approximately $100 a year. India should therefore pay young parents who agreed to sterilization $250. Seeing a smile play across the face of Oscar Harkavy, the head of the Ford Foundation's new population office, Enke said the foundation should "go to it."[36]

Before Harkavy could reply, Raina said he was "very much shocked." India had already made tremendous progress in its family planning program. He did not think people would agree to be sterilized for money, and if they did it would be the kind who did not understand what they were agreeing to. But this seemed to provide another reason to "go to it." One AVS board member, Lester Cushing Rogers, CEO of an international construction firm, said he had found like-minded people in his travels through Latin America, equally concerned that a growing proportion of the population was "entirely uneducated, many so ignorant that they scarcely know the language of the country. They may be from Indian tribes," he offered, "they may simply be semi-moranic [sic]. The slums on the hillside are simply frightful." Rogers therefore agreed that for such people only money talked.[37]

Other conference participants proposed an experimental program in which mass communications would be "beamed" at the "heartland of Appalachia" to encourage sterilization. In Virginia and North Carolina eugenic laws were already "propitious." Success could demonstrate that compulsory programs were unneeded. In fact, in the late 1950s and early 1960s, legislators in nineteen states proposed measures to sterilize unwed mothers. Even without them, most North Carolina counties already targeted women on welfare for sterilization. With the appearance of Daniel Patrick Moynihan's report on high fertility and illegitimacy rates among African Americans, the clamor to use population control to cure "pathological" poverty would only increase. As Hugh Davis, the inventor of the Dalkon Shield, would later explain, the popularity of birth control pills among middle-class women presented a "real hazard," because other women with less motivation ("the individual who needs birth control the most—the poor, the disadvantaged, and the ghetto-dwelling black") were more likely to become pregnant. Similarly, when he introduced incentive payments, the chief secretary of Madras explained that only sterilization would work for "the large mass of the people who will not space their pregnancies or limit their number except as a result of Governmental action."[38]

What was contemplated was not, therefore, just targeting poor countries, but rather the sterilization of poor people worldwide, including in the United States. And for some, in India as much as America, the reason was differential fertility and its eugenic consequences, not just population growth per se. But as C. P. Blacker argued at the AVS conference, given India's relatively greater need, it had to provide leadership. The conference

This fashionably dressed Thai woman describing the IUD was the model for the hand-drawn "happy family" poster in the background. It often proved impossible to overcome the evident class difference between those who promoted family planning and their "targets." IPPF Archives.

led to the creation of an *International* Association for Voluntary Sterilization. India's minister of health and family planning, Sripati Chandrasekhar, eventually accepted the chairmanship.[39]

In the meantime, the industrious Enke prepared analyses for USAID on how paying for IUD insertion and sterilization could be more effective than conventional kinds of development aid. In April 1965 one such study landed on the desk of a senior National Security Council staffer, Robert Komer. It was excellent timing. Since Lyndon Johnson had succeeded JFK in November 1963, he had grown increasingly unhappy with the U.S. foreign aid program. Its two largest recipients, India and Pakistan, were on the brink of war. Neither country supported U.S. policy toward China and Vietnam, yet both seemed to expect this aid would continue indefinitely. India had become dependent on American grain shipments and had just requested another fourteen million tons over two years. Johnson was concerned that back-to-back visits by President Ayub and Prime Minister Lal Shastri—Nehru's successor—would spark new criticism in Congress, and perhaps jeopardize all U.S. foreign aid. He therefore canceled both visits, sending the message that neither country could take U.S. support for granted.[40]

Until this time Johnson had shown little inclination to make population control a larger part of U.S. aid. In his state of the union address he had promised to "seek new ways to use our knowledge to help deal with the explosion of world population and the growing scarcity of world resources." But he agreed to this only because he did not want to accede to pressure from the population lobby for a presidential commission. Johnson would not even meet with Rockefeller and Draper, despite repeated requests and a personal recommendation from Dean Rusk. While they were gaining allies in Congress, and Senator Ernest Gruening would shortly commence thirteen days of hearings to publicize the cause, population control still seemed like a tar baby. A close aide, Jack Valenti, advised that it was "not a matter that the President wants to visibly touch at this time."[41]

If proponents of population control were to bring the president around, they needed a new argument that would appeal to his political instincts. That is just what the Enke study provided. Komer passed it on to McGeorge Bundy, Johnson's national security advisor. "Here's a little flank attack that I think might just penetrate LBJ's defenses," he wrote. "It's a hard dollar and cents argument for taking a more serious view of birth control in the [less developed countries]." Considering the impact Enke's calculations had on U.S. policy, they require a closer look.[42]

Enke posited that any increase in GNP was the product of more workers, increased capital stock, or new innovations that boosted productivity. He assumed that every country that simply added workers to the labor force would reap diminishing returns. But Enke believed that countries already "overpopulated" would gain even less from new manpower. For him, such countries were also distinguished by the absence of innovation. With these assumptions it was easy to show how, in a hypothetical case, a 12 percent increase in labor would result in only a 6 percent increase in output, and hence lower per capita GNP. But was it really true that countries like India did not produce innovation, and that more workers would be less productively employed? Enke did not say, nor did he try to show any correspondence between his mathematical formulae and real-world data.[43]

In fact, Enke showed scant understanding of "overpopulated" countries. As he further refined his calculations, he assumed that from the time they were born until the age of 15 people were simply "too young to be anything but consumers," aside from "a few chores around the family dwelling." This was hardly true of most agrarian societies, where children worked from an early age, and not just to take out the garbage. And if they did spend more time in school, would they not be more productive and innovative when they entered the workforce? In fact, Enke conceded that returns on investment in health and education "may be very great." Rather than pursue this, he simply assumed "misallocation."[44]

The crux of Enke's argument—what made it possible to argue that fifteen years of consumption during childhood would outweigh a lifetime of adult employment—rested on the concept of discounting. Economists point out that a gain is worth more to us if we can have it now rather than later, and a loss is felt less keenly as long as it is deferred. For certain kinds of analysis, future gains and losses are therefore discounted, just as a bond costs less than its face value until the day it is redeemed. For "overpopulated" countries, Enke used the high discount rate of 15 percent. When compounded yearly, it steadily reduced the present value of a child's consumption, even though a growing child would be consuming more food, clothing, and so on. Nevertheless, the amount of consumption that would be "released" had this child never been born still came to $279. In fact, it would have been even more if one did not also have to factor in the one-in-five chance that a child in an overpopulated country would die before consuming all these resources. As for the children who survived and worked all of their adult lives, their production was so far in the future that its discounted value was practically nothing. Enke did not even bother to calculate it.[45]

All of this might be dismissed as a statistical sleight of hand. Some argue that, however valid for individual decision-making, discounting is inappropriate for public policy. The factors that explain our preference for short-term gains, including immediate gratification, should not apply when we are also thinking about our children and grandchildren. Everyone who has a child and sends him or her to school accepts a lower "per capita GNP." And with a sufficiently high discount rate, no one can justify the child's existence economically, least of all the far more needy youth of rich countries. Using the same logic, one could make an economic argument against vaccinating for childhood diseases. After all, diphtheria and whooping cough also "release" consumption that might be profitably invested.[46]

Nevertheless, assigning a negative value to an individual life allowed Enke to argue that a $4 vasectomy would have an impact on per capita

GNP equivalent to $1,000 invested in industry or infrastructure. Enke thought that if people did not volunteer, governments should pay incentives. They could target harder cases and calibrate the payment according to age, religion, and occupation in order to prevent the maximum number of births. For Enke, this was merely a transfer payment that imposed no cost on society as a whole (never mind the political price paid to collect taxes, and the opportunity costs of not investing in health or education). In fact, such a payment rewarded those who did not commit the "anti-social" act of bearing children. He suggested as much as $325 for sterilization, or $30 a year for those who agreed to forgo childbearing with an IUD. According to Enke, any objection to buying and selling the right to bear children was "atavistic."[47]

Although it featured the mathematical formulae of cutting-edge economic modeling, it was Enke's modest proposal that could be seen as a throwback. For decades eugenicists had offered hard dollars-and-cents arguments for paying people not to reproduce. Here is how Sanger put it when addressing the international birth control conference of 1925:

> If the millions upon millions of dollars which are now expended in the care and maintenance of those who in all kindness should never have been brought into this world were converted to a system of bonuses to unfit parents, paying them to refrain from further parenthood, and continuing to pay them while they controlled their procreative faculties, this would not only be a profitable investment, but the salvation of American civilization.

The idea had changed little, but proponents were now far closer to real power.[48]

When Komer made his case to the president, he scaled back some of Enke's claims, at the same time underscoring them. "While you're thinking about foreign aid," he began, "here's a fascinating statistic. A recent study claims that if economic resources in many [less developed countries] were devoted to retarding population growth rather than accelerating production growth, *these resources could be 100 times more effective in raising output per capita!*" The figures were "just one good economist's," Komer conceded. "However, even if they're off somewhat, there's no doubt of the rapidly declining cost of population control because of new devices." This could have "immense significance" for India, Pakistan, and other recipients of U.S. aid. "The process of getting these countries to the stage of self-sustaining growth, *and thus reducing the longer term foreign aid burden on us*—could be greatly foreshortened."[49]

Komer did not mention Enke's proposal to pay individuals incentives to undergo sterilization. It had given him even bigger ideas: "the relevance of the figures like the above to the achievement of our foreign aid goals is so striking that you may want to consider ways and means of gradually using our foreign aid more as an incentive to major efforts in this field by the less developed countries themselves." Johnson said that he wanted to hear more. We don't know what more he was told. But the next day Komer wrote to Bill Moyers, also known to be "a bug on this," that he was "delighted with what you told me, and I'll keep mum."[50]

Two months later, at the twentieth-anniversary celebration of the United Nations in San Francisco, Johnson had this to say:

> Let us in all our lands—including this land—face forthrightly the multiplying problems of our multiplying populations and seek the answers to this most profound challenge to the future of all the world. Let us act on the fact that less than five dollars invested in population control is worth a hundred dollars invested in economic growth.

While further discounting Enke's claims, and without necessarily understanding their justification, the president had signed off on the idea that children in poor countries were a burden to the world. Even more important, he now insisted on personally approving every new food shipment to India, typically a month's supply, in a policy that came to be known as "the short leash."[51]

Indian diplomats—and even Johnson's own aides—were left wondering what he expected them to do. A complex and interlocking set of issues divided the two countries. India's perennial dispute with Pakistan—which broke out into fighting along their disputed border in May 1965—was driving America's longtime ally into the arms of Communist China. Washington also wanted to prevent Delhi from developing nuclear weapons, which would make the situation even more explosive, and persuade it to lend support for America's war in Vietnam. But "wise men" like Dean Acheson advised Johnson that India could not be starved into submission on such issues. Instead, the United States would use the short leash only in matters where their interests were linked, but where Delhi had to be pulled in the right direction. The president and his advisors therefore began to focus on the idea of "self-help," compelling India to develop an economic program that would reduce its need for U.S. aid, and that included population control.[52]

The United States had to coordinate with a "consortium" of other donors, led by the World Bank. Its president, George D. Woods, also wanted many things from India, including devaluation of the rupee and easing of import controls. But he was keenly interested in population and sent an advisory mission. Even the UN Bureau of Social Affairs was eager to give advice. India's representative to the Population Commission, deputy chairman of the Planning Commission Asok Mitra, worked with his U.S. and Swedish counterparts to obtain authorization for another mission. It would be headed by IPPF director-general Colville Deverell and include Leona Baumgartner of USAID. Nine more Ford Foundation consultants assisted a Planning Commission study. It too would offer recommendations. Soon Delhi would be crawling with foreign experts.[53]

As one Ford consultant noted, "Much time is spent telling visitors and one another what is wrong with the program. Everyone has a diagnosis!" Officials had hardly begun to implement all of the changes recommended after the last major evaluation, in October 1963. Many in the Ministry of Health still did not agree with the idea of extending family planning services beyond medical clinics. Even with total unity of purpose, just hiring enough personnel to reach the remotest areas was a mammoth task. The plan called for training 49,000 auxiliary nurse midwives by 1967 along with a commensurate number of doctors, health visitors, educators, and so forth. Some state programs were already cutting corners. In the state of Kerala, for instance, doctors received only two days of training before they started performing sterilizations. In a follow-up study, "a substantial percentage" of their patients reported complications like pain, weight change, or lessening of sexual desire. It was not possible to conclude whether sterilization was responsible, because no thorough medical exam had been done beforehand. After the Health Ministry received reports of fatalities, Nayar reminded states that only trained doctors should undertake the procedure. By the end of the third five-year plan, in 1966, 42,000 people had received some kind of family planning training, including 7,000 doctors. But many areas were still woefully understaffed. Given India's federal structure, officials in Delhi could do little if state health departments did not share their goals.[54]

Federal officials were themselves shorthanded. Though responsible for a

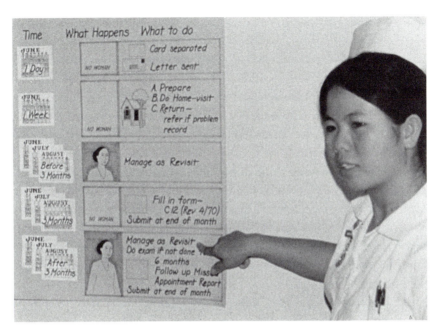

As population programs ramped up, hiring and training sufficient staff posed a growing problem. Hundreds of thousands of workers had to be taught to keep careful records and relentlessly pursue "targets" until they became "acceptors." IPPF Archives.

budget that was three hundred times larger than in 1957, the family planning staff in Delhi had grown hardly at all. They had virtually no modern office equipment—only desks overflowing with old files. New personnel, the review by the Ford Foundation and the Planning Commission noted, seemed to "sink in the murky waters of papers which should long ago have been disposed of." The entire office was weighed down by the traditions of the Indian civil service: valuing seniority over competence; demanding prior approval for any improvisation; requiring detailed accounting of even the tiniest expenditures; and insisting on equal treatment for all states—such that no one could receive research funds or supplies or transport if it was not made available to all. No request, however small, was answered quickly. Supplying audiovisual equipment to thirty-three training centers, for instance, took more than two years. The World Bank team requested many reports on the family planning program during the month it spent in India. Commissioner Raina was unable to produce even one.[55]

Everyone had a different diagnosis for what was wrong, because there were so many reasons to choose from. But there was little disagreement among these experts about what should be done. Outside the Ministry of Health, "everyone begs that we should not be too polite and gentle," Deverell reported, and most accepted the need for "changes at the top." In essence, this meant creating within the ministry an independent power center that would control budgets and staff and concentrate solely on meeting targets. As Deverell's UN team put it, "the programme may otherwise be used in some States to expand the much needed and neglected maternal and child health services."[56]

The expert reports all emphasized the need to leave behind the medical model of family planning. The team from the Ford Foundation and the Planning Commission advised that India should move training programs out of medical colleges and undertake a "radical revision" of the old curriculum, with its talk of "'prescribing' contraceptive methods to 'patients.'" Everyone liked the use of camps and mobile clinics. Of course, some methods, like female sterilization, still required doctors, many doctors still worked in hospitals, and hospitals still had to treat sick people. So consultants from both the Ford Foundation and the UN called for a "strong policy" requiring large hospitals to reserve beds for sterilization. This would "avoid delay and consequently possible loss of motivation." Considering that in most Indian hospitals the small number of maternity beds were the only ones available to women, this would further reduce the scant resources devoted to women's health.[57]

While they advised against an overreliance on any one method, and the UN team called for a tenfold increase in the rate of sterilizations, all of the expert committees insisted on immediate adoption of the IUD. This was a foregone conclusion, considering that the Population Council had coordinated with both Ford and the World Bank in selecting all these consultants. Whether the Indian Health Ministry would agree was another matter. Sheldon Segal had had to smuggle IUDs into India disguised as Christmas tree ornaments in order to begin clinical testing. When Nayar got wind of it, she had threatened to have the responsible Indian official brought before a disciplinary council. Guttmacher and Baumgartner managed to bring her around. Two-thirds of the Indian clinical studies had only begun in 1964, and nearly half of the investigators had not yet submitted any data. But the Ford Foundation paid for Indian medical research as a way to win support from doctors for family planning, not to prove the safety or appropriateness of the IUD in different settings. The objections of the testers were quickly overridden.[58]

The Population Council sent more than one million IUDs to India before a factory was constructed on-site. They came in nonsterile form, and the Council provided only one inserter for every twenty IUDs. Personnel would therefore have to be trusted to sterilize both the device and the inserter before every procedure. The UN team did not mention the issue. Instead, it recommended that "initial training for the Reinforced Programme should be reduced to the bare minimum, and staff should be sent into the field to gain experience, and return for further training later on." For Deverell, the IUD's major advantage over other contraceptives was that it "can be inserted once and then can be forgotten about for an unlimited period," relieving women of "a great number of agonising decisions during the course of any one month." The experts all agreed on the need for performance targets for every district in every state. And to meet them, they also accepted the "emergency need for promotional incentives," as the Ford–Planning Commission report put it, especially considering that every "birth averted" represented a "saving to the nation."[59]

The Indian government had a tremendous incentive to accept this advice, which came from representatives of USAID, the World Bank, the United Nations, and the Ford Foundation. Together these organizations provided most of India's annual $1.5 billion aid package. By September 1965, when India and Pakistan went to war over Kashmir, officials in Delhi were unnerved by their vulnerability. Daily food rations in Calcutta had already been cut. "Right now 40 million Indians, most of them low income people living in large cities, are wholly dependent upon US foodgrains," Ambassador Chester Bowles reported. Any interruption of supply threatened famine. But Johnson insisted on putting his signature to every 500,000 tons of grain—"using food as leverage," as Komer put it, "by only dribbling it out slowly."[60]

A more effective population control program was only one of a number of things that Washington and the World Bank wanted from India. But devaluing the rupee, easing import controls, and shifting investment from industry to agriculture all required agonizing reappraisal of national plans and priorities. It was widely suspected that Washington really wanted to use its leverage to force India to settle the Kashmir dispute. Action on population seemed merely to require reshuffling the staff and budget of a single ministry so Colonel Raina could take charge. Minister of Finance T. T. Krishnamachari, Minister of Agriculture Chidambaram Subramaniam, and Minister of Planning Asoka Mehta all favored a more forceful population control policy. Now was their moment to move decisively against Health Minister Nayar.

The cabinet first created a committee on family planning, where in monthly meetings Krishnamachari, Subramaniam, and Mehta could gang up on her. The Planning Commission took the lead in spelling out what Nayar had to do. Prompted by an advance look at the World Bank report, Asok Mitra noted that "none of the recommendations and targets are either new or unanticipated." To meet these targets, Mitra emphasized, "the guts of the matter is administration."

> Where the Planning Commission should insist would be to hold the Ministry to its proclaimed time and physical targets. To be able to fulfill them, very large scale expansion of the entire machinery all down the line, an enormous widening of the base, and real stiff-ening of the administrative machinery will be required. The [Family Planning] Commissioner's writ must run swiftly and unimpeded all down the line.[61]

Though the ministry had only begun IUD insertions a few months ear-lier, by 1970–71 Mitra wanted twenty million women to be using them. India would also have to perform over a million sterilizations annually. To achieve these goals every Primary Health Centre would have "a realizable physical performance target." But mobile units and camps would be the mainstay of the program. "It should be possible for [the] IUCD campaign to forge ahead of the [Rural Health Centre] programme and not depend upon it," Mitra wrote. Perhaps anticipating the consequences, he noted that, while studies had shown some women would spontaneously expel IUDs or request their removal, "with the expansion of the programme, these rates will be higher."[62]

Waging war became the favored metaphor for India's new approach to population control. Officials in Punjab, which already had the highest rate of sterilization, announced that they considered themselves "on a war foot-ing." While fighting raged across the border with Pakistan, IUD insertions continued, totaling sixty thousand by December 1965 in just this one state. Martial metaphors also meant that some would be sacrificed. As Mehta put it, population growth was "the enemy within the gate. . . . It is war that we have to wage, and, as in all wars, we can not be choosy, some will get hurt, something will go wrong. What is needed is the will to wage the war so as to win it."[63]

Spending on family planning reached 110 million rupees in 1965–66, an increase of more than two-thirds over the previous year, while the rest of the health budget grew not at all. It already swallowed up nearly 13 percent of the ministry's expenditures and would keep on growing, totaling 255 million rupees by 1967–68. Population control was imperative in time of war, Nayar explained, because "it is the quality of people that decides the result of a battle and not mere numbers."[64]

Program strategy also shifted to reflect the new line. While the ministry paid lip service to the "cafeteria approach," giving people a choice between several methods, it also told state health ministers that the superiority of the IUD and sterilization was now "quite obvious." The Health Ministry directed states to set targets and concentrate on densely populated areas. For instance, all maternity homes, maternal and child health centers, and outpatient departments with women doctors were to start inserting IUDs, "giving it to at least 50 per cent of cases after delivery." Targeting women immediately after childbirth had long been a favored tactic of Howard Taylor's. More than a decade earlier, he had described his success in sterilizing women in Puerto Rico by explaining that they "are most susceptible to the influence of physicians at the time of delivery." IPPF director-general Deverell did not see any problem with targeting postpartum women. "If we can set up some form of simple reporting agency," he speculated, "which will identify the target as it were and bring the target into the scope of the introducing machine then much of our problem will be over."[65]

But in the summer of 1965, even before India had assembled its "introducing machine" and started targeting millions of Indian women, major problems with the IUD were coming to light in the Population Council study in Taichung. In the 13–24 age group—girls as young as 13 were apparently inserted with IUDs—28 percent had asked that they be removed after just a year because of pain or excessive bleeding. Within two years, 23 percent expelled them spontaneously. By this point fewer than half of those in all age groups still retained the devices, and almost 10 percent had become pregnant—pregnancies that would now be significantly riskier.[66]

Because the Taichung experiment had started earlier, it provided forewarning of the problems that would soon become widespread in the national programs in Taiwan and South Korea. In both countries, "acceptors" had accepted willingly. They had received no incentive payment and, unless they were indigent, had to pay half the cost. Moreover, health services were well organized compared with those in India. Extrapolation of the

Taichung data indicated that tens of thousands of women would now be coming to these clinics for pain or bleeding, pelvic inflammatory disease, ectopic pregnancies, or septic abortions. In India, where mobile teams would bribe or coerce millions of women into accepting IUDs, the number would be orders of magnitude greater. How could a family planning program so understaffed and poorly trained, having "forged ahead" of rural health services, possibly cope with the consequences?

Most Indian women had no idea what risks IUD insertion posed. In Punjab, for instance, village leaders were instructed to tell people that "contraceptives are harmless." Even top officials seemed completely unprepared for what was coming. Mukherjee described early reports of bleeding as a local administrative problem. His boss professed not to believe that there was any problem whatsoever when she addressed the first meeting of the new Central Family Planning Council, even though some of the state health ministers present reported the same difficulty. "Rumours regarding its harmful effects are baseless," Nayar insisted. She had to produce bimonthly reports for the Planning Commission showing that her ministry was meeting targets. The new IUD plant—built according to Population Council specifications and funded by the Ford Foundation—was ramping up production to twenty thousand devices a day.[67]

At the time, most people were less worried about the family planning program than an impending food crisis. Some anticipated a shortfall of as much as twenty million tons, and there were already sharp price rises, hoarding, and black marketeering. Komer told Johnson that starvation was now inevitable. LBJ agreed to supply another one and a half million tons of wheat and to meet with Agriculture Minister Subramaniam when he came to Washington. The United States, Johnson told his visitor, was "not interested in disciplining anyone, in becoming the masters of anyone, or in dominating anyone." America had its own problems, he admitted. But he wanted to achieve new results in food and population both at home and abroad: "We would exercise whatever persuasion we could toward these ends. We wanted to provide incentives too."[68]

Johnson's advisors were now afraid to approach him with any new requests for food aid. "Frankly, what worries me, as a planner," Walt Rostow wrote, "is that a good many human beings may starve in those critical months before the next harvest." In his memo to the president, Rostow noted that the last major famine, in 1943, left three million dead. On January 11, 1966, Prime Minister Shastri died of a heart attack in Tashkent at the conclusion of a successful peace summit with President Ayub. A week later, the day Parliament elected Indira Gandhi to take his place, Secretary of Agriculture Orville Freeman suggested that Johnson might pledge another one and a half million tons of food as a goodwill gesture. The president told Komer "to get Freeman to quit giving stuff away."[69]

Unlike Shastri, the new premier had a personal interest in family planning. When repeatedly pressed on *Face the Nation,* her predecessor had explained, "I have six children, so I am in no position to advise others on that." Gandhi, on the other hand, nearly died after the birth of her second son, Sanjay. She had wanted to donate her family's ancestral home in Allahabad so that it could become an Institute for Family Planning. As information minister, she had pressed a plan to distribute hundreds of thousands of radios across rural India to transmit family planning information. And Gandhi together with Rama Rau was also among those who had been pressuring Nayar to pay women to accept IUD insertion.[70]

Gandhi's interest in family planning was apparent in her first meeting with Ambassador Bowles. So too was her evident need for American help. Bowles told her that she could "relax," that she had many friends in the United States, but that good relations required three things: peace with Pakistan, genuine and positive neutrality in the Cold War, and "pragmatic economic policies . . . giving high priority to agriculture, education, and population planning." Gandhi replied that managing relations on this basis would be "an easy matter," promising to "press hard on such programs as family planning." On January 25, 1966, the day after she was formally sworn into office, the Ministry of Health was renamed the Ministry of Health *and Family Planning,* including a separate department with its own permanent secretary and minister of state.[71]

The president considered Bowles too easy on India, and he would have liked much more than neutrality vis-à-vis Vietnam and China. What Johnson did insist on, with an eye to how all of this would play with the public and Congress, was that India help itself. One of his advisors, Joseph Califano, suggested the United States commit more food aid before Gandhi came to Washington. "Johnson exploded all over my memo," he recalled. "No, *Hell* no." Califano got a call that afternoon, and before he could even say "Yes, Mr. President," Johnson yelled, "Are you out of your fucking mind?" He would not ease the pressure until India demonstrated real determination in population control: "I'm not going to piss away foreign aid in nations where they refuse to deal with their own population problems."[72]

For Komer, who had been the first to suggest that Johnson use food as leverage, and had just been promoted to national security advisor, Gandhi's visit was the most important international summit since Kennedy faced Khrushchev. "We finally have the Indians where you've wanted them ever since last April . . . coming to us asking for a new relationship on the terms we want." Through skillful management they had even avoided a backlash in the Indian press against U.S. pressure, "when it's been just that for almost a full year." Better still, "That tough-minded George Woods and the World Bank are with us." Woods would be "a great ally" in conveying the clear message that *"from now on we hinge aid to performance."*[73]

In all the papers that Johnson's advisors gave him to plow through before Gandhi's arrival, population control was only one subject among many. But it was always there, and it was one subject that always counted in Gandhi's favor. Thus, the local USAID administrator noted that, under Gandhi and Mehta's leadership, "more punch in very recent weeks is being added to the Central Government's family planning program." This was an essential part of any deal. Asking "What do we seek?" Rusk listed India's obligation to reorient its economy as the first of five points, and that required "a massive effort to *control population growth.*" Johnson did not have to insist. As Rusk noted, "*She* knows and *we* know that without tangible and continuing American interest in the future of the Indian Union, that Union does not have much of a future."[74]

There is no record of the conversation between Gandhi and Johnson when they met alone on the morning of March 28, 1966. But afterward Johnson was apparently satisfied. When he sent a message to Congress two days later requesting that it approve food aid for India, he reported, "The Indian government believes that there can be no effective solution of the Indian food problem that does not include population control. The choice is now between a comprehensive and humane program for limiting births and the brutal curb that is imposed by famine."[75]

In fact, India would suffer from both famine *and* a brutal program to curb population growth—the famine only added to its brutality. Shortly after Gandhi's return, Nayar agreed to a set of recommendations from a special committee under Mukherjee. Doctors would now receive hardship pay for working in mobile units, but also minimal performance standards: 150 vasectomies or 300 IUD insertions per month. State doctors based in clinics would also receive a bonus if they performed 75 vasectomies or 150 insertions. As for private practitioners, they were paid piecework-style: 10 rupees per vasectomy, and 5 rupees per IUD insertion. Where states were short of women doctors, the ministry would dispatch "loop squads" to ensure they met their targets.[76]

The Mukherjee report made only oblique reference to the program's growing problems, affirming that "systematic follow-up of the cases is of utmost importance." Yet like all of the foreign expert reports, there was no provision to ensure such follow-up. The UN committee was positively disingenuous on this point: "In the initial stage of the programme," its report declared, "it is especially important that nothing should go wrong. Women accepting a loop should be fully prepared for the possibility of *complications*. Arrangements should be made for a health visitor to call on them about one week after insertion." Considering that the UN encouraged camps, mobile clinics, and millions of insertions, and that there was only one health visitor for every forty thousand people, this was a ludicrous suggestion. The ministry's method of paying for state family planning programs discouraged better care. Thus, when officials in Uttar Pradesh asked whether they might receive funds to treat those who, because of counterindications like pelvic inflammatory disease, were found unsuitable for the IUD, they were told to absorb the cost out of the three rupees they received for each IUD insertion.[77]

Rather than providing some incentives to encourage follow-up, the Mukherjee report recommended paying every woman five rupees at the time of insertion. The minister agreed, in principle, though it was several months before this was communicated to the field. The report's authors would not, however, wait for the development of more adequate health services, citing the World Bank recommendation to "forge ahead." Govind Narain, the new permanent secretary for family planning, emphatically agreed that they "should not waste any more time."[78]

Indian officials were proud of the dynamism and boldness with which they pursued population control, and proudly told the world about it. "India's population explosion" was "a threat more serious than any military invasion," according to pamphlets distributed by the embassy in Washington. In a White House interview in May 1966, Asoka Mehta regaled President Johnson with their achievements and aspirations: "In 1965 there were more vasectomies than in the preceding 10 years. In five states targets for 'the loop' had been reached within five months. Twenty-nine million IUD's would be fitted within the next five years."[79]

The Population Council's senior staff and advisors, above all Notestein and Guttmacher, were in the best position to know that these targets were not merely unattainable, but positively reckless. As the main backer and co-ordinator of IUD programs all over the world, the Council was receiving regular reports of mounting problems. In June 1966, for instance, a Singapore postpartum project discovered in follow-up exams that 20 women out of 3,400 inserted with IUDs had suffered a perforated uterus—a rate fifteen times higher than anticipated. This was one of the most dangerous complications. These women had better care and diagnostic procedures than most, so investigators were "sure that there must be many cases of undiagnosed perforations in other programs." The next month Guttmacher learned that the rate of IUD insertions in Hong Kong had fallen off "rather shockingly" because of side effects like heavy bleeding and ectopic pregnancy. By August it was obvious that higher than anticipated rates of spontaneous expulsion or removal due to side effects were a systemic problem, common to IUD programs in the United States, Puerto Rico, Taiwan, South Korea, and Pakistan.[80]

The Population Council privately circulated this information to program administrators. They suggested that doctors do a better job educating their patients and perhaps be paid for follow-up visits. "The strange thing," Guttmacher remarked, "is that Nayar claims such magnificent results in India. Perhaps it is because follow-up is less complete." In fact, the monthly rate of IUD insertions in India had fallen by half since March—from approximately 120,000 to 60,000. In June Delhi received reports that in some areas nearly half of all those inserted were complaining of prolonged bleeding, "creating a very bad reaction." By October 1966 the pace was only a tenth of that required to meet the annual target, fewer even than the number of sterilizations, though that too had begun to level off. India's family planning program was turning into a fiasco.[81]

A few states seemed to show a way out of the impasse. Punjab, for instance, had been paying IUD acceptors, and it achieved 277 percent of its target for 1965–66. Madras instead concentrated on sterilization, with higher incentive payments for both acceptors and motivators than any other state—and the highest performance per capita, totaling almost three hundred thousand that same year. On October 27, 1966, Delhi finally accepted what Ford Foundation, UN, IPPF, and World Bank consultants had been recommending all along: everyone who agreed to be sterilized or inserted with an IUD could get a cash payment. Rather than set a nationwide pay scale, it provided states 11 rupees for every IUD insertion, 30 per vasectomy, and 40 per tubectomy (later increased to 90 rupees). Out of this sum, they could pay whatever incentives appeared necessary, whether to individuals, to staff, or to freelance "motivators."[82]

For individuals, 3 to 7 rupees was typical for an IUD insertion, and sterilization earned anywhere from 12 to 45 rupees (some states, like Madras, supplemented central funding to provide higher incentives). India's Family Planning Association also received compensation for every procedure, and local clinics sometimes made larger payments. Some Indian industrialists also got into the act. In 1967, Tata Industries, one of the country's largest conglomerates, began paying 200 rupees to every employee who agreed to sterilization. Even this relatively generous pay package, equivalent to $26.70—or $161 in today's money—might seem trivial. But at the time, 2 or 3 rupees was a decent wage for a day's work, and many people earned less.[83]

More significant than the amount of the incentives was the timing of their introduction. Just a few weeks earlier, the monsoon rains had failed to arrive in Bihar, Rajasthan, Madhya Pradesh, and parts of Uttar Pradesh. Over one hundred million people now faced famine. Bihar was particularly hard hit—it was the third year of drought. In affected areas, annual per capita income would range from 74 to 112 rupees a year (that is, $10–$15). The possibility of even a modest cash payment would take on extraordinary importance.[84]

At no point did anyone state as a matter of policy that poor people would starve if they did not accept IUDs or sterilization. Even when Johnson signed a "Food for Peace" law requiring that a country's family planning efforts be taken into account before granting food aid, he kept insisting that population programs be "freely and voluntarily undertaken." At the same time, USAID officials were told "to exert the maximum leverage and influence" to ensure that governments were meeting their obligation to "control population increases." Indian officials were often reminded not to use the word *incentive,* maintaining that the payment was for travel and lost wages (even when there were no travel costs or lost wages). Incentive payments were subtly coercive even in the best of times, since many Indians were always at risk of malnutrition. Now some people in Bihar were subsisting on less than 900 calories a day.[85]

Immediately after the incentive payments were announced, there was a spike in the number of sterilizations and IUD insertions. Bihar had previously had the lowest rate of sterilization per capita of any state or union territory in India, and had only met 12 percent of its IUD target. But in 1966–67, with some people living on leaves and bark, nearly 100,000 "acceptors" suddenly came forward. The next fiscal year's performance was even better: almost 200,000, with fully 78 percent opting for sterilization (and higher incentive payments). As a Ministry of Health and Family Planning analysis concluded: "The large number of sterilizations during 1967–68 was due to drought conditions."[86]

In Madhya Pradesh it was much the same story. Its rate of sterilization and IUD insertion had been considered "very poor." But the number of "acceptors" swelled almost fourfold to 130,000 in 1966–67, then to 230,000. "The year 1967–68 was the third year of continuous drought in the State of Madhya Pradesh," the ministry analysis explained, "and that was one of the main reasons for the performance in that year being very good." Uttar Pradesh and Orissa were also hard hit, and had a similarly dramatic increase. If it was not for these states, there would have been no increase in the number of "acceptors." Because of them, and because of their plight, more than 300,000 more Indians submitted to IUD insertion or sterilization in 1966–67, or 1.8 million altogether.[87]

The chief minister of Uttar Pradesh was not satisfied, because India's most populous state was still far short of filling its quotas. He had family planning officials report directly to district magistrates. They tried to explain that they were not hired just to bring in cases. They were also supposed to educate people and build support in the community. But their new bosses also had to answer for their performance in controlling population growth. Failure could lead to termination or suspension of pay. Meeting targets therefore led to "constant whipping of the staff," as magistrates put the whole weight of the state behind population control drives, sending out block development workers and revenue collectors. They promised higher payments, free fertilizer, and even land grants. As promises were made and broken, motivators started to bring in the aged and infirm, and poorly trained medical staff botched operations, the whole program fell into disrepute.[88]

Studies by the Indian Planning Commission found much the same pattern in Punjab and Maharashtra. Though neither state was hit by drought or famine, even here the population control campaigns were often coercive and sometimes appalling. Officials in Maharashtra abolished the positions of field-workers and educators in order to free up more money for incentive payments for sterilization. All kinds of people took on the role of "motivator," including businessmen who set up their own camps and started leaning on employees. This spirit of "catching cases" developed even among doctors in Punjab, who competed for incentive money. Conversely, in May 1967 Delhi demanded disciplinary action against government doctors who did not meet their quota.[89]

Sterilizations were performed on 80-year-old men, uncomprehending subjects with mental problems, and others who died from untreated complications. There was no incentive to follow up patients. The Planning Commission found that the quality of postoperative care was "the weakest link." In Maharashtra, 52 percent of men complained of pain, and 16 percent had sepsis or unhealed wounds. Over 40 percent were unable to see a doctor. Almost 58 percent of women surveyed experienced pain after IUD insertion, 24 percent severe pain, and 43 percent had severe and excessive bleeding. Considering that iron deficiency was endemic in India, and would have been still worse in famine-affected areas, one can only imagine the toll the IUD program took on the health of Indian women.[90]

Of course, some of these same men and women desperately wanted to avoid pregnancy, with or without any incentive payment. Most had not even heard of state-sponsored family planning until 1966. They received a very poor first impression. In Maharashtra, for instance, three-quarters of husbands were initially happy with their wives' decision to use the IUD. More than half changed their minds. When monthly performance fell short, new "family planning fortnights" were launched with higher incentives, only to bring diminishing returns. People who might have willingly participated learned to wait on the chance that they might earn more later. Many of those rewarded for sterilization would never have had any more children. A study from Uttar Pradesh found that the ages of those undergoing vasectomies had been systematically falsified in official records. On-the-spot verification showed that almost half were over 50 years old. Moreover, 63 percent were either unmarried or separated, or had wives aged 45 or older. With villagers openly showing their distrust or even contempt, family planning officials began to see their assignment as a punishment. In one study, 69 percent said that they would happily take another job if it were offered to them.[91]

Oblivious to all of this, in January 1967 Lyndon Johnson urged Gandhi to "take the lead in inspiring and urging all nations—rich and poor alike—to join a truly world wide effort to bring population and food production back into balance." He viewed India's struggle against famine as emblematic of a global crisis. "We count on the Government of India to become an example of what a determined people can do for themselves." Five months later, he wrote that Gandhi must "find satisfaction" with her government's population program.[92]

In fact, India was falling further and further behind, inducing desperate officials to devise a scheme that would come to signify everything wrong with population control in India. Someone in the Ministry of Health and Family Planning recommended that, instead of still higher incentive payments for sterilization, the Ministry give people transistor radios. It would make manifest what had only been a promise: the idea that family planning, by itself, could make people modern, with all the modern accoutrements. As Mitra argued, it was essential to "present before the common man the choice of an entire range of readily purchasable goodies like a bicycle, a watch, a transistor radio." Posters promoting the "happy planned family" would often show family members with a radio or television set. The unhappy unplanned family, on the other hand, was shown alone with their misery—except, perhaps, for the stick with which the father beat his children, or the children beat each other.[93]

Some officials instead called for the state to punish those who would not submit to IUD insertion or sterilization. At the end of 1966 both Kerala and Mysore had begun denying maternity leave to government employees with three or more children. The Central Family Planning Council formed a "Small Family Norm Committee" to study incentives and disincentives and make recommendations. Before it could report, the cabinet of Maharashtra took what it admitted were "radical decisions." In June 1967 it agreed that India should not only deny free medical treatment and maternity benefits to those with three or more children, but should make sterilization compulsory. To demonstrate its seriousness, Maharashtra announced that in fourteen months all state employees that elected to have three or more children would be denied government scholarships, grants, loans, and maternity and housing benefits. Haryana and Uttar Pradesh soon followed with similar measures. In a conference of the chief ministers of India's states, all but two said that they were favorable to the idea of mandatory sterilization of prolific parents.[94]

Maharashtra's cabinet members ostentatiously included themselves when specifying who would face loss of benefits, though one suspects that few ministers would need free medical care, much less maternity leave. Part of

the ethos of family planning in India was that it included everyone, reproducing the nation as a nation. Given India's unstable borders, this was part
of the appeal. Officials therefore set the same targets everywhere—such as
20 sterilizations per 1,000 rural population—despite the tremendous diversity of India, not to mention regional differences in actual demand.[95]

Now that India's government was finding it impossible to persuade the
nation to reproduce itself according to plan, underlying concerns about differential fertility resurfaced. The chief minister of Maharashtra, V. P. Naik,
insisted that compulsory sterilization should apply to "all citizens irrespec-
tive of caste or creed." But he expressly included polygamous men. "This
meets the objection of a certain section," he explained, "which feels that
any group, whose personal law allows more than one wife, may overwhelm
the rest of our population."[96]

A year earlier, the Central Family Planning Council had taken up the
sensitive question of whether Muslims were participating in the program
and just as quickly dropped it. Virtually everyone present agreed that religious differences were not necessarily an impediment, but also that they
had to try harder. This included meeting with Muslim leaders and publishing fatwas endorsing birth control. But in one study in Uttar Pradesh far
fewer Muslims agreed to sterilization, and some political leaders encouraged their followers to out-reproduce everyone else. It did not help matters
that well over 90 percent of senior family planning officials interviewed
were high-caste Hindus.[97]

The cabinet committee on family planning was warned that such "rum-
blings" might "snowball into large scale opposition." It was agreed that
some minorities were seeking to take advantage of family planning to gain a
"larger say in the affairs of the country." Some were prepared to target particular groups, beginning with India's scheduled castes—that is, the Dalits
once deemed "untouchable," who were eligible for scholarships and other
assistance to integrate into Indian society. They could start by stripping
them of these benefits. Maharashtra and Uttar Pradesh announced that
scholarships would be barred to families with more than three children, except for those awarded on individual merit. At the grassroots level, population control already focused on outcastes. In Uttar Pradesh it was found
that, while they made up 29 percent of the population, they constituted 41
percent of those vasectomized. They were an even larger proportion of
those brought in by revenue collectors and block officials. Typically the

most impoverished and powerless in any community, Dalits were the most vulnerable to local notables intent on achieving targets and reaping the rewards.[98]

Nayar emphasized education as the solution, but she was finally eased out. The new minister, the demographer Sripati Chandrasekhar, wanted to make sterilization compulsory for every man with three or more children. Violators would only have to pay a fine, so the measure would be compulsory only for those who could not afford to pay. Chandrasekhar justified his position by paraphrasing Oliver Wendell Holmes's judgment in *Buck v. Bell:* "The principle that sustains compulsory vaccination is broad enough to cover cutting the vas." As for the three generations of imbeciles, that apparently went without saying.[99]

After a prolonged debate the cabinet judged Chandrasekhar's proposal for compulsory sterilization to be impractical. Even if legislators agreed, family planning services were unequal to the task. The question remained whether the government would adopt more limited measures to penalize large families. Six states had issued orders to deny maternity benefits to government employees with three or more children. It was pointed out that stripping scheduled castes of scholarships would cause hardship, and that withdrawing free medical care and maternal leave would harm women and children. Others would continue to press, including Asoka Mehta, who now held the combined portfolio of Social Welfare, Petroleum, and Chemicals. "This has an element of inhumanity in it," he admitted, but Malthusian growth could be even more inhumane. "Here we have to wield the surgeon's knife. It may hurt a little, at a point, for a while, but it will help to impart health ere long."[100]

India was the cutting edge of the population control movement. Scientists and activists worldwide had agreed that high fertility was to be treated as a disease, and that birth control for nations made individuals expendable. The hasty and incomplete safety testing, the determination to "forge ahead" of health services, the insistence on setting targets, and the idea of paying for performance and penalizing resistance all derived from what Draper characterized as a general consensus. As he put it in testimony before Senator Gruening's committee, "the world population explosion represents a serious and imminent threat," and "something has to be done to dispel that threat."[101]

Even in failure, India was a leader, because it convinced population control proponents that crash programs were now all the more necessary. Thus, in August 1967, Draper wrote to Rusk that it was time to "push the panic button." He suggested once again using food aid as leverage to force Delhi to create a fully independent agency and train thousands of paramedical personnel to insert IUDs, bankrolled with up to $100 million from the United States. Otherwise, India's failure would set a "horrible example" for other countries just beginning to mount programs. Notestein, on the other hand, chalked it up to a learning experience. "Mistakes have been made," he admitted, "but the overall gains have been valuable and highly significant." Indeed, he bragged that the Population Council had "bought" five years' advance in contraceptive technology with the IUD. It was now fully behind the idea of a "crash program" to find something better.[102]

Until this point, the Ford Foundation's money had given it a uniquely influential position. The $26.3 million—$163 million in today's money—Ford allocated to assistance in population control in 1966 was more than had been provided by every other foundation, foreign aid agency, and international organization combined.[103] Yet even the resources of the world's wealthiest foundation seemed insufficient compared with the scale of the problem. Moreover, Draper considered both Ford and the Council too conservative in their methods. He would join Hugh Moore in creating the Population Crisis Committee (PCC) to drive home the need for drastic action in Washington. In November 1967 they persuaded Senators Gruening, Fulbright, Joseph Tydings and other supporters in Congress to earmark $35 million of USAID's budget for family planning. While the administration's request for technical cooperation and development grants had been slashed by a quarter, they now had to spend ten times more on family planning. USAID administrator William Gaud had insisted that they would not be able to do it. But the new head of the population program, Reimert Ravenholt, was full of ambition, and he said that he could spend all of it and more.[104]

This money would become a windfall for all the NGOs with ongoing programs in family planning. Even before the new budget was finalized, the IPPF was promised $3 million. Following on its first government grant in 1966, $368,000 from Sweden, the whole character of IPPF funding changed almost overnight. By 1968, more than half of its $6 million budget came from foreign aid agencies. Within the IPPF, people like Joan Rettie, regional secretary for Europe, had waged a "battle" for years to fend off Draper's demand that "population control" should be the overriding priority. Now that Draper was securing most of the IPPF's budget, this battle was over.[105]

The Population Council was the second leading recipient of USAID money, with $675,000 in 1967, then $2.7 million in 1968. As a downturn in the Ford Motor Company's share prices forced the foundation to rein in spending, USAID support was crucial in sustaining the Council's relentless expansion. Its $11.3 million budget in 1968 was enough to employ 170 full-time staff. In 1962, the Ford Foundation, the IPPF, and the Population Council were the only significant sources of international support for population control, and they spent $4.2 million. By 1968, these three together with USAID, the National Institutes for Health, Sweden, the United Nations, and the Rockefeller Foundation would commit $77.6 million.[106]

The UN provided only a small fraction: just $2.2 million. But at the end of 1966, the General Assembly had finally agreed that the UN should help states that were developing population policies and programs. Indeed, compromise language had made it unanimous, including a clause stating that childbearing "should be the free choice of each individual family." In 1967 quiet discussion began that would eventually lead to a UN Fund for Population Activities, and the only thing that could set every UN agency in motion: a large pool of money open to bidding.[107]

Notestein became accustomed to a lifestyle he could hardly have imagined only a few years before. As he traveled the world promoting the IUD, with introductions from John D. Rockefeller 3rd and World Bank President Eugene Black, this once obscure academic found doors opening to him before he even touched the handle. Heads of state and cabinet ministers sought his counsel, and everywhere family planning workers awaited his approval. With obvious relish, Notestein recorded the hospitality he received in diaries with datelines from such places as Cairo, Taipei, Karachi, and Seoul. They were true banquets, from which he "staggered away from the table my soul content within me," only to face "another day of tremendous eating." One evening in Dacca he was idly watching the river traffic when an acolyte commented on how "we had come a long way since we started thinking about this ten years ago," and that "Notestein had the most exciting job in the world." Notestein could not help but agree.[108]

Among his peers Notestein was not particularly conspicuous in his consumption. Alan Guttmacher was in the habit of beginning letters to the Planned Parenthood membership with comments like "This is written

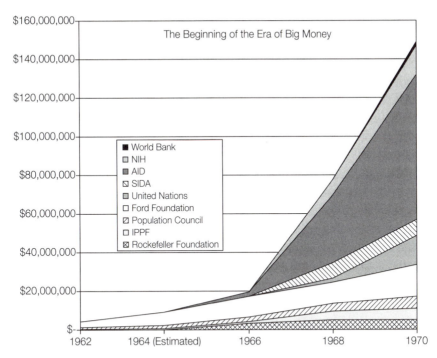

With the rise of concern about the "Population Bomb," the field was suddenly awash with money. Government agencies like USAID gained preponderant influence over NGOs that had previously been the main source of funding. Data are from "The Foundation's Work on Population," September 1970, Ford Foundation Archives, New York City, file 001976.

31,000 feet aloft as I fly from Rio to New York." He insisted on traveling with his wife, first class, with the IPPF picking up the tab. Ford officials flew first class with their spouses as a matter of policy. One wonders why Douglas Ensminger ever left his residence in Delhi—he was served by a household staff of nine, including maids, cooks, gardeners, and chauffeurs. He titled this part of his oral history "The 'Little People' of India."[109]

Ensminger insisted on the need to pay top dollar and provide a plush lifestyle to attract the best talent, even if the consultants he recruited seemed preoccupied with their perks. One of these strivers ran his two-year-old American sedan without oil just so that the Ford Foundation would have to replace it with the latest model. The fund-raising consultants the IPPF hired advised that it had to spend money to make money, not least because it was hoping to get more from corporate donors. Thus, while the new headquarters in London had "an air of quality," it still needed a

director-general with international standing and the qualities of a businessman. Such a personage would need a salary and staff support costing
twenty thousand pounds (equivalent to more than $450,000 today).[110]

The lifestyle of the leaders of the population control establishment reflected the power of an idea whose time had come as well as the influence of
the institutions that were now backing it. But technological changes had
also made it far more practical and pleasurable to manage such a far-flung
enterprise than even a decade before, and this benefited midlevel professionals as well. In 1953, one of the lead investigators of the Khanna study
had traveled from London to Delhi together with his family. His propeller-
driven plane was first held up by twenty-four hours of engine trouble,
caught fire in the air, filled with smoke, lost altitude over the Arabian
desert, then hit a sandstorm out of Karachi. It was diverted to Calcutta and
finally landed on an emergency strip when it ran out of fuel. At every stop,
his wife had had to boil formula for their 3-month-old baby. Since no one
had ever attempted this kind of program before, the Khanna researchers
had to spend years there with their families developing new methods. As
much as they were criticized for failing to understand rural Punjabi society,
it was not for lack of trying.[111]

Not so with this new jet set of population experts. Some did not even
try to adapt their methods to different countries. When the Population
Council's man in Taipei flew into Karachi for three days in 1965, he immediately concluded that "Pakistan needs a simple program, standardized, like
the Model T Ford." Later that year, a four-man Population Council team
landed in Kenya to prepare a report for the Ministry of Economic Planning
and Development. It described a step-by-step program: the government
should declare a policy to reduce population growth; set up a family planning council with representatives from different ministries and NGOs;
start the program where success was most likely; and "rely heavily on the
intrauterine device." There was also something in it for the demographers:
"We recommend that early attention be given to improvement of the collection and analysis of information, particularly vital statistics"; they should
conduct a Knowledge-Attitude-Practice survey to show support for the policy; "immediate attention should be given to training personnel for the
program both locally and abroad"; and because "well-trained personnel in
the population field" are "scarce," "we would recommend that Government
give early attention to the need for foreign advisors and explore their availability from various potentially interested agencies."[112]

One year later, a completely different team landed in Tehran and prepared a report for the Ministry of Health. It was virtually identical, with all the quoted language repeated verbatim. One might have expected some nuance, considering that one country had a president and the other had a shah, one was animist-Christian and the other was Muslim, one was in Africa and the other in Asia. Instead, the only significant difference was to explain to the Iranian health ministry that "the basic purpose of a family planning program is to improve the health and welfare of mothers and children." Kenya's Ministry of Economic Planning and Development, on the other hand, was promised "savings to Government in reduced maternity care expenditures." Other than that, even Notestein's cover letter was a cut-and-paste job:

> You and your government are to be congratulated. . . . We believe that [name of country here] is in the fortunate position of having recognized the implications of population growth at a stage in the country's development when there is still time and opportunity for effective action.[113]

As for the fertility surveys these reports invariably demanded, demographers wanted uniform data sets suitable for ever more sophisticated statistical analysis. Locally hired field-workers would therefore carry the same surveys across the African Sahel, over the mountains of Nepal, and through the back alleys of Bangkok. The surveys arrived in places like Ann Arbor or Chapel Hill in the form of neatly stacked IBM punch cards. In just one country, Ghana, this required translating the same question, such as how often respondents had sex, into twenty-nine languages. Did such questions, and the responses they elicited, mean the same thing to all concerned—including the children who overheard their mother asked whether she would have preferred that they had never been born?[114]

Critics would later marvel at how these consultants worked: "no need to speak the language, or even to meet a non-Ph.D.-holding native. Visits to the country, if required at all, could be confined to short stays in Western luxury hotels." And for those who were willing to stay, they were told that after just two years abroad they could pick and choose among several job offers from U.S. population centers. Doubtless many sincerely wanted to help, but this was not always how they came across to their counterparts overseas. In one postmortem, forty-three Indian officials were asked what motivated these foreign consultants to come help them. Monetary gain was considered the main reason, followed by desire to write a book.[115]

For population experts this was the beginning of an era of constantly expanding opportunities. The budgets, the staff, the access were all increasing even more quickly than the population growth their programs were meant to stop. There was "something in it for everyone," Population Association of America President John Kantner later recalled: "the activist, the scholar, the foundation officer, the globe-circling consultant, the wait-listed government official. World Conferences, a Population Year, commissions, select committees, new centers for research and training, a growing supply of experts, pronouncements by world leaders and, most of all, money— lots of it."[116]

There was only one problem. Population growth still appeared to be accelerating. The experts could not agree why, and family planning—even crash programs—did not seem capable of stopping it.

7

BEYOND FAMILY PLANNING

In the 1940s, when leaders of new UN agencies embraced the cause of family planning, they promised it would help perfect "the global family." Instead, by 1968 the global family seemed increasingly dysfunctional. Family planning did not seem to have slowed the rate of population growth. In fact, the most planned generation in history was also the largest and most misbehaved. Young people were leading violent protests not just in Chicago, Paris, Mexico City, and Prague, but in more out-of-the-way places, like Bihar, which witnessed 132 student riots between 1967 and 1971. Labor disputes and communal violence increased apace all over India. More educated children of affluent parents, the baby boomers who gave hope to "reform" eugenicists, were often in the vanguard in demanding revolutionary change. And it was assumed that the young people of the 1960s were about to give birth to a still larger, hungrier, and angrier generation. Researchers and administrators began to question whether their favored strategy and slogan would ever work. As the new Population Council president, Bernard Berelson, put it in 1969: it was time to look "beyond family planning."[1]

Birth rates were actually falling in the 1960s in most of the world, at virtually the same rate as literacy was increasing among women. One reason campuses were so restive was that so many students were crowded inside of them. In Europe, Asia, and the Americas, fertility had peaked by 1965 and began declining even before most government-sponsored family planning programs were up and running. But it took time before censuses confirmed this trend, and Draper was not the only one inclined to "push the panic button." USAID administrator William S. Gaud assured Lyndon Johnson, "We are pushing this as hard as we can." But the president let it be known that he did "not think the Government is doing enough or doing it effectively enough." World Bank president Robert McNamara complained that "there has been damn little accomplishment in the reduction of age specific fertility rates in the major countries.'" Indira Gandhi was advised that "the whole Family Planning and Health Programme is in a bad shape."[2]

India could be written off as a special case, "frequently terrifying and wasteful of money and spirit," as one donors' meeting concluded. Yet even where fertility was falling, it was difficult to claim credit for family planning campaigns. "With the exception of Hong Kong and possibly Taiwan," the IPPF's Colville Deverell admitted, "I know of no cases where it can be truthfully said that voluntary organizations have of themselves succeeded in operating programs which have brought down the birthrate on a national scale." Deverell insisted that only governments could achieve this outcome, and the partnership between Taiwan's public and private agencies provided a model for other countries. But here, as in Hong Kong, South Korea, and Singapore, parents had already begun having fewer children by the time it got started. Even when government officials and volunteers worked hand in hand—and usually they did not—they might only accelerate a preexisting trend.[3]

This was not population control as proponents understood it. Providing contraception was supposed to reverse fertility trends, since it was assumed that people needed help planning smaller families. For decades this belief had provided a basis for cooperation among the most disparate factions. Whether they wanted to lower fertility to liberate women, improve population quality, preserve the environment, or promote economic growth, there was no need to argue if parents merely needed help to do it themselves, for their own reasons. This consensus was codified at a 1968 UN conference, which proclaimed that "parents have a basic human right to determine freely and responsibly the number and the spacing of their children." Pointing out the potential conflict between freedom and responsibility would risk a schism, and insisting on "population control" as the official goal might provoke a backlash. But when insiders spoke to one another, they made their motives plain. As Lyle Saunders of the Ford Foundation put it to a meeting of USAID officials, "In this company it is not necessary to argue that our primary purpose is to reduce the rate of population growth." The population establishment had committed to family planning as "its major vehicle." But improving the health of mothers and asserting their right to birth control were, to Saunders and many like him, "means, not ends."[4]

By the end of the 1960s, the force of global population growth and the object of childbearing as a private choice finally collided. Kingsley Davis kicked off the debate in 1967, writing from the very epicenter of student protests in Berkeley. In an article in the prestigious journal *Science,* he declared, "The things that make family planning acceptable are the very things that make it ineffective for population control. By stressing the right of parents to have the number of children they want, it evades the basic question of population policy, which is how to give societies the number of children they need."[5]

Davis had both the credibility and the evidence to expose the deep fissures beneath the growing weight of the population establishment. After Notestein blocked his promotion at Princeton and the Population Council refused support for a new center at Columbia, he also had a motive. He pointed out that the "Knowledge-Attitude-Practice" surveys that had been used to demonstrate "unmet need" for birth control in such places as South Korea, Tunisia, India, and Indonesia showed that the average respondent still wanted at least four children. At this rate, even if family planning programs met all the unmet need, the world was still doomed to disastrous overpopulation. For some, it was already too late. "The battle to feed all of humanity is over," Paul Ehrlich declared in his 1968 bestseller, *The Population Bomb.* "In the 1970's the world will undergo famines—hundreds of millions of people are going to starve to death in spite of any crash programs embarked on now." John Rock, who might have been more optimistic, considering his success with the pill, agreed that a massive famine was inevitable: "There is nothing we can do to stop it."[6]

The cause of population control had always attracted more than its share of worriers, eccentrics, and extremists. But even establishment figures now began to debate whether family planning would ever work, or whether instead they would have to resort to either outright compulsion or broader social changes, such as measures to "make children expensive," as one Ford-sponsored meeting suggested. In part, this was a difference over whether to focus on supplying contraception or increasing the demand for it. But there was also a more fundamental debate over the very idea of treating individuals as "populations," as well as over the interests served by such an agenda. When new, more invasive and manipulative techniques were directed at not just poor countries, but poor people in the United States, the resulting grassroots opposition proved impossible to ignore. The ensuing controversy could not be contained within any establishment, however big it had become, especially when longtime leaders like John D. Rockefeller 3rd repudiated past policies and called for an entirely new approach.[7]

In this period no authority, however august, went unchallenged. Graduate students accused their professors of prostituting demographic research; women's health advocates disrupted U.S. Senate committee hearings that would listen only to men; and dissident UN officials attacked the hidden agenda behind UN conferences. The Catholic Church was also rocked by dissent, as laymen and -women protested that "natural family planning" was unnatural and unworkable, and demanded to have a voice in reexamining Church teaching. More quietly, but nonetheless effectively, academic economists cast doubt on the core assumption that population growth in poor countries caused hardship, something even Pope Paul VI had been prepared to accept. Leaders of many newly independent countries insisted it was all a distraction from—if it did not exemplify—neocolonial exploitation. In all these ways, going "beyond family planning" also meant going beyond the old politics of birth control, as new constituencies came to the fore and old coalitions began to break under the strain.

Family planning was always meant to be a flexible concept, so going beyond family planning could mean many different things. In the late 1960s and the 1970s it often signaled a shift from voluntarist to coercive measures. But there were other axes that pivoted on different conceptions of family planning and led to different policies. For many, family planning was a way to reduce, if not eliminate, the incidence of abortion. For them, it was beyond the pale to campaign for women's right to make this other choice, much less to promote abortion as a means of controlling population growth. Going beyond family planning could also mean pointing out how the family itself was a political institution and how rigid sex roles reproduced inequality. The alternatives were potentially liberating. But in the context of debates about population control, promoting women's access to education and paid work, changing how society valued childrearing, and accepting homosexuality and alternative lifestyles were considered indirect means to reshape fertility preferences. Still others went to extremes in their single-minded commitment to flooding the world with contraceptives.

Among the most flamboyant figures during this period was Dr. Reimert Ravenholt, head of the USAID population program, which provided well over half of all international family planning aid, whether public or private. USAID shipped so many contraceptives that it struggled to manage its own inventory and keep unused condoms and pills from spoiling in the warehouses. Strikingly tall and self-confident, Ravenholt personified the spirit of expansiveness, often provoking a reaction by personally demonstrating USAID's pump-action abortion kits and handing out condoms with his business card printed on the wrapper. He was the kind of official who, rather than merely inspecting a Filipino sterilization clinic, donned surgical scrubs and joined the operating team.[8]

Yet neither Ravenholt nor his principal patron, William Draper, believed there was any need to go "beyond family planning" if that meant losing focus on meeting the demand for birth control. They remained committed "supply-siders," pointing out that in most of the world conventional family planning had hardly been tried. Social scientists had just begun to study the reproductive behavior of poor people in poor countries. With ever more fine-grained analyses of earlier fertility declines, it had become clear that highly motivated people in agrarian societies had lowered their fertility with unreliable or onerous methods—including delayed marriage, abstinence, and withdrawal. It was at least possible that less motivated people might opt to limit the size of their families if they were shown how easily they could have sex without having children.[9]

The need to find out led USAID to provide a quarter of a billion dollars for the largest international social science survey ever attempted: the "World Fertility Survey," which began in 1972 and reached some 330,000 women in sixty-two countries by 1984. In 1977 USAID commissioned Westinghouse to conduct additional surveys designed to determine how convenient contraception would have to be before people would start using it. If insufficient motivation hindered efforts to reduce fertility, this research would show how to reduce the importance of individual motivation to the absolute minimum.[10]

Part of this "supply side" approach was to develop contraceptives that, like the IUD, did not require daily diligence, only without all the side effects. Together with the National Institute of Child Health and Human Development, USAID became a major player in the development of new technologies. Even as other countries began to assume a larger role in financing service programs, the proportion of research funding from public and private sources in the United States actually increased: by 1979, Americans paid for over 70 percent of research in reproduction and contraception. This included major corporations like Upjohn, which developed the first mass-market injectable contraceptive, Depo Provera. UNFPA and the IPPF adopted it with alacrity, and over a million women were injected by 1978. Government grantmakers, foundations, and pharmaceutical companies collaborated in bringing the next generation of devices to market, including Norplant, a subdermal implant that had been first envisioned in the late 1960s. The Population Council began field-testing it within the decade.[11]

Many questioned the safety of injectable and implanted contraceptives. USAID was barred from sending them abroad as long as the Food and Drug Administration withheld approval for domestic use. Instead, it experimented with new ways to flood countries with condoms and pills, a policy that came to be known as national "inundation." Rather than wait for people to come to clinics, program administrators kept extending their reach through commercial marketing and "community based distribution." Street vendors were supplied condoms below cost to maximize their profit motive, while in other cases door-to-door distribution made contraception "cheaper than free," as one economist put it. Everywhere they sought ways to reduce the need for training and infrastructure, and thus eliminate the kind of management problems that had undermined the Indian program. Nonmedical personnel were permitted to dispense pills without prescription; illiterate midwives were taught to insert IUDs; and doctors learned how to further simplify and speed up sterilization and abortion procedures.[12]

Proponents continued to argue that, considering the high rate of maternal mortality in most developing countries, even risky contraceptive methods imperfectly administered would save lives. But most health professionals insisted they must "first do no harm." World Health Organization director-general Marcolino Candau of Brazil issued a stark warning at the May 1966 World Health Assembly. Addressing an audience of doctors and health ministry officials, he attacked those "promoters going around the world" who were seeking a "reversal of priorities," diverting funds from public health to population control. He vowed that WHO would provide no "moral support." Indeed, "an international organization could not be involved in these highly political decisions."[13]

No issue would prove more politically divisive than abortion. If people wanted fewer children, but could not or would not use contraception, helping them terminate their pregnancies seemed to provide the solution.

Depo Provera gave many women the chance to use contraception without depending on their partner's permission. But programs that made reducing population growth the overriding priority threw up barriers to proper medical screening and follow-up—in this case quite literally. IPPF Archives.

Abortion had been largely responsible for Japan's rapid fertility decline after World War II, and it was also the "silent partner" to the IUD and the pill in the later declines in Taiwan and South Korea. Keeping abortion illegal, moreover, was a silent killer. Many of the leaders in the field—such as Guttmacher and Helena Wright—knew from personal experience how poor women paid the ultimate price for abortion bans.[14]

But while most people in the field opposed laws that would prosecute these women, even the IPPF's own secretary-general, Colville Deverell, thought it "quite a different thing to contemplate the adoption of abortion by a Government as a deliberate instrument of population control." Notestein worried that unless Population Council staff "constantly and firmly take the anti-abortion stance," they would give credence to the claim that they were "against life." But whereas Notestein nonetheless considered it a personal right, Council board member John T. Noonan, a Catholic historian, believed abortion was a "moral evil." George Zeidenstein, the new president, threatened to resign in 1976 if he were not permitted to reverse Council policy. Noonan left instead.[15]

Some governments—such as the United States, Sweden, and Norway—began to promote abortion in poor countries before they provided or even permitted abortion at home. USAID along with UNFPA began to fund programs on the subject of "Law and Population," which focused on abortion and sterilization restrictions along with what they considered overly stringent safety regulations for the IUD, the pill, and newer contraceptives. This initiative helped activists coordinate efforts to remove these and other laws that stood in their way. Between 1967 and 1978, forty-two countries made it easier to terminate pregnancies, such that by the end of the decade only 20 percent of the world's population lived in places where it was still prohibited. In several countries, such as India, South Korea, Tunisia, and China, a desire to reduce overall fertility had played a major role in the change. Between 1970 and 1978, seventeen countries, including Indonesia, the Philippines, and six more in Latin America, also broadened access to sterilization.[16]

Ravenholt, for his part, saw no ethical distinction between these different means of controlling population growth. He believed that new research on a pill to induce abortion would make contraception obsolete, because women could simply take such a pill whenever they missed a period. He was quoted observing that abortion was especially appropriate for poor people, since they lacked the foresight to use birth control. The potential for abortion would be greater still if parents were able to discover the sex of a fetus. In June 1967 the head of research at the PPFA, Steven Polgar, called for new methods of sex determination to help parents have fewer children, an idea that Berelson also backed. The prospect of sex-selective abortions may be why, in 1973, Ravenholt said that eventually even the poorest people would pay for abortion. In the meantime, he wanted to distribute millions of "menstrual regulation" kits worldwide, providing a manual vacuum aspirator designed by a California abortionist that would be so cheap and so simple to use that no government could stop its spread.[17]

The supply-side approach reached its apotheosis in Ravenholt's USAID. But it was still premised on the consensus view, according to which parents needed help in planning smaller families, even if they wanted only sons. Others wanted to go "beyond family planning" by changing people's desire or ability to have children. In the 1969 article that coined the phrase, Berelson described a host of measures, including proposals for indoctrina-

tion via satellite television broadcasts, the reversal of tax and housing benefits for families, and adding sterilizing agents to water supplies. Simply
compelling people to stop having children appeared to be the clearest departure from family planning, one that critics focused on when defending more conventional programs. As one senior World Bank official put
it in 1969:

> It is extremely dangerous for us to slide into the habit of referring to
> "population control." What we and other concerned groups and in-
> dividuals are talking about is family planning—programs that will
> enable people who wish to limit the size of the families to do so.
> "Population control" carries an altogether different image, which
> those who are opposed to or suspicious of family planning are all too
> ready to exploit. To put it bluntly, it is the image of the rich white
> north controlling the growth of the poor dark south by putting
> chemicals in the water, forced inoculations, etc.[18]

But the same official went on to note that senior people in the field,
such as Ernst Michanek—the first director-general of the Swedish International Development Authority (SIDA)—believed that "the time will come
when we will have population control in the strict sense of the term."
Ensminger of the Ford Foundation thought India would be justified in prohibiting parents from having more than three children if incentives failed.
Another Ford official thought failure might be inevitable absent a dramatic
technological breakthrough: "An annual application of a contraceptive aerial mist (from a single airplane over India), neutralized only by an annual
antidotal pill on medical prescription." While Berelson was skeptical about
the ethics and effectiveness of such measures, he observed that "there seems
to be [a] 'natural history' progression in family planning from softer to
harder methods." He favored "research on a mass involuntary method with
individual reversibility."[19]

Among the major funding agencies, Ravenholt's population office was
virtually alone in its policy of refusing support for programs to create demand for contraception. He argued that supplying "unmet need" would be
enough to solve the problem of population growth, or was at least worth
trying before trying anything else. Many of his superiors and subordinates
disagreed, and pressed Ravenholt for experiments with incentives. Even
leaders of Planned Parenthood admitted that incentives and disincentives
might be necessary. Alan Guttmacher told reporters that compulsion would

be immoral and that he doubted it would ever provide a practical solution—after all, "Just who is going to round up 200 million Latin-American men and sterilize them?" But in 1969 he decided that if they could not cut the rate of increase in world population by one-quarter in ten years it would be time to "get tough." Thereafter, they should use taxes, incentive payments, and new—unspecified—contraceptive technologies. By 1973, the idea that it was enough merely to help parents plan their families was, according to a National Security Council official, a "school of thought about population which has few respectable adherents left."[20]

It is difficult to say, even now, exactly when people began to go beyond family planning, not least because so many leaders always considered it to be a means to other ends. For some, voluntarism and coercion were not conflicting but complementary strategies. Incentives and even disincentives could be ethical, officials at both the Ford Foundation and UNFPA agreed, but only if people were given a choice of alternatives—above all, safe and convenient contraception. Some, like Saunders and Freedman, suggested that small payments merely overcame inertia. But the bitter experience in India made it difficult to avoid ethical questions. At what level did such a payment become a bribe? Conversely, how desperately poor did someone have to be before even the smallest "incentive" became coercive?[21]

In this period SIDA, the World Bank, the Food and Agriculture Organization, and UNFPA all funded programs that distributed food as a way to increase use of contraception. In a country like Haiti, where one in eight children were severely malnourished, was it wrong for the FAO to distribute food "as an incentive for mothers to regularly visit the health centre where notions of family planning are taught"—especially if, as seems likely, using fertility reduction was a rationale to secure more funding for nutrition programs? Administrators issued clear instructions that acceptance of contraception was not to be a condition for receiving aid. But in Bangladesh, field-workers fearing loss of salary or dismissal would ignore them, and deny food to destitute women who did not help them meet sterilization targets. Other programs offered incentives to leaders of a community, or the community as a whole, to adopt contraception. Was this kind of peer pressure more acceptable than government pressure on individuals?[22]

Considering incentives begged the question of disincentives, because those forgoing these payments would incur opportunity costs and those resisting community pressure would also pay a price. In the late 1960s a number of Indian states went further still: denying maternity benefits to state employees with three or more children. In the 1970s Indonesia followed suit. Singapore withdrew tax and housing benefits from anyone who had more than three children (and this in a country where all but the wealthiest typically lived in public housing).[23]

Even in affluent countries, some were already calling for constraining reproductive choice without equivocation or apology. Because of America's influence on population control programs all over the world, debates there had much wider implications. When, in 1965, Lyndon Johnson began backing family planning both at home and abroad, he presented it as part of a global war on poverty, explaining that his vision of a great society "did not stop at our boundary." While USAID supplied contraception abroad, Sargent Shriver's Office of Economic Opportunity (OEO) funded the first programs at home. In the United States as in other countries, this often meant helping Planned Parenthood provide services in poor neighborhoods.[24]

The people who worked in these clinics knew that they were wanted and needed, especially at a time when American women were still dying from illegal, botched abortions. Less educated people tended to want larger families, but by 1965 there was overwhelming approval in every socioeconomic group for birth control. Without money for doctors or birth control pills, poor people tended to make do with the least reliable methods, such as sponges, douching, and foam. After federally funded family planning programs began, the number of people enrolled grew rapidly, from 450,000 in 1965 to over three million in 1973, more than half with incomes below the federal poverty line.[25]

But this work had scarcely begun before some declared it was too late. "Biological anarchy in zoological tenements" was already an issue in the 1964 presidential election according to Pulitzer Prize–winning journalist Theodore White. Dwight Eisenhower, honorary co-chairman with Harry Truman of the PPFA, complained in 1965 that the United States was "spending money with one hand to slow up population growth among responsible families and with the other providing financial incentives for increasing production by the ignorant, feeble-minded or lazy." The biologist Garrett Hardin had worried about dysgenic trends in U.S. population as early as 1952. But in a classic 1968 article, "The Tragedy of the Commons," he managed to harness old and new "quality of life" arguments by drawing an analogy between childbearing and abuse of public lands. "In a welfare state," Hardin asked, "how shall we deal with the family, the religion, the race, or the class (or indeed any distinguishable and cohesive group) that adopts overbreeding as a policy to secure its own aggrandizement?" Only the most conscientious would respond to appeals for more responsible parenthood. Over time, Hardin warned, they would have fewer children, and the thoughtless and selfish would have more. Family planning would not only fail to stop population growth, wrecking the environment, it would lead to "the elimination of conscience from the race." To Hardin, it was a "tragic ideal" that had to be abandoned. People should instead "reexamine our individual freedoms to see which ones are defensible."[26]

"The Tragedy of the Commons" formed part of a broader trend in American intellectual life and is still in the canon of environmental studies. Many more people were turning to biologists and biological concepts to make sense of social problems. This was emphatically the case for C. Lalor Burdick, a major contributor to the PPFA and head of a foundation that contributed millions more for research in reproduction. He complained to Hugh Moore that welfare programs "provide breeding pads and free sustenance for the proliferation of the kind of people that hate us and would destroy us, if they could." It was a mistake for Moore to proceed with plans to campaign for the two-child family as a universal standard. "Would it not be better if the truly useful families should have three or four children and that those who have demonstrated themselves as feckless should not have any?" (Though for the people of India and Africa, also characterized by "fecklessness" and "hopeless inabilities," the UN should "just go away and leave them to work out a survival of their fittest in their own way.")[27]

Burdick admitted that Draper "would shoot me for this kind of approach." He represented a faction that the population control movement was finding increasingly hard to control. Journalists reported that in 1969 the question of voluntarism vs. coercion was "boiling up" in nearly every meeting of the PPFA board. An emphasis on opening clinics in poor neighborhoods suggested that "we want the poor to stop breeding," as one dissenting member put it, "while we retain our freedom to have large families." Guttmacher himself said that much of the concern about population growth was a smoke screen for something else, including white supremacy and anticommunism. The PPFA found it even more difficult to deal with radical ecologists, who rallied to a new group, Zero Population Growth (ZPG), that was dedicated to the idea that population control begins at home and therefore focused on reducing U.S. fertility rates. ZPG's first executive director, Richard Bowers, authored a "model penal code" that called for compulsory sterilization of parents with five children. A senior PPFA staffer, Frederick Jaffe, found himself surrounded by ZPG activists in one appearance later that year at the University of New Hampshire. A botany professor suggested the solution was "a contagious virus to sterilize all of mankind," though saving other primates would require administering them with an antidote. "With friends like these," Jaffe concluded, "we don't need enemies."[28]

At the outset, family planning was intended to be the solution for both qualitative and quantitative population problems as well as the political problem of obtaining consent for policies intended to shape how a society reproduced itself. The original concept did not preclude preventing the "unfit" from having children. But "reform" eugenicists like the Myrdals, C. P. Blacker, and Frederick Osborn had argued that coercion would be exceptional or even unnecessary if benevolent welfare states helped parents plan their families. With free health care, day care, and education, fitter parents would have more children and family planning programs would persuade the unfit to plan fewer. Some criticized what they considered the unscientific conflation of nature and nurture, but it had far more political appeal than Mendelian genetics. For instance, in a 1967 message to Congress defending his Great Society programs, LBJ argued that "bad environment becomes the heredity of the next generation," quoting the progressive-era reformer Jacob Riis.[29]

Johnson was probably unaware of how, at the turn of the century, Henry Cabot Lodge and Francis Walker had used Riis's reports on the lives of recent immigrants to argue that such poor "stock" could only degrade the quality of the U.S. population. But he should have known that linking environment to heredity, no less than speculating about innate inferiority, risked stigmatizing whole communities—and that their leaders would bitterly resist. Daniel Patrick Moynihan discovered this after the press picked up his Labor Department report, *The Negro Family: The Case for National Action*. Moynihan had noted that an increasing proportion of the U.S. population was being raised in impoverished, broken homes. He wanted to provide all families with an allowance rather than continue subsidizing only single mothers and their children. Most industrialized countries already followed this practice, many for the purpose of maintaining population growth. Moynihan would never understand why he was accused of "blaming the victim"—he himself came from a broken home. But even sympathetic commentators immediately realized that his report could not help but have pernicious consequences.[30]

As much as Moynihan insisted that the "pathologies" he described resulted from a legacy of racism, he had helped white people, racist or not, see African Americans not as individuals, nor as families—since fatherless households were "disorganized" by definition—but as a "population," a population that, according to Moynihan, had a lower IQ, committed most violent crimes, and was growing ever larger by reproducing itself in the most unnatural fashion. More than any racist caricature, framing African American people as a population legitimated policies that treated them as such—and given the history and context, they had ample reason to worry about what might follow. Across the United States, state legislators continued to introduce bills calling for mandatory sterilization for recipients of Aid for Families with Dependent Children (AFDC). As the sociologist Charles V. Willie argued:

> It would seem that whites are concerned about the size and stability of the black family now only because the number of black men who are dying prematurely is decreasing and the number of black children born who survive is increasing. If you can understand the basis of the alarm among white liberals about this situation, then you can understand the basis for a charge of genocide which is made by black militants.[31]

For a time Moynihan was persona non grata among liberal Democrats. He later remarked, "If my head were sticking on a pike at the South West Gate to the White House grounds the impression would hardly be greater." But inside, White House officials continued to ponder whether and how they should explore the "relationship of mounting welfare rolls to family planning and population stabilization." Others spoke more vaguely of the need to address issues of "quality" and not just quantity, both for domestic and foreign population policy. They worried that welfare programs and foreign aid promoted fertility rather than responsible parenthood, permitting both poor people and poor countries to continue having large families.[32]

In fact, U.S. vital statistics showed that fertility had peaked in 1957 at 123 births per thousand women aged 15–44, then fell by almost half, to just 65, by 1976. The proportion born to unmarried mothers increased among both whites and African Americans. But overall fertility in the two groups declined at virtually the same rate. The birth rate among poor non-white people in the period 1965–1970 fell faster than any other group. These were harbingers of a global trend, for any who cared to see it. Instead national media continued to stoke fears of black fertility. "In the process of creation right now," Cornell professor Andrew Hacker warned, "are rioters and rapists, murderers and marauders who will despoil society's landscape before the century has run its course."[33]

Those who felt pressured to adopt contraception began to perceive such efforts, not as part of any Great Society, but as utterly unlike any welfare program they had known before. African American leaders in Pittsburgh forced the Office of Economic Opportunity to shut down its family planning program. "What U.S. hospital has a policy of visiting sick people who skip appointments?" asked a local NAACP official. "What welfare group sends volunteers to the homes of people who miss getting their check or the chance to get welfare food supplies. Do they have 'volunteers' to go out and tell people about good jobs?"[34]

These population control techniques had been developed by American consultants in places like Punjab and Taiwan. In spite of the Pittsburgh controversy, in 1969 the OEO provided a major grant to the Population Council to bring U.S. hospitals into its international program promoting contraception among new mothers. The premise, as Deverell described it after a Council-sponsored meeting, was that "the most effective procedure is usually to attack women in the post partum stage, and there is virtually no opposition to the view that family planning is necessary to sustain the quality of human life." The economic rationale for reducing fertility, on the other hand, was "not one which can often publicly [be used] to advance a government programme."[35]

In fact, many congressmen and senators, including George H. W. Bush of Texas, focused on welfare mothers when they pushed the OEO to make family planning a priority, mandating that part of maternal and child health funds in the Social Security budget go to family planning. The House version of the 1967 bill prohibited increased federal assistance to states in which children on welfare made up a growing proportion of the population, illustrating how populations—instead of individuals—could become the objects of policy. Though this provision was dropped in conference, the final version required that welfare case workers tell AFDC recipients that birth control was available free of charge. Contrary to OEO guidelines, several participating hospitals in the postpartum program also

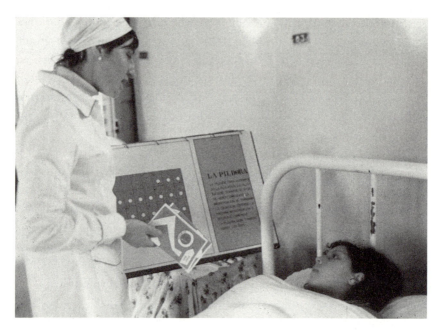

Postpartum programs this one in Colombia, could enhance maternal health. But workers often pressured women when they were most vulnerable. IPPF Archives.

provided abortion and sterilization, provoking new accusations of genocidal intent from community leaders in Harlem and Chicago.[36]

The director of the Johns Hopkins project in Baltimore, Hugh Davis, insisted that they were "losing the population war globally" and argued for programs at home and abroad that emphasized IUDs, abortion, and sterilization. "The quality of life," he concluded, "may indeed be preserved by the prevention of excessive human life." To that end he tested a new IUD, the Dalkon Shield, on his own Johns Hopkins patients. It featured rows of prongs that made it resistant to expulsion but also made it painful to insert or remove, and all too likely to pierce the uterine wall. The most fatal flaw was a multifilament "tail" that provided a path for infection. When the manufacturer, A. H. Robbins Corporation, began getting hit with lawsuits, it offered Dalkon Shields in bulk, unsterilized, to USAID at nearly half off. Ravenholt accepted the offer, and 440,000 women in forty-two countries would be using them by the time the recall order was issued in 1975.[37]

As "the population war" escalated, both governments and nongovernmental organizations concluded that they had to shore up the home front, if only to continue making advances abroad. The Ford Foundation backed another Baltimore program that used "indigenous" field-workers to discourage teen pregnancy. Foundation staff concluded that, here too, they needed "systematic study of economic incentives that might be introduced to lower fertility." They had to be weighed against moral and political considerations, but the United States could not be exempt:

> As attention is turned to questions of social control of fertility in developing countries, emphasis should also be given to population policy in the United States. While all but a relatively small segment of our population has fertility under voluntary control, insistent questions are being asked about possible long-term effects of individually determined fertility performance on the quality of life of our nation. As we prescribe for Delhi, so we must take account of Detroit.

A Population Council report agreed that "the United States has problems similar to those of many 'less developed' countries in providing family planning for its 'less developed' population."[38]

In the UK, an interdepartmental committee originally charged with examining policy on world population determined that it could not ignore Britain's own problems. Calcutta, they agreed, could bear lessons for London. Even France finally repealed its 1920 law against contraception, in part because of complaints from elected representatives and government officials worried about high fertility in the overseas departments of Guadeloupe, Martinique, and Réunion.[39]

Opponents, for their part, began to suspect that family planning was really just part of a larger plan to promote the interests of white people everywhere. Those who represented transnational, diasporic communities were particularly outspoken. Spokesmen for the Black Panthers and the Nation of Islam as well as Cesar Chavez, leader of the National Farm Workers Association, concluded that they should find strength in numbers. Family planning leaders pointed to such arguments as proof of the irrationality of the opposition, and noted that the clinics had considerable support among African American women. At a 1971 National Conference on the Status of Health in the Black Community, male participants "had to come to grips with the fact that many Black women want and accept family planning services." But participants agreed there was a clear distinction between family planning and population control. All too many programs ignored the role of men and the needs of children. Rather than focus only on preventing births, they demanded that clinics also provide infertility treatment and prenatal counseling. Even mainstream civil rights leaders like Julian Bond worried that racists would stoke population anxieties to justify genocide against black people.[40]

Such worries grew during the Nixon administration. Nixon had been elected in 1968 on a "law and order" platform, warning about "the city jungle" and the prospect that "the brutal society that now flourishes in the core cities . . . will annex the affluent suburbs." In public he argued that riots and street crime resulted from a lack of personal responsibility, not inequality. But in private he shared LBJ's concern that poverty and population growth were increasing the prospects for social breakdown both at home and abroad. "Because of the growing gulf between our wealth and that of most countries, and the shrinking of the world through modern communications, people in the [less developed countries] will not stand for continuation of the status quo," he told the head of a new task force on international development, Rudolph Peterson. "It will develop like our own urban problems."[41]

For Nixon, population growth was making individual liberties increasingly untenable. More than a decade earlier he and his wife had barely escaped death at the hands of a crowd of protestors during a visit to Caracas. Upon his return, he observed that the United States might be "running against the tide" during a discussion about population trends. "It was a genuine question whether or not the U.S. could continue to try to promote democracy and free enterprise, in the form we understand these systems, in the underdeveloped countries." Now that he was president, Nixon told Peterson—whom he would nominate to head the UN Development Program—that democracy could not be a condition for allocating foreign aid. On the other hand, "population control is a must . . . Population control must go hand in hand with aid," he insisted. "The U.S. has finally bitten the bullet on this issue and made it a top priority national policy."[42]

Nixon hired Moynihan as a domestic policy advisor and asked him to join Secretary of State William Rogers in bolstering their efforts at home and abroad. A presidential commission was appointed to "show other countries that we are prepared to attack our own population problem as well as theirs and thus make credible our efforts at world-wide population control." Officials at the Department of Health, Education and Welfare (HEW) appeared to be dragging their feet in funding family planning efforts, so the United States adopted the same solution American consultants had long urged on other governments. In a special message to Congress in

July 1969, Nixon called for consolidating federal funding for domestic family planning in a new unit of HEW, the "Office of Population Affairs." Administration officials let it be known that Nixon favored population control, and not just family planning. According to the first head of the HEW office, "most of us hope government population policy will be based on voluntarism," but "long delays in the formulation of a governmental population policy can dissipate the last hope of a voluntary solution. The planners of involuntarism are already at their drawing boards."[43]

The atmosphere of alarm, even hysteria, surrounding the population issue made coercive policies seem inevitable. Much of it was driven by fundraising, since growing programs had to be fed with cash infusions. Soliciting six-figure donations, Draper warned that if the population of Asia, Africa, and Latin America doubled, "the resulting conflict, poverty and general misery will be such that present-day civilization, as we know it, will be seriously threatened and perhaps largely destroyed throughout most of the world." Pharmaceutical corporations needed little inducement—much of the money would come back to them anyway. Ortho, G. D. Searle, and Syntex all joined the "honor roll" of IPPF contributors by pledging $150,000 each. But the competition for money and attention from the uncommitted led to ever more extreme appeals.[44]

Leaders of poor countries sometimes stoked these anxieties, especially during aid negotiations. "Asia is in an explosive state," Gandhi told Johnson, because the poor would not put up with deprivation indefinitely. Johnson came to believe that America's overseas bases constituted the first line of defense. "There are 3 billion people in the world and we have only 200 million of them," he told troops guarding the Korean demilitarized zone in November 1966. "We are outnumbered 15 to 1. If might did make right they would sweep over the United States and take what we have." The very next day, the State Department's special assistant for population, Philander Claxton, called for a worldwide program costing $150 million, or double or triple that with the use of incentive payments and pills. Otherwise, he predicted, there would be "more intense and widespread social dissatisfaction, armed insurrections, and political upheaval which will directly affect the safety and welfare of the American people in the most serious way." Draper and Rockefeller prepared charts for the president demonstrating how, if trends continued, famine would come to India, Pakistan, and China by the early 1970s, Egypt, Indonesia and Peru by the mid 1970s, and most of the rest of Asia, Africa and Latin America by the 1980s.[45]

But even the most rapid population growth could never be proven to have caused any particular crisis or emergency. And even if people went hungry, did that mean they would march on America? In other countries too, analysts struggled to explain how large families led to anarchy, and why poverty threatened anyone but its victims. "Are we really liable to have a breakdown in law and order in some of the large metropolitan areas in the poorer countries?" asked William Clark, director of the Overseas Development Institute. "This would involve some very strange changes in the present power structure." Canadian prime minister Pierre Trudeau warned vaguely that the growing populations of the poor would create "pressures" on the rich to "make decisions of frightening moral consequence." He quoted Chateaubriand's observation that a poor man would not tolerate gross inequality once he became aware of its magnitude: "in the last result you would have to kill him."[46]

Seeking vivid images, authors and publicists tried out different techniques to make population growth seem more "explosive." Harrison Brown happened upon a solution when he asked an audience to imagine what would happen if population growth continued to accelerate. Eventually, people would not only cover the entire planet, but would form a "sphere of humanity" expanding at the speed of light. Such rhetorical devices demanded an extremely long-range perspective—5,300 years, in this case. In the meantime, how would wealthy Americans and Europeans—who were expected to pay for population control—be threatened if Asians and Africans continued to have large families?[47]

An alternative approach was to recall that "population is people," and that continued growth would cause human suffering. But if population growth was people, and it was also a problem, some drew the implication that certain people were inherently problematic. In 1967, for instance, the Bihar famine and a new initiative in family planning inspired Oxfam to try a new kind of publicity campaign. Posters on some nine thousand billboards told Britons that "Oxfam HATES hungry children," and asked them to "Help Oxfam STOP feeding hungry children." It was clever marketing, but staff were taken aback by the number of letters advising that they "sterilise the bastards."[48]

Propagandists for the Population Crisis Committee decided to put the issue in personal terms. Full-page ads in the *New York Times* and the *Washington Post* featured an out-of-focus shot of an elderly man throttled by a young attacker, asking readers: "Have you ever been mugged? Well, you may well be!" The text explained that "city slums—jam-packed with juveniles, thousands of them idle—breed discontent, drug addiction and chaos." It linked population growth to rising rates of forcible rape. A "*crash program* for population stabilization" could not only reduce crime—as well as improve the environment and "quality of life"—it would save spending on welfare.[49]

Yet to portray the population issue in such a way risked confirming the suspicion that it only concerned affluent white people. Draper was appalled by the ad, and Planned Parenthood publicly disassociated itself from it. When most of the movement's leaders said "population is people," they did not mean muggers and communist insurgents, but needy people, preferably small children. As Robert McNamara put it in his first major address on population as president of the World Bank: "They are not mere statistics. They are human beings. And they are dying, now; at this very moment; while we are speaking." Indeed, those dying were "fortunate," because millions more would "live languidly on," though "stunted in their bodies, and crippled in their minds."[50]

Everyone understood that vivid imagery, not statistical data, is what created the requisite sense of urgency. The IPPF therefore dispatched an experienced photojournalist from the *Observer* to South Korea and Thailand, suggesting as possible subjects "abandoned children" and "the despair of a woman with a large family." The man spent several weeks traveling to clinics and even backstreet abortionists, taking hundreds of shots. But none seemed satisfactory. As the IPPF information officer tried to explain, they needed photos that were "dramatic and full of human drama. . . . I do hope you will bear in mind our need for drama!" After another unsatisfactory consignment arrived by post, she spelled it out for him: "I would welcome photographs showing through facial expressions the despair of parents with too large a family, and their relief at receiving family planning information." Even when officials had artists paint such images of happy and unhappy families, they could not control how they were seen. In Kerala, for instance, people pitied the handsome family because it had only a few children.[51]

An unanswerable argument remained to population control proponents: if crash programs failed, even worse methods would follow. As explained by the director of the family planning program of Church World Service, "This generation is probably the last which will have the opportunity to attempt democratization of voluntary family planning. Failure will mean ei-

ther mass starvation amid holocaust or the institution of pregnancy police."
Rather than revealing the bankruptcy of the crisis management approach to
family planning, the failure of crash programs was used to justify ever more
drastic measures, since someone could always imagine something worse.[52]

But the most effective propaganda for population control in the period
did not threaten or cajole, or invoke poor victims. It played on anxieties
about crime, contagion, and mass migration, but without actually naming
them. It made people feel, viscerally, that it was already too late, and that
they were living a nightmare. It began in 1968, when the executive director
of the Sierra Club, David Brower, heard a Stanford entomologist speaking
on the radio. Paul Ehrlich had first become interested in population when
he noticed how habitats for butterflies were being paved over and DDT
spraying made it harder to raise them. Brower realized that Ehrlich had the
credentials and the media presence to insert population into the ecology
debate. He commissioned a short paperback and, after barely a month, was
pleased to receive "a first-rate battle tract." It was called *The Population
Bomb*.[53]

There were many more books like it, with titles like *Hungry Nations,
The Hungry Planet, Born to Starve,* and *Breeding Ourselves to Death*. Ehrlich
stood out because he made himself—and his family—part of the story,
threatened by the "population bomb" in a way that his readers could readily
understand. They had been in India in the summer of 1966 to study butterflies. Ehrlich seems to have been utterly unaware of the IUD and sterilization campaign that was unfolding all around him. But the experience of
riding a flea-ridden taxi on a "stinking hot night in Delhi" provided the
ideal device to open his book:

> As we crawled through the city, we entered a crowded slum area. . . .
> the streets seemed alive with people. People eating, people wash-
> ing, people sleeping. People visiting, arguing, and screaming. People
> thrusting their hands through the taxi window, begging. People defe-
> cating and urinating. . . . People, people, people, people.

Ehrlich wrote that he had understood the problem of population growth
intellectually, but now he could "understand it emotionally." For the first
time he had experienced "the *feel* of overpopulation."[54]

Ehrlich could have encountered far larger crowds on a hot night in New
York or London. What seemed to disturb him was not their numbers, but
their quality—that is, their race and poverty. Ehrlich considered himself a

liberal and was taken aback by this criticism. He would go on to write books condemning racism. But he probably connected precisely with those readers who had imagined getting lost in a large city and ending up in the wrong neighborhood—not Delhi, but Harlem or Watts. Only Ehrlich invited readers not just to imagine a wrong turn, but to recognize that America—all of it—was turning into a bad neighborhood. Thus, every time in the following pages he described population growth in poor countries, he called on readers to imagine what it would mean for America. How, for instance, could the United States manage if 200 million people were "dumped" on it in thirteen years, as he said would occur to India.[55]

When Ehrlich finally suggested that population growth would disturb more than his readers' consciences, that indeed poor people were likely to "attempt to overwhelm us in order to get what they consider to be their fair share," readers would have been ready for drastic solutions both at home and abroad. He called for population control for the United States, "hopefully through a system of incentives and penalties, but by compulsion if voluntary methods fail." For other countries he adopted the concept of "triage." He urged the United States together with other developed states to abandon areas of the world that were beyond help by cutting off food aid. As for the others, the first step would be to distribute transistor televisions for communal viewing. Each community would in this way receive the warning that food aid depended on using contraception.[56]

The book sold well. But it took off only when Johnny Carson invited Ehrlich to appear on his show in January 1970, overriding the objections of his own production staff. Ehrlich's appearance prompted more calls and letters than any other guest during the preceding months. When Carson invited him back three weeks letter, he let Ehrlich give the address of Zero Population Growth. By March 1971, ZPG had thirty-two thousand members. As for *The Population Bomb*, by 1974 it had gone through twenty-two printings and sold two million copies. It provided practical instructions in how to proselytize friends and pressure political leaders, including sample letters. Ehrlich even urged readers to write letters to editors and "complain bitterly about any positive treatment of large families" in the media. Congressmen and senators reported receiving multiple copies of the book. Gerald Ford found it "quite startling," and several more voiced their support, including William Proxmire and Claude Pepper.[57]

At this point, Congress was only prepared to increase funding for family planning, and it was careful to specify that it must be voluntary. But Paul

McCloskey (R-CA), who would run against Nixon in 1972 on a peace platform, concluded that if voluntary family planning failed, "we will have to
confront head-on the question of legal sanctions against individuals who
intentionally or negligently sire or bear children they cannot or will not
support." And just as Ehrlich called for a federal department to oversee
population and environmental policy and research—including "development of mass sterilizing agents"—Congressmen Emilio Daddario and Tom
Mosher introduced a bill that would reorganize the Department of the Interior to deal with population problems. Daddario acknowledged that a
policy on procreation had "gruesome potentialities," like those envisioned
by Aldous Huxley and George Orwell. "But we are reaching the point
where we have no choice but to curb and disperse, somehow, the accelerating crush of a crawling, sprawling humanity which is voraciously stuffing
its collective maw with more and more of the irreplaceable resources of this
planet."[58]

For those who wanted to regulate population numbers and "quality,"
controlling borders usually seemed easier than controlling the fertility of
fellow citizens. In this period, Europeans and Americans began to organize
against migration from the Global South. The trend reflected such factors
as the disintegration of empires, rising unemployment, and women's entry
into the workforce. But in Britain and France as well as the United States,
concerns about the high fertility of both migrants and the countries from
which they came was a common refrain. Here again, domestic and foreign
population policies intersected and started to go far beyond family planning.

This trend began with the regulation of movement within decomposing
empires. The United Kingdom ended the old tradition of free entry in
1962. By 1968, with a rising popular clamor against continued entry by
nonwhites, the Labour government barred the door to British citizens of
Asian descent who were being expelled from Africa. Even so, the Tories'
shadow defense secretary, Enoch Powell, warned that Britain's growing minority communities would eventually provoke the same civil unrest then
occurring in American cities, and called for promoting their expulsion in
terms never to be forgotten:

> As I look ahead, I am filled with foreboding. Like the Roman, I seem
> to see the river Tiber foaming with much blood. That tragic and in-
> tractable phenomenon which we watch with horror on the other side

of the Atlantic but which there is interwoven with the history and existence of the States itself, is coming upon us here by our own volition and our own neglect.

Four years later Britain's leading environmentalists issued a "blueprint for survival" that demanded governments commit to policies to stop population growth, including "an end to immigration."[59]

France had long favored both immigration and high fertility, albeit with reservations about particular ethnic and religious minorities. Muslim families who loyally opposed independence for Algeria were prevented from entering the metropole even when, at war's end in 1962, they faced death. In fact, de Gaulle conceded independence in part because it provided France with a legal basis for stopping Muslim immigration. It continued because a growing economy required still more manpower. In 1967 de Gaulle reconvened the Haut Comité de la Population, which noted that unassimilated immigrants and their offspring constituted a growing proportion of the population. Rather than continuing to subsidize every birth, the government elected "to improve the quality" by legalizing contraception and favoring medium-size families. In July 1974, France finally ended the policy of recruiting North African workers. That same year, the countries of the European Economic Community agreed on the need to coordinate immigration policies, with most states calling for a common approach to stopping illegal entry.[60]

In Washington, supporters of the Immigration Act of 1965 had promised that dismantling discriminatory quotas would not lead to a major increase in immigration, or even change its predominantly European character. The family reunification provision was expected to benefit those already related to the "American family," that is, Europeans. When, instead, the process of "chain migration" dramatically increased the numbers of immigrants, especially from Latin America, Africa, and Asia, Zero Population Growth lobbied for restriction. Most American environmentalists did not want to make immigration a conservation issue, and perhaps only a few had any interest in crypto-eugenics. But both Hardin and Kingsley Davis—who advised ZPG—wanted to work toward a day when people would accept constraints on their fertility, and stopping immigration seemed like a good place to start. Davis judged that "human genetic control seems bound to occur, unless all progress is halted." Hardin, for his part, admitted that *coercion* was usually considered a dirty word, but it need not be—it could

be "cleansed" by "saying it over and over without apology or embarrassment."[61]

These debates had global implications because participants understood
issues of population increase, quality, and migration to be parts of a larger
whole. Rich countries could become like India, in other words, if they
too failed to contain population growth, if that growth compromised the
"quality of life," if too many poor people moved north, or some combination of the three. And for those who were thinking globally, local measures might not be enough if the rest of the world became like India—more
hungry, less healthy, and increasingly discontent. When coercive population control came to be implemented on a massive scale in India in 1976,
it would not, therefore, be a merely local development. It would be the
culmination of a worldwide campaign calling for ever more extreme measures.

Yet going "beyond family planning" meant more than just reconsider-
ing individual rights—whether of prospective parents or potential immigrants—and it did not necessarily have to lead to more coercive forms of
population control, much less to what an earlier generation would have
called "world eugenics." It began to inspire a searching examination of the
whole range of political institutions and cultural practices that shaped populations. In a sense, this was a return to the early, more sophisticated articu-
lations of family planning. It was only with the shift in focus from differential and declining fertility in the industrialized West to high fertility in the
Global South that "family planning" became synonymous with supplying
contraception, rather than shaping the demand for contraception. Demographers might have liked to do more, but they did not have great expectations. As Ronald Freedman put it in 1958: "the crucial research question is:
what minimum change in the social environment is necessary to make a
perceptible change in motivation toward fertility?" Ten years later, demographers were still uncertain about the answer, but they were finally in a position to demand more than minimal changes. Education, health care, and
even infrastructure projects would be designed with an eye to shaping fertility preferences.[62]

Some countries always took a broad view of "population policy." This
often meant giving priority to maternal health care or even urban planning.
The demand to "populationize" all kinds of development policies and projects was, in part, just bureaucratic politics, as officials protective of their
budgets scrambled to rebrand programs. Those committed to population

control had long recognized that improving health and nutrition was more appealing than merely reducing fertility—what some Latin American administrators derided as the "veterinary approach." And evidence of the link between women's education and preferred family size continued to accumulate.[63]

But officials intent on slowing population growth were wary of losing focus. Sam Keeny of the Population Council argued that "most talk of 'integrating' [family planning] into [maternal and child health] as WHO does, or Ceylon has done, often means losing the ballgame." Administrators not primarily concerned with population control could not "get the job done." McNamara said he was reluctant to finance health care "unless it was very strictly related to population control, because usually health facilities contributed to the decline of the death rate, and thereby to the population explosion." Even when administrators acknowledged the correlation between women's access to education and lower fertility rates, they asked whether less costly measures might achieve the same goal.[64]

The World Bank and other donors backed many programs that had only a tangential connection to controlling population growth because agencies with different interests did not give them good alternatives. But as competition increased, donors could be more demanding. As Howard Taylor put it, the essence of "integrated" projects was to determine "how much health will yield how much family planning." Ravenholt cut off funding for the Pan American Health Organization because it assigned an "unduly large emphasis on the introduction of maternal and child health activities into family planning programs, *rather than the reverse.*" His representative in the Philippines observed that, with more requests for aid, "it becomes increasingly important to show progress in the only terms which ultimately matter—births averted (or declining growth rate)."[65]

The search for more subtle means to shape reproductive behavior would make population control even more ambitious as an experiment in social engineering. At its peak, these programs amounted to only 2.3 percent of total development aid. But that was also when aid administrators began to ask how they might use the rest as leverage to demand a whole range of indirect measures to "show progress in the only terms which ultimately matter." What came to be known as "integration" could include raising the age of marriage, legalizing abortion, and encouraging girls to stay in school, but it could also mean calibrating access to health care and public housing according to parents' reproductive behavior.[66]

In many cultures, the very idea of separating sex from childbearing was subversive. This included the California suburban communities that revolted in 1968 when they discovered what children were learning in sex education classes. Even the most conventional family planning programs included an "information-education-communication" component to counteract pro-natalist cultural norms. Much like Ehrlich and ZPG, the lead Ford consultant in India wanted "to make people ashamed of having more than three children." All over the world, posters, films, flip charts, and folk performances depicted the "unplanned family" as unclean, unhealthy, violent, and ugly. This included the 1967 Disney film *Family Planning*, commissioned by the Population Council, which was translated into twenty-four languages. It gives Donald Duck the task of crossing off unwanted children. Planned families were not merely smaller, more beautiful, and athletic, but surrounded by consumer goods, and sometimes explicit promises of better housing, health care, and education—a veritable garden of earthly delights. Family planning promised miraculous results, to the point that the planned/unplanned family resembled nothing so much as a medieval altarpiece, with viewers offered the choice between heaven and hell.[67]

For those who believed that too many children were a source of poverty, the depiction of planned and unplanned families conveyed an obvious truth. Family planning propaganda could be read as empowering, part of a broader social trend that encouraged "self-actualization." But appearing to replace children with transistor radios also encouraged a consumerist attitude toward something that many people considered sacred and mysterious. When respondents in fertility surveys said that it was for God to decide how many children they would have, family planning workers persuaded them to think otherwise. The Disney film, for instance, insists that "deep down in their hearts all men are the same—they want the same things for themselves and their families." It depicts an "everyman" effortlessly flying through the air. The ability to plan "may someday make man master of all he surveys." Audiences are shown that it is their decision, not just when and how they would have children, but whether their families would be rich or poor, beautiful or ugly. Researchers even considered a skin-absorbent contraceptive that could be worn in a ring or applied with cosmetics. This was meant to underscore the association of family planning with a new aesthetic alien to most societies, which associated beauty with fertility, rather than the reverse. And whereas households in many cultures were multigenerational, and fosterage and adoption practices could involve the extended family in "family planning," all of these efforts posited that the nuclear family should be the global norm. American economists even likened parents' decision to have a child to the purchase of a washing machine. The government of Singapore instructed citizens to think of the third child as "a luxury." At the same time, it warned that "the fourth and fifth are anti-social acts."[68]

In one crucial sense family planning was fundamentally conservative, at least for those who assumed that the two-parent family really was the world standard. The Disney film, for example, instructs fathers that "when he uses family planning, each man is more of a master in his own house. He can now have the freedom to decide when to have children." The wife lets the man speak for her, whispering in his ear. As Davis pointed out, this kind of family planning reinforced the assumption that such families were the building blocks of any society, whether locally, nationally, or globally:

> It stresses parental aspirations and responsibilities. It goes along with most aspects of conventional morality, such as condemnation of abortion, disapproval of premarital intercourse, respect for religious teachings and cultural taboos, and obeisance to medical and clerical authority. It deflects hostility by refusing to recommend any change other than the one it stands for: availability of contraceptives.

The population establishment had set out to persuade poor people to stop making poor choices. It fretted when they did not comply, and considered offering parents abortion and incentives, or coercing them through disincentives and involuntary contraception. "All of us have a responsibility toward the family of man and especially to those we love," Disney's narrator intones, as Donald Duck points at the audience, "including you!"[69]

Going "beyond family planning" could mean going farther still, to the point that critics began to argue that the patriarchal family stood in the way of progressive change. In *The Feminine Mystique,* Betty Friedan had posited that the pressure to marry young, drop out of college, and have large families was all part of "the problem that has no name," a problem that was also causing "the population explosion." Judith Blake, a colleague of Davis's at Berkeley, argued that the standardization of sexual roles, and particularly women's status as wives and mothers, was increasingly dysfunctional. Like Davis, she pointed out that most people, and especially the poor and uneducated, wanted relatively large families. But whereas Davis thought that a set of policies to dissuade them "reads like a catalogue of horrors," she argued that it required removing penalties against anti-natalist behavior. This included ending sexual "indoctrination" in schools, abolishing legal and social sanctions against homosexuality, and eliminating tax breaks and housing policies that rewarded procreation. She urged "those who have the power to guide our country toward completing the vital revolution."[70]

Blake's intellectual inquiry recalled the earliest roots of the birth control movement, when pioneers like Elise Ottesen-Jensen had critiqued patriarchy and championed progressive sex education. These roots had grown tenuous through a series of political compromises, culminating with the consolidation of a well-funded and well-connected establishment committed to the most conservative version of family planning. Now that it seemed likely to fail, there was an opportunity to renegotiate. It required many years of activism, and not just appeals to "those who have the power," before Blake's critique of gender roles entered the mainstream. In 1970 Notestein provoked laughter at the National War College when he noted with exasperation that "one even sees homosexuality defended on the grounds that it helps curtail population growth!" Yet the mere fact that such an audience had to consider this perspective is more noteworthy than the alacrity with which they dismissed it.[71]

With hindsight one can see more clearly the significance of what Blake and others like her were attempting. But one can also overlook the complexity of this struggle in the context of its times. It might appear, for instance, that there were really just two ways to go beyond the family planning consensus that had prevailed up to this point—one favoring responsibility and social order, the other rights and empowerment; one side coercive, manipulative, and patriarchal, the other liberated, principled, and tolerant. Yet in this period the politics of population were particularly fluid. Indeed, Davis and Blake—who might seem to personify these different tendencies—were married to one another. Some of the same people who advocated increasing women's access to education would also pay the poor to be sterilized and penalize those who resisted. The technologies that helped women control their bodies without having to answer to men, such as injectable contraceptives and "menstrual regulation kits," also appealed to those who simply wanted to control populations more effectively.

These apparently opposing tendencies often proceeded side by side, and it was not always clear to contemporaries how to tell them apart. What now seems indivisible, reproductive rights and health, appeared to present tradeoffs. Did women's right to contraception, for instance, make it wrong to insist on medical screening? Raising the legal age of marriage seemed like a comparatively painless way to both limit childbearing and improve women's status. But some Sri Lankans perceived it as a form of coercion, and resented it as such. India initially intended to legalize abortion only on condition that one of the parents also accept sterilization. Sex-selective abortions would make the notion of "choice" even more complex, if not deceptive. After all, how did controlling reproduction empower women in a context in which virtually all the fetuses aborted were female?[72]

For the likes of Friedan and Germaine Greer, empowering women and promoting tolerance were not just more subtle means to reduce fertility. They were good in and of themselves. But such ideas might not have gotten nearly so much attention if they did not also offer a more effective way of controlling population growth, especially among poor people. Rather than seeing them as mutually contradictory policies, people with command of budgets and staff considered them different paths to the same end. Rather than deciding based on which was right, they wanted to know which would work.

Yet in this period many people began to change their minds and started to think anew. The most dramatic example was the crisis in the Catholic Church. In the early 1960s, theologians and bishops from Germany, Belgium, and Britain became alarmed at the declining numbers coming to confession and mass, and asked whether so many Catholic couples could be in a state of mortal sin. Paul VI admitted that the issue raised "extremely complex and delicate questions," that he needed expert advice, and that his conscience might compel him to change the Church's position.[73]

In an unprecedented move, but in keeping with the spirit of Vatican II, in 1965 he expanded a Pontifical Commission for the Study of Population, Family and Births to include a majority of laymen and -women from different walks of life. Among the new members were Pat and Patty Crowley, leaders of the Christian Family Movement, who had carried out a survey of about three thousand Catholic couples in eighteen countries. For many decades, Rome instructed priests and the Catholic media to warn the faithful of eternal torment in hell if they used birth control. The Church had publicly endorsed "natural" family planning after 1951, and the basal temperature method made it more reliable (if more onerous). Sex within marriage was increasingly accepted as a positive good, not a necessary evil, and sermons blaming women for men's failings were becoming rare. But the survey showed that even members of the Christian Family Movement had found the new teaching to be contradictory and increasingly frustrating. Patty Crowley had to explain to the assembled cardinals and bishops that daily recording of one's rectal temperature and abstaining during what some considered the peak period of sexual arousal was not only a cross to bear, it did not even work for all too many who tried it. Her testimony had a major impact.[74]

The papal commission also included demographers, including some closely tied to the population establishment, who argued that population trends might compel change. They worried that, if they continued, "governments which formerly had not observed any demographic policy will suddenly find themselves, over-night, obliged to adopt brutal measures." But they also advised that any papal message should be a moral pronouncement directed at all married couples, and "avoid any reference which might be interpreted as the Church taking a stand in favour of systematic limitation of the growth of population in the under-developed countries."[75]

In June 1966 an overwhelming majority of the commission, including most of the bishops, approved a report that endorsed "responsible fruitfulness" and specified criteria by which Catholics could assess particular contraceptives. But more than a year later, the pope had still not issued a new encyclical. Commission member John T. Noonan had shown in the course of their deliberations how Church teaching on contraception had changed and changed again over the centuries. Many Vatican officials were concerned that if the pope now changed once more, it would undermine papal authority. We cannot know for certain, but critics believe that Paul "reconverted" for this reason. He was quoted at the time as being worried that, otherwise, theology would "become the servant of science." Instead, he reaffirmed that the Vatican answered only to God.[76]

Like his predecessor, when publicly explaining his position Paul stressed the corrupting influence of sex without procreation, not merely on individuals but also on states and societies. He made a particular point of alerting the faithful to what might happen if he ceded his role as patriarch to a powerful government:

A dangerous weapon would thus be placed in the hands of those public authorities who take no heed of moral exigencies. Who could blame a government for applying to the solution of the problems of the community those means acknowledged to be licit for married couples in the solution of a family problem? Who will stop rulers from favoring, from even imposing upon their peoples, if they were to consider it necessary, the method of contraception which they judge to be most efficacious?

What was overlooked at the time, and since, is how the pope accepted that population growth was indeed a problem, one that would impose hardships on the poor. Unlike Pius XI, he did not describe contraception as a "mortal sin," and he enjoined priests to be patient with their parishioners. But for most people the bottom line—announced in banner headlines—was that the pope still banned birth control.[77]

Ironically, rather than upholding Church authority, the encyclical provoked unprecedented dissent. Lay members of the commission openly disagreed with the pope's decision, and insisted on their right to do so. Hundreds of theologians in the United States and Europe signed petitions declaring that the pope was not infallible and that Catholics could exercise their individual conscience. One of them, Father John A. O'Brien at Notre Dame, worked with Hugh Moore in collecting thousands more signatures from scientists, publishing their protest statement in major newspapers, and sending translated versions to Church leaders all over the world. Bishops in thirty-seven countries felt obliged to convene national conferences and offer their own commentary. None publicly challenged the pope's authority, but many—including those in Austria, Belgium, Canada, the Netherlands, England, and Wales—offered support for the idea that Catholics should also be guided by their conscience. When analyzed in terms of the 1,556 dioceses represented, only 17 percent offered unequivocal support, 28 percent were ambiguous, and fully 56 percent of the bishops backed statements that mitigated the force of the encyclical. Patty Crowley continued to organize opposition among the faithful. "It's my Church," she explained, "no one's going to kick me out." Three New York women went on to found Catholics for a Free Choice in 1973 to protest Church efforts to ban supporters of the *Roe v. Wade* ruling from communion.[78]

The Church would increasingly focus on abortion, since it could rally support even among those who ignored the ban on birth control. Even some feminists remained ambivalent about it, mainly over differences in tactics—reform vs. repeal—but perhaps also because population controllers with their own agendas were initially among the most determined backers of full legalization. Hardin and ZPG lent crucial support in the early days of the National Abortion Rights Action League. When Friedan had helped convince the National Organization for Women (NOW) to back legalization, some delegates resigned. "Feminists for Life" argued that abortion exemplified society's failure to meet women's needs. It was only gradually that grassroots activists made the right to end an unwanted pregnancy a unifying cause for the great majority of feminists.[79]

The growing complexity of reproductive politics was evident in the controversy over contraceptive safety, beginning with continuing questions about the pill. By January 1970, when U.S. Senator Gaylord Nelson called hearings on the issue, concerns about the health risks divided even those committed to reproductive rights. Users had been found to have an elevated risk of thrombosis, leading British authorities to issue warnings against high-dose pills. In a book called *The Doctors' Case against the Pill,* Barbara Seaman related painful, firsthand testimony from women who had never been warned about side effects, as well as from doctors concerned that there might be a still greater risk of cancer. Though Seaman helped Nelson's staff line up most of the witnesses, neither she nor any other women were invited to appear.[80]

Seaman came to the Senate hearing anyway, covering the first day as a reporter. "All of a sudden," she later recalled, "these women started standing up and yelling . . . I heard my name, 'why isn't Barbara Seaman testifying?' And then somebody else was saying, 'why isn't there a pill for men?' And someone else was saying, 'why aren't there any patients testifying?'" Senator Nelson was sympathetic to the protesters' concerns, and eventually succeeded in compelling pharmaceutical companies to insert warnings about potential side effects in every package—a landmark in the history of consumer rights. But he chided the "girls" for the disruption, and they were ordered out of the hearing as television cameras rolled. It was just the beginning of a long struggle that made it impossible to ignore women's voices when making policy affecting women's health. Seaman, who until this point merely sought reform within the medical establishment, decided to join the activists in their efforts to hold it accountable, going on to cofound the National Women's Health Network.[81]

Questions about the safety of contraceptives provoked discord even in closed-door meetings of groups that otherwise agreed on the need to give priority to controlling population growth. In a 1971 discussion of a Population Council program to support promising research, IPPF secretary-general Henderson worried that it would concentrate too much on physician-administered methods.

> She also feared that prestigious committees would back off too quickly from methods that entailed certain risks and noted that IPPF has 75 nongovernmental clinics that might be prepared to take on testing of somewhat more risky compounds than those that would be tested by government programs.

Segal denied that they would neglect self-administered methods. "He did state forcefully, however, that they are not prepared to push forward on methods that entail substantial medical risks." Berelson added that "it is usual to trade off individual risk against social benefit, [but] there is now a movement to compare individual risk with *individual* benefit."[82]

It was already becoming clear that such trade-offs might be subject to public scrutiny and legal action. Lawsuits eventually bankrupted A. H. Robins over wrongful deaths and injuries from the Dalkon Shield. Medical doctors also faced malpractice claims from women, and even teenage girls, who had been sterilized without their consent. Some of the most notorious cases occurred on Indian reservations and among African Americans in North Carolina. Doctors on the payroll of the Office of Economic Opportunity threatened families with loss of welfare benefits and medical care much as their counterparts were doing in India.[83]

In the late 1960s and the 1970s, many people became aware of the dark chapters in the history of eugenics and medical experimentation, and aware too of how that history was hardly just a thing of the past. It was only then, for instance, that sustained debates began about the unique nature of the Holocaust—a term little used before—not just in the United States, but also such countries as France, the Netherlands, and Germany. In films, television, and investigative journalism, people began to relate past and present in ways that undermined the political legitimacy of population control. In July 1972 the revelation of a government experiment on syphilitic black men in Tuskegee—who were studied for forty years but never treated for their condition—provided startling new evidence that eugenics posed an ongoing threat, if only because so many professionals had been trained in racist science and medicine. In a sign of the times, *Eugenics Quarterly* was renamed *Social Biology* in 1969, and three years later the American Eugenics Society became the Society for the Study of Social Biology.[84]

This was hardly the end of eugenics. In many countries, it was still considered a legitimate basis for adjudicating social rights and responsibilities. In 1972, for instance, the Council of Europe's Consultative Assembly narrowly defeated a recommendation for easing national legislation restricting abortion "to allow for surgical interventions for therapeutic and eugenic purposes." That same year India made eugenics a criterion for access to abortion. In this context, most proponents probably had in mind genetic counseling, in which prospective parents would be helped to make more informed choices. But many still understood eugenics as a way to engineer qualitatively better populations. In 1975 the chairwoman of the Indian Council of Child Welfare called for sterilization of the feebleminded. Eugenic improvement would also be an important motive behind China's one-child policy. As long as eugenics was defined broadly enough to encompass different concerns about "the quality of life" and alternative cures for social problems, it would continue to shape the way people thought about population.[85]

But in the United States, eugenics was becoming synonymous with racism and "pseudoscience," discrediting the larger cause of population control. This came about not just through the work of opponents, but also supporters like Nobel laureate William Shockley. In the 1960s, he suggested that people with below-average IQs might be given incentive payments for sterilization, and that girls could be injected with a sterilizing capsule so that the government would decide who could have children. Berelson included his ideas among the proposals to go beyond family planning. Shockley's own training was in physics, but he became interested in the subject after traveling to India—here again, not unlike many other population controllers. But whereas more established leaders of the movement now denied any interest in racial differences, Shockley began to demand that the National Academy of Sciences study why blacks scored lower in IQ tests. The NAS finally empaneled a committee under Kingsley Davis in the hopes that it would put an end to the controversy. Instead, their report refused to rule on the merits and called for additional research. They insisted on the "freedom to pursue scientific thought and investigation without hindrance because of fear that the results may be misused or because the conclusions may be unpalatable."[86]

This kind of controversy corroded the popular faith in science that had been essential to the cause of population control. It was not just guilt by association with a shady past, but the growing suspicion that research always reflected political agendas of one kind or another. Davis went on to argue that family planning had become "a vested interest seeking government funds." Demographers faced the same accusations from their own research assistants. This new generation condemned the establishment, not because it had failed to control population growth, but because they considered the whole enterprise to be tainted with racism and classism. In 1969, graduate students at Wisconsin, Michigan, and Cornell founded the journal *Concerned Demography*, complaining that "demographers are being used as administrators and public relations men for government family planning programs around the world."[87]

At the 1970 Population Association of America conference virtually the entire membership closed ranks against the students. They could be dismissed as radical malcontents, and Davis written off as a gadfly. It was more difficult to ignore the accumulating body of research showing that high fertility was not, after all, correlated with poverty. No one was more persistent and provocative in arguing this position than Julian Simon. In the late 1960s, he had backed payments for sterilization and published analyses like Enke's purporting to prove that poor countries could get rich by reducing their population. But together with Simon Kuznets and Richard Easterlin, he found that actual data did not support this conclusion. Instead, he was persuaded by work like that of Ester Boserup showing how population growth could actually spur innovation. Simon also had a flair for publicity, challenging Ehrlich and two Berkeley physicists to a bet in which, after a decade, either he or they would pocket the difference in the price of a thousand dollars' worth of commodities like copper and tin. Ehrlich was betting that, because supplies were finite and population continued to grow, metals would become more expensive. In fact, they cost less in 1990 than in 1980. He had to send Simon a check for $576.07.[88]

While economists typically worked with large data sets at the macro level, anthropologists like Mahmood Mamdani used ethnographic research to show why poor people continued to have large families, and why demographers misunderstood them. It was not just more fine-grained research that led Mamdani to go on the attack in *The Myth of Population Control*, which became an instant classic when published in 1972. He resented the way demographers jumped to the conclusion that those who resisted their efforts did not value human life. He came to agree with one of his interview subjects, who one day told him that they "were enemies of the smile on this child's face. All they are interested in is war or family planning."[89]

Similarly, Simon was converted, not just by a different reading of the data, but also by a realization that life is worthwhile to those living it, for poor people in India no less than affluent people in the United States. Hearing Paul Ehrlich on TV, he later recalled, "absolutely drove me out of my skull." Yet Ehrlich himself had become disturbed by some of his fans, "who seem mainly interested in controlling other people's populations." In 1970, he co-wrote with his wife, Anne Ehrlich, a statement titled "Population Control and Genocide," in which they argued that African Americans had ample reason to resent those who focused on their fertility. Because affluent white Americans did far more damage to the environment, any government program should aim first to reduce *their* birth rate.[90]

Population debates "beyond family planning" became unpredictable, because the issue of who would do the planning, and for whom, was recognized as moral and political, and not merely logistical. It was no longer a simple matter of "Catholics" versus "feminists," for instance, when "Catholics for a Free Choice" squared off against "Feminists for Life." Similarly, the protest of black militants against what they viewed as racially motivated population control was answered by women from the same communities asserting their right to control their own bodies. Even those who agreed that controlling world population growth was a priority—such as environmentalists—bitterly debated whether stopping immigration was a legitimate means, or just a racist canard. None of these debates could be settled by asserting professional expertise. Instead, the public was now directly and sometimes emotionally engaged in exploring the history of eugenics, the ethics of medical experimentation, the biology of genetic differences, the economics of population planning, and the theology of conception.

Leaders of the population control movement responded to these attacks by defending their record and fighting back. They lined up heads of state, major corporations, and international organizations behind a global strategy to slow population growth. But they also worked more quietly to insulate their projects from political opposition by co-opting or marginalizing critics, strengthening transnational networks, and establishing more freestanding institutions exempt from normal government oversight. They provided states willing to participate with unprecedented powers to surveil and control their citizenry. But activists and aid agencies were also prepared to work around governments reluctant to go along. The population control movement was not going to be stopped by outside critics or internal discord. Instead, growing disarray at the top, grassroots opposition from below, and a continuing tendency to remove all checks and balances would send it careening out of control.

8

A SYSTEM WITHOUT A BRAIN

In January 1969 the chair of the IPPF's Management and Planning Committee, George Cadbury, tried to explain to an important contributor why population control could not be pushed through the "corridors of power" by an advisory council of eminent men. For Cyril Kleinwort, one of London's leading bankers, the need seemed self-evident, and Kleinwort had a history of threatening to withdraw support when he did not get his way. "We do indeed need help in bringing our influence to bear in places where it will count," Cadbury allowed. "This is, in fact, our main purpose in life, and much more important than running clinics, or distributing contraceptives." But that task differed in different countries. Such an organization "would have to be as international as we are," he explained, "and that would bring together a strange bunch of bedfellows, including Latin American generals and African revolutionaries now in power, Communist officials and western business men. I should prefer to deal with each of them on his own home ground. As a group I fear they would be ineffective."[1]

Cadbury's message captured the dilemma of a global movement at the apex of its power. An astonishing array of world leaders had now agreed, in the words of a joint declaration organized by John D. Rockefeller 3rd, that a "great problem threatens the world . . . the problem of unplanned population growth." Ferdinand Marcos and Josip Broz-Tito, King Hussein of Jordan and Gamal Abdul Nasser, Mohammed Ayub Khan and Indira Gandhi, Park Chung Hee and Harold Wilson, were among the thirty who urged other heads of state to join in recognizing "that family planning is in the vi-

276

tal interest of both the nation and the family." By 1970, twenty-seven countries had committed to cutting birth rates.[2]

The population control movement was trying to control the population of the world, country by country. Success required working with and through very different constituencies who could agree on little else. Governments that had committed to reducing fertility resisted external supervision. Others would tolerate a private association promoting "maternal health," but not if it was subsidized by the U.S. government, the main source of aid. Even within the United States, family planning in poor communities had provoked a backlash, such that the Planned Parenthood Federation of America redirected funds from the international campaign to shore up the home front. Many more governments, especially in communist and Catholic countries, opposed international aid for birth control—to say nothing of going "beyond family planning"—and attacked it at every opportunity.[3]

Up until this point, proponents had advanced their cause by creating a constellation of organizations, each tackling a different part of what all agreed was a global problem. The division of labor was both geographical and functional: the UN Population Division helped design the censuses and derive projections that made "world population" a measurable problem for both policymakers and propagandists; the Ford Foundation gave researchers the world over financial incentives to focus on fertility; the Pathfinder Fund and the IPPF organized adherents into associations that started clinics and lobbied for official action; the Population Council created new contraceptives and, together with university and UN centers, trained new experts to run standardized programs; finally, well-connected individuals like Rockefeller and Draper, together with elite groups such as the Population Crisis Committee and the Japanese Parliamentary Federation on Population, worked the corridors of power to claim a place at the top of the international agenda.

Once population control had achieved this conspicuous position, and in some places started to become a mass movement, it became more difficult to work in the same way. It was impossible to reconcile all of the environmentalists, feminists, and anti-immigration activists who demanded a place within it. And since their disagreements were often very public, opponents could see how the movement worked together, and how it might be driven apart. The problem was already intractable in the United States. Worldwide, it was mind-boggling. As Cadbury had anticipated, the first time population control proponents managed to assemble political leaders from almost every country—the 1974 World Population Conference in Bucharest—the proceedings escaped their control.

Leaders in the field therefore sought to devise institutional arrangements that put population control above politics. They had long favored free-standing agencies with the autonomy to ignore health ministries and other potential opponents. In this period, with the UN Fund for Population Activities (UNFPA), they took it to a whole new level. This was an international agency that could operate independently of the member nations, recycle population assistance to recipient countries that would not accept it from the United States, and support a host of NGOs that sometimes skirted or ignored local laws. When UNFPA could not manage this global strategy, leaders looked to more informal means to coordinate their efforts and overcome recalcitrant states.

The Catholic hierarchy, buffeted by dissent, was also compelled to adapt. Paul VI was stung by the "revolt and challenge" provoked by *Humanae Vitae*. The Vatican did not initially try to mount a concerted response to population control campaigns, perhaps because it could not count on support even within the Church. While it gradually suppressed internal opposition, efforts to keep birth control and abortion illegal became more decentralized. In the United States, grassroots Right to Life committees were encouraged to take the lead, with the expectation that they might go farther than would the Church itself. Indeed, they were encouraged to "fight emotional fire with emotional fire." The bishops felt justified by the threat that coercive population control would be imposed even in the United States, and that new pills would make abortion routine. But such tactics, in turn, and the more general fear of a gathering backlash by religious conservatives worldwide, encouraged their opponents to resort to population control measures that did not require government oversight or even permission. They disseminated low-cost abortion kits and tolerated the diversion of contraceptive supplies to black marketeers. If the goal was to reduce fertility worldwide, it mattered little who was using these contraceptives, or how they came by them.[4]

Divided from within and besieged from without, leaders created a "system without a brain," setting in motion agencies and processes that could not be stopped. The idea of a "population crisis" provided the catalyst. But this was a system that ran on money. Earmarked appropriations greased the wheels of balky bureaucracies, and lavish funding was the fuel that drove it forward. But so much poured in so fast that spending became an end unto itself. The pressure to scale up and show results transformed organizations ostensibly dedicated to helping people plan their families into tools for social engineering. Leaders pushed "crash programs" farther and faster than ever before, even though the risks entailed were now palpable, including the long-range sustainability of programs that really were devoted to advancing reproductive rights and health. Rather than accept constraints or accountability, they preferred to let population control go out of control. It finally culminated with the Emergency Period, when over eight million Indians were sterilized in a single year. A century after Annie Besant first made India an example of a worldwide population emergency, it marked the beginning of the end of the campaign to control it.[5]

Foreign aid and philanthropic giving are usually tokens of good intentions. To understand how they could lead to more coercive population control requires tracing this money back to its source. In this period, most international aid for population programs came from American taxpayers by way of the U.S. Agency for International Development. Even if the checks were written by UNFPA, or the IPPF, or any number of other international and nongovernmental organizations, they were largely underwritten by the U.S. government. And international aid was more important in shaping contraceptive delivery programs than any other development activity. In such countries as South Korea, Pakistan, Thailand, and Tunisia, it provided two-thirds or more of the family planning budget. That was one reason why it was easy for leaders of these countries to agree with Rockefeller that population control should be a priority: they did not have to pay for it.[6]

Rockefeller was the international ambassador of family planning. But it was William Draper who continued to raise most of the money that turned words into action. Draper was a tireless worker with tremendous charm and persistence. Against USAID's continuing opposition, Draper had persuaded allies in Congress to earmark another $50 million for population control in fiscal year 1969, $75 million in FY 1970, and $100 million in FY 1971 (or $498 million in today's dollars). All this came during a period of fiscal retrenchment. Between FY 1969 and FY 1971, the agency had to lay off 30 percent of its personnel. Even while the rest of the Technical Assistance Bureau, more fortunate than most, cut 10 percent from its bud-

William Draper raised billions for family planning through the international network he developed working in government and banking. Ryoichi Sasakawa, who made a fortune during the Japanese occupation of China and later through organized gambling, was one of many who succumbed to his blandishments. IPPF Archives.

get—including all health initiatives—Ravenholt's Office of Population kept growing. By the end of 1969, USAID had eighty full-time staff working on population. Three years earlier, Ravenholt could not even hire a secretary.[7]

It was not just the size of these appropriations that made the Office of Population powerful. McNamara at the World Bank also wanted to make population control a priority. But the Bank could provide loans only to governments that agreed to take them. Ravenholt, on the other hand, enjoyed extraordinary autonomy in deciding how to spend his budget. Along with the ubiquitous Draper, he was fortunate to have members of his large family occupying strategic positions in Washington. His brother and sister-in-law were on the staff of Vice President Hubert Humphrey before they moved to the office of Senator Daniel Inouye (D-HI). With help from his allies, Ravenholt outmaneuvered opponents in USAID in winning the authority to provide hard-dollar grants outright. He also pioneered a new procedure that permitted recipients to use a portion to hire staff and develop project proposals, with Ravenholt deciding whether they merited the release of the remaining funds. The beneficiaries, including pharmaceutical companies and research universities, helped lobby Congress to continue increasing Ravenholt's earmarked appropriations. In this way, with lawmakers and lobbyists behind him, Ravenholt became a law unto himself. As a sign of his status, he became the only person in the agency to have the "GS-18" pay grade. And whereas the rest of USAID occupied a shabby wing of the State Department building in Foggy Bottom, Ravenholt's staff moved into a high-rise in Rosslyn.[8]

But Ravenholt's budget—and the power it entailed—would continue growing only if his office found ways to spend money as quickly as Congress appropriated it. When the authorizing legislation became law in January 1968, they had only six months to program the first $35 million (approximately $203 million in today's dollars). Within weeks Draper was urging the IPPF to ask USAID for more money. He delivered the same message directly to the Pakistan and India family planning associations, bypassing the IPPF's central office. Colville Deverell was enraged. He warned that the IPPF would "court disaster" if it encouraged family planning associations to expand any faster.[9]

Deverell worried that the IPPF could not control or even account for how associations spent the money. Jealous officials in USAID were in a good position to point out that the central office had no idea where it was going. A program officer in Lagos described "a bad situation" with the Nigerian association, which had been allotted $60,000 in USAID money even though no one from London had come through "in a very long period of time." Much larger sums had been negotiated for Caribbean and Latin American associations—for instance, $450,000 for Colombia's PROFAMILIA—because of the "tremendous pressure and emergency created by A.I.D.'s Latin American Bureau."[10]

After *Humanae Vitae,* the IPPF became even more important for USAID's global strategy. Support for population programs in Catholic Latin America was so sensitive that references to it were deleted from the *Congressional Record.* Draper reported that funds available to the IPPF in 1969–70 were "beyond 'what any of us dreamed of.'" But even the minimal control and accounting USAID initially required went against a guiding principle of the IPPF—indeed, of the whole population control movement. Only by accepting the independence and diversity of its member associations could the IPPF operate in so many different developing countries. Deverell said he would resign if USAID did not accept this, and the risks it entailed. Cadbury shared this concern about how the "crisis mentality" could wreck the IPPF: "Is a sudden access of money," he asked, "especially if it has to be spent quickly, necessarily an unmixed blessing?"[11]

Even in terms of fund-raising, the long-term risks of crash programs were already apparent during this "era of 'big money'"—as the president of the Family Planning Association of India, Avabai Wadia, later called it. The industrialist J. R. D. Tata suggested using some of it to create an endowment for the day when fund-raising became more difficult. But Draper said no. If the whole point was to stop the growth of world population—as opposed to ensuring the long-term availability of family planning services—spending every last dollar as quickly as possible made sense. For the chairman of the IPPF's Governing Body, Cass Canfield, there was no slowing down: "we face a very great crisis in solving the problem of over rapid population growth." The IPPF's leadership had to "drive our engine at high speed, the highest possible in terms of negotiating dangerous curves, and to overcome roadblocks."[12]

The IPPF therefore took all of the money USAID offered and then went back for more. It had originally proposed that in order to preserve an independent identity Washington should underwrite no more than 40 percent of its budget. In 1968 the Swedish International Development Authority gave $486,000 and the UK Ministry of Development provided $130,000. Even so, this was a pittance compared with U.S. fund-raising, private as well as public. Draper obtained contributions from some two hundred U.S. companies, including Du Pont, Standard Oil, Texaco, AT&T, IBM, and Aetna. Altogether, American money accounted for more than 90 percent of the IPPF's 1968 budget. As for private fund-raising abroad, all of it put together, including a $40,000 donation from Oxfam, now amounted to less than 1 percent.[13]

Draper professed to be astonished by these figures. "This is really shocking," he told Canfield, ". . . really pitiful." If it became widely known, Draper warned, Planned Parenthood of America might not go on sharing half its revenue. He therefore had a strong hand when he pressed the IPPF both to meet USAID halfway and to seek other sources of revenue. A new initiative would promote private fund-raising in Canada and the UK. At the same time, Draper developed a plan to target the twenty to thirty richest families in every Latin American country as well as large companies owned locally or linked to multinationals.[14]

A Ford Foundation representative, William O. Sweeney, participated in one such effort. He explained that they wanted to counteract the idea that family planning was "a 'gringo' plot." But in this case, involving the Salvadoran Demographic Association, Sweeney thought it could have been called "Operation Takeover." Three of Draper's American associates flew to San Salvador together with Luis Leite, director of the IPPF's Western Hemisphere office in New York. They used a local contact, Mrs. Francisco de Sola, to arrange a meeting with the rest of the board. "After we all agreed that the lingua franca of the meeting would be English," Sweeney recalled, "in a very fast series of moves it was agreed to establish a fundraising committee." De Sola was made the chair, Leite promised a professional fundraiser, and they agreed to raise $100,000 by year's end.[15]

Even with such heavy-handed direction, non-U.S. contributions could never keep pace with USAID's largesse. Some senior IPPF leaders, like Agnete Braestrup of Denmark, soon to become president, bristled at Draper's demands. In Europe she thought government aid should be the mainstay. In 1969, Sweden increased its support to $729,000, the UK contributed $243,000, and Japan provided its first $100,000. This made possible a concomitant increase in USAID funding, but it was not enough to satisfy Draper.[16]

In April 1969 Deverell resigned. He could not be talked into staying. The new director-general, Sir David Owen—formerly co-administrator of the UN Development Program (UNDP)—quickly agreed with Draper that USAID-supplied contraceptives should be excluded from the 40 percent formula, which would therefore come on top of another $7.3 million in aid. But USAID's director, John Hannah, did not want the IPPF to seem like "a tool of the US government." Excluding commodities was a transparent ruse. "Any critic could easily smoke this out."[17]

Owen also went to the Ford Foundation for a grant to cover the federation's growing overhead expenses. The evaluations Ford requested from its field staff around the world show why, here too, he met with frustration. They gave the IPPF credit for having started family planning programs that spurred government action in Singapore, Barbados, Trinidad, Jamaica, El Salvador, and Costa Rica. Associations in the Philippines and Indonesia were also well regarded. Powerful private organizations in South Korea, Hong Kong, and Taiwan continued to complement or even substitute for official programs.[18]

There was tremendous variation among the IPPF's many members—some existed largely on paper, while others, like BEMFAM in Brazil, PRO-FAMILIA in Colombia, and the FPA of India, operated large networks of clinics. The problem was that all too many were struggling to find a role after governments began providing contraceptives through national health care systems. In some cases the associations claimed to set an example by operating model clinics. But the clinic and poster model itself seemed outmoded. Others tried to shift to commercial marketing, community-based distribution, or more sophisticated information-education-communication (IEC) programs designed to stimulate demand. But all this required new expertise, and even experts disagreed on how it should be done. Family planning associations were expected to expand their budgets, move in new directions, while simultaneously giving way to government programs.

IPPF staff in London and the regional offices were not yet able to provide much guidance, and Ford representatives were scathing in assessing their efforts. Ford's man in Tunisia, for instance, noted the "rather casual and sometimes lavish way in which IPPF representatives have floated around this area and others." Ford representatives described Luis Leite as a strange, even "psychopathic" person who refused to cooperate with them. USAID officials also noted that the New York office's relations with some national associations, such as Venezuela's, were marked by bitterness, which they attributed to IPPF's Byzantine committee structure.[19]

The problem went beyond particular personalities. Leite himself thought that many associations suffered from an "empire building neurosis," which made them eager for more USAID funds and reluctant to cede responsibility to governments. PROFAMILIA was a notorious example of an association that took on large projects even in areas, like IEC, where they "didn't know what they were doing." Ford's review concluded that "these problem affiliates might be less apt to cling to their pill-dispensing careers for dear life if they were steered much more professionally on how to begin relating their activities to budding interests of governments."[20]

In fact, even the most professional staff could not actually "steer" the associations, even with the "power of $" (as Ravenholt put it, when he too grew exasperated after having given the IPPF so many millions). Donors only gradually learned that decisions about how to spend it were all made by committees composed exclusively of volunteers. Owen could not even replace senior staff without permission from the board. To some informed observers, the IPPF seemed like nothing more than "a loose federation of tribal chiefs."[21]

As money continued to pour in, it laid bare the weakness at the very core of the IPPF's volunteer model. In the 1950s and 1960s, it began in most countries by organizing local elites into an association, typically including expatriates and dominated by doctors, which then sought out a prominent politician or his wife as patron. Most associations seemed content to set up urban clinics staffed part-time by volunteers, who rarely ventured into slums or rural areas. Mobile clinics appeared to provide a solution. But because of poor roads they often broke down and became white elephants. If they actually made it to a remote rural area, the affluent volunteers who emerged gave the impression of stepping onto the surface of the moon. They had little understanding of the people they were trying to help.[22]

In the early 1970s, as its budget continued increasing by more than 25 percent every year, the IPPF would undertake a wrenching transition. After his untimely death, Owen was succeeded by another senior UN official, Julia Henderson. She turned the IPPF's "central office" into a true headquarters staffed with professionals; established a system of planning, budgeting, and evaluation; and began to use "the power of $" over member associations. In the meantime, USAID used many more vehicles to expand its population control program, rapidly expanding the Pathfinder Fund, funding a new international arm of the PPFA, and starting up the International Fertility Research Program—"Ravenholt's toolshop" for testing new contraceptives.[23]

But it was becoming clear that USAID could not mount a population control program on the scale required by partnering only with NGOs. A truly global campaign could be carried out only under the auspices of the United Nations. USAID needed the United Nations as "a further channel through which (additional) American dollars may be provided and cleansed for assistance in areas like Latin America," as one British official put it. Similarly, the Swedes thought that a WHO program could "sanitize US money"—that is, remove from it the taint of appearing to serve a hidden agenda. It would also provide a way to funnel still more money to the IPPF without appearing to exceed the 40 percent formula.[24]

Yet the UN system, even more than the IPPF, would prove to be an unwieldy instrument, and for much the same reason. Its heterogeneous and international character made it more acceptable in some countries than USAID, but also rendered it impossible to control. Development agencies with overlapping mandates like UNICEF and the FAO fiercely competed for business. Every one of them ultimately answered to governing bodies of member states, many of whom—like France and the USSR—were hostile to population assistance. UN officials in the best position to organize a program, especially at WHO, viewed family planning as a distraction from their core mission. A sympathetic secretary-general like U Thant could work through his own Secretariat, which included the demographers of the Population Division. But they had no practical experience in running family planning programs.

U Thant's first proposals for an expanded UN program focused on training and hiring more staff for the Population Division and the UN's regional commissions. In 1967 he projected incremental growth in UN expenditures, rising from $1 million to $1.7 million by 1971, including a special "Trust Fund"—the UNFPA—to which countries could contribute for projects to distribute contraceptives.[25] Draper criticized the UN effort as piddling compared with the exponential growth at USAID. He wanted a $100 million program. Together with Rockefeller, Hannah, Woods, and other heavy hitters, Draper published a tough report in May 1969 for the United Nations Association of the United States. After two years, the UN agencies had only just agreed on procedures for drawing on the fund: "The question of mandates," the report concluded, "of which agency should undertake what activity, has been used as a classic delaying tactic by a United Nations system which, taken as a whole, is reluctant to make a more impressive commitment." They called for a population commissioner with wide-ranging powers to administer UNFPA on behalf of the secretary-general, answering only to a committee appointed by donor and receiving countries.[26]

The report appeared like a golden opportunity to one of U Thant's top aides, Chakravarthi Narasimhan, the undersecretary-general for General Assembly affairs. The United Nations had the chance to assert leadership in a vital field with "vastly increased contributions" from the United States. He did not have any new ideas as to how to get UN agencies to develop a coordinated program. He did, however, suggest some novel ways to insulate their work from potential opposition. First, UNFPA field activities could be removed from the control of the UNDP Governing Council, giving the proposed commissioner "more or less complete freedom of action." A board that would be "representative of, but would not necessarily represent," donor and recipient countries could offer "advice and guidance." In fact, it was later explained, it would "meet only once a year and function primarily as a fund-raising group."[27]

A plan thus began to form to make UNFPA an entirely new kind of United Nations agency—one that was not under the control of member nations. The board would consist of family planning stalwarts like Rockefeller and serve at the pleasure of the secretary-general. UNDP administrator Paul Hoffman appointed the first director, Rafael Salas of the Philippines, at Rockefeller's recommendation. When drawing up the new organization's principles and procedures, Salas consulted only governments and individuals favoring population control. He did not submit them to the UN Economic and Social Council (ECOSOC), explaining that he did not want "his wings being clipped." Even representatives of friendly states, like the United Kingdom, found this "to say the least, odd." At the board's first meeting, members encouraged Salas to fund nongovernmental organizations even in countries that had not given prior approval.[28]

In an angry speech to the Population Commission, Alfred Sauvy pointed out that UNFPA had left them completely out of the loop. It was unprecedented for an agency to act in the name of the United Nations without having to answer to member nations, free to dispense money to pharmaceutical companies and family planning associations anywhere in the world. Brazil and the USSR also protested, but for the moment could do little else.[29]

Draper set about raising $15 million. The United States promised to match all other contributions—indeed, Hoffman remarked that he was "sure that he could obtain all the money that he wanted." But it remained vital that other countries, including developing countries, contribute. Subject to Draper's importuning, British officials noted that he and others were "doing their best to play off one donor against another in a good cause." Already a "powerful and influential figure," he was now "insinuating himself into the position of an international lobbyist."[30]

Draper surpassed his own goal, with contributions from twenty-four countries. Some were symbolic, such as $3,000 from the Dominican Republic, and $240 from Cyprus. Other countries, such as Indonesia, gave more only on condition that it fund local programs. But the symbolism was important to counter what a State Department official described as a "widespread feeling that the Fund is a US front organization." Hoffman chose Salas in part simply because he was "Catholic and brown." Donors were well aware that he was ill-prepared to manage the money that started to pour in. When the British minister of overseas development first met with Salas and his deputy, she was informed that they constituted the entire UNFPA. Neither one had any experience organizing a family planning program. They could not explain how they would find help without hiring personnel away from national programs. They were a funding organization, mainly for other UN agencies, but admitted they had no capacity to evaluate how the funds would be spent.[31]

A year later UNFPA had only expended $4 million of the $15 million raised. Nevertheless, Draper was already well on his way to raising $28.6 million more. Donors competed with one another to show that they were contributing to the solution of the "population crisis." A British official observed that standing aside "would be represented as a characteristically mean and carping British approach to such matters, [and] we would find ourselves in uncomfortable isolation." In 1970, Norway's aid agency, Norad, decided to devote fully 10 percent of its appropriations to family planning. At the time, it had only one employee working on this issue. Her budget increased almost tenfold by 1977, to over $36 million. Japan and the Netherlands increased population aid eightfold, to $15.4 and $9.7 million, respectively, and they too had to turn most of it over to UNFPA.[32]

The Swedish International Development Authority (SIDA) had an experienced staff of eighteen managing bilateral population aid projects in more than a dozen countries. But when the government committed to spending 1 percent of its annual gross national product on foreign aid, and increasing family planning assistance at the same rate, they decided to give priority to countries that could "easily 'swallow' large efforts . . . e.g. India." For the same reason, half the budget—which would grow from $7.4 to $31.4 million by 1977—would now go to multilateral organizations. As for Ravenholt's Office of Population, by 1978 it was spending over $200 million a year, or $643 million in today's dollars. The 40 percent it gave to international and nongovernmental organizations made it unique within USAID. It was the largest contributor to both UNFPA and the IPPF, and provided almost the entire budget for smaller NGOs such as the Pathfinder Fund.[33]

"There were times when it was kind of embarrassing," Ravenholt later recalled. "When I would go to a meeting of the Population Association of America and these various universities would sic their youngest, most nego-tiable women on me, to ensure that they would get a larger share of the grants. I was sitting on a $150 million budget. It was wild." To Norad's Karin Stoltenberg, the endless succession of meetings on the population crisis seemed like "an international traveling circus." No one knew what to do with all the money. "I traveled around with tens of millions of kroner in my pockets, and I had to find a way of spending them."[34]

Partly by default, partly by design, donors were creating something completely new in the "development" field: a system to recycle money through international and nongovernmental organizations that, unlike the United States or even Sweden, could claim to work on behalf of all human-ity. In 1971, 62 percent of population assistance was bilateral, going from one country to another. By 1975, 45 percent was bilateral, 37 percent was multilateral, and 18 percent funded NGOs. Some of this money went through not just one but several intermediaries—for instance, USAID con-tributed to UNFPA, which contributed to the IPPF, which then turned the funds over to local affiliates.[35]

But even with all the money in the world, most of it "sanitized" by the World Bank, UN agencies, or NGOs, how could they stop world popula-tion growth? The problems were both political and practical, and equally intractable. By UN standards, the UNFPA approach was straightforward: "primary emphasis will be placed upon operational programmes and proj-ects assisting efforts to moderate fertility rates where such assistance is de-sired." Carl Wahren of SIDA advanced the "rather novel" idea of support-ing a project in a country like the Netherlands to show that rich countries

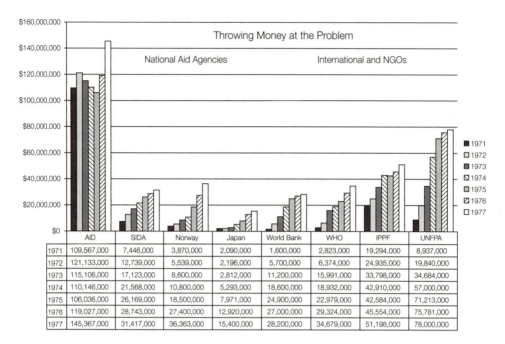

	AID	SIDA	Norway	Japan	World Bank	WHO	IPPF	UNFPA
1971	109,567,000	7,446,000	3,870,000	2,090,000	1,600,000	2,823,000	19,294,000	8,937,000
1972	121,133,000	12,739,000	5,539,000	2,196,000	5,700,000	6,374,000	24,935,000	19,840,000
1973	115,106,000	17,123,000	8,600,000	2,812,000	11,200,000	15,991,000	33,798,000	34,684,000
1974	110,146,000	21,568,000	10,800,000	5,293,000	18,600,000	18,932,000	42,910,000	57,000,000
1975	106,036,000	26,169,000	18,500,000	7,971,000	24,900,000	22,979,000	42,584,000	71,213,000
1976	119,027,000	28,743,000	27,400,000	12,920,000	27,000,000	29,324,000	45,554,000	75,781,000
1977	145,367,000	31,417,000	36,363,000	15,400,000	28,200,000	34,679,000	51,198,000	78,000,000

As foreign aid budgets for population doubled and doubled again, harried officials passed the money along to international and nongovernmental organizations, many of them no better prepared to spend it sensibly. Data from Stanley P. Johnson, *World Population—Turning the Tide: Three Decades of Progress* (London: Graham and Trotman, 1994), 100.

had population problems too. The U.S. and UK representatives strongly objected, and no one supported the idea ("the Dutch representative did not comment"). But everyone recognized that an exclusive focus on reducing fertility in poor countries risked undermining UNFPA's already dubious legitimacy. In a General Assembly debate at the end of 1970 on whether to declare 1974 "World Population Year," many African and Latin American countries asserted that they were *under*populated.[36]

Salas signed agreements with eight countries by mid-1972 to develop comprehensive programs. All eight agreed on the need to reduce population growth. But UN agencies developed 75 percent of the projects, so the programs reflected *their* priorities. For the Population Division, better censuses were key, so UNFPA paid to train more demographers. WHO insisted that contraception was part of maternal and child health, so UNFPA gave them money to expand their maternal and child health programs.

UNESCO emphasized the fundamental importance of education, so UNFPA supported the development of curricula that educated schoolchildren about population. Because Salas could not wrest control over "population programme officers" from the Population Division, UNFPA did not have any field staff of its own.[37]

By all accounts Salas was "too gentle" with both the agencies and the receiving countries. Donors were frustrated by the way some, especially India, played one off against another, or adopted a "take it or leave it" attitude when proposing joint projects. Britain funded UNFPA with the idea that international agencies were ideal for pressuring these governments. According to Ulla Lindström, Salas needed to deal on an equal footing with heads of state. The head of the advisory board, Alberto Lleras-Camargo of Colombia, wanted the UNFPA to invoke the authority of the UN to develop a world strategy to reduce fertility. Even a tough bureaucratic infighter with battalions of experienced staff would have had difficulty meeting these expectations.[38]

UNFPA, like the IPPF, struggled to grow into a role that required it to discipline recipients. It was designed to give Salas maximum autonomy in dealing with them—in this case, both UN agencies and sovereign governments. But that did not give him the power to make them do anything. New programs could provide incentives, but it was difficult to exert any leverage when Salas needed everyone's help to spend the money as quickly as it poured in.

India, to take only the most important case, was not about to be pressured by an upstart UN agency. Indeed, after the IUD debacle, there was a groundswell against foreign advisors. Gandhi decided to appoint a new and formidable secretary for family planning. "Shri K. K. Das has been accused of a variety of things," her top aide noted, but he had "a very healthy attitude in the matter of indiscriminate use of foreigners in various branches of our administration." In short order, Das sent the Ford Foundation consultants packing. He also refused to help USAID spend the $20 million it had budgeted for India's program. The United States "tilt" toward Pakistan during the 1971 war over Bangladesh made the closure of the USAID office inevitable. But disenchantment with foreign aid for family planning was deeply rooted and applied to the entire donor community. Early that year, it appeared that India would reject proposed projects by the World Bank, Norway, Sweden, and Denmark.[39]

At last, in the summer of 1971, Indian officials indicated that they would consider accepting new aid, but on their terms. SIDA's Wahren was startled upon reading the Indian conditions:

> As far as I can remember, we have never previously been encouraged in such a direct manner to just hand over a cheque—and that's that! To place a ban on "special reporting, joint reviews, and technical consultancy" and directly stipulate this as a condition for co-operation . . . is unusual, to say the least.

Norad's Stoltenberg, on the other hand, was relieved to be given the chance to provide 22 million kroner (over $15 million in today's dollars) for a postpartum project, and would have given much more. Wahren relented and joined the World Bank in agreeing to a seven-year, $31.8 million project (nearly $160 million in today's dollars).[40]

Because there was so much money to give away, and so little time to do it, donors were led to support whatever projects appeared plausible. Of course, they had to satisfy constituents intent on population control. India asked several donors for more money for maternal and child health, for instance, arguing that if children lived longer their parents would not have so many of them. But even Norway turned them down, even though it was a leading proponent of the "integrated" approach. Norad's population budget, like USAID's, was built on warnings of "a catastrophe of unknown dimensions," of a "hunger crisis or war," as two MPs put it during parliamentary debate. "When one went to India to find a family planning programme," Stoltenberg explains, "one could not come back with a vaccination programme."[41]

The interaction between donors and Indian government officials therefore reinforced a tendency to focus only on reducing fertility. The postpartum program, first developed by the Population Council, was still premised on the idea that it was easiest to "capture" women while they remained bedridden in maternity wards. Only now, with Norway's help, staff appealed to the entire surrounding community through educational films, public meetings, and radio programs. Norway also paid for sterilization wards at participating hospitals—over 500 by 1974—and paid staff incentives to meet performance targets.[42]

Similarly, the Swedish–World Bank project in the states of Uttar Pradesh and Mysore included many "components," including expanded training, mobile IEC teams, and supplementary food for pregnant and lactating women. But all of this was focused on reducing fertility—even the nutrition program was intended to "assess the value of food as a direct incentive for family planning." In the first year, they projected a 60 percent increase in the number of tubectomies in Mysore and a doubling in Uttar Pradesh. Nineteen annexes were added to hospitals for these procedures, along with twenty maternity-sterilization wings at primary health centers. As for UNFPA, nearly half its funding was a direct grant for an "innovative vasectomy program."[43]

Some foreign experts were now cautioning their Indian counterparts about relying on targets, incentives, and sterilization (even while admitting that previous advice "was not always helpful"). But the messages were mixed. Ford's Ensminger advised that the program was not "crisis-oriented" enough, and should consider higher incentives for sterilization. Saunders of the Population Council insisted that "the principle of incentive payments has not been thoroughly or systematically tested yet." Money spoke louder than words, and foreign donors obviously wanted a vast expansion in India's sterilization program. They got what they paid for.[44]

It was impossible to ignore how foreign funding was once again spurring the Indian program in a dangerous new direction. Oxfam's local representative warned headquarters that paying providers incentives for sterilization would lead to the same abuses and backlash as had occurred with the IUD campaign, and there was the "prospect of more draconian measures on the horizon." In November 1971, Don Lubin, the IPPF's new controller, reported that "India is going wild on vasectomy and tubectomy camps and the FPA is following the lead." He advised that the IPPF send no more money to the Indian FPA, since it had $60,000 sitting in the bank and would have had $300,000 if it had not "undertaken some unprogrammed vasectomy and tubectomy camps." Over the previous year, the state of Kerala had demonstrated that, by offering incentive payments of one hundred rupees in cash or kind, it could sterilize more than sixty thousand people in a month. Staff performed as many as fifty vasectomies at a time in a vast operating theater. The payment was equal to a month's wages for half of those who came forward.[45]

The state of Gujarat also boosted incentives to achieve what one official there called "a new world record": in just sixty days, 223,060 people were sterilized. No less important than the incentive payments, though more difficult to document, was the role of these local officials—especially a powerful figure like the district collector, who controlled the police and taxes.

In the Kerala sterilization camps of 1970–71, organizers made an example of a large, poor family, at the same time giving cash payments and prizes to volunteers. IPPF Archives.

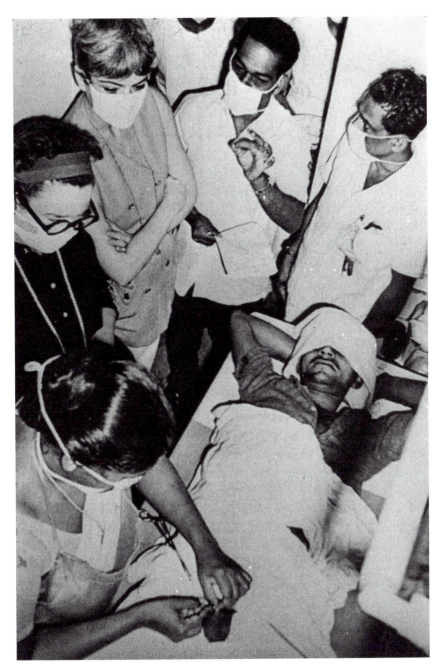

Fifty-bed operating theaters afforded little privacy to those who agreed to be sterilized, especially when foreign donors were permitted to inspect the results. IPPF Archives.

Officials in Gujarat were given awards of up to a thousand rupees, which was more than the annual income of most of the people who agreed to the procedure. The Planning Commission advised that Kerala's experience should be "used as a model." The Ministry of Health agreed and established guidelines for the creation of camps across India, setting a national target of 5.6 million sterilizations—or double the number in the previous year. Kerala officials themselves abandoned the approach when they discovered that a fifth of "acceptors" were over 45 and few came forward after the camps disbanded. When others tried to sterilize even more people in a poor district of Uttar Pradesh, tetanus from dirty instruments took the lives of eleven men.[46]

International aid contributed to a chaotic situation in which pressure to spend money and show results led to ever larger and riskier programs. But there was widespread agreement among population control proponents worldwide that, as one Gujarati official put it, family planning was on a "warfooting and it is high time to consider it as an emergency programme." Similarly, Rockefeller remarked that a 1970 meeting of top administrators—including McNamara, Hoffman, Owen, Chandrasekhar, and WHO director-general Candau—had confirmed him in the belief that "the situation facing the world in relation to population is of such overwhelming magnitude, urgency and importance as to justify a wartime type approach."[47]

In a period in which the United States was fighting a war in Vietnam while also waging a "war on hunger," such language might seem only metaphorical. But it set a tone and structured the way people considered population control programs: "We programmed for success," Ravenholt later explained, "much as army quartermasters must do when girding an army for battle: making sure enough ammunition is made available in advance so that the troops can defeat the enemy." The concept of contraception as "ammunition" also helped the executive director of the Association for Voluntary Sterilization explain to its president, Hugh Moore, how they could take advantage of the problems with the IUD:

> Although the IUD is another weapon in the war against hunger, its effective firepower in destroying the enemy is limited by its 40% failure rate. As a former artillery officer, you recognize the significance of this. . . . It is hoped that we Americans will not lose the war on hunger by supplying the troops with blank cartridges, instead of calling for artillery strikes.

Using such language did not necessarily predetermine how policymakers made decisions about this or that contraceptive or program strategy. Draper, a former Army major general, admitted that all available techniques had problems. But shortly after McNamara moved from the Pentagon to the World Bank, Draper advised him that "the war must be fought now, with the best weapons available, without waiting—and it can be won." Military language could justify different approaches to the same problem, but it always conveyed the idea that population control required risks and sacrifices, and that there was no substitute for victory.[48]

But who was the enemy? Even if it was "the population problem," and not people, particular individuals were being asked to make sacrifices in order to solve it—in fact, the very "clients" whom these programs, at other times, were said to "serve." In Kerala, a local official described the sterilization camp as "a sacred, concerted, and concentrated effort involving the sacrifice of the participants and the public at large at the altar of the future welfare and prosperity of the nation." Describing the Population Council's approach, Deverell wrote that "the most effective procedure is usually to attack women in the post partum stage."[49]

The women in these offices were amazed at how their superiors talked about their work. Adrienne Germain was one of a handful of professional women working at the Population Council, and then the Ford Foundation's population office, in the early 1970s. "It's as though women weren't human," she now recalls. Senior professional staff "could walk the corridors and be in meetings and talk about 'users' and 'acceptors' and write about users and acceptors and have absolutely no interest in who these people were." Women were excluded from discussions about contraceptive technology and the ethics of research trials. When Nafis Sadik arrived at UNFPA in 1971, she already had sixteen years of experience as a gynecologist and public health professional, culminating with her nomination as director-general of the Pakistani General Council for Family Planning. But in her new workplace, where she was the only professional woman, "it felt as if I didn't exist." Any substantive suggestions she made "fell on deaf ears." Even in the IPPF, women were underrepresented on the key budget and finance, medical, and scientific committees. As for the World Bank, it did not have even one professional woman involved in population programs.[50]

Worse yet was the sexual harassment many women experienced. "The way some women were treated by some of the topmost of the senior leaders was despicable," Germain recalls, "whether it's how they treated their graduate students, or how they behaved at the huge number of conferences. . . .

especially by Northern men vis-à-vis women in foreign countries."[51] One senior official bragged to the author about his escapades at these meetings, including having sex with one woman in a conference room after other participants had left. Asked whether he used protection, he replied, "I let her worry about that."

Leaders in the field gradually became more sensitive to how they spoke about such issues, as well as the near total absence of women in their deliberations. Even while calling for a "wartime type approach," Rockefeller "recognized that use of words such as birth control and population control has given an unfortunate negative ring to the subject." He would go on to promote women to leadership positions and listen to their ideas. But while donors worried about perceptions, few did anything to stop practices that targeted poor people, and especially women, practices that alarmist rhetoric both subsidized—because of the contributions it elicited—and justified.[52]

Of course, international assistance was supposed to help improve programs in places like India, in some cases by ensuring that family planning was "integrated" with health care, better nutrition, and so on. SIDA's board was told that it was "first of all a research project." But the Swedes soon discovered that their Indian counterparts had no intention of helping them identify flaws in the program. They "would never dare let their data reveal this," as SIDA's local representative explained, "as they would probably lose their positions immediately." This did not stop Sweden from funding what the World Bank also described as "an experimental program," albeit an experiment involving twenty million people.[53]

The overriding concern for donors—including even many of those who favored a more "integrated" approach—was how to make their programs more effective in reducing fertility. In November 1971, the Population Commission agreed on the need to consider "a global population strategy." But observers came to view UNFPA as merely providing "a weak 'umbrella'" for uncoordinated efforts by a motley assortment of UN agencies. In Pakistan, for instance, the UNFPA program had already come to a grinding halt. Recruitment proved difficult, training was poor, and morale and performance deteriorated. After twenty-eight months, Pakistan estimated it had received barely $500,000 of the $1.7 million pledged just for the first year (though UNFPA estimated it had delivered more than $900,000). A UN review noted that the program had become "a significant source of government employment," provoking corruption allegations.[54]

When UNFPA began negotiating a much larger program in Indonesia, McNamara was determined to help it do better. The director of the population projects office, Dr. K. Kanagaratnam, persuaded Salas to let the World Bank serve as the "executing agency" and decide whether and how other UN agencies might be permitted to take part. UNESCO had proposed a "massive" project in population education, despite having neither relevant experience nor approval from the Indonesian government. Similarly, UNICEF won the contract to procure and maintain a fleet of vehicles, even though it did not have any facilities to do so.[55]

The Bank therefore played the heavy in helping Salas fend off pork barrel projects. Candau protested to McNamara that UNFPA had cost WHO business in Indonesia. The undersecretary-general for economic and social affairs, Philippe de Seynes, was even more indignant, and insisted that the Population Division still had responsibility for evaluating and coordinating UN programs. Salas responded by threatening to cut the Secretariat out of the Indonesia project. De Seynes finally went along, even while insisting that it should not serve as a precedent. When negotiations concluded in February 1972, UNFPA and the Bank agreed to each provide Indonesia with $13.2 million. This was four times more funding than any previous UNFPA program. But it had taken almost two years, and earned the enmity of Candau and de Seynes, who between them had almost four decades of experience as senior UN officials.[56]

Two months later, Draper proposed a series of initiatives to the new secretary-general, Kurt Waldheim. He urged him to appoint his predecessor, U Thant, as his personal representative in organizing both the World Population Conference in Bucharest in 1974 and the other activities that would mark World Population Year. This world population czar and his staff would supervise both the Population Division and UNFPA. Since it would all be paid for by contributors to the Fund, Draper suggested that it be exempt from "any normal budgetary limitations or ceilings." Altogether, he estimated that population programs merited at least three billion dollars annually, a sum equivalent to more than a third of all international humanitarian assistance.[57]

But Draper had miscalculated. Waldheim was a conservative Catholic who would one day be knighted by the Vatican. De Seynes seized the opportunity to cut Draper down to size. In a personal note to Waldheim, he laid it on the line:

> There is no doubt that General Draper and other private individuals have done a magnificent job in fostering the role of the United Nations in the population field and raising considerable sums of money.

However, I believe that *an excessively active involvement on their part* in the development of the United Nations activities and in matters of administration and management, as well as in the elaboration of United Nations population policy, *could conceivably become compromising and lead to serious embarrassment.*

De Seynes recommended temporary appointment of a lower-ranking official under his own authority to organize the conference. To take charge of drafting the global population strategy, he called for choosing the new leader of the Population Division purely on competence, setting aside the usual UN practice of letting countries divvy up the top jobs.[58]

Waldheim agreed with de Seynes and accepted his nominees, both of them respected demographers from Catholic countries: Carmen Miro of Panama would organize the conference, and Leon Tabah of France would take over the Population Division. De Seynes then informed Salas that he had changed his mind about the Indonesia deal: the Population Division would have nothing to do with it. By this point, the UN General Assembly itself had weighed in, calling for "improvements in the administrative machinery of the Fund" and a full report at its next meeting.[59]

Salas obviously needed help, but simply abdicating to the World Bank was not a sustainable strategy. The Advisory Board assembled a review committee under Ernst Michanek, the head of SIDA. In May 1972, Michanek asked Berelson and the Population Council to host a two-day meeting of the top administrators. They included Harkavy of Ford, Henderson of the IPPF, Kanagaratnam of the World Bank, Ravenholt of USAID, and Gille of UNFPA—seemingly everyone except Salas himself. Together they would lay out a new strategy to control the population of the world.[60]

Of all the problems they confronted when they came together at the Population Council office on Park Avenue, the top leadership focused on what they called a "backlash." Mass IUD programs had provoked resistance, and not just in South and East Asia. In Tunisia, for instance, an effort to push the IUD without proper training or alternative forms of birth control led to "a strong backlash of antagonistic feeling and anti-propaganda." Much the same thing had happened in Haiti. The failure to persuade "acceptors" to continue accepting had become the bane of programs all over the world.[61]

The meeting also noted troubling incidents in which normally passive resistance had suddenly become very vocal. In December 1971, for instance, a UN-sponsored meeting of African demographers in Accra was commandeered by a group of "Young Turks" who accused the senior researchers of pursuing neo-Malthusianism on behalf of the countries that dominated the world economy. At the first UN conference on the environment in June 1972, an unprecedented gathering of activists and other nongovernmental representatives challenged the official program for focusing on overpopulation among the poor rather than conspicuous consumption among the rich. Indira Gandhi, of all people, lent her support to this critique, demanding a more integrated approach.[62]

On Park Avenue they called it a "Third World syndrome," but it was much more. In the United States as well, opponents of population control would not be bought off with tokenism. When, for instance, Rockefeller and the Nixon White House carefully selected what they considered a representative commission to study "Population Growth and the American Future," minority members protested that the research staff was "lily white." American graduate students who wrote in the pages of *Concerned Demography* offered much the same critique as the "Young Turks" in Accra. In the Scandinavian countries too, there was growing opposition to aid for population control programs abroad.[63]

Most worrisome of all, Nixon himself decided to reject his own commission's recommendation that the United States should aim for "population stabilization." He declared that he had "a basic faith that the American people themselves will make sound judgments regarding family size and frequency of births." In fact, several cardinals and bishops had gotten to Nixon in a coordinated letter-writing campaign. The president went out of his way to pledge "admiration, sympathy and support" for Cardinal Terence Cooke's campaign to prevent legalization of abortion in New York. U.S. policy changed little after Nixon's announcement, which was aimed at Catholic voters in the 1972 election. But it still had ramifications. As proponents had repeatedly stressed, pressing other countries to reduce growth would be more difficult if the United States did not accept this policy for itself.[64]

The top administrators agreed that "the battle for the hearts and minds of 'key' leaders in the developing world (and of course in the developed countries as well) was far from won." These leaders needed both "more facts" about population, as well as "'objective' analyses of the implications of these facts for social and economic development and the environment." UNFPA "was not yet equipped to do either of these jobs." It was not just a matter of expertise. "Objective" meant, first of all, less vulnerable to "the Third World syndrome":

It was argued that developing countries were perfectly capable of see-ing UNFPA for what it was: a new name attached to an old face. USAID might push its money through UNFPA but these were still dollars and [less-developed countries] were not really in doubt as to their provenance. Until UNFPA acquired a constitutional structure that was representative of governments (and preferably all govern-ments), the "advantages of multilateralization" would remain sus-pect.[65]

"Objective" also meant analyses that would make clear the *implications* of demographic facts, and this included what might happen if countries re-fused to recognize them:

The key relationship here would be the partnership between UNFPA and the World Bank. For the World Bank made a continuing ap-praisal of member countries' economic prospects and performance; dealt with key planning and finance ministries and was in a position to draw the attention of governments to the effects of demographic trends.

In effect, the top leaders agreed that the World Bank should make govern-ments work with UNFPA. Rather than waste time with health ministries, "the crucial dialogue" would now be conducted with finance and planning ministries. In some cases that required stressing health and welfare, espe-cially in Africa and Latin America—that is, "go on 'doing' population via family planning, etc., but not to talk about 'doing' population." Otherwise, they would need new mandates, and no one in the room wanted to risk re-opening the debate over population control before all the UN governing bodies. Instead, they agreed that the World Bank and UNFPA would form the "dominant axis for population policy assistance." This seemed like the only way out of the "historic dilemma" of the UN system: making it greater than the "sum of the parts."[66]

These ideas were not entirely new. This is how foreign donors had worked with allies in India to bring about a more forceful population con-trol program five years earlier. But the administrators present were now in a position to adopt the same approach all over the world. Michanek's com-mittee therefore recommended measures that would, on the one hand, make UNFPA more appealing to governments, while, on the other hand, enabling it to develop its own identity and capacity for independent action. To dispel the impression that it was a front for certain developed countries or a slush fund for UN agencies, Michanek's report advised that most of the Fund should support programs operated by receiving countries themselves. It repeatedly called for a "dialogue" with governments, though one that would contribute to "a strategy to reduce population growth" by changing "the motivation of the remaining 'target population' in the reproductive age groups." This would require cooperation with the World Bank and continued funding of non-UN bodies like the IPPF. Oversight would still be kept to the bare minimum: Salas would answer only to the secretary-general and the UNFPA advisory board.[67]

Waldheim accepted only some of the committee's recommendations. He said that he agreed with his own advisors—almost certainly de Seynes—that UNFPA was now too big to be his sole responsibility. In fact, the General Assembly resolved that Salas would have to report to *both* ECOSOC *and* the governing body of the UNDP. But Michanek, speaking "as one of the Fund's principal donors," won agreement on two key points: UNFPA would shift support to governments prepared to execute their own programs, and it could advise these governments to use nongovernmental organizations like the IPPF instead of UN agencies.[68]

This was just the beginning of a long struggle, but over the period 1970–1983 UNFPA would give $111 million to NGOs and $245 million to governments—more than twice as much as it granted to WHO, the biggest recipient among UN agencies. Henderson had already resolved that the IPPF should develop an "*operational* partnership" with UNFPA and the World Bank, telling member associations to join their country missions. Of course, Ravenholt's USAID was still far and away the largest source of funding, especially for UNFPA and the IPPF. But its very size meant that it would have to continue working with and through these other organizations. Ravenholt therefore took on the role of ensuring that the World Bank–UNFPA "axis" actually had some pull and started moving. In meetings of the international agencies, he began to upbraid the Bank for its slowness in disbursing loans. And he backed UNFPA by threatening to cut funding if subgrantees like WHO did not make population control a priority.[69]

In 1972 UNFPA and the World Bank—with USAID cracking the whip—were harnessed to the axis of an emerging system that could pull national and international programs beyond family planning. They did not undertake another joint program like that in Indonesia, but did agree to provide "parallel" financing for complementary projects in Malaysia. In Kenya the top official in the Ministry of Finance announced that the World Bank and UNFPA would jointly coordinate external aid. And Tunisia shifted its family planning office from the Ministry of Health to the Ministry for Economic Affairs.[70]

To outside observers, the clearest case of World Bank pressure was Mexico, which in June of that same year abruptly reversed its earlier policy of favoring population growth. Mexico depended on Bank loans, and the announcement coincided with a visit by McNamara. But the archives have not yielded much evidence that he pushed for the change. In fact, it appears that Draper was the catalyst. He had deep roots in the country, having made a fortune as CEO of an electric utility years before. He won over Mexico's minister of finance and brokered $1.2 million in UNFPA funding for the family planning association. But here again a National Population Council headed by the secretary of the interior—not the health minister—would direct the government program.[71]

The Vatican was alert to the inroads population control proponents were making even in Catholic countries. In 1973, members of the Roman curia together with laypeople came together in a Committee for the Family to coordinate a response. The secretary of state, Cardinal Jean-Marie Villot, ordered the national episcopal conferences to begin preparing for the World Population Conference in 1974, ensuring that the Church view was represented in national delegations. Rome would back the demands of the nonaligned movement, which had just held an historic summit in Algiers to demand the creation of a new international economic order. "Any population policy must be part of a sound development policy," Villot insisted. "It proceeds from the principle that development is for man and for the whole man, and not vice versa. . . . There remains the task of achieving greater justice in the division of goods, as much within individual nations as at an international level."[72]

Marathon meetings of the Population Commission worked over endless drafts of a World Population Plan of Action (WPPA). The United States and Sweden took the lead in pressing other countries for an explicit commitment to control fertility, while UNFPA and the World Bank fought the UN agencies' efforts to shape a final document that would merely call for more money for more of the same. The Vatican and its allies objected to the very mention of contraception. Villot warned the bishops not to permit well-meaning Catholics to agree to any compromise: "some subjects are not open to discussion." Fresh from the fight, IPPF chairman George Cadbury told the Management and Planning Committee that they should not even try to reach out to liberal Catholics, since "appeasement rarely pays." The IPPF and its allies must therefore continue to insist that "there is an urgent and pressing population problem in the world," and the WPPA should say so. If they were to prevail, they must not underestimate their enemy: "the oldest and most successful political organization in the world."[73]

While arguments continued over the World Population Plan of Action, the informal strategy of working through finance ministries and national coordinating boards continued to unfold. It took shape, not in UN working groups and regional conferences, nor in drafts of preparatory documents, but in shifting institutional structures that aligned national and international agencies toward a global goal of reducing fertility. It was also evident in the means marshaled to achieve this end. The UNFPA–World Bank project in Indonesia was positively breathtaking. It called for building more than fifteen hundred structures and hiring some thirty thousand workers. In addition, no fewer than 115,500 community leaders would be "oriented" to support the program. Here, too, donors negotiated not with the Health Ministry, but with the National Development Planning Agency. Overall management was entrusted to a National Family Planning Coordinating Board that answered only to Suharto.[74]

Walking through an obstetric ward in Central Java, a World Bank official was startled to hear tape-recorded "family planning propaganda" being broadcast to a captive audience of new mothers through the hospital's intercom system. Among them "female sterilization was being practiced systematically." The biggest innovation was the *Banjar* system, named after the Balinese term for small communities governed by male heads-of-household. In regular public meetings, each was required to report on whether their families were using contraception and explain noncompliance. The results were compiled in color-coded maps that were publicly displayed to facilitate "community re-enforcement and even pressure to use contraception." Some leaders went so far as to beat drums daily to remind women when to take their pills. A USAID official described it as "one of the great social engineering feats of modern times." The United States paid to expand the system and tried to export it to other countries.[75]

UNFPA, the World Bank, and USAID were also the major donors in another blockbuster program in Bangladesh, this one managed by a National Population Council. It began with "several tens of thousands of workers," as the Population Council's Paul Demeny described it. "Besides the personnel—administrators, doctors, nurses, midwives, lady welfare visitors, supervisors, female family welfare workers, drivers, peons, and so on—there will be the bewildering paraphernalia of *things*—from buildings to flip charts, from battery operated slide projectors to four wheel drive vehicles, from motor launches to surgical equipment and supplies of pills, IUDs, injectables, and all the rest." Almost all of this was paid for with international funding. Bangladesh accepted it but provided almost nothing itself.[76]

These programs were organized on a national basis. But just as its architects intended, the global population control campaign did not depend on government support. With USAID money the Pathfinder Fund was able to blaze a trail through countries hostile to family planning, such as Bolivia and Paraguay. Ravenholt also helped the Association for Voluntary Sterilization start an international project, and provided $34,647,000 by 1980. And he funded a Johns Hopkins program that trained thousands of doctors from more than seventy countries and sent them home with equipment, such as new "gun-style" devices that could shoot plastic bands around the fallopian tubes. With just one laparascope, a physician working in otherwise "primitive settings" could perform more than ninety female sterilizations in a day (though women suffering complications in these settings would presumably have to manage on their own). In Mexico, where the United States could not provide bilateral aid, Ravenholt nevertheless sent a "forward spotter for our artillery" to the U.S. embassy. When he "called for a strike," USAID provided training and supplies through the Johns Hopkins program. AVS subgrantees were supposed to respect local laws, but were told, "In those localities where sterilization may not be 'acceptable,' they may be supplied under the euphemism to 'detect and treat abdominal disease.'"[77]

Ravenholt planned to promote abortion in much the same way with mass distribution of low-cost "menstrual regulation" kits. In 1971, while abortion was still against the law in India, a USAID grant enabled the Population Council to provide 120 such kits for use in its Post-Partum Project. Ravenholt ordered 10,000 more for distribution at a Menstrual Regulation Conference in Hawaii in December 1973. Three hundred conferees took them home to countries all over the world, many of which still banned abortion. Encouraged by the results, he ordered 100,000 more kits. The idea was "to make abortion so easy to perform and so widely available that legal restrictions would be meaningless."[78]

Seasoned observers, like Notestein, marveled at the way very different kinds of institutions—ranging from the Pathfinder Fund to the Rockefeller Foundation to UNFPA—were coming together to form a quasi-organic system, one that not only controlled populations but produced new global norms:

> The situation is almost like the links of a food chain. The personally led special-purpose foundations experiment for and nourish the larger and more deeply institutionalized foundations and universities. These, in turn, experiment for and nourish the governments, which now show signs of experimenting for and nourishing the international organizations. The same activity which is viewed as improper, if not downright wicked, at the beginning of the chain is transformed by the end into an essential constituent of virtue if not a basic human right.[79]

This system ran on money, and the competition to discover new ways to spend it continued to propel it forward. Ravenholt's man in Manila staged a photo op in which he pitched condoms and pills out the door of a government helicopter. Another USAID staffer confided, "A core value for the Office of Population is showing that you can spend money faster than other organizations in the field." In essence: "Look how quickly we act, how outrageous we dare to be."[80]

In fact, Ravenholt would not go as far as some—sensing the opportunity—wanted to push him. In 1973 an influential congressman, Otto Passman, cut his budget because Ravenholt refused to fund the Family Health Foundation (FHF) of Louisiana, which gave kickbacks to the governor and his cronies. The FHF's clinics served a poor community that included many African Americans and Catholics but nonetheless claimed to have the highest "continuation rate" in the world—thus appearing to refute Judith Blake and other doubters. For a time its president, Joe Beasley, was the hero of the population control movement, which may be why Passman was able to push Ravenholt in this way. Beasley was awarded a professorship at Harvard, was made chairman of the PPFA's executive committee, and received grants from Ford, Rockefeller, and Tulane University together with over $50 million in matching federal funds. This money paid for a private jet, two apartments, and countless parties in the French Quarter, as well as envelopes of cash to buy off "Black Power" leaders. On the other hand, fifteen mobile clinics turned out to be figments of his imagination. Even after Beasley was indicted—he would be sentenced to two years in prison—Passman continued pressing USAID to support his program as a model for export to Latin America.[81]

If they had thought of Beasley as the avatar of family planning, it must have occurred to these administrators that programs in places like Pakistan, Indonesia, and the Philippines might also be out of control, or at least beyond *their* control. In 1973, Rockefeller was sufficiently disturbed to hire an outsider, Joan Dunlop, to provide a fresh perspective. "There's something wrong with the population field," he said. He told her to "go to meetings and listen to people and tell me what you think is wrong." Blanchette Rockefeller took Dunlop aside and expressed relief that her husband had finally hired a woman. "He's not being told the truth," she said. "And in order for you to tell him the truth, you must consider yourself to be his equal."[82]

Dunlop was amazed at what she heard in gatherings of population experts and administrators, where they talked about using technology to "attack" the birth rate. For these men, almost "sexless" in their approach, "it was all very theoretical." Notestein proudly reported how demographers were employing the same mathematics to track birth, death, and migration as the Manhattan Project had used to analyze neutrons. Many had spent so much time traveling abroad, they seemed oblivious to the way other Americans had been challenged to think harder about race and gender, and to acknowledge that fertility was more than a mathematical exercise. For others, making crude jokes and harassing women seemed to be part of the job. "We used to laugh at these guys when we weren't crying," Dunlop recalls, "because of the way they were treating other women and the way they were dealing with the issue." Dunlop finally came back to Rockefeller and told him the problem was simple: "A very small number of men control all the money and the ideas in this field. They have a stranglehold on it and there is really no innovation."[83]

Rockefeller's own meetings with these same leaders confirmed what Dunlop was telling him. In this period they regularly gathered to review the latest research and coordinate their efforts at the Rockefeller estate, Villa Serbelloni, overlooking Bellagio on the shore of Lake Como. If the system had a brain, it would have been on display here, in a conference room small enough to exclude all but two dozen key officials and invited experts. At the 1973 meeting the president of the International Development Research Center, David Hopper, asserted, "We're dealing with a species response . . . [we] can and must view [the] population problem independent of cultures." But Berelson warned that the "backlash" was continuing to grow. Moreover, new research on earlier declines in fertility in Europe had shown that it "looks more complex than we used to think.—And it looks less applicable to [less developed countries] than we had thought." There was debate but no agreement about how family planning might be "integrated" with health, education, and employment policies. McNamara found it all "depressing," complaining that "data are inadequate, we don't know how to analyze cause + effect relationships. Don't know how to advise countries on what to do."[84]

If someone did figure out a solution, it was important that the finding be rapidly communicated to the rest of community. But Michanek warned that "the U.N. bodies are too official. They are too afraid of government. They are not in a position to give a strong enough position to nongovernmental institutions. . . . [UNFPA] isn't strong enough to take on this particular job, and we are now trying to find somebody who is." Ravenholt felt much the same way, and said so, even though UNFPA's number two official, Halvor Gille, was sitting right in front of him. Salas was humiliated upon hearing of this discussion—another one he missed—and sought to reassure Ford Foundation vice president David Bell that he was not "beleaguered."[85]

Salas had more help now, including forty-five professionals in New York and seventeen overseas. But many lacked experience—their average age was 35. They had to oversee nine hundred global, interregional, and country projects involving nearly one hundred countries. The largest of them, in Indonesia, would still not be fully up and running three years after the original agreement. Nearly all the management consultants went home "weary, frustrated, and sometimes angry." Meanwhile, in Pakistan, UNFPA's failure to provide committed funds in a timely fashion "undercut any leverage the international donors might have had," as one experienced observer noted. Together with the World Bank's own difficulty in getting countries to accept and spend its loans, the population control "axis" was under increasing strain. Salas had to improve UNFPA's performance, or at least increase the rate at which it allocated and expended the contributions that continued to pour in, or the whole effort would stall.[86]

This helps explain why, in July 1974, UNFPA awarded its largest grant ever to the government of India. It could not have come at a more critical time. The 1973 oil crisis had struck India hard, diverting export earnings and driving up inflation. The government was forced to cut expenditures across the board, including about $26 million, or more than 28 percent, from the family planning budget. Gandhi announced a five-year plan shifting resources to nutrition and health care. She had to publicly upbraid officials who complained about the change in priorities (family planning had been swallowing up 59 percent of the health ministry budget). The sterilization camps and extra incentive payments were suspended. In 1973–74, the number of sterilizations nationwide fell more than 70 percent.[87]

A UNFPA grant of $40 million—$164 million in today's dollars—would help change all of that. It was the usual dog's breakfast, only bigger, with vehicles from UNDP, a training program from UNICEF, and so on. But the single largest part, $14 million, was for "expansion of sterilization program." Singh told officials that he had decided to *increase* payments from 45 to 70 rupees for tubectomy operations (25 for acceptors, the balance to reimburse providers), and keep paying 35 rupees for vasectomies (20 for acceptors). He did not bring back mass camps, but the number of sterilizations in 1974–75 still rebounded to 1.4 million, up 44 percent from the year prior.[88]

The UNFPA grant received little attention at the time. In the summer of 1974 all eyes were trained on the World Population Conference, where the World Population Plan of Action would finally be presented. With Draper in the lead, population controllers sought backing for a plan that would be time-bound and target-oriented, only now on a global scale. But only a system without a brain would have selected Bucharest, of all places, to hold the conference. Romania was the only country in the world with a time-bound, target-oriented policy to *increase* its population. Eight year earlier, Nicolae Ceausescu had discouraged contraception and banned abortion. By the time conference participants gathered in August 1974, an epidemic of botched abortions had caused maternal mortality to soar.[89]

More than a thousand official delegates representing 133 countries came to Bucharest. There were also fourteen hundred participants in the Tribune of NGOs, four hundred in the "Youth Conference," and over nine hundred accredited journalists to report on the proceedings. The nonaligned countries, fresh from a special session of the UN General Assembly, planned to use the conference to renew their demands for a new economic order. They wanted to regulate multinationals operating on their territory, control their own assets, and work together to improve terms of trade.

It was standing room only in the Great Hall of the Palace of the Republic on the first day, August 19, 1974. Everyone was grateful that here, at least, the air conditioning was working. The first thing many of them noticed was that the man in charge of it all, Antonio Carrillo Flores, was also the father of six children. In his opening address, Carrillo reminded delegates that it was a population conference, not an economic conference, and it could not cover everything.

For those who wanted to control population growth, talk of neocolonialism was irrelevant: they were just helping parents plan their families to ensure a better future. But their cause was deeply rooted in the history of imperialism and decolonization, a history that was still far from over—as the four delegations representing national liberation movements made clear. This included Americans' own long—and long-denied—experience of exercising power over other people without having to answer to them. The senator who had crusaded throughout the 1960s for U.S. support for population control abroad, Ernest Gruening, had been director of the Division of Territories and Island Possessions in the 1930s. In Puerto Rico he implemented a program to reduce population growth, the first of its kind in any colonial possession in the world. When the Johnson administration began to support a global campaign, it decided to set an example by focusing on America's own island territories, Native American reservations, and what Moynihan called "the urban frontier." The IPPF's first secretary-general, Sir Colville Deverell, was a thirty-year veteran of Britain's colonial service. He often turned to former colonial officials when hiring staff. A member of the IPPF's Western Hemisphere regional council observed that the central office seemed to be "an extension not only of British philosophy but virtually of that of the British Civil Service or Colonial Office."[90]

Family planning associations redolent of the colonial era, typically including expatriates, had long had difficulties in dealing with newly independent governments in Africa and Asia. The most dramatic clash occurred with Singapore, which in 1968 seized control of the local association's assets. A high-ranking official explained that it had been run for too long by members of "the imperial generation." Newly independent governments could not so easily dismiss the UNDP—the UNFPA's parent agency—which was also represented disproportionately by former colonial officers from the UK, France, and the Netherlands. Observers noted how, like the

IPPF, it hired these men to run population programs even though they had no relevant experience—except, that is, experience dealing with "indigenous people." The same might be said of much of USAID's Office of Population. Ravenholt's most loyal operatives came from USAID's Public Safety Division, who would otherwise have been out of a job after evacuating from Vietnam. In their prior line of work, *population control* was the term used for setting up constabulary forces and identity card systems to combat communist insurgency.[91]

Even the Ford Foundation described its men in Asia as a "thin red line," recalling Rudyard Kipling's image of the defenders of empire. And even the World Bank's man at the Bucharest conference, representing an institution dominated by former colonial powers, was "appalled" by the "patronizing attitude of the mostly white anglosaxon representatives" of organizations like the IPPF and UNFPA. The "imperial generation" included all kinds of people, and not all were WASPs. For instance, India's representative in Bucharest, Karan Singh, was heir to the last maharajah of Kashmir. More nouveau royalty with imperial pretensions, such as the shah of Iran, represented in Bucharest by his twin sister, were also confirmed supporters of population control. The most prominent proponent in Japan, former prime minister Nobosuke Kishi, had been a top official in Japan's imperial adventure in Manchukuo and was indicted as a war criminal. Even Carrillo, secretary-general of the conference, had personal reasons to dismiss the issue of neo-colonialism out of hand. When he was Mexico's finance minister in the 1950s, he had helped foreign-owned utilities by changing tax policy and authorizing rate increases. One of the biggest beneficiaries was the Mexican Power and Light Company, then under the chairmanship of one William Draper.[92]

Colonial officials had experienced the demise of empire as at least partly a function of population growth and their inability to stop it. Many now feared a further redistribution of wealth and power between rich and poor, in which metropoles would lose access to oil and other natural resources, or even suffer reverse "colonization" by migrants from former possessions now grown overpopulated. Was it so far-fetched to imagine that a "World Population Plan of Action" to reduce the number of poor people was really a form of imperialism by other means? After all, what is empire but the pursuit of unaccountable power, even if it now operated in the guise of international and nongovernmental organizations? A system without a brain might have seemed an unlikely successor. But was it not said—rightly or wrongly—that the greatest empire ever was acquired "in a fit of absent-mindedness"?

The World Population Conference of 1974 thus witnessed an epic battle between starkly different visions of history and the future: one premised on the preservation of order, if necessary by radically new forms of global governance; the other inspired by the pursuit of justice, beginning with unfettered sovereignty for newly independent nations. The Vatican was able to profit from this clash, which made it unnecessary to push its theological objections to birth control. But the U.S. delegation would be in the thick of the fight. In a high-level review, policymakers agreed that Americans must lead efforts to establish a global strategy to reduce fertility to the replacement rate by the year 2000. Otherwise, food riots and revolution would close markets to U.S. investment, and countries exporting raw materials would be led to form more cartels just to feed their people. The Western industrialized economies were still reeling from the effects of the OPEC embargo attending the 1973 Arab-Israeli war. Draper together with Maxwell Taylor—former chairman of the Joint Chiefs of Staff—warned of "a Holy War of the Have-nots against the Haves, possibly in alliance with the oil-rich developing states. . . . The resulting conflict may well be the ideological schism which will split the world."[93]

But by the time the U.S. delegation arrived in Bucharest, Nixon had left office in disgrace, leaving its leader, Secretary of Health, Education, and Welfare Caspar Weinberger, in a weak position. Gerald Ford felt it necessary to warn countries against taking advantage of U.S. vulnerability. China was the first to declare its opposition to the plan, insisting that the future of mankind was "infinitely bright." Algeria was also a leader in the push for a new international economic order, and their representative chaired the critical working group. Argentina introduced over a hundred amendments to the WPPA, with support from other Catholic states such as Peru, Italy, and Ireland. Weinberger reported that the Vatican "seemed to have representatives in every committee, working group and sub-group meeting." The USSR and its East European allies opposed numerical targets, but not family planning. When China's representative condemned the two superpowers as equally imperialist, the Soviets in the audience turned around and shook hands with their American counterparts.[94]

As preparation of the WPPA deteriorated into a war of attrition, with every change of wording requiring another vote, more and more people began to drift over to the Tribune. Pro-choice activists handed out buttons

proclaiming "My Body Belongs to Me," while pro-life organizers distributed handbills depicting aborted fetuses. NGOs supportive of population control tried to hold on to the spotlight with panels on themes like
"Threats of the Future." The headliner, Lester R. Brown of the Overseas
Development Council, claimed the world had less than a month's supply of
food and that the United States might soon become arbiter in deciding
which nations starved. But Brown had been saying much the same thing,
or worse, for a decade already. The session was poorly attended, and question time brought ridicule. An African woman delegate demanded to know
why people from rich countries did not "listen to us for a change."[95]

Stalled elevators, broken telephones, and ninety-degree temperatures
raised tensions. As opposed to the official conference in the main hall, po-
larized over whether the WPPA would have numerical targets, the Tribune
was fragmented in multiple ways. Germaine Greer—"her head a mass of
wet ringlets from the heat"—together with Betty Friedan and Margaret
Mead led a revolt after discovering that the draft plan had only a single
paragraph on women, which only suggested that including them in development might reduce fertility rates. Greer had already challenged the Ro-
manian foreign minister, elected president of the conference, to explain
why his country forced women to bear unwanted children. He appeared
not to understand her question. Eighty percent of the official representatives were men, and men were in charge of 127 out of the 130 national delegations. The few women began working together with their counterparts
in the Tribune to obtain recognition for gender equality in education, employment, and development planning.[96]

Another surprise came with the strong reception for a group heretofore
unheard of, the "Non-Malthusian Coalition." A week into the conference—when only fourteen paragraphs of the WPPA had won approval—
they hosted a large press conference to publicize abuses in vasectomy camps
and contraceptive research funded by the Population Council. The Ford
Foundation was irked to discover that the Coalition included employees of
UNFPA. Together with like-minded delegates, these renegades pressed for
a WPPA that would address issues of social inequality and urge more attention and resources for public health.[97]

The biggest shock of all came that same day, when Rockefeller walked
up to the front of the plenary session of the NGO Tribune. The room was
packed and charged with anticipation. More than anyone, Rockefeller rep-

resented the establishment, even if his manner was always diffident and un-assuming. But he was now ready not just to listen to the concerns of dissenters like Dunlop and Germain but to champion their cause. He began by describing how much had changed in the forty years since he first decided to take up birth control, something that formerly could not even be discussed in polite society. All along, he thought of it as a way "to help make it possible for individuals everywhere to develop their full potential." But "in certain other respects," Rockefeller admitted, "I have changed my mind." He announced that he had come to Bucharest to issue "an urgent call for a deep and probing reappraisal of all that has been done in the population field." Family planning programs could not simply target fertility, but had to be made an integral part of a development program with a moral purpose: meeting basic human needs. Any such program had to give "new and urgent attention to the role of women," and recognize that women themselves should decide what their role would be.[98]

The next day, in a decisive vote, a majority of delegates rejected a U.S. proposal that the WPPA recommend reducing the average size of families. But the coalition included many who would have supported goals for establishing voluntary family planning services, as opposed to population control. Nevertheless, the Americans continued pressing for a plank that would set targets for reducing the rate of population growth by 1985. The final result—"rammed down the throats of substantial opposition," as Notestein put it—just recognized that some countries had set population targets and invited others to consider doing the same.[99]

Because of the failed push for population control, the WPPA did not give strong backing for family planning services. The principle of national sovereignty was explicitly recognized no fewer than five times. The "international community," on the other hand, was given just two priority assignments: reducing mortality and boosting food production. As for nongovernmental organizations, many delegates displayed open hostility. NGOs, mentioned only once in the final document, were directed to work "within the framework of national laws, policies, and regulations."[100]

"It was a humbling experience for everyone," Notestein remarked. "I've seldom been so blue." Bucharest was a particular setback for the programs of UNFPA and the IPPF, and a victory for those who opposed population programs as a distraction from fundamental problems of poverty and inequality. Henderson noted that the IPPF stood accused of being a front for

U.S. interests. Its image was tarnished among many Latin American, African, and Asian countries, and its very survival would now depend on working out a new partnership with national governments.[101]

Bucharest was the Waterloo of the population control movement. But it also provided a preview of how a different kind of movement, one that did not aim to plan other people's families, might regain the moral high ground. The WPPA called on all countries to "respect and ensure, regardless of their over-all demographic goals, the right of persons to determine, in a free, informed and responsible manner, the number and spacing of their children." Postmortems noted how well women had been able to advance their agenda where others failed, in part because women from the Third World were permitted to take the lead. It pointed up the fact that almost everyone who spoke against family planning was male. Even these men tended not to oppose planks endorsing gender equality. The Bucharest conference rejected population control not just out of deference to national sovereignty but also in recognition of the inalienable rights of individuals to education and comprehensive health care.[102]

Mike Teitelbaum of the Ford Foundation's population office had earlier argued for "steadfastly continuing our focus on population matters and viewing much of the rest as peripheral." But after coming back from Bucharest, he concluded that making women's role in development more explicit "might have the direct effect of increasing political support for population policies and programs." The Ford Foundation and others in the field began funding programs to improve the status of women, including the first UN Conference on Women in Mexico City the following year.[103]

Because of all the baggage they carried, it was difficult for groups that had long pushed population control to establish their bona fides as advocates of women's rights. "Too many short-term opportunists from the population field are jumping on the women's bandwagon," Dunlop told Rockefeller. "I am beginning to fear that the demographic imperative may do to the 'status of women' what it did to family planning." Critics pointed out that a new Women's International Network and its calls for "feminist unity" were subsidized by the Population Crisis Committee. A panel in Mexico City on "Population and Family Planning" seemed to be an exercise in tokenism. The IPPF organizer simply stipulated the agenda—which did not include burning issues like contraceptive safety and coercion—and then went looking for a "male from a developing country" and a "Ghanaian or Nigerian woman market leader to appear at the dias." When audience members accused the panel of serving a neocolonial agenda, IPPF assistant secretary-general Fred Sai of Ghana complained, "We didn't expect to go back to all the political claptrap."[104]

A broader vision of development that encompassed gender equality and contraception as human rights could give groups like the IPPF a new mandate. But they had to stop subordinating health care, education, and employment to the achievement of population targets, and recognize that improving the "status of women" was not just a means to some other end. Those still committed to population control were contemptuous. As the medical director of the IPPF, Malcolm Potts, put it, "'integrated' development is the most recent diversion of funds from the obvious. It is jargon." For Berelson, Bucharest was "the latest fad in a faddish field." They might need to use some "new rhetoric," even if that meant having to exchange "new clichés in place of old clichés." But the conference was "best seen as a political battle in a larger war," and Berelson was not about to surrender. It was important to remain "steady-at-the-helm," with "no failure of nerve." Population control proponents knew that leaders like Suharto, Marcos, the shah, and many more besides backed population control, whatever their emissaries said in the limelight of a UN conference.[105]

For someone like Karan Singh, minister of health *and* family planning to the world's largest democracy, this debate posed an acute dilemma. He had to address both international and domestic audiences, and reconcile a continuing commitment to reducing fertility with a new determination to address "minimum needs" in nutrition and health care. This inspired him to offer the most memorable line from Bucharest: "the best contraceptive is development." But other than his formal statement, Singh remained largely silent—like the life-size cardboard images of Indian villagers that Oxfam set up throughout the conference site.

Back home, flesh-and-blood Indians were fighting over their country's future. In Bihar and Gujarat, populist movements were calling for Gandhian revolution. The Congress Party, which had once provided a broad tent that included large numbers of the Dalits, Muslims, and landowners, as well as the landless, was riven by factions. It now functioned only to win elections and dispense patronage. In June 1975 Prime Minister Indira Gandhi herself was found guilty of violating electoral law. When the leader of the Bihar movement, Jayaprakash Narayan, called for her resignation, she invoked emergency powers and began to arrest more than a hundred thousand of her opponents.[106]

The population control establishment had long treated their work in India as "research." It now became a vast laboratory for the ultimate population control campaign. Singh declared in the Lok Sabha, India's lower house, that his statement in Bucharest had been misinterpreted: population control was a priority. In October 1975 he wrote to the prime minister advising a "crash programme," noting that the Emergency provided "an appropriate atmosphere for tackling the problem." With the constitution suspended, they could now employ any means necessary. Singh advised beginning with more incentives and disincentives. But in January 1976 he said that the government would not oppose those states, like Maharashtra, that wanted to jail parents with three or more children if they did not accept sterilization. Conversely, officials went to Bihar for "some plain talking," describing its failure to meet targets as "a criminal and anti-national act." All understood the implied threat. The prime minister admitted that such steps were "drastic." "Some personal rights have to be kept in abeyance," she explained, "for the human rights of the nation, the right to live, the right to progress."[107]

Population control during the Emergency was not just about compulsion. It was an experiment in "integration," in which every branch of government would take part. In April 1976, after a detailed cabinet discussion, Singh announced a comprehensive program. It called for raising the age of marriage, increasing women's literacy, and improving child nutrition. But it also raised incentive payments for sterilization and calibrated them to maximize fertility reduction—150 rupees for those with two children, 100 for those with three. And it gave a green light for states to introduce "disincentives" as well as compulsory sterilization. Gandhi personally signed off on every point in the program.[108]

One man came to personify population control during the Emergency, especially when people later on tried to distance themselves from it. It was said that Sanjay Gandhi had always caused problems for his mother. His birth had nearly killed her. He was spoiled as a boy, struggled to find a career, but finally succeeded in enriching himself through control of a failed government car factory. Though Sanjay held no official position, he coopted or intimidated much of the cabinet, placing sycophants in top jobs at the key ministries. Indira was so afraid of losing his affection that she allowed him to do almost anything—according to one story, even slapping her one evening in front of guests.[109]

A man with no formal title who answered to no one—not even his

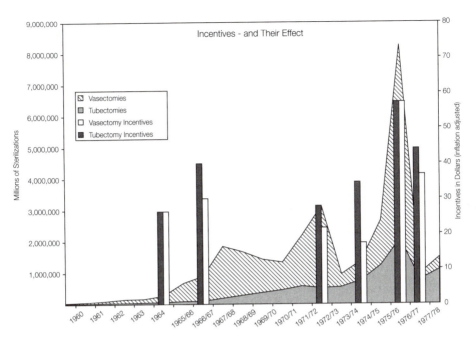

When Delhi raised the level of incentive payments to providers and "acceptors"—or reduced it, as in 1973 and 1977—it had a direct and sometimes dramatic effect on the number of vasectomies and tubectomies. Data from Alaka M. Basu, "Family Planning and the Emergency," *Economic and Political Weekly* 20 (March 9, 1985): 423; and K. Srinivasan, *Regulating Reproduction in India's Population: Efforts, Results, and Recommendations* (New Delhi: Sage, 1995), 52.

mother—was just the kind of person to lead a population control campaign. It suited both Sanjay's politics and his temperament. He was an outspoken anticommunist, favored foreign investment, and cultivated contacts with the Americans (much to his mother's embarrassment). As she explained, "Sanjay isn't a thinker—he's a doer." But he could have done little if he did not find a whole infrastructure to control population already in place along with international agencies and local officials prepared to advise him on how to proceed. Sanjay even spouted talking points prepared for him by the Washington-based Population Reference Bureau. Officials learned that the way to win his favor was to triple targets for sterilization. The worst abuses occurred not through legal sanction—the legislative process still moved too slowly—but when they acted on their own authority out of fear, greed, or ambition.[110]

Sanjay had an integrated four-point program of his own: family planning for two children, increasing adult literacy, abolition of dowries, and slum clearance. It was the linkage of slum clearance and family planning that made him notorious. But here again, Sanjay reflected and amplified a trend long building in population control circles: a concern to control the seemingly chaotic growth of cities. Contemporaries considered this a global problem, affecting the United States as well as Asia, Africa, and Latin America. For some governments, such as Tanzania, Bangladesh, and Indonesia, "population control" included ambitious schemes to redirect migration to what they considered underdeveloped lands. UNFPA also funded projects to persuade peasants to stay put. Kingsley Davis considered such policies impractical, because people would not stop having large families until they adopted an urban lifestyle. After Gandhi declared the Emergency, Davis offered his support: "The only force capable of managing such cities appears to be a strong government that stands in contrast to the populace in skill as well as power."

> In short, the likely way the massive urbanization projected for Asia can take place is by a totalitarian government, highly competent and rigorously committed, ruling a docile mass of semi-educated but thoroughly indoctrinated urbanites existing at a low level of consumption, working very hard, and accepting passively what is provided for them.[111]

Sanjay took up the challenge. He announced plans to demolish whole neighborhoods, including centuries-old residences around Turkman Gate in old Delhi. In April 1976—just after Singh had announced the new population policy as part of "a frontal assault on the citadels of poverty"—one of Sanjay's associates, Ruksana Sultana, opened a sterilization camp nearby in the Muslim neighborhood of Dujana House. She too lacked any official position but earned enormous sums by "motivating" others to be sterilized, striding into slums in chic sunglasses with the police at her side. When demolition squads began tearing down homes by Turkman Gate, residents beseeched her aid. According to some, she agreed to help only on condition that they produce three hundred people prepared to be sterilized.[112]

As poor beggars were rounded up and loaded onto a van, a burqa-clad woman blocked the road. Police intervened, arrested one man, provoking people to begin fighting back. Sultana had to be rescued from women who surrounded the sterilization camp. Residents joined together to battle the demolition squads, who were armed with bulldozers and pickaxes. When security forces responded with tear gas and beatings, protesters stormed the police post. Reinforcements finally reestablished control with live ammunition. No one knows how many were killed, but the bulldozers were soon back at work, clearing the way for a new commercial complex.[113]

Looking at a map showing the areas demolished across Delhi during this period, an astute surveyor found it looked "more like a bombardment plan than a development plan." Some seven hundred thousand people were driven from their homes. Many more cities were subjected to the same treatment. Gandhi herself was shocked when a close friend confronted her with photographs of a neighborhood of half-wrecked houses in the holy city of Varanasi, which "looked as if a bomb had fallen on it." Those with three or more children had to produce a sterilization certificate to be eligible for new housing.[114]

This war against the poor also swept across the countryside. In one case, the village of Uttawar in Haryana was surrounded by police, hundreds were taken into custody, and every eligible male was sterilized. Hearing what had happened, thousands gathered to defend another village named Pipli. Four were killed when police fired on the crowd. Protesters gave up only when, according to one report, a senior government official threatened aerial bombardment. The director of family planning in Maharashtra, D. N. Pai, considered it a problem of "people pollution" and defended the government: "If some excesses appear, don't blame me. . . . You must consider it something like a war. There could be a certain amount of misfiring out of enthusiasm. There has been pressure to show results. Whether you like it or not, there will be a few dead people."[115]

Foreign donors closely watched these developments. They responded by increasing their support. A World Bank official was in Delhi when Singh announced the new program. When he got back to Washington he reported that "facilities are needed to support the GOI's main drive for sterilization in its new, more vigorous FP campaign." India proposed a second, $26 million project, which included both literacy programs and "sterilization annexes" to clinics covering a population of fifteen million people in Bihar, Andhra Pradesh, Jammu, and Kashmir. The scale was deemed "disappointingly conservative." The World Bank wanted to triple it.[116]

When asked by UNFPA, a Bank official said they had taken no formal position on compulsory sterilization. The IPPF management and planning committee also decided that "it would be premature for the IPPF to issue an official statement at the present time." Most of the top leaders of the

Family Planning Association of India, including Rama Rau, initially supported compulsory sterilization. But the president, Avabai Wadia, persuaded them it was "too drastic." Instead, she quietly advised the government that "the reinforcement of pressure points may prove more effective and this is where incentives and disincentives can be of great assistance." The FPAI, which received more than 90 percent of its budget from London, had already conducted more than one hundred sterilization camps in 1974–75. Henderson agreed that "the present environment in India is favorable towards a major expansion of voluntary sector effort." FPAI was granted a nearly 60 percent increase in funding, to $1.6 million. It decided to expand both the sterilization camps as well as nutrition and maternal health programs, "as they have a high potential for drawing acceptors into the family planning programme." FPAI leaders complained the government was not giving them enough credit for their contribution. Fulfilling government targets was their "first priority."[117] With funding and encouragement from London, FPAI branches sterilized eighty thousand people in 1976, more than the world total for the entire IPPF in the preceding year.[118]

Sterilization became a condition not just for land allotments, but for irrigation water, electricity, ration cards, rickshaw licenses, medical care, pay raises, and promotions. Everyone, from senior government officials to train conductors to policemen, was given a sterilization quota. This created a nationwide market in which people bought and sold, sometimes more than once, the capacity to reproduce. Of course, for the very poorest, with no money and nothing else to sell, sterilization in such conditions was not really a choice. But some were themselves agents in the process—such as the men who made up the demolition teams, a low-status job that typically went to Dalits. They demolished one another's homes.[119]

By October 1976, when SIDA sent one of its staff, Peter Hegardt, for a joint inspection of the Swedish–World Bank projects in Uttar Pradesh and Karnataka, the oppressive nature of the program was plain for all to see. "Obviously the stories . . . on how young and unmarried men more or less are dragged to the sterilization premises are true in far too many cases," he reported. There were many "shocking stories." In fact, in the two weeks he was in India, there were three incidents in Uttar Pradesh of police killing people for protesting the population control program. Nevertheless, in December 1976 SIDA decided to fund a second India project to the tune of 75 million kroner ($17 million, or $60.2 million in today's dollars).

Hegardt advised that they would not accept compulsion, but that "civilized and gentle pressure should be used."[120]

McNamara visited India in November, and Singh told him that incentives for sterilization would remain necessary "for quite some time." States were urged to punish instances of coercion, but "a certain number of incidents were probably unavoidable." McNamara was encouraged by what he heard and saw. "At long last," he wrote, "India is moving to effectively address its population problem." He commended Singh for providing "dynamic direction." In a joint meeting with World Bank staff, UNFPA program director Nafis Sadik said that any country that used compulsory sterilization should not receive UN assistance. She also "expressed her personal view that compulsion may be needed at the expense of human rights, but that people should be provided with all types of contraceptive means. Without such a provision, compulsory sterilization seems 'unethical.'"[121]

USAID no longer had any office in India, and its policy was to not be involved with incentives. But Ravenholt attended meetings of the AVS where he was informed of its contributions to a program that was sterilizing a quarter of all eligible couples in some areas. By October 1976, Ravenholt was planning a "rapid expansion" of support for voluntary sterilization worldwide. Two months later, he met with WHO officials and signaled that USAID wanted to support the Indian program. Working through NGOs, USAID had already sent sixty-four laparoscopes to India, each one capable of as many as two hundred sterilizations a day.[122]

Altogether, in the course of one year, the government would record more than 8 million sterilizations: 6.2 million vasectomies and 2.05 million tubectomies. Rather than registering any concern, the Swedish economist Goran Ohlin advised the World Bank that there were still too many empty beds in sterilization clinics. But with continued funding, they had the opportunity "to follow at close range a great deal of important experimentation in the world's biggest family planning program." Indeed, "for the future, the Bank should probably try to achieve even more experimentation." As for whether the project should be affected if some states adopted compulsory sterilization, Ohlin wrote that it was "not for me to say."[123]

Ohlin wanted more experimentation in nutrition and literacy programs, not just sterilization. But, as might have been anticipated, the rest of the "integrated" program fell by the wayside when the whole government was incentivized to sterilize. There were more immunizations of expectant mothers and children during the Emergency. But they achieved less than half the targeted number. Performance in the nutrition program actually declined. Indian critics, like Debabar Banerji, asked why, if the government was inclined to use compulsion, it did not enforce the minimum age of marriage. It did not even manage to update the toothless, half-century-old Sarda Act, contrary to Singh's pledge.[124]

Now that draconian population control was being practiced on an unprecedented scale, the most senior people in the field agreed that they must not interfere. Berelson, Ronald Freedman, Fred Sai, and A. Chandra Sekhar—secretary of India's Health and Family Planning Ministry—concurred with Ohlin's assessment. Appraisals of national programs seem "overly confident that now, in the mid-1970's, the outsiders know better than the insiders what to do, whereas more modesty and humility, more response and less aggressiveness, is called for." Asked to advise the Bank on its population program worldwide, they argued that "there is no current reason to think that the outside community necessarily knows better, and the Bank would do well to listen carefully to local ideas, and to encourage them." After all, they noted, the major innovations were coming not from international agencies but from the countries themselves, especially India, Indonesia, and Communist China.[125]

The people of India, however, had had enough. Hundreds were being killed from botched sterilizations—according to official statistics, 1,774 of them. There is no way to count the number who were being hauled away to sterilization camps against their will. One indicator is the number of those sterilized who had only one child or none at all—unlikely to volunteer in a country where parents depended on sons to support them in old age, and a quarter or more did not survive infancy. In Andhra Pradesh there were 21,653 such sterilizations; in Gujarat, 7,834; in Karnataka, 10,244; in Maharashtra, 6,958; in Orissa, 19,237; in Punjab, 19,838; in Uttar Pradesh, 11,434; in West Bengal, 8,098. Many states did not report such data but provided other measures of both the degree of coercion that was being used and the resistance it was arousing. In Delhi, for instance, 100 people were arrested for opposing population control; in Haryana, 428; in Madhya Pradesh, 161; in Rajasthan, 283.[126]

The prime minister initially dismissed reported abuses as mere rumors. But one of her top aides, P. N. Dhar, sensed that she was growing uneasy about the arbitrary power Sanjay wielded—that "things were getting out of control." Gandhi had still not decided whether states like Maharashtra would be permitted to impose compulsory sterilization. One day, in November 1976, Dhar passed along a report describing how schoolteachers were treated when they failed to meet their quota. Teachers, like everyone else, could be demoted, fired, or threatened with arrest. They, in turn, sometimes expelled students when their parents did not submit to sterilization.

Gandhi seemed saddened and remained silent for some time after reading the report. At last, in a tired voice, she asked Dhar how long he thought the Emergency should go on. It was the first in a series of conversations, but shortly thereafter Singh began issuing circulars to all states demanding that sterilization camps maintain proper medical standards and promptly compensate families of those who had died. The next month a cabinet discussion on Maharashtra's compulsory sterilization bill was canceled. And in January, Indira overruled Sanjay and dissolved the Lok Sabha. She was calling an election.[127]

Most observers, like Gandhi herself, expected the Congress Party would win again, as they had won every election since independence. Opponents were released from jails, but they had only two months to reorganize. Moreover, much of Gandhi's program during the Emergency was broadly popular, including a moratorium on repayment of rural debts and the abolition of bonded labor. The government gave 1.7 million acres to the landless and three million sites to the homeless—though many were landless and homeless because of slum clearance. It was thought that they would be grateful to have relief from strikes, a crackdown on smugglers and black marketeers, price controls on essential commodities, and a tax amnesty.[128]

Opponents cobbled together a coalition and agreed to field their strongest candidate in every district. India's newspapers were finally free to report the abuses in the family planning program. More than half the election coverage explicitly mentioned the issue. Gandhi's cabinet rejected the Maharashtra law, withdrew a proposal for "disincentives" for government employees, and closed the sterilization camps. Nevertheless, as Gandhi campaigned across India, the crowds were disappointingly small at her campaign rallies. When some women literally turned their backs to her, the prime minister waded into the audience and tried to turn them around. Key allies began to desert her, including the leading spokesman of the Dalits in her own cabinet. Even her aunt, Vijaya Lakshmi Pandit, who backed compulsory eugenic sterilization as a member of the National Planning Committee almost a half century earlier, decided it was time to keep her distance.[129]

At last, in the largest democratic election in history, the people of India produced one of history's greatest political upsets. The prime minister herself was routed in her home district. So was Sanjay. The Congress Party was defeated all across northern India. They lost 141 out of 142 seats in the states that had registered the largest increases in sterilization (in the previous election, Congress had carried 80 percent in these areas). In Delhi the crowds stayed up through the night to cheer as the results came in. Passing cars honked their approval. "The skinny, ill-fed, semi-clothed, so-called illiterate villager of India refused to be seduced by promises of food and fuel in exchange for their basic human rights," the solicitor-general observed. "To the spurious question of whether the poor man would rather have food or freedom, the resounding answer which he gave was: 'we will have both.'"[130]

Something even more powerful, even more implacable, had finally defeated the ideology of population control: People voting, one by one.

9

REPRODUCING RIGHTS,
REPRODUCING HEALTH

The 1974 World Population Conference, followed by India's repudiation of Indira Gandhi three years later, exposed for all to see that the population control movement had no mandate. It was rejected by a majority of governments and the most populous democracy. Together, the effect was like a body blow followed by an uppercut. The "system without a brain" would never be the same.

All of the most important international and nongovernmental organizations in the field entered a period of agonizing reappraisal. Facing staff and budget cuts, population controllers could only take bitter satisfaction in receiving confirmation that fertility rates had begun to fall in almost every region of the world. Together with unfulfilled predictions of global famine, it only made their work seem less urgent, and their excesses all the more unforgivable. Continuing debates about whether government programs were reducing fertility rates—in most places, it started without them—were becoming matters of merely academic interest.

Some people working from the inside had always resisted the idea that they needed to plan other people's families. But there was too much invested in population control for these institutions to transform themselves overnight. In 1980, about $2 billion was being spent on population programs in poor countries, including some $490 million in international aid. For the most part, it was not pledged to promote gender equality or maternal health. With funding already down relative to inflation as well as other kinds of international aid, downplaying a long-standing commitment to reduce fertility might lead to further losses. Careers and reputations depended on the proposition that it remained both practical and urgent. The status quo was also buttressed by the ideological detritus of decades, which continued to attribute war, famine, disease, and degeneration to "overpopulation."[1]

Those who genuinely wanted to empower people, and not control them, struggled to disentangle themselves from all of this. It was not obvious how to make a clean break—that is, how to stop coercion and advance a different agenda that could work in dozens of different countries. Hundreds of millions of people had come to depend on family planning programs, for all their faults, and many more were still left to their own devices. If this was not to be a self-serving exercise in exculpation, their needs had to come first. Radical reform was all the more difficult at a time when a new "pro-life" movement was seeking opportunities to divide its opponents and discredit every last one of them. A popular American president, a charismatic pope, and swelling legions of orthodox Muslims and evangelical Christians worked not just to finish off the increasingly feeble vestiges of population control, but to strangle the renascent but still vulnerable movement for reproductive freedom.

The central battleground in this struggle was China, which witnessed the largest crash program of all. It developed its one-child policy in isolation and owed little to any foreign organization. But UNFPA and the IPPF, conflicted about their future and pressured by donors, decided to extend their aid despite repeated warnings about what they were walking into. Their enemies quickly realized that China was the weakest point in the entire system, one that now made everyone connected to it vulnerable to attack. These attacks were disingenuous, often defamatory, but they compelled those committed to reproductive rights to close ranks and insist on a more principled platform. A new consensus, the fruit of long struggle, would finally displace population control at the last United Nations population conference in 1994.

The torturous road to Cairo began two decades earlier, when delegates at the Bucharest conference turned for home. As Henderson reported to the IPPF's Governing Body: "It is getting increasingly clear that there are many FPAs and quite important personalities within the Organisation who are not completely conversant with the structure, organisation, and policies of the Federation. A major internal education within the system is necessary." The IPPF began to fund more "integrated" projects. At the same time, Henderson fought to create a unified secretariat from regional staff.

When her successor, Carl Wahren, challenged the New York office's hold over the Western Hemisphere associations, they threatened to form a Pan-American Federation and split the IPPF in two.[2]

The Population Council was also facing a total breakup, in part because of unprecedented budgetary pressure. There was still plenty of money for family planning programs—USAID was offering to fund even more than 40 percent of the IPPF's budget, on the condition that it expand its sterilization programs. But many in the Council, led by the demographer Paul Demeny, thought they should walk away from this business. Between 1974 and 1977, the number of staff would fall from 275 to 174. Among those leaving was Bernard Berelson. More than two years later, the board of trustees had still not found a new president.[3]

In 1974–75, four trustees under David Hopper conducted a wide-ranging review. When staff members asked why the board did not get on with appointing a new president or have younger, more committed members, Hopper turned on them. In a tumultuous meeting, he shouted that it was they who had failed in their jobs. Privately, he said that the Council's involvement in national family planning programs and neglect of issues like migration was a historical accident. After seriously considering total liquidation, his committee recommended that it be broken in three. The biomedical and technical assistance divisions would go their own way while a much-reduced Council would do basic demographic research.[4]

John D. Rockefeller 3rd and the rest of the board decided to reject the Hopper committee's recommendation. Rockefeller agreed with the Ford Foundation that, as David Bell argued, it "would signal that population [is] not a problem—ec + soc development is all you need to work on." This would be "a major blow to the world population movement." But the two continued to cast about for ideas. Bell permitted Adrienne Germain to travel to Asia—at her own expense—to look into opportunities for projects to help women. One of the people she met in a follow-up visit with Dunlop was George Zeidenstein, head of the Ford Foundation office in Bangladesh, and formerly country director for the Peace Corps in Nepal. He was not a product of the population establishment. In fact, he grew his hair long and impressed people with his patience and willingness to listen. Dunlop agreed with Germain that this was just the person who could transform the Council.[5]

Rockefeller convinced the trustees to make Zeidenstein president despite Berelson's and Notestein's bitter opposition. "The anger and the resentment and the resistance against George were enormous," Germain recalls. In a series of contentious meetings, senior staff implied he was imposing a political agenda. Zeidenstein reminded them that the Council had been founded with a political agenda and had a record of promoting risky contraceptives. He questioned whether there could be any such thing as value-neutral research.[6]

The Ford Foundation itself was undergoing a period of self-doubt and deep budget cuts. In a June 1977 strategy paper, Ford staff asked questions of the trustees that would scarcely have occurred to an earlier generation: "To what extent is it appropriate for us as a Western Foundation to support activities which profoundly affect traditional mores and value structures? Can we not be charged with seeking to impose on others our views on what is good for the world?" But they assured the trustees that they worked to reduce fertility only because their partners all over the world agreed it should be a priority. The trustees kept on cutting their budget. The Rockefeller Foundation's board was also said to be "shying away from field experiments + fam. pl."[7]

What had happened in India during the Emergency was not always acknowledged in these reappraisals, but its impact was obvious nonetheless. Just five days after Indira Gandhi was thrown out of office, the IPPF Management and Planning Committee finally resolved that "no sterilization procedure should be performed unless the person concerned has given voluntary unpressured informed consent." It also stressed proper medical standards, advised against paying providers piecework-style, and urged that incentives and disincentives should not infringe on basic human rights.[8]

A month later, Henderson faced close questioning in a congressional hearing. She maintained that the IPPF opposed incentive payments for individuals and had never funded them. Another witness, the national director of the U.S. Coalition for Life, Randy Engel, tried to argue otherwise. She was able to document how USAID had used intermediaries in India. But she did not actually connect all the dots, which might have led from U.S. support for the IPPF to the eighty thousand sterilizations that its Indian affiliate carried out in 1976. Congress was assured that USAID guidelines prohibited support for incentives or involuntary sterilization.[9]

In Sweden, SIDA began to prepare another population project for India in spite of protests from its own officials. Together with Indian informants, these officials helped journalists launch an investigation that concluded that the Emergency Period was "Sida's Watergate." A series of newspaper reports and radio broadcasts argued that, after growing "completely out of control," Sweden's program in India had provoked riots and massacres. SIDA was forced to cancel the hundred-million-kroner project.[10]

The long-feared backlash against population programs was gathering strength and had already spread far beyond India. In the Philippines a program that distributed contraceptives through grocery stores touched off what newspapers called the "condom war." The Catholic Women's League won agreement that condoms would not be advertised or given to minors. In Indonesia, it was "simmering Moslem opposition" that gave administrators pause. In March 1976 government officials decided to accelerate the sterilization program anyway. Four months later, reports that the U.S. ambassador had resisted demands to accept delivery of more contraceptives led to scathing editorials in the Indonesian press. One argued that the family planning program had already led to underage prostitution. Another, titled "Condom from the U.S. and Sex from Indonesia," warned that distributing contraceptives would "destroy our tradition and morals. Sex will become the most important thing as in the Western countries and the U.S."[11]

In March 1977, the same month the Congress Party was routed in India, an opposition alliance across the border in Pakistan attacked Prime Minister Zulfikar Ali Bhutto for promoting family planning, "a filthy business and against the spirit of Islam." The alliance promised to convert all family planning clinics into health dispensaries. Bhutto, for his part, used program personnel and vehicles to turn out the vote. He claimed victory at the polls, but opponents took to the streets. Four months later General Muhammad Zia-ul-Haq seized power with the support of the Muslim League. At the end of the decade, new revolutionary regimes were refusing international family planning assistance. The Sandinistas of Nicaragua instead called on mothers to have more children in order to repopulate the war-torn country. Ayatollah Ruhollah Khomeini attacked the shah's family planning program from exile and shut it down when he took power. In Iran as well, it was called an instrument of Western imperialism.[12]

When the leaders in the population field decided to meet once again in June 1977, they did not return to the romantic shores of Lake Como. "The image of 'Bellagio' is inappropriate," it was explained, "given the politics of the world population scene." This meeting would also have more representatives of developing countries, including Carmen Miro and Fred Sai. When they convened in the sober setting of Ulvshale, Denmark, the discussion was noticeably different. "Note how revolutionary we all are,"

Bell exclaimed, "—decentralized, poverty-oriented. Note how powerful are [the] obstacles—not simply national leaders with vested interests but middle class with vested interests" (that is, physicians reluctant to let others provide family planning services).[13]

But even those who agreed on the need to be more progressive and "poverty-oriented" differed about how to proceed. How, exactly, did one improve the "status of women"? If through education, then what kind? If by increasing access to paid work, might that not simply add to all the work women were already expected to do? There was not even a consensus about whether incentive payments were coercive. For many officials, all of these measures were complementary parts of a more "integrated" program. Robert McNamara would defend Indira Gandhi in a major policy speech at MIT:

> No government really wants to resort to coercion in this matter. But neither can any government afford to let population pressures grow so dangerously large that frustrations finally erupt into irrational violence and civil disintegration. That would be coercion of a very different order. In effect, it would be nature's response to our own indifference.

At the same time, McNamara pointed out that reducing infant mortality, improving nutrition, and increasing women's access to education and paid employment could also reduce fertility rates. This speech prompted Bell to write that McNamara and everyone else at the Ulvshale meeting were "ready to accept a powerful, sweeping set of ideas which were hardly on the horizon four years ago."[14]

In fact, the link between education and fertility had been discussed for decades, beginning with eugenicists who worried that educated women did not contribute to the gene pool. It was catching on now not because of new research but rather because the political environment had changed. A 1975 Population Council review argued, "The linkage of fertility with educational attainment, employment status, and role within the family doesn't hold up very well. . . . It seems a pretty reasonable prospect that further research along this line will not turn up any more significant direct relationships, particularly of policy relevance." But Zeidenstein decided to proceed anyway. The focus of such research should be human welfare, he said, not demography per se. McNamara might well have replied, "Not on my dime."[15]

Some World Bank officials in the field did not think the status of women was any of their concern. As the head of the Bank office in Bangladesh, Leonard Weiss, explained to Dunlop and Germain, at the same time impatiently looking at his watch: "This was interfering in the social structure and the role of the family. He seemed somewhat taken aback," Dunlop reported to Rockefeller, "when Ms. Germain asked him what he thought family planning was, if not intervening in the most private and personal of decision-making in the family's life."[16]

People like Dunlop, Germain, and Zeidenstein insisted on examining the underlying values and larger social goals of family planning. But this did not necessarily lead to a new consensus. Few leaders in the field would make an unconditional commitment to advancing reproductive rights. Nafis Sadik later recalled how she could not yet even talk in such terms within UNFPA. Dunlop herself, perhaps to put Weiss at ease, said that the cause of gender equality was not necessarily relevant to poor, rural societies. For most, it depended on the context. As the steering committee of the Ulvshale meeting put it: "international stress on certain human rights and values as universal may retard national action that could result in attractive policy trade-offs when measured by local standards."[17]

Even within the United States, interpretations of "local standards" differed, as the increasingly volatile debate over abortion made clear. Was it a fundamental right, or merely a good "policy trade-off"? In June 1977 the Supreme Court upheld the constitutionality of bans on federal funding for abortion. The next month Dunlop asked Frederick S. Jaffe, head of the research arm of the Planned Parenthood Federation of America, if he agreed with an editorial that had just appeared in the *New Republic*. It argued that "there clearly is no logical or moral distinction between a fetus and a young baby; free availability of abortion cannot be reasonably distinguished from euthanasia. Nevertheless we are for it." The social costs of compelling a woman to bring an unwanted child into the world were too great. Just the first year on Medicaid amounted to $2,200, compared with less than $200 for abortion. "And of course," the editors noted, "in most cases the first year is not the end of it."[18]

The editorial raised the question Sanger had faced more than half a century earlier: Was it enough to leave women alone to make their own choices? Giving them real options might require making it a compelling public interest, even if that meant assigning a negative value to their potential offspring. Jaffe made the same choice Sanger did:

Government should pay for abortions for the poor, not primarily because it's their constitutional right to control their own bodies or even because government should enable the poor to have what the rich have. Government should pay for such abortions because the cost (all kinds) of denying them abortions outweigh the benefits, for both the individuals concerned and for society as a whole.[19]

The very next day, President Jimmy Carter was asked in a press conference about federal funding for abortions. He said that he opposed it, except for medical reasons, rape, or incest. Carter was asked whether this was fair to the poor. "Well, as you know," he replied, "there are many things in life that are not fair, that wealthy people can afford and poor people can't." But the government could not remedy every inequality, "particularly when there is a moral factor involved." Planned Parenthood leaders came to the White House to complain that Carter's position was "racist, sexist, classist." Phyllis Piotrow of the Population Crisis Committee insisted that access to abortion was "a basic human right." If the United States regressed it would set back their work abroad. Even so, it still seemed worth pointing out that New York State saved $135 million by paying for abortions rather than caring for unwanted children.[20]

Planned Parenthood leaders could not decide whether to defend access to abortion as a human right or an "attractive policy trade-off." How could they agree on a rights agenda abroad, where the trade-offs were far more complex, and where they had to work with organizations still committed to population control? Rockefeller might have provided leadership, but he died in a car accident near his family estate in 1978. Some reproductive rights advocates believe this delayed reform for a decade. In fact, the leaders of the Population Council and the foundations were now inclined to oppose coercion, and the Ford Foundation in particular was beginning to promote a more comprehensive approach. The problem was that their relative weight was much diminished. The IPPF and the UN agencies had to listen to the main sources of their funding. That is why, amid all the revolutionary rhetoric at Ulvshale, Henderson made "a conservative plea: let's show some consistency in striving for the goals we've set—we still need earmarked funds for population questions." Similarly, Miro suggested that they "keep the money in fam. pl. *Add* other programs."[21]

The World Bank, USAID, and UNFPA were now reluctant even to show up for Bellagio-type meetings, wherever they were held. When sounded out about another such gathering, they did not bother to be polite: "They criticized the foundations, Ford and Rockefeller as well as the Council, saying that they are no longer as important in the field as they have been, and that if they mount the conference in the same old way they will seem to be trying to regain the leadership that they can't have anymore." After convincing them to come to Ulvshale, Bell was pleased with the outcome, though he suggested "a deliberate effort to include more women in the gathering." McNamara, by contrast, complained that there were "too many speakers who wasted time."[22]

The consensus Bell thought he discerned was illusory, resting on philosophical differences and unexamined assumptions. The World Bank, in particular, was utterly opposed to a fundamental reexamination of population control. It was not just McNamara. The Bank's vice president for South Asia agreed that "given India's problem, significant incentives seem to us an essential ingredient of any program." A new "Operational Manual" declared that "population projects should be designed to do what is necessary to reduce fertility." This included "demand-influencing activities" such as measures to improve women's opportunities and lower infant mortality, but also "community incentives that may create peer-pressure for limiting fertility or that may establish differential privileges in education or housing." The end goal was unambiguous: "general health benefits that cannot be presumed to influence fertility, while certainly welcome, should not be used to justify project expansion." Some officials protested. James A. Lee warned that "the withholding of necessary health services to achieve leverage for 'population' would be politically dangerous and potentially embarrassing to the Bank." He was overruled.[23]

Ravenholt was even less inclined to admit any doubt about the priority USAID gave to reducing fertility. He complained, "We have been subjected to an extraordinary gush of rhetoric from a number of people with very limited knowledge of what actually is happening in the field." Ravenholt explained that he had tried to use whatever medical services existed, and they simply could not keep up. He also innovated by including iron supplement pills in oral contraceptive packets. He firmly believed that nothing would improve women's health and opportunities more than helping them to have fewer babies. "When they speak of 'integration' of family planning with health," Ravenholt concluded, they "are really seeking to obliterate or at least weaken the family planning element."[24]

But Ravenholt could no longer ignore his critics, not when there were mounting problems with his programs that had to be explained to Congress. In 1975, for instance, he discovered that the Filipino program was "floundering." The head of the Population Commission, Rafael Esmundo, assured him "that we are all determined to saturate the rural communities." When they could not distribute enough contraceptives, Ravenholt demanded to know "what the hell is going on" in what he called the "Philippines family planning swamp." The U.S. ambassador met with Ferdinand Marcos, who called Esmundo in. Mercedes Concepcion complained to a World Bank official that, if they were not able to show "favorable results," they would be forced into "choosing between professional integrity and the donor agency's pressure."[25]

When the new military government in Pakistan suspended its population program, Ravenholt grew even more irate. The U.S. ambassador had already complained to Bhutto about how it had become a source of patronage for incompetent hacks. A mountain of contraceptives, enough to supply the whole country for three years, sat in warehouses and fed the black market, but never seemed to reach rural clinics or shopkeepers. When USAID's representative, Steven Sinding, could not get permission from the new regime to launch a household distribution study, Ravenholt complained, "You simply don't yet know how to get your utilitarian wheels down."[26]

Ravenholt knew that he was driving his people to distraction and provoking opposition in the countries where they had to work. But he felt he had no choice. As he explained to Esmundo, "I live with the apprehension that unless we accomplish the essential work within minimum time that the oppositional forces will gain the upper hand and obstruct both our action and yours." Ravenholt had already experienced a major setback in 1973, when he tried to ramp up production of his "menstrual regulation" kits. Instead, Senator Jesse Helms was able to introduce an amendment to the Foreign Assistance Act that prohibited any such use of federal dollars. Ravenholt responded by brokering deals in which the IPPF obtained the kits and distributed them to its own affiliates as well as other NGOs. Opponents subjected USAID grantees to increasingly intrusive congressional oversight and government audits, and caught some of them using U.S. money to pay for abortions.[27]

Ravenholt suspected a Catholic on his staff was leaking incriminating information. When Jimmy Carter became the first "born again" president, and political appointees tried to rein Ravenholt in, he became even more suspicious. In fact, evangelical Christians were still either divided or indifferent about family planning, in part because opposition was so closely identified with Rome. The senior Catholic hierarchy had come close to endorsing Gerald Ford in the 1976 election, and only backed off when younger bishops and staff at the U.S. Catholic Conference protested against making abortion their only issue.[28]

Some of Ravenholt's subordinates concluded that neither covert nor overt Catholic opposition was his main problem. They considered Ravenholt to be his own worst enemy. He seemed to seek out conflict. This included a shoving match with Halfdan Mahler, director-general of WHO. Ravenholt also punched the director of his research division when a study "didn't turn out the way he thought it should." Ravenholt would apologize for such outbursts. In other cases it was impossible to undo the damage. In 1977, faculty and staff at Washington University protested a USAID-funded program that trained foreign doctors to perform sterilizations but did not address what they viewed as the underlying causes of poverty. The *St. Louis Post-Dispatch* reported after interviewing Ravenholt that the United States had a plan to "provide the means by which one quarter of the fertile women in the world can be voluntarily sterilized." The reporter doubtless hyped what Ravenholt told him. But the president of the university, members of Congress, and the foreign media all protested against his all-too-honest explanation that the United States was not acting just from humanitarian concern: "Without our trying to help these countries with their economic and social development, the world would rebel against the strong U.S. commercial presence."[29]

USAID's assistant administrator Sander Levin asked Ravenholt to "move on to another challenge." When he refused several more such requests, his boss vowed to "destroy" him. The two waged their battle through lawyers and newspaper leaks. Ravenholt lost a key ally when Hubert Humphrey passed away, and a committee reassignment made it tougher for Senator Inouye (and his staff) to defend him. Embarrassing reports appeared in the *Washington Post* describing how Pakistani children made balloons from surplus condoms. In another case, the U.S. mission in Nepal requested $50,000 to burn condoms that were past their shelf life. Levin shifted authority over country programs from the population office back to the regional bureaus. Ravenholt was demoted and finally driven from office.[30]

USAID's population budget kept increasing, but Congress no longer supported the inundation approach. "The view that all we want to do is control population rate[s] for our own purposes, and it's not related to any compassionate concern for overall health matters," Rep. James Scheuer (D-NY) explained, "is crippling efforts to get family planning programs going." Ravenholt protested that demand far exceeded supply wherever they succeeded in delivering contraception to people's doorsteps. For Scheuer, this was missing the point: Freestanding contraceptive programs unconnected to health services provoked opposition. "It's the setting, Ray, it's the setting."[31]

Even Ravenholt's critics had to acknowledge at least one apparent success. In February 1978, a triumphant cable arrived from Jakarta with the subject line "They Said It Couldn't Be Done." It reported that in Bali some 60 percent of eligible couples had started to use modern contraceptives. Despite continuing high rates of poverty, illiteracy, and infant mortality, fertility fell 30 percent between 1971 and 1976, indicating that the decline was due to the family planning program, not to broader development. The success of the Indonesia program was reported far and wide, and seemed to form part of a trend. Thailand and Colombia also reported declines in fertility after beginning family planning programs. For the leaders in the field—and not just Ravenholt—it vindicated their work.[32]

The problem was that, even according to the most favorable contemporary studies, family planning efforts explained less than 5 percent of fertility levels in developing countries. As Paul Demeny pointed out, in country after country, "from Malaysia to Mauritius, from Taiwan to Trinidad," fertility was declining *before* family planning programs had even begun. Brazil had experienced a major decline with no official program whatsoever. Moreover, it could not be shown that even the 5 percent effect was actually caused by such efforts, or whether instead broader socioeconomic or cultural changes explained *both* the decline in parents' preference for large families *and* government willingness to provide them with contraceptives (what economists call the endogeneity problem). After all, national leaders also had families, and many were married to women deeply committed to family planning (not to mention those leaders—like Gandhi—who were mothers themselves).[33]

Of course, Demeny observed, a government could, like India's, sterilize a million people in a month. It could do anything "within the bounds of physics." As for Indonesia, this was no simple contraceptive distribution program—not with local authorities, backed up by a military dictator, monitoring and even mapping the fertility behavior of every household in their jurisdiction. Ravenholt's field representative judged that the "institutionalization of [family planning] at the sub-village level is awesome—perhaps the most socially directed family planning program outside of red China." It was not the kind of approach that could be standardized, replicated, and exported around the world.[34]

By the time the leaders in the field reassembled for another Bellagio meeting in April 1979, they confronted a depressing agenda. Carl Taylor, who had a quarter of a century of experience in the field, warned Bell that UN officials, program administrators, and politicians were "scared stiff because of *growing religious crusades*" against family planning, with Muslims and Catholics on the move in the Middle East, North and South America, and parts of Europe. Others noted that, despite their rhetorical opposition at Bucharest, several countries started programs to limit population growth after the conference. But there was no denying that the sense of urgency was fading. Despite considerable pressure exerted by UN officials, more than half of all governments did not even respond to repeated queries about what they were doing to implement the World Population Plan of Action. The agenda for this meeting included a series of tough questions: "Looking to the 1980's, what is the population field after? Is it to be broadened to include all the development issues? . . . If the population field is 'drifting,' with no clear set of objectives, what can be done to revitalize it? Where does it go from here?" No good answers were forthcoming.[35]

It was with this sense of disillusionment and uncertainty that the leaders in the field approached a fateful decision: whether to help Communist China implement an audacious new policy to control population growth. There had long been tantalizing reports of "barefoot doctors" dispensing oral contraceptives like postage stamps. But reliable information was scarce. In Bucharest, Beijing's delegation did not even permit the organizers to publish Chinese demographic data. Gradually Chinese officials began to open up, and it emerged that they were determined to reduce fertility rates. From 1970, as per capita grain production stagnated, contraceptives began to be provided free. In 1973, economic plans started to include population targets. The following year, Mao told foreign affairs officials that it was "unacceptable not to control population." The people responsible for implementation suggested a sunny slogan: "Marry a little later, space a little more, have a little fewer, raise a little better." But no one could be relaxed about deviating from any new course set by the Great Helmsman.[36]

The first IPPF delegation to China arrived in 1977. During a seventeen-day visit to six cities and two communes, they discovered that provincial family planning committees decided the permissible number of births for each factory and agricultural production team. "If persuasion fails to work," they apprised the *International Herald Tribune*, "draconian measures may be used. A woman pregnant for the third time may be pressured to undergo an abortion and, if she insists on continuing the pregnancy, the child will receive reduced rations and find his educational opportunities limited." According to Jaffe's confidential report to Carl Wahren, they did not observe anything that "smacked of direct coercion."[37]

Up to this point, China's program had developed without any help from international and nongovernmental organizations. One might have thought that they would be wary of involvement in another "crash program" with coercive tendencies. But after being attacked as neocolonialists—and Chinese criticism was the most vituperative—suddenly being courted by the largest, most revolutionary country of all led to a very human reaction: it made their heads spin. Moreover, much of what they observed would have seemed familiar. This included use of mobile IUD and sterilization teams, incentives and disincentives, mobilization of peer pressure, and an interministerial committee to ensure that the entire government worked to achieve time-bound targets. China's program was evolving into a demonstration of total "integration" even more ambitious than India's under the Emergency Period. People like Kinsley Davis and Bernard Berelson were eager to give it another go, even if it required coercive sterilization.[38]

China may have been looking abroad as it developed this program. As early as 1963, Zhou Enlai had called for sending experts to Japan. Ravenholt would later suggest that China learned from USAID's own strategy. After the crackdown on "Malthusian" heresy in the late 1950s, and the Cultural Revolution that followed ten years later, China had few population experts of its own. But the most important foreign influence on China's program was entirely fortuitous. In 1978 a missile scientist named Jian Song went to a conference in Helsinki and learned about the Club of Rome, a group of European scientists, industrialists, and officials who met regularly to grapple with what they called "the world problematique." In 1972 they had commissioned a team at MIT to run simulations modeling the effects of increasing population and resource use. The results were published as *The Limits of Growth,* which warned of "overshoot and collapse," including the exhaustion of known oil reserves by 1992. The book sold more than ten million copies in dozens of different languages. The limits of this kind of approach were apparent by the time Song had heard of it. But it provided a way for him to apply his training to social problems. After all, if one ignored the uncertainties of predicting reproductive behavior, a growing population could be analyzed with partial differential equations much like the velocity and thrust of a missile in flight. In fact, by controlling reproduction, hundreds of millions of people could be guided to a single target.[39]

It only took about five minutes of borrowed time on missile ministry computers to run all the computations. The results appeared to provide a precise forecast of ultimate doom. Song's team projected that, if Chinese women had an average of three children, the population would grow to more than four billion by 2080. A large cohort was about to enter its reproductive years. If China could not reduce the fertility rate to 1.5 or even 1—that is, one child per woman—the resulting depletion of resources spelled disaster not only for China, but for the entire world. Conversely, if they were able to contain this growth, the Chinese could become prosperous and assume their rightful place among the world's leading nations.[40]

In China, like so many other countries, proponents of population control ignored the possibility that people might choose to have fewer children even with less coercive measures. In fact, the fertility rate was already falling rapidly: from 6.4 to 2.7 over the preceding decade. But by using future projections as evidence, Song helped convince China's senior leadership to launch a campaign to halt all population growth by the year 2000. The one-child family was one policy on which Mao's successors could agree. Deng Xiaoping, now preeminent among them, was particularly concerned about rising unemployment and the need to improve per capita consumption. Here too, demographers began to calculate each birth as a net liability to the state. And to override all opposition, the government created an independent bureaucracy and freestanding clinics specifically charged with controlling population growth.[41]

Implementing China's one-child campaign required processing census data to determine national and local birth quotas. That is why the Chinese wanted international assistance with improving this data and acquiring computers to process it. This would ultimately produce something population controllers had long dreamed of but never beheld: individual birth permits. "Based on the nationally issued population plan targets," one such permit stated, "combined with the need for late marriage, late birth, and fewer births, it is agreed that you may give birth to a child during the year [198–]; the quota is valid for this year and cannot be transferred." Brigade, production team, and street committee leaders were instructed to closely monitor women of childbearing years. Some of these cadres made women submit to monthly gynecological exams and posted reports on their menstrual cycles.[42]

Women who instead opted for an abortion earned 14 days of paid vacation—40 days if it occurred in the second trimester of pregnancy and was promptly followed by sterilization. Other incentives and disincentives varied from province to province, but those in Hubei were typical: If parents had only one child, they were to be given subsidies for health care, priority in housing, and extra retirement pay. The child was also favored with preferred access to schools, university, and employment. But if the parents had another child, they were required to repay these benefits. As for those who had more than two children, both mother and father were docked 10 percent of their pay for a period of fourteen years.[43]

In rural areas, where bearing a son was crucial to security in old age, it proved difficult to collect fines. The community would conceal offenders, others would temporarily move, or the baby would be given to relatives. Where compliance was high, on the other hand, there was not enough money to pay all the promised benefits. Authorities responded by unleashing "crash drives" in which "shock teams" led by senior officials moved from village to village. They reinforced local medical personnel to carry out more IUD insertions, abortions, and sterilizations. They also browbeat cadres and singled out offenders. A Stanford graduate student, Steven Mosher, witnessed one such meeting while doing field research in a Guangdong village. Eighteen women who were at least five months pregnant, "red-eyed from lack of sleep and crying," were told they would have to undergo abortions, in two cases by caesarean section. Through such means, China registered 7.9 million abortions, 13.5 million IUD insertions, and almost 7 million sterilizations in 1979, a 44 percent increase in total procedures over the previous year.[44]

Unlike Mosher, most visitors to China went on tightly orchestrated tours, typically for two or three weeks. Even so, the nature of its program was no secret. In late 1979, after returning from one of the first UNFPA missions, Carmen Miro reported that in one Beijing neighborhood of 22,000 no fewer than 1,700 people were working to enforce the one-child policy. She confirmed that decisions as to who could become pregnant were made collectively. But UNFPA nevertheless invited China to develop a joint project, signing the first agreement in December 1979, and increasing the grant to $50 million in September 1980—the same month the one-child policy was announced to all party members. It would help pay for more consultant visits, an international conference in Beijing, fellowships and study tours in the United States and Japan, and eleven new population institutes.[45]

The first UNFPA resident advisor in China, Walter Mertens, envisioned his work as helping his Chinese counterparts understand the long-range implications of the one-child policy. Ansley Coale was brought in and warned them about the risk of a grossly distorted age structure. But how could UNFPA avoid becoming implicated in China's program when most of its grants went to training and equipping the people who would carry it out? Mertens noted that the Chinese seemed to want to spend all the money on computers—crucial in calculating birth quotas.[46]

Senior IPPF staff had their eyes wide open when they began to negotiate with Chinese officials. In January 1980, IPPF information officer Penny Kane informed Carl Wahren that those who had recently returned described "very strong measures being taken to reduce population growth—including abortion up to *eight* months." BBC radio had relayed reports of suicides and attacks on would-be parents. Kane had long favored developing closer links with China, but felt compelled to issue this warning:

> I think that in the not-too-distant future all this will blow up into a major Press story as it contains all the ingredients for sensationalism—Communism, forced family planning, murder of viable fetuses, parallels with India, etc. When it does blow up, it is going to be very difficult to defend.

This program was entirely governmental, unlike India's, where the IPPF insisted its Indian affiliate could not be blamed for the abuses of the Emergency Period. Kane therefore repeated the point for emphasis: "We might find it extremely difficult to handle the press and public if there were a major fuss about the Chinese methods."[47]

China's program was not just intended to stop population growth. As a UNFPA-funded conference concluded in 1980: it was "also imperative to devise measures to raise the quality of population from the point of view of eugenics, education and ethics etc." Medical genetics, like Malthusianism, was making a comeback, and was designated a priority area for research.

REN KOU PU CHA YOU LI YU KONG ZHI REN KOU ZENG ZHANG

人口普查有利于控制人口增长

Foreign donors emphasized that much aid to Beijing was technical in nature. But Chinese officials viewed their one-child policy as an integrated whole. As this 1982 government poster emphasized: "The Population Census Will Help Control Population Growth." IPPF Archives.

Authorities took a broad approach, including prevention of consanguineous marriages, improved nutrition, and prenatal care. But the Chinese press called for laws to prevent reproduction among those likely to give birth to babies with defects, "including bisexuality," and even "elimination measures when abnormal babies are discovered."[48]

After a late 1980 visit, Kane reported hearing firsthand some "extremely simplistic" ideas about how to reduce the numbers of mentally and physically handicapped children, ideas that sounded "like Western eugenists did in the early 1900s." But she also provided reassurance. While some "overenthusiastic cadres" still went too far, she said, authorities were backing off from the most coercive measures. As for eugenics, "probably luckily, the country is in no position at the moment to do much about it. Foetal testing is virtually unobtainable: intelligence tests not yet developed and nobody has the knowledge for genetic counseling."[49]

In time, Chinese authorities would use international aid to acquire training and equipment to implement a eugenics program. Kane herself provided copies of eugenic laws when the Chinese asked for them. But whatever their qualms, both the IPPF and UNFPA faced insistent demands that they help control China's population growth from an unexpected quarter: Japan. Officials in Tokyo made clear that otherwise they would cut their contributions. UNFPA was able to obtain $12 million from Japan by pledging to spend it in China, and the IPPF was encouraged to do the same. Japan was now the second-biggest contributor to population programs in developing countries. But Kazutoshi Yamaji, president of the leading NGO, the Japanese Organization for International Cooperation in Family Planning (JOICFP), warned that "a cold wind appears to be blowing across UNFPA/IPPF relations." Both organizations were now facing budgetary shortfalls. For Tokyo, the whole point was to reduce population growth. As Yamaji explained: "If Asian population programmes do not succeed then stability in Asia may be endangered." When the one-child policy began arousing criticism in the media, Japanese aid officials focused even more intently on channeling their support through third parties.[50]

Helping China—while at the same time reducing the IPPF's exposure—required ingenuity. Representatives explained to Beijing that they were a federation of voluntary associations and that China should form one of its own. An agreement with a new Chinese FPA was duly signed in March 1981. The Chinese were interested in international seminars, study tours, and scholarships. London "gave them a great deal of attention, much more than we would normally do," as one senior IPPF official noted after arranging a twelve-day tour of India, including a meeting with Indira Gandhi. But the first secretary-general of the Chinese FPA was also the deputy director of the state family planning commission. In seeking more IPPF grants, he was unable to provide a budget independent of the government program, or even explain any distinction between them. Don Lubin, the IPPF's deputy secretary-general, was finally driven to ask: "Is the China FPA to be treated differently from any other grant-receiving organization of IPPF?"[51]

As the IPPF and UNFPA stepped up support, China's program became ever more coercive. In November 1981, the *Wall Street Journal* relayed reports in the Hong Kong press of how vehicles transporting Cantonese women to hospitals for abortion were "filled with wailing noises." Some pregnant women were reportedly "handcuffed, tied with ropes or placed in pig's baskets." The same reporter cited an anonymous Chinese source describing how women with unauthorized pregnancies were receiving injections resulting in stillbirths or early infant death. "Every day hundreds of fetuses arrive at the morgue." By 1982 reports were appearing regularly in the Chinese press, soon reprinted abroad, of how the birth of girls had led to abuse of mothers and infanticide. They predicted a long-term shift in the ratio of women to men.[52]

The very fact that such abuses were being reported in China—along with the punishment meted out to offenders—supported the contention that they were deviations from a policy that depended on voluntarism and improved the status of women. China increased the legal age of marriage, encouraged new husbands to join their in-laws' households, and sought to improve women's ability to care for themselves and their parents. Nevertheless, by December 1982 Wahren reported to a visiting Chinese delegation that some IPPF donors had become uneasy and needed a full accounting. The Chinese replied that "mistakes had been made, and one had to be open about them to find solutions." They invited Wahren to visit and asked him to advise authorities on how to better inform foreign opinion.[53]

Wahren and other top IPPF officials spent twelve days in China during April 1983. Preliminary census data were said to show a normal male–female ratio, and the delegation was reminded that publicity given to female infanticide was intended to stop it. Wahren was reassured that "it does not seem to be the policy of the Government to consider direct compulsory measures in family planning." He allowed that "in some areas there appears to be some compulsion [but] . . . immediate steps are taken to put an end to such practices." In fact, demand for sterilization was said to exceed capacity. Wahren seemed untroubled at the use of incentives, disincentives, and peer pressure to generate this demand. By this point, UNFPA had helped establish three training centers. The IPPF, for its part, was presented with a $9 million plan for equipping 290 "publicity, education, and technical guidance family planning centres." This would include equipment "for providing ante-natal check-ups of a specialised nature and genetic counseling." The amount was beyond the IPPF's means, but upon returning to London, Wahren called for a significant increase in support.[54]

This was the most coercive phase in the whole history of China's one-child policy. The campaign was led by Xinzhong Qian, a Soviet-trained former major general in the People's Liberation Army. All women with one child were to be inserted with a stainless-steel, tamper-resistant IUD, all parents with two or more children were to be sterilized, and all unauthorized pregnancies aborted. While usually expressed in euphemisms— abortions were called "remedial measures"—Shanxi Province's new regulations explicitly stated that "under no circumstances is the birth of a third child allowed." Qian was quoted as stating that compulsory abortion and sterilization were "understandable and reasonable."[55]

In the internal circulars and official directives, there was not even a pro forma injunction to avoid coercion—something that was standard in previous campaigns. For cadres, this was a clear signal. UNFPA, on the other hand, while worried that a new policy might pose "serious problems for the United Nations," specified that it would have to be a "a very explicit regulation that all couples with a second child must be sterilized." Similarly, Bernard Aluvihare, the IPPF's assistant secretary-general, pledged that if this came to pass the IPPF would "express our concern through appropriate lines of communication." But he claimed that London "views the relationship of the [Chinese Family Planning] Association as separate from the Government as we do with all other Associations." As in India during the Emergency, a voluntary association could play "a very important watchdog role." This may well have been, considering that by this point China had trained no fewer than ten million "volunteers." But as Aluvihare knew, all of the leaders of the Chinese FPA were either active or retired government officials.[56]

In 1983 more than 16 million women and more than 4 million men were sterilized in China, nearly 18 million women were inserted with IUDs, and over 14 million underwent abortions. Population control seemed poised for a comeback, and not just in China. In India, for instance, after a virtual collapse following the Emergency Period, both the Janata government and the Gandhi cabinet that replaced them backed ambitious new targets. Following the advice of a high-level working group, "the entire Government apparatus" was to be mobilized from 1983 to stop all growth by 2050. The policy "integrated" measures to improve the status of women and maternal health care with higher cash payments for sterilization and lower priority in housing for everyone else. Though relatively few Indian men would come forward, almost 4 million women were persuaded to undergo sterilization in 1983–84 and over 2 million accepted IUDs—an almost 270 percent increase overall since 1980–81. The cabinet once again backed the idea of denying maternity leave and housing loans to public-sector employees who elected to have more than two children. They backed off only when employee representatives refused to go along. Foreign donors supported India's tough new program with over $53 million in aid in 1984–85—about $103 million in today's dollars—a nearly fivefold increase over 1980–81.[57]

In the same period, South Korea launched an intensified campaign to promote a one-child norm. Although fertility rates had already fallen precipitously, during the 1982–1986 economic plan period the government introduced an array of new incentives and disincentives to stop population growth by 2050. Access to housing, health care, and education were all calibrated to encourage contraception, especially sterilization. At the same time, the government revised laws to diminish the cultural preference for sons. But in South Korea, just like India and China, a policy that purported to raise the "status of women" could have the very opposite effect. Caught between the government and their families, women were under mounting pressure to have fewer children but still bear sons.[58]

Women were also the target of a new eugenics program in Singapore. In 1983, Prime Minister Lee Kuan Yew declared in a National Day speech that if university-educated women continued having half as many children as those with little formal schooling, the quality of the population would suffer. Poor women—almost all from the Malay minority—were offered $5,000 if they agreed to sterilization. At the same time, Singapore gave university graduates tax breaks if they had three or more children, children who would now have priority in admission to the best schools.[59]

Donors used international aid as an incentive for more population control programs, focusing on recalcitrant countries in Africa. For instance, a

1980 State Department cable instructed the U.S. embassy in Burundi to insist on "performance," despite the obvious political sensitivity of "a Tutsi-government sponsored population control program over the Hutu majority." (A hundred thousand people had been killed in communal conflict just eight years before.) The government could receive $85 million in U.S. aid. "Without performance," it would be $6 million. Some U.S. officials even wanted a policy in which "highly specific performance tests in population" would determine aid levels. When another USAID official, Carole Henderson Tyson, protested that this would violate assurances given to Congress, it was agreed that the Burundi cable "goes too far." Commitment to a population program was "only one among several key factors."[60]

In other cases, like Kenya, donor pressure was unrelenting. McNamara repeatedly brought up its "frightening" rate of population growth in discussions about a possible loan with President Daniel Arap Moi and other Kenyan leaders. In 1981 a donor consortium presented a plan for a National Council on Population and Development (NCPD). Its executive committee and staff would have to be selected with World Bank approval and include NGO representatives. This Council would be in charge of the budget and direction of "demand creation" activities, eventually focusing on "national laws and policies on fertility, and pilot schemes of incentives to reduce fertility." The Kenyan Ministry of Health opposed the initiative, but the World Bank made it a condition for the release of a structural adjustment loan. Kenya finally received the loan in August 1982 on the very day it agreed to the NCPD.[61]

By this point, Bangladesh was one of the biggest recipients of international population assistance, which paid for two-thirds or more of its national budget. In 1983 UNFPA, USAID, and the World Bank called for the prime minister to lead a National Population Control Board with emergency powers and an independent budget. They wanted to create demand for contraception by reducing infant mortality, raising the age of marriage, and reserving places and scholarships for female students. But higher incentives would also go to those who accepted sterilization or IUDs. Public-sector employees who chose to have more than two children would be denied raises, and their children would have lower priority in schools and hospitals. This prompted one UNFPA official, Alan Keller, to warn that the proposed program would "punish innocent bystanders."[62]

In fact, abuse was already rife. That same year, the Bangladesh army would round up hundreds of people for forcible sterilization. Emergency aid from the World Food Program meant to feed people made destitute from floods was denied to those who would not agree to sterilization. Just as in India, the elderly and the childless were caught up in campaigns driven by incentive payments, "motivators," and targets. But even while agreeing that poverty drove people to accept sterilization, and they often regretted it, a World Bank review argued against abolition of incentive payments because it would "discriminate against the poor." UNFPA's man in Dhaka, Walter Holzhausen, argued, "It is time for donors to get away from too narrow an interpretation of voluntarism." Governments willing to use pressure deserved support. In fact, he reported, "Most donor representatives here greatly admire the Chinese for their achievements, a success story brought about by massive direct and indirect compulsion."[63]

In this larger context, it is not surprising that both the IPPF and UNFPA decided to help China implement the one-child policy. UNFPA even awarded Qian with the first United Nations Population Award, complete with diploma, gold medal, and a monetary prize of $12,500. Indira Gandhi was the co-winner. When the honorees came to New York to receive their awards, Secretary-General Javier Perez de Cuellar congratulated them: "Considering the fact that China and India contain over 40 per cent of humanity, we must all record our deep appreciation of the way in which their governments have marshaled the resources necessary to implement population policies on a massive scale."[64]

One of the advisors to the prize committee, the Nobel-winning economist Theodore Schultz, resigned in disgust, calling it a travesty. But the Chinese media gave it extensive publicity. Along with the one-child propaganda posters printed in English as well as Chinese—"the U.N. will like to see how its money is being spent," the Chinese explained—it confirmed that Beijing had international backing. Salas maintained that China's "official policy is family planning." To say that it had led to infanticide was "an inference." Qian, for his part, pledged in his acceptance speech to continue the policy of "controlling population quantity and raising population quality."[65]

In fact, Qian was increasingly isolated back home. All along, critics like Zhongtong Liang had been calling for more flexibility, only to lose their jobs. By 1983 the All-China Women's Federation was demanding an end to infanticide and abuse of women. Together with grassroots resistance in rural areas, the opposition finally convinced the government to soften its stance. Qian was removed from office, notwithstanding his UN gold medal, and in 1984 a new directive called for implementing China's policy in a "fair and reasonable" way more "acceptable to the peasants."[66]

At the same time, the IPPF European Regional Council began gathering evidence to determine whether the Chinese association had violated IPPF policies. Members were increasingly critical of Wahren's willingness to look the other way. "The day may soon come when we have to do more than say, as UNFPA has said, that our agreement with the Chinese FPA prohibits coercion," one IPPF official predicted. "We shall need to take the further step of demonstrating that our FPA is not involved in cases of coercion, whether these cases are just occasional abuses or not. In short, I think there is a time bomb in this issue and we would do well to pay attention to it before it explodes."[67]

It was already too late. The UN award, together with the rest of UNFPA's and IPPF's support for China's one-child policy, had given the pro-life movement the perfect opportunity to attack family planning worldwide, beginning with its center of gravity: the USAID population program. After Ronald Reagan became president, it had kept running like a juggernaut, with supporters in Congress continuing to protect its budget. This was exasperating to pro-life activists, many of whom opposed contraception and not just abortion. They were frustrated about having had so little impact after helping to give Republicans control of both the White House and the Senate. The main focus, of course, was reversing the *Roe v. Wade* ruling. Thanks in part to the efforts of Dr. C. Everett Koop in raising awareness about abortion among evangelical Christians, pro-life organizations were proliferating, and many were calling for direct action. But confrontational tactics led to internal divisions, as did the question of whether to countenance exceptions in cases of rape and incest. Liberal Catholics insisted that being pro-life also required opposing the death penalty and defending everyone endangered by dire poverty and government indifference.[68]

In 1981, divisions among pro-lifers erupted into a protracted and public brawl over dueling legislative initiatives. The next year both the Human Life Bill and a constitutional amendment went down to defeat. The Supreme Court would also strike down most restrictions on abortion and explicitly reaffirm *Roe v. Wade*. The Catholic bishops began to accuse the Reagan administration of using disunity among pro-life groups as an excuse for inaction. Attacking USAID's support for the IPPF and UNFPA promised a way out of this impasse. It could come through executive order, and need not await court rulings or legislative majorities. Any connection to abortion or coercive sterilization, no matter how tenuous, could be used to justify defunding organizations that supplied condoms, pills, and IUDs to millions of people—something that could never be achieved by attacking contraception head-on.[69]

In January 1983, on the tenth anniversary of the *Roe v. Wade* ruling, pro-life groups came to the White House to talk about something different. They told Reagan that he should dismiss the director of USAID's population office, Joseph Speidel. In fact, the whole program should be "dismantled and defunded." To begin with, they provided a hit list of organizations receiving USAID funds. Speidel left USAID just a few weeks later. But if this was a direct result of the pro-life demands, it appears to have been the only one. They did not yet have a single issue to dramatize the need to reverse U.S. policy.[70]

A year later, after the worst year in the history of the one-child policy, pro-life groups returned with the ammunition they needed. They compiled press clippings on Chinese abuses, the UNFPA awards, and editorial condemnations in a thick briefing book they presented to Reagan. They thanked him for Speidel, then called for another scalp: the State Department coordinator of population affairs, Richard Benedick. They wanted the IPPF to be audited (again) and demanded that population funds instead support "natural family planning." And for the second time, they called for sending a pro-life delegation to the August 1984 World Population Conference in Mexico City.[71]

It was a relentless campaign, one based in the United States but branching out all over the world. One of the more important leaders in internationalizing the cause was Paul Marx, a pugnacious priest and sociologist based at St. John's University in Collegeville, Minnesota. He traveled the world haranguing audiences and local media while screening graphic anti-abortion films. Marx even went after U.S. bishops whom he thought were too complacent about the IUDs and pills that were also "killing millions of future Catholics." In 1981, Marx moved to Washington and founded Human Life International (HLI). All the time he worked around the American hierarchy by developing contacts with Church leaders abroad—in some cases, by sending cardinals checks for a hundred to a thousand dollars, which Joseph Ratzinger refused but others were happy to accept. In Latin America Marx found many allies angered over USAID and UN programs. In 1984 Cardinal Alfonso Lopez Trujillo of Colombia—who would become the president of the Pontifical Council for the Family and a key advisor to John Paul II—condemned a Pathfinder-supported sterilization project as a type of "global castration," prompting the government to launch an immediate investigation. That same year, the Vatican began to fund Marx's campaign, which helped to stop the legalization of abortion in Ireland and Mexico. By 1987 HLI already had eighteen branches abroad, employed thirty-five people at its Maryland headquarters, and shipped propaganda to over one hundred countries.[72]

Reagan had compelling reasons to agree to pro-life leaders' demands. They provided many of his most loyal campaign workers. To underscore the point, they reminded him that pro-choice Republicans might be vulnerable to pro-life Democrats in the next election. Aides instructed Reagan that his own reelection required friendly relations with Catholic leaders. They advised that he focus on abortion, because contraception caused "anxiety and guilt among Catholics," surmising that it had "something to do with medieval philosophy."[73]

John Paul II, for his part, was critical of many aspects of U.S. policy, especially the nuclear arms buildup. But he also made it a top priority to uphold *Humanae Vitae,* which he had helped to draft for Paul VI. Reagan needed his support in maintaining sanctions against the communist regime in Poland and combating revolution in Central America. The pro-life briefers made a point of telling Reagan that the IPPF had criticized Catholic influence in the Solidarity movement. Because Latin American bishops were backing opposition to the Sandinista government of Nicaragua, Reagan had good reason to be responsive to their complaints about USAID. He not only assured them his administration was redressing past problems but asked the bishops to report any new transgressions.[74]

Jack Svahn of the Office of Policy Development drafted the statement to the UN population conference, what came to be known as the Mexico City policy. It would cut off international and NGO family planning programs that provided abortions, even if the abortions were funded from other sources. The IPPF's top lobbyist in Washington learned that it was being served up as "a sacrificial lamb." But officials in the National Security Council, USAID, and the State Department fought for three months to defend U.S. support for UNFPA, and with it America's tradition of leadership in efforts to control world population growth. Many still considered it a national security priority. Ten years earlier the vice president himself, George H. W. Bush, had helped protect UNFPA as ambassador to the

United Nations. And in January 1984 a bipartisan presidential commission under Henry Kissinger backed USAID's efforts to reduce birth rates in Central America. Criticizing China's one-child policy would also risk complicating U.S. efforts to deepen the anti-Soviet alliance with Beijing.[75]

Competing drafts of the statement gave more, less, or no credit to the idea that population growth remained a problem, and that family planning programs—and not just free markets—provided a solution. When pro-life groups insisted that the U.S. delegation be led by James Buckley, a former senator committed to their cause, they gave him leverage to insert crucial language in the official statement. Chief of Staff James Baker finally agreed that the United States would call for "concrete assurances that the UNFPA is not engaged in, *or does not provide funding for,* abortion or coercive family planning programs." This implied that any support for a program like China's would disqualify the entire UNFPA for U.S. funding.[76]

The U.S. statement at Mexico City was a startling turnaround, so much so that McNamara warned that the Americans would be "laughed out of the conference." In fact, while some of the U.S. delegates, notably Alan Keyes, seemed to enjoy the verbal fencing required to defend an unpopular stand, the position they represented was politics of a high order. Its two feet were planted firmly in the two sides of Reagan's ruling coalition—social conservatives and Wall Street Republicans—and by surrounding the pro-life thrust with free-market rhetoric, it threw critics off balance. In the heady atmosphere of a United Nations conference, where most attendees were committed to spending large sums to persuade poor people to have fewer children, it seemed like an outrageous provocation to suggest that "population growth is, of itself, a neutral phenomenon" and that "more people do not necessarily mean less growth." It rankled to hear the United States—once the world's leading worrier—argue that population growth had provoked an "overreaction." Rather than focus on the hidden agenda, many opponents followed the waving of the invisible hand.[77]

The debate about the "voodoo demographics" of the Reagan administration distracted attention from the revival of hard-core population control in Asia as well as the pressure being put on African states to follow suit. The U.S. delegation was obviously playing to a domestic audience—the conference came just weeks before the Republican national convention. They would not discuss how they defined "coercive," perhaps because USAID would have had much to explain. After all, in contravention of U.S. law, it continued to cover 85 percent of the costs of incentive payments for sterilization in Bangladesh. And notwithstanding his anti-abortion stance, Surgeon General Koop had approved expanding a Department of Health and Human Services cooperative research program with China in human genetics, focusing on prenatal screening. When pressed by reporters about which countries used coercion, Buckley refused to name a single one.[78]

The Indian and Chinese delegations were therefore allowed to pay lip service to a declaration that disavowed coercion. "The U.S. may be concerned about abortion," an Indian delegate argued, "but it cannot impose its views on the free world." The principle that was once considered an obstacle to solving the global population problem, state sovereignty, now served to protect national policies, no matter what their nature.[79]

The Mexico City conference marked the moment when population growth was no longer treated as a global problem requiring a global solution. U.S. funding for family planning would continue for those organizations, like Pathfinder, willing to submit to the new guidelines. But the IPPF refused to stop supporting affiliates that provided abortion. This required laying off 60 of its 217 staff, canceling the opening of five field offices, and shelving a plan to start new programs in Africa. Some of the strongest links that once gave the population movement its strength began to fall away.[80]

At first the Reagan administration deemed UNFPA in compliance with its policy. But still more articles appeared detailing the abuses that had occurred in China over the previous two years. In the summer of 1985 pro-life senators and congressmen specifically directed the president to deny funds to any organization that "supports or participates in the management of a program of coercive abortion or involuntary sterilization." In November 1985, USAID director Peter McPherson cut off all support.[81]

The Mexico City policy as defined by Congress set the terms of much of the subsequent debate on international family planning assistance. The IPPF and UNFPA indignantly denied that they "actively promote abortion" or funded "coercive family planning." They disputed what it meant to "participate in the management" of a program like China's. Family planning advocates could not win this debate as long as it would be decided in a Republican White House. But they demonstrated that there were many ways to lose. Their position depended on subtle distinctions that seemed like dissembling compared with the passion of their opponents. It was perfectly true, for instance, that they supported China's program, not its policies. It might even be said that they opposed the one-child policy before anyone else outside China, pressing officials to stop abuses and warning that it could drastically skew the age structure. They relied on the reassurances of their interlocutors and could not know what was happening in every factory and hamlet.

Yet UNFPA and the IPPF did not oppose setting targets and using incentives and disincentives to generate "demand." By providing training to no fewer than seventy thousand people, UNFPA made it easier for China's leaders to implement any policy they chose, defended their right to make the decision for themselves, then applauded the results. China's FPA continued to grow until it had no fewer than twenty million members in five hundred thousand local associations, the largest mass organization in the country. But IPPF officials continued to insist it was an affiliate like any other. True, their private urging led officials to expressly forbid abuses. But press reports showing the persistence of infanticide and physical compulsion showed that the "incentives" and "disincentives" inherent in the one-child policy had unleashed forces no one could control.[82]

After 1983, stubborn resistance in rural areas gradually led cadres to allow farmers with one daughter to try to have a son. Those who had prospered with the coming of market reform could afford to pay fines or move to China's growing cities. The policy of granting exceptions was gradually formalized. A key element in this mutual accommodation was a new technology that began to arrive in rural areas in the early 1980s: ultrasound machines. They could be used to determine whether an IUD was still in place or to detect birth defects, thus serving both the quantitative and the qualitative goals of the one-child policy. But they could also be used to determine the sex of a fetus by the fifth month in order to abort females for parents who preferred sons. Authorities forbade the practice, to no avail.[83]

Since the late 1960s it was understood that sex determination would reduce fertility rates—that is precisely why Planned Parenthood and the Population Council had backed research in this area. Some of the first ultrasound machines in rural China came through international assistance. The second half of the 1980s marked the peak period of imports, with 2,175 ultrasound machines arriving in 1989, though it is not clear how many came through international aid. In 1990, for instance, the Australian Agency for International Development shipped 200 ultrasound machines to China as part of a $4 dollar grant. Foreign Minister Gareth Evans was asked whether he would seek assurances that they would not be used for coercive abortions. "I am not," Evans replied, "going to ask anybody anything," retorting that the unregulated export of coat hangers could also lead to abortions.

In 1994, a guide to doing business in China listed ultrasound machines as one of the "HOT items," and advised exporters to "monitor the medical research programs of the World Bank and other multilateral agencies." (For a decade already the World Bank had been providing hundreds of millions of dollars in interest-free loans for "Population-Health-Nutrition" projects in China.)[84]

China gradually gained the capacity to make as many as ten thousand of its own ultrasound machines every year. With prospective parents paying as much as fifty dollars to determine the sex of their fetus, the machines could pay for themselves. In its very first joint venture in China, General Electric set up a plant to produce still more ultrasound machines. By this point, the combination of ultrasound and late-term abortions was already known to be shifting the sex ratio all across China, from approximately 6 percent more boys than girls—within the normal range—to nearly 17 percent more in 1995.[85]

Defenders would argue that UNFPA had to stay in China to improve services and change the way their counterparts provided family planning. In January 1989 officials grew hopeful that China was ready to end abuses once and for all. Beijing promised that offenders would be expelled from the Party or sent to jail. A new $59 million UN grant would end involvement in the census program and shift to enhancing maternal and child health. But once again, UNFPA had the worst possible timing. With the Tiananmen Square crackdown in June 1989, China also cracked down on unauthorized pregnancies. For four long years, birth-planning officials launched wave upon wave of new "shock attacks." People who once managed to bribe or evade authorities were subject to strict surveillance, heavy fines, and sometimes beatings. The head of the State Family Planning Commission spoke of the need for "crack troops" to ensure victory. Local family planning associations identified the offenders, and the people's militia "mobilized" tougher cases. This included confiscating stored grain, taking livestock, and demolishing homes. Those targeted, in turn, sometimes killed cadres and informers.[86]

China's program has indeed changed over the years. But far more important than any foreign influence has been the resistance of the Chinese people, especially women, many of whom bear the scars from years of what one analyst has called "reproductive combat"—IUD insertions and surreptitious removals, hidden pregnancies and late-term abortions, child abandonment and sometimes infanticide. In the early 1990s, toddlers were

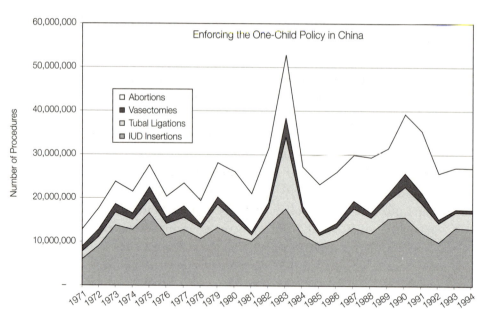

The number of abortions, sterilizations, and IUD insertions rose and fell in unison, not with changing demand, but with a series of campaigns in which officials cracked down on resistance. Data from Tyrene White, *China's Longest Campaign: Birth Planning in the People's Republic, 1949–2005* (Ithaca: Cornell University Press, 2006), 136.

found with notes inside their clothing in which parents expressed shame and complained bitterly about having to abandon them.[87]

Without the international meetings, study tours abroad, and model UN projects, the plight of these people might have been even worse. It is impossible to know. But one thing is clear: by failing to protest in their defense, demonstrate, and demand that China end the one-child policy, family planning organizations ceded moral high ground to their opponents. It was not much, but it was enough of a perch to permit pro-lifers to pile calumny upon calumny on China's program and all who could be associated with it. True, they never recognized, much less encouraged, efforts to make it less coercive and improve the quality of care. Attacking aid for China's one-child policy was a way to undermine family planning worldwide. But the more outrageous their claims, the more irresistible it was to debate pro-lifers on this, their chosen ground.[88]

In this same period, a renascent movement for reproductive rights defied physical assaults, clinic firebombings, and serial killers to protect American women's freedom to choose. Why did they remain silent or defensive about efforts to force IUDs, abortion, and sterilization on Chinese women? This would have required much more than just conceding that pro-lifers were right to protest China's one-child policy, albeit for the wrong reasons. It would have required confronting the whole historical trajectory of population control that led organizations like the IPPF and UNFPA to China in the first place. A new generation of activists were often innocent, even oblivious to the earlier history of the movement that they were joining, and changing. In a sign of the times, the IPPF adopted strict new rules in 1983 disallowing acceptance of funds from pharmaceutical companies. Many more women were now taking up leadership positions at all the major institutions, though in 1987 they were still only about 30 percent of the staff at UNFPA. By that point, UNFPA and even the World Bank were no longer promoting incentives and disincentives. The major institutions really were different from what they had been ten years or even five years before.[89]

Veteran leaders, those who had long used family planning as a vehicle for population control, assured everyone that there was nothing to answer for. The UNFPA denied not merely that it sanctioned coercion in its program in China, but that it had ever sanctioned coercion in any program. Salas claimed not only that it did not provide help with abortion, but that it had "never funded an abortion of any kind." Senior Population Council officials insisted that they had never accepted, much less recommended, incentive payments. In the IPPF, a legend grew up that it had been a watchdog, rather than a lapdog, during India's Emergency Period. Over time, in the face of unrelenting attacks, it became ever more difficult to acknowledge that helping China implement the one-child policy had been a terrible mistake.[90]

As a collective amnesia descended over the field, the rank and file were told to focus resolutely on the very real threat posed by their opponents. Pro-lifers, on the other hand, discovered in history another field in which they could put advocates of reproductive rights on the defensive, in some cases by combing feminist critiques for ammunition. By 1986, IPPF staff and volunteers were said to be "in a state of distress and confusion about the onslaught of attacks on everything they do." In one of the first meetings of a new IPPF Task Force on Opposition, participants viewing a film called "The Great Population Hoax" had "a strong feeling that the makers were out-of-date and ill-informed about IPPF and its aims." One of those present, Julian Heddy, had to point out that the film was simply quoting statements from the early days of the movement, "and it was not possible to deny their existence." He suggested that the IPPF compile an in-house glossary because "not everyone took enough care to avoid unfortunate terms like 'overpopulation.'" They were told that they need not "jump to the defense of policies [IPPF] does not stand for," such as had occurred when abuses in Bangladesh's sterilization program came to light. And most of all, they were not to shut out all their critics: "people we thought were supporters—in particular some feminist groups. . . . Enemies may turn out to be friends after all."[91]

Friends indeed. Feminist critics of population control, long marginalized and belittled, ridiculed and harassed, would finally redeem the cause of reproductive rights, and not a moment too soon. All over the world, they were organizing behind a revolutionary new agenda. The practices that had once been used to control populations—especially transnational networking, philanthropic funding, international conferences, and a discourse of "unmet need" and crisis—would now be used to help people stand up for themselves and address real and ongoing afflictions, not just future threats. Beginning in 1981 a series of international meetings of activists condemned *both* abusive population control programs *and* efforts to force women to bear unwanted children. In 1984, at the suggestion of women from poor countries of the Global South, the group was christened the Women's Global Network for Reproductive Rights.[92]

That same year, feminist researchers from poor countries gathered in Bangalore and went on to found Development Alternatives with Women for a New Era (DAWN). The group, which included Peggy Antrobus of St. Vincent, Lourdes Arizpe of Mexico, Carmen Barroso of Brazil, Fatima Mernissi of Morocco, and Gita Sen of India, redefined "reproduction" as "the process by which human beings meet their basic needs and survive from one day to the next." It could not be isolated from questions of how, for instance, international institutions were forcing poor countries to cut spending on education, public health, and food subsidies—even while insisting on population control—much less be reduced to a problem of "unplanned unhappy families." Women's many contributions, usually unremunerated and invisible in economic statistics, could determine whether and how societies survived. The conditions of their lives, not abstract claims of sisterhood, would define feminism, and making it a political force required a willingness to listen to discordant voices.[93]

Many of these women remained wary of working with institutions that once stood for population control, certain it would lead to a sell-out. But others, working from the inside, began to put resources behind the rights and health approach, showing how it could make a real difference. The Mexico City policy was a catalyst, breaking connections in the old population control network and demonstrating the need to forge new alliances. In Bangladesh, for instance, local NGOs that provided abortion services were suddenly cut off. Adrienne Germain, now head of the local Ford Foundation office, knew that it was a death warrant for poor women. She insisted that Ford take a stand. For the first time, the foundation agreed to support abortion services. During her four years in Bangladesh, Germain also oversaw grants to improve the quality of family planning services, include maternal and child healthcare, and undertake research on sexually transmitted diseases—another first for the foundation.[94]

Even when working to ensure access to abortion, Germain was not one to simply "inundate" countries with abortion kits and hope for the best. In one case, a personal inspection required chartering a small plane to fly to a remote area of the Amazon even though Germain was still recovering from a car accident. After the plane crash-landed, she had to walk seventeen hours through the jungle to find help. A hallmark of the new approach to family planning was going the extra mile to ensure that clients had decent quality of care.[95]

By this point, Germain was working for Joan Dunlop and the International Women's Health Coalition. It was originally founded by the PCC to distribute abortion kits. The Hewlett Foundation wanted to use it as a vehicle to help more groups that would lose USAID funding because of the Mexico City policy. Dunlop was committed to the cause—she herself had had an illegal abortion many years earlier. But she too insisted that, if she were to assume leadership, the IWHC had to address reproductive health and rights more broadly. She brought Germain over from the Ford Foundation, and the two worked together to promote research and advocacy on sex education, sexually transmitted diseases, and the growing prevalence of reproductive tract infections, a major cause of sterility.[96]

By beginning this dialogue and broadening their agenda, feminists were in a stronger position to confront pro-life forces head-on. An important battle occurred in Nairobi in 1985 at the third UN World Conference on Women. When Germain and Dunlop arrived, they discovered that Marx and HLI were preparing to stage a replay of the Mexico City conference ten years earlier, when the media reduced all their work to a fight between American women in favor of family planning and Third World women suspicious of population control. The local papers had banner headlines claiming the women had come to Nairobi to promote abortion. HLI was already on the ground organizing Kenyans from local churches for a mass demonstration to demand a pro-life plank in the official statement.[97]

Germain began to search out all of the people she knew from her years with the Ford Foundation, notably Peggy Antrobus, Carmen Barroso, and Joan French of Jamaica. They were scattered in different delegations and meetings, but she finally managed to get them together in one room: "Listen, I basically never asked you for anything, any of you," she began, "but I need to ask you now. I know you've got a lot of sessions of your own, but if we don't mobilize to push back against these people we're just going to be drowned. It's going to be the end."

Germain's confederates fanned out and began to organize other women to stop pro-life demonstrators from taking over the NGO Tribune. The struggle came to a head on the central lawn of the University of Nairobi. The women formed a big circle, and two of them stepped forward: Joan French, statuesque and elegant, together with a more diminutive but no less courageous new ally, Mercedes Sayagues of Uruguay. Together they declared, "We need to craft here and now a statement on how women see the right to life." Standing in the midst of the crowd, they set to work, and kept working even when pro-life provocateurs tried to stop them. Their supporters were too many, and they pushed back. Together they created the first United Nations statement that recognized not only that women had "the basic right to control their own fertility," but that this was the basis for all their other rights.[98]

Victory in Nairobi resulted from an "inside-outside" strategy, in which women in NGOs, international institutions, and official delegations all worked together. The ultimate insider, one who could turn the family planning field upside down, was Nafis Sadik. In 1987, after many years as Salas's deputy, she finally took over as executive director of UNFPA. From the first staff meetings, she insisted that improving the lot of women would be a top priority in every UNFPA program. She recognized in the spread of HIV/AIDS another compelling reason to broaden their mandate beyond reducing population growth. Prevention programs would help revive interest in barrier methods and male responsibility for contraception.[99]

Sadik decided that she could now speak out on all the things she had always wanted to say. She was the first leader in the field to take on Beijing, something particularly difficult for the head of a UN agency. She publicly criticized efforts to target mentally handicapped people for compulsory sterilization and abortion. In 1993 she demanded that a project intended to demonstrate the effectiveness of voluntary family planning not be subject to the national policy of birth quotas and penalties, threatening otherwise to withdraw UNFPA support.[100]

If family planning really was about reproductive rights and health, and not just a means to control population, the next UN population conference in Cairo in 1994 would pose a crucial test. Sadik, designated secretary-general, was not originally planning to give NGOs any particular role. At the 1992 Conference on Environment and Development in Rio de Janeiro, the Vatican had exploited suspicion of population control once again to weaken all references to family planning. Germain and Dunlop together with Bella Abzug convinced Sadik that, with foundation backing for a series of national, regional, and international meetings, they could marshal support behind a truly different agenda. Sadik urged them to work together with environmentalist and mainstream family planning groups: "We must act as if our lives depended on what we do in the next decade." In the run-up to Cairo, feminist NGOs won representation in many official delegations, formed a women's caucus, and began pushing a common program.[101]

The battle at Cairo began with a series of skirmishes in the preparatory meetings, as contending forces maneuvered for position. The draft program of action acknowledged that rapid population growth was unsustainable, but repeatedly insisted that reproductive rights and health were essential in addressing this and every other development challenge. Poor people had many "unmet needs," and not just contraception. If there was truly a crisis, it was in the needless death of hundreds of thousands of women every year from botched abortions and substandard maternal health care. Where abortion was illegal, these women deserved better care.

The backing of the Clinton administration might have seemed assured. It had already restored funding to the IPPF and UNFPA. But the men who would lead the U.S. delegation, State Department counselor Tim Wirth and Vice President Al Gore, were mainly concerned with the environmental ramifications of rapid population growth. Intense debates about how to define a post–Cold War U.S. foreign policy led environmentalists to link high fertility to civil wars and refugee flows. Population controllers were only too happy to agree that it was contributing to a catastrophe that might one day make coercive measures unavoidable. But feminists were more persuasive in arguing that pushing Malthusianism risked a repeat of what happened at Rio. Making contraception part of a rights and health agenda was the only legitimate way forward.[102]

As establishment figures began to realize that they were being relegated to the margins, some fought back. They resented the way their earlier work was criticized and population goals deemphasized. Robin Chandler Duke, for instance, a wealthy and influential leader of the Population Crisis Committee, came to one of these preparatory meeting and never came back. The PCC attacked the emerging consensus for failing to make population control the top priority. Princeton demographer Charles Westoff told *New York Times* readers that this was the work of "extremist feminist groups." Even without taking on a whole new agenda, family planning budgets were inadequate to meet all the unmet need for contraception, undertake more "motivational efforts," and thus "reduce the likelihood of more drastic programs later." In a withering rebuttal, Ellen Chesler pointed out that "cavalier bureaucrats, misguided strategies, and poor services" had already taken a terrible toll on both women and men. Women insisted on being makers of policy, not just acceptors, and they were finally succeeding. They had the backing of not just hundreds of grassroots organizations worldwide, Chesler wrote, but the IPPF, the Ford and Rockefeller foundations, and donor governments like the United States and Sweden.[103]

By this point Nafis Sadik was tirelessly traveling the world to win over opponents, even the pope. After all, she had much to offer. The draft program called for reducing the incidence of abortion, did not specify "targets," and condemned the use of incentives and disincentives. Most of all, this would be the first UN conference on "Population *and Development.*" All of this was consistent with the Vatican's own critique of population control. But when pushed on the issue of unsafe abortion, Sadik was unyielding: "Over my dead body will that come out."[104]

After almost a year of negotiations, Sadik went to the Holy City in March 1994 for a private audience with the pope. It was the most momentous meeting on the subject since Sanger had debated Gandhi sixty years before. There was no prospect of complete agreement. But there was at least a chance that the Church would not oppose Sadik's efforts to end manipulative and coercive population control. She arrived wearing a sari and hoping for the best. But Vatican officials rushed up to tell her to get ready for a clash. She found the pope alone in his office, dressed in his white cassock. He rose slowly from behind his desk.[105]

It was supposed to be the Year of the Family, he began, but now seemed set to become the "Year of the Disintegration of the Family." A debate immediately ensued about whether there was only one kind of family. John Paul insisted that only couples had needs and rights, not individuals. Sadik pointed out, like Sanger before her, that women could not abstain, as the Church instructed, if their husbands insisted. The pope, according to Sadik, suggested that women were responsible for the irresponsible behavior of men. Seeing her reaction, he tried to change the subject. "Excuse me," she interjected, "I must respond to your statement about the behavior of women. In most of the developing countries men look on marital relations as their right, and the women have to comply. . . . Violence within the family, rape in fact, is very common in our society. The most upsetting thing about all this is that only women suffer the consequences."

Time and again the pope exclaimed that it was materialistic and immoral culture that was to blame, especially in wealthy countries, which exported their values to the rest of the world. For him, the Cairo program was a case in point. Sadik agreed responsible education was essential, but she pointed out that the Catholic countries of Latin America had some of the highest rates of unsafe abortions. As a doctor she followed a different code of ethics: "Even if you don't approve, you still have to treat patients for the consequences of their actions. You might even disapprove, but it's not for us simply to sit in judgment. We have to help if we can." When the pope insisted that the UN could not include compulsory sterilization in its program, Sadik began to think that he had not actually read it.

The meeting ended abruptly and Sadik was ushered to the door. There would be no smiling photograph. A week later, 140 papal nuncios arrived in Rome to receive marching orders. They were to launch a worldwide diplomatic offensive, a campaign unprecedented in Church history. The pope personally wrote every head of state, suggesting that the draft program was being imposed by "certain fringes" in rich and secularized societies. His letter was accompanied by a study alleging "contraceptive imperialism." The pope pressed Clinton by telephone and again in a tense meeting at the Vatican. But the president would not relent. The pope issued a new encyclical, *Evangelium Vitae* (the Gospel of Life), and vowed that Church teaching on birth control would never change. Visibly angry, he pounded home his arguments in virtually every public appearance. In one of them, he insisted that the Church supported equality for women, but "perfection for woman is not to be like man, making herself masculine to the point of losing her specific qualities as a woman."[106]

The Vatican could no longer hope to succeed by joining those who at-tacked population control as a poor excuse for development. By redefining reproductive rights to include education, health care, and employment, feminists forced John Paul to rely on Muslim conservatives. Whereas Is-lamic law is flexible on birth control, they could agree to oppose any lan-guage that appeared to accept abortion, contraception for the unmarried, and a more inclusive definition of "family." Vatican emissaries worked with diplomats from such countries as Iran and Morocco in demanding that these passages be bracketed for further negotiation, then set about mar-shaling more support. The brother of the Iranian president, Mohammed Hashemi Rafsanjani, predicted "future war between the religious and the materialists" after meeting with the papal nuncio. "Collaboration between religious governments in support of outlawing abortion is a fine begin-ning for the conception of collaboration in other fields." The Vatican was not above horse-trading: Libya's official news agency announced that an-other papal emissary had promised to help end the dispute over Muamar Qadaffi's role in the 1988 Lockerbie bombing.[107]

In Cairo itself, Muslim clerics condemned the conference as "a Zionist and imperialist assault against Islam." President Hosni Mubarak had ex-pected that hosting a UN meeting would demonstrate that he had won his struggle against Islamist militants. Instead, the preceding week brought a deadly machine-gun assault on a tourist bus, followed by a formal warning from the hard-line Gama'a al-Islamiya to stay away. Sudan, Saudi Arabia, and Lebanon urged other countries to join them in a boycott. Sadik hoped for a show of support from Muslim women prime ministers. Instead, Tur-key's Tansu Ciller and Khaleda Zia of Bangladesh backed out, sending lower-level delegations in their stead. In Pakistan street protests led most observers to expect that Benazir Bhutto would also withdraw, but she de-cided to defy them. "If I am going," Sadik recalls Bhutto saying, "I want all my ministers there with me." They were joined by Prime Minister Gro Harlem Brundtland of Norway and fifteen other heads of state.[108]

The Cairo conference would have a much larger international audience than the typical UN colloquy, in part because Catholic leaders all over the world kept insisting that nothing less than the future of humanity was at stake. The mere discussion of women's rights made for high drama when it required protection by fourteen thousand police and soldiers. But support-ers of the conference had also been maneuvering behind the scenes to en-sure that the most high-profile media were there to see it. Sadik won a pledge from Ted Turner that CNN would publicize her agenda, scheduling a series of documentaries and news features on such issues as female genital mutilation. Delegates tuning in to CNN from their hotel rooms would not be able to miss them. The *New York Times* did not typically send its own correspondents to such events. But Deputy Foreign Editor Steve Weisman was alert to the growing controversy after his wife, Elisabeth Bumiller, investigated the persistence of population control targets and sex-selective abortions in India, as well as the dire need for decent reproductive health care. The *Times* dispatched its UN bureau chief, Barbara Crossette, and carried nearly twenty stories from Cairo. Many quoted members of the NGO network cultivated by Dunlop and the IWHC.[109]

Most important of all, when almost twenty thousand people began to arrive at the conference, journalists discovered they had stories that were worth listening to. Tours of well-attended family planning clinics in Cairo, to begin with, belied the myth that devout Muslims were opposed to birth control. Some two-thirds of Iranian women were using contraception distributed free by the state, which also provided sex education in schools. The birth control pill was available over the counter in Saudi Arabia. Some of the most Catholic countries in the world, such as Spain, Italy, and Brazil, had some of the highest rates of contraceptive use. So who, exactly, did the Vatican emissaries represent? Rather than just repeating tired clichés about religious opposition and "overpopulation," journalists could report how people all over the world were fighting for their rights and—in some cases—fighting for their lives.[110]

In the opening session in Cheops Hall, Brundtland gave an unexpectedly tough speech, taking direct aim at the Vatican: "Morality becomes hypocrisy if it means accepting mothers suffering or dying in connection with unwanted pregnancies and illegal abortions." When she suggested that access to safe abortion was a human right, many in the audience rose to their feet. But Vice President Gore wanted to take the issue "off the table," denying that the United States supported this stance. In one of the most keenly anticipated speeches, Bhutto embraced the cause of family planning as part of her dream for Pakistan, but warned that it could not mean the same thing for every society. She had tremendous credibility in demanding consensus, considering that Islamists had helped to overthrow and execute her own father: "Leaders are not elected to let a vocal narrow-minded minority dictate an agenda of backwardness."[111]

In closed-door meetings that evening, the European Union representative proposed language that Sweden called a "rock-bottom compromise." It made clear that governments should try to minimize the need for abortion, and "in no case should abortion be promoted as a method of family planning." But unsafe abortions were "a major public health concern," and women must have access to "reliable information and compassionate counseling." The language won assent from Pakistan, Iran, and holdouts in Latin America and West Africa. When the Vatican representative called for more negotiation, it brought a chorus of boos.[112]

To those who believed that a fetus is a human life, no abortion can ever be safe. What is harder to explain is why the Vatican would continue to claim that the Cairo program was just another amoral effort to control the population of poor countries. But Rome was no less opposed to the idea that women should have the right to control childbirth—even through contraception. "The Catholic-Islamic counter-offensive is not really about abortion," a keen observer concluded; "it is against uppity women and policies that are designed to make women even more uppity." As such, Cairo represented "patriarchy's last stand."[113]

Old-guard population controllers began to criticize the conference organizers. "The way this (U.N. plan) was written makes you wonder what people were thinking," Wahren declared. "The language in some of the sections was almost guaranteed to inflame people, and not just here in the Middle East." But reproductive rights supporters did not lose their nerve. In tough negotiations, they defended the key points in the abortion paragraph and even strengthened it. Women deserved reliable information about abortion even where it was illegal. Those who suffered complications needed not just medical treatment but prompt access to family planning services. When the papal delegates persuaded their few remaining allies, including Malta and Guam, to help postpone a resolution once again, an Egyptian minister suggested they should never have come to Cairo. Some diplomats began to question the bona fides of the Vatican delegation. After all, Catholicism was the only religion with permanent observer status at the UN. If it was truly a state, it counted the smallest population in the world, the only one with no mothers and no children. If the Holy See actually succeeded in wrecking the conference, it risked losing its capacity to act as both a transnational force and an international actor, unique in world politics, and now all too conspicuous.[114]

The pope finally decided to cut his losses, and his delegation ended the five-day standoff. The Cairo Program of Action was endorsed unequivocally by 162 states. Seventeen others voiced reservations over specific passages. The head of the Vatican delegation, Archbishop Renato Martino, associated the Holy See with some parts, including the basic principles and even the chapter on "Gender Equality, Equity and Empowerment of Women." As one conservative theologian gamely explained, such empowerment might "lead to a revitalization of the traditional family and a reaffirmation of the distinctively maternal power of women."[115]

At the time many worried that the Vatican had succeeded in focusing all the world's attention on a single debatable paragraph in the 113-page document. In fact, by employing a wedge strategy aimed at the most controversial point and still failing to stop the emergence of a global consensus, it merely underscored its isolation. All over the world, the Cairo program came to represent a new era, both reflecting and reinforcing efforts to end population control once and for all. That same year a high-level advisory committee in India called for the abolition of all acceptor targets and a new focus on quality of care. In China as well, new government plans and white papers emphasized increasing girls' access to education, respecting the rule of law, and elevating individual rights. And in Bangladesh a new and comprehensive reproductive health strategy based on the Cairo program helped reduce maternal mortality by almost 20 percent by 2000, saving thousands of women's lives every year.[116]

Women had finally won international recognition of the most basic fact of life: they had always been held responsible for reproducing society—by their families, by their governments, and even by a "world community" anxious about overpopulation. It was only right that they should be able to choose freely—free of manipulation and coercion, free of ignorance and prejudice, free of hunger and preventable disease. Of course, none of this was cheap. Many governments failed to follow through, and some retreated. But no one could claim any longer that when governments tried to plan other people's families, they were acting on behalf of all humanity. Population control as a global movement was no more. The Cairo program constituted the instrument of surrender.

Conclusion

THE THREAT OF THE FUTURE

The growth of world population is now slowing and a whole generation of demographers and activists is passing from the scene. Some of them look back in anger, resentful of how the people who now lead the family planning field denigrated their historical contributions and drove them to the margins. They are proud of having alerted humanity to an oncoming catastrophe that would have made the world's many problems all the more intractable. Their work must be seen in the context of the times, when no one could be certain how population growth would begin to slow. And while they admit to mistakes, they point out that reducing fertility did much good. It was a catalyst for economic growth and helped women most of all. As Frank Notestein said not long before he died, "I don't think we did so damn bad."[1]

This is an ironic turn of events. When people were still trying to control world population, they had little use for history. Since everyone was living on borrowed time, proponents of each new campaign dismissed prior experience—even very recent events—as irrelevant or misleading. As societies met the challenge of feeding larger populations, Malthusians old and new claimed that this had only set up a still greater calamity. Immigrants did not turn out to be inveterate criminals or culturally unassimilable. But nativists invoked their accomplishments to point up the defects—and danger—of the new arrivals. If improved nutrition, sanitation, and education made people physically fitter and measurably smarter, eugenicists warned that this only masked the genetic deterioration that would manifest itself in generations to come. Though demographic projections consistently failed to anticipate change, new projections were still taken as irrefutable signs of inevitable doom. And when family planning took the form of crash programs to control the fertility of poor people and poor countries, the resistance provoked appeared to prove that more ruthless measures would be necessary.

Empathy, imagining oneself in the place of another, is a core value for historians. Without it one cannot begin to make sense of the past. One can well understand why future threats seemed so compelling to those who felt that only they could recognize and defeat them. But people who have little appreciation of history when they still have the power to change it realize too late that it is not just a source of alibis. History also has something to say about the future. By reexamining the experiments of the last century, whether containing Asians in their own continent, or breeding better people, or paying the poor to stop having children, we can consider more critically the kind of population control now being contemplated. The greatest threat of the future, now as ever, is that focusing only on what may be will lead us to overlook the lessons, and legacies, of the past.

There is no denying that population controllers confronted head-on a challenge without precedent, as humanity doubled and doubled again during their lifetimes. The rise in world population was an epochal event, perhaps of more long-term importance than any other in the extraordinary twentieth century. But partisans claim much more, suggesting that their work has stopped and will ultimately reverse this trend, saving humankind from untold misery. Since the 1960s, the total fertility rate has fallen by more than half, with the average woman now having fewer than three children instead of six. Where population programs were most successful, especially in East Asia, once-poor countries are said to be reaping handsome rewards from a "demographic dividend." Swelling workforces no longer struggle to feed large families. Persistent high fertility in Afghanistan and much of Africa, on the other hand, is blamed for making them breeding grounds for terrorism and genocidal violence. And with new challenges like global warming, "population stabilization" is more important than ever.[2]

These are the two strongest claims population controllers make for their long-term historical contribution: that they raised Asia out of poverty and helped keep our planet habitable. If true, the subject of this book would be one for the ages. The dismantling of the population bomb would inspire endless historical studies, colloquies, and debates. Many scholars might well conclude that the end result was worth the price, especially as the sacrifices recede into the past and we continue benefiting in the future. Over-looking it all in equanimity would be statues erected to honor the leaders of the great population control campaigns of the twentieth century, including Draper, Ravenholt, Qian, and Sanjay Gandhi.

But these claims cannot both be correct. After all, if Asians have 2.1 children, but also air conditioning and automobiles, they will have a much greater impact on the global ecosystem than a billion more subsistence farmers. Throughout the twentieth century, poor people were promised that they would have more things if only they had fewer children. All the posters, flip charts, and films displayed happy, planned families surrounded with conspicuous consumption, to the point that even Nigerians were shown a suburban home with a car in the garage. And if promoting the nu-clear family as well as alternative lifestyles reduced the rate of population growth, it also increased the number of homes in which people live—the number is now increasing at almost twice the rate of world population. When people move from multigenerational households to live by ones and twos, they tend to consume more of everything per capita, whether fuel, or water, or wide open spaces.[3]

If increasing prosperity magnified environmental challenges, it might nonetheless provide ample justification for controlling population growth. People who made their careers in the field assign great importance to eco-nomic theories that linked demography and development. But political leaders had many other, more mixed motives, and less sophisticated eco-nomic analyses—such as cost/benefit analyses of "births averted"—often had greater influence. Moreover, economists were unable to show that a de-cline in fertility was correlated with increasing per capita GNP, despite all the money and reputations that were riding on it. Population control pro-grams proceeded regardless.[4]

A new version of this argument no longer claims that declining fertility will free capital for more profitable investment, but instead stresses how it has changed the age structure. Increasing the number of workers relative to the dependent population of children and the elderly is said to have spurred the Asian tigers. Ironically, the population bomblets of the 1960s and 1970s—the babies economists once judged to be worse than worthless—are the very people who are now credited with making Asia boom. But oth-ers point out that this one-off effect does nothing for countries that do not also encourage people to be productive and invest in "human capital." Moreover, these same workers may yet pay a price for this "demographic dividend," because their children's generation may be too small to pay for all their pensions and prostate surgery.[5]

Both the ecological and the economic arguments for population control rest on the same debatable premise—that these campaigns actually succeeded in controlling populations. This would appear hard to deny. The global decline in birth rates has coincided with the increasing use of contraceptives. Yet fertility typically began to fall even before family planning programs—much less coercive population control—really got going. Conversely, pro-natalist policies did not have more than a slight effect. It turns out that about 90 percent of the difference in fertility rates worldwide derived from something very simple and very stubborn: whether women themselves wanted more or fewer children. People were planning smaller families, in other words, and they would not be bribed to have more babies. Nor did they wait for mobile clinics, sterilization camps, or door-to-door "inundation" campaigns to provide them with the means. This can be statistically demonstrated, but it is also consistent with both historical experience and common sense. After all, French peasants did not need Napoleon to provide them with pessaries. Even then, avoiding childbirth was less expensive and troublesome than unwanted children.[6]

If the population control movement did not succeed in controlling populations—except during the darkest episodes in its history, which proponents are quick to disavow—this is hardly the only measure of its impact. Hundreds of millions of people—over 60 percent of married women in their reproductive years—now use contraception, nine of ten choosing "modern methods" like pills, IUDs, implants, and injectables. Many more people are staying single, and not all of them are remaining chaste, even if these data are more difficult to collect. Trying to stop population decline led some countries to improve maternal health, maternity leaves, and day care. And to reduce birth rates, many others raised the age of marriage and legalized abortion. If all those who wanted to regulate their own fertility were still left to their own devices, the world would be a very different place, even if the number of people in it would not be greatly different.[7]

To determine *how* different that number might be, analysts have tried to compare national and local programs displaying more or less intensive efforts to reduce fertility. This can present philosophical and not just methodological problems. Should measures of "unmet need"—a critical variable—include women who do not want more children but will not use birth control for religious reasons? Societies that mount family planning

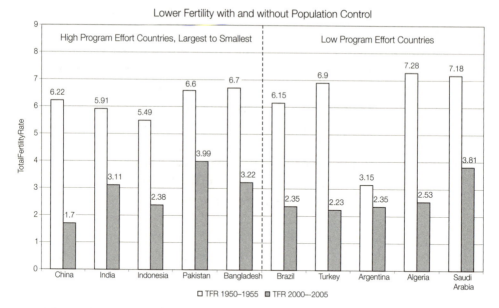

Lower Fertility with and without Population Control

☐ TFR 1950–1955   ■ TFR 2000—2005

The number of children per woman fell between 1950 and 2000 in countries with strong population control programs. But it also dropped dramatically during the same period in countries that made little effort to stop population growth or even encouraged it. Fertility data from UN Population Division; program effort data from W. Parker Mauldin and John A. Ross, "Family Planning Programs: Efforts and Results, 1982–1989," *Studies in Family Planning* 22 (1991): 354.

programs are also likely to be societies in which ordinary people, and not just national leaders, desire smaller families. Setting these problems aside, a generous estimate is that the impact of mounting a massive population control program like that in Indonesia is to reduce the average size of families by about one child.[8]

Is this reason enough to make this kind of investment? By all means, if it actually helps people exercise their right to control their own bodies. But we should not expect that it will lift them out of poverty, especially if it costs more than the per capita GDP to avert this one birth, as occurred in one particularly profligate program in Bangladesh. A far more important factor in fertility rates is whether women have had an education, and thus more opportunities to accomplish things besides bearing children. This effect has been shown to reduce average birth rates from an average of six or seven children for women with no education to three to five for those with seven years or more. And quite apart from its impact on population growth, or even per capita GNP, educating and enfranchising more than half of humanity made an incalculable contribution to general welfare.[9]

It is therefore the emancipation of women, not population control, that has remade humanity. In the family planning community, this link between education and fertility is treated like a revealed truth. It enabled feminists and environmentalists to forge a new, more enlightened consensus that promoted empowerment, not population control. In fact, it was something that many people knew all along. But for much of this history, lest we forget, everyone from the Catholic hierarchy to the Population Council to India's Congress Party agreed that the preference educated women had for fewer children was a problem, even a threat. It was not until more of these professional women won a place in international debates that promoting education became the solution.

Until then, population controllers preferred to deal with the high fertility of poor and uneducated people with increasingly blunt instruments. If the "dumb millions" did not plan their families in ways experts found intelligible, contraception had to be dumbed down. If people were too poor to afford it, they could be paid for using it, with group pressure providing further inducement. When many people would not accept IUDs, sterilization, implants, or injectables, population controllers dreamed of something that could be diffused through the air or the water, making everyone sterile without an antidote only authorities could provide. When no such technological fix was forthcoming, they denied parents maternity leave, housing, and health care, or simply dragged people to abortion and sterilization clinics. Even while proponents proclaimed that their aim was to help individuals master their fate, experiments involving tens of millions treated them as if they were bacteria in a petri dish, exhibiting regularities in reproductive behavior that could be measured and manipulated.

This history still weighs like a millstone on family planning organizations, if only because they have not unburdened themselves of it. Decades of warnings about "population crises" have made it more difficult to sustain support for reproductive health care now that it is not the fate of the earth, but the lives of individual women and children that hang in the balance. The legacy of "crash programs" to reduce fertility was to make many poor people mistrust sexual and reproductive health care, such that in India the very term *family planning* can no longer be used without implying population control.

Too often, the IPPF, the Population Council, UNFPA, and other orga-

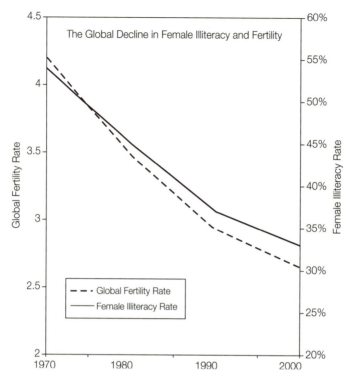

The worldwide decline in fertility rates corresponds far more closely with the worldwide decline in illiteracy among women than with population control programs. Data
from UNESCO Institute for Statistics and UN Population Division. Graph adapted
from FEWER, copyright © 2004, by Ben J. Wattenberg, by permission of Ivan R. Dee,
Publisher.

nizations in the field respond by asserting that they have always promoted
reproductive rights and never backed coercive policies. In effect, nothing
has changed. Ironically, their pro-life adversaries agree with them: nothing
has changed. Except the pro-lifers argue that these same organizations have
always been pursuing a hidden agenda and have never stopped trying to
control the population of poor people and poor countries.

To assert that nothing has changed, no less than to focus single-
mindedly on the future, is to turn one's back on history. The present becomes eternal, bounded by a Manichean conflict without end. But whereas
some in the pro-life community really do see their struggle as otherworldly,
this way of thinking is debilitating for people who are genuinely committed

to progressive change here and now. Because every time pro-lifers find ex-amples from the past of coercive practices or questionable motives—much of it taken from feminist critiques—it provides more grist for the mill.

At their most tendentious, opponents of birth control reduce history to a vast conspiracy theory, in which rich countries practice eugenics on the rest of the world. True, calls to stop population growth in poor countries came even before there was evidence that such growth was occurring. Time and again population controllers spoke out of both sides of their mouths—promising maternal health to some audiences, "crypto-eugenics" to others. Moreover, much of this discourse was conducted in English, and American dollars dominated international aid for the early population control cam-paigns.

But the Catholic hierarchy was no less concerned with "race suicide"—and not just salvation for sinners—when they fought to keep birth control and abortion illegal. The Church worked hand-in-glove with population controllers when their interests coincided, including a form of "positive eu-genics" that promoted higher fertility, and "natural family planning" under the supervision of priests. The one constant was a tireless defense of patriar-chy, in which the pope was father to the Church, men ruled their families, and women made babies. But the bishops, like their opponents, preferred indirection when they entered the political arena, letting others make their arguments rather than risk identifying the cause too closely with Rome.

The people who pushed different forms of population control in rich and poor countries are really part of the same story, a story far too impor-tant and complex to reduce to a conspiracy. In such countries as Italy, France, Argentina, Mexico, and Brazil, pro-natalist leaders found it oppor-tune to cooperate with the Church's campaign against contraception. Con-versely, elites in India, China, and many other countries worried about the size and "quality" of their populations, and willingly collaborated in proj-ects to reproduce whole nations according to plan. But whatever the plans of population controllers, whether pro- or anti-natalist, events proceeded in the most paradoxical fashion. Contraceptives like the pill, initially devel-oped as a "foolproof" means to reduce fertility, helped spark a sexual revo-lution that swept the globe. Fund-raising campaigns and population con-trol techniques directed at Asia and Africa changed the politics of family planning in Europe and the United States. The reaction provoked when population control came home brought into focus the question of how contraception could empower people, rather than control them, and whether women, in particular, had a stake in defending reproductive rights and health wherever they were threatened.

It is important to commemorate the struggle for reproductive rights, which emancipated men as well as women. But if we do not also recognize the full magnitude of what leaders like Margaret Sanger were up against, we cannot convincingly explain why so many of them, including Sanger, compromised and became complicit with population control. The temptation to plan other people's families was pervasive and persistent, and proponents often had both good intentions and deep pockets. Those who resisted therefore had to wrest control of fertility, not just from church and state, but from many people and institutions who professed that they only wanted to help. By acknowledging how hard it was to reform organizations like the IPPF and UNFPA, we may finally put to rest the canard that they are still pursuing the same hidden agenda.

The great tragedy of population control, the fatal misconception, was to think that one could know other people's interests better than they knew it themselves. But if the idea of planning other people's families is now discredited, this very human tendency is still with us. The essence of population control, whether it targeted migrants, the "unfit," or families that seemed either too big or too small, was to make rules for other people without having to answer to them. It appealed to the rich and powerful because, with the spread of emancipatory movements and the integration of markets, it began to appear easier and more profitable to control populations than to control territory. That is why opponents were correct in viewing it as another chapter in the unfinished history of imperialism.

The link between population control and imperialism is not merely conceptual, but historical. The ambition to control the population of the world emerged directly from the travails of territorial empires. Struggles over settlement colonies inspired the first practical programs. By the close of the nineteenth century, regulating migration was conceived as a means to maintain both the fertility of European settlers and the mortality of Asian "hordes." When birth rates continued to fall among the European peoples in the first decades of the twentieth century—at the same time fears of degeneration became pervasive—increasing the size and "quality" of nations provided both the rationale and the human resources to renew colonial expansion. But in the 1930s and 1940s imperial authorities began to realize that population growth was accelerating in North Africa, Korea, India, Indochina, and the West and East Indies. Only the Nazis made it a matter of policy to systematically attack the fertility of conquered peoples, at the same time applying colonial practices like wholesale expropriation, forced labor, and exemplary terror in the heart of Europe. The reaction this provoked reinforced the reluctance of other imperial powers to risk measures to reduce the growth of subject populations. This was true even when there was local support for promoting contraception. In India under the Raj, and in occupied Japan, local elites recognized in population control a means to regenerate nations and reclaim their independence.

This was the historical conjuncture that gave birth to the movement that sought to plan "the global family." Family planning was expected to eliminate differential fertility between rich and poor both within nations and around the world. Former colonial officials staffed new international and nongovernmental organizations and, together with leaders of newly independent states, devised ways to reproduce nations according to plan. After a few false starts in the 1940s, and the consolidation of a population establishment in the 1950s, a transnational network began to organize massive social engineering experiments. As Gunnar Myrdal remarked at the time, UN agencies had inherited the mantle of the *mission civilisatrice.* Poor countries were pressed to accept population programs and rich countries were expected to pay for them. A majority finally agreed to create a United Nations Fund for Population Activities. Lobbyists and UN officials worked behind the scenes to shield it from government oversight or at least ensure that it could funnel money to NGOs, which operated even more independently. When some states openly accused proponents of neocolonialism, powerful aid agencies like USAID and the World Bank convinced them to reconsider.[10]

In recent years, as the institution of sovereignty has seemed increasingly shaky, observers have focused on the fate of "failed states" like Yugoslavia, Rwanda, and Afghanistan. International and nongovernmental organizations exercise stewardship on behalf of the "international community," creating a kind of "Empire Lite," albeit with the best of intentions. The population control movement demonstrated decades ago that "global governance" can work not just in war-torn territories, but domains into which few colonial officers would ever have dared enter. What Michel Foucault called biopolitics, the power to regulate both individual and social bodies, presented infinite opportunities to substitute "governmentality" for state sovereignty. Preaching rules that regulated how people reproduce nations was far more subversive of sovereignty than restoring a state's capacity to control its own territory. Proponents were trying to remake humanity to fit global norms.[11]

The world is now full of transnational movements responding to a range of different "crises," including epidemic disease, refugee flows, and climate change. Many see global governance as the hope of the future. It appears to be a welcome alternative to the sovereign state at a time in which the most important problems cannot be contained by national borders. But it is worth recalling how many of those who sought to control the population of the world also dreamed of world peace, world government, or at least a closer union of like-minded peoples. The longtime leader of the Malthusian League, Charles Drysdale, did not just want to fine overly prolific parents, he wanted a federation to unite Germany, France, and the United Kingdom. Corrado Gini called for a Euro-American union, if only because he was concerned about how miscegenation was making the two sides of the Atlantic drift apart. Harry Laughlin drafted and redrafted schemes for an international parliament in which the best stocks could rule the earth. Julian Huxley and his fellow "reform eugenicists" fought against this kind of racial prejudice, but they too wanted a "federation of the whole world" to undertake a comprehensive program of genetic improvement. And the One, Holy, Catholic, and Apostolic Church has always claimed to represent all God's people both in this world and the next.[12]

Why should we presume that would-be governors of the globe will be essentially different from other kinds of governors? And why should we expect that politics will improve merely by pouring it into different containers, especially when these new vessels also divide people and keep them unequal? The nation-state, for all its faults, at least provides a structure in which governors and the governed are presumed to belong to the same polity. This has not precluded terrible atrocities, the worst typically inflicted on stateless people after they have been stripped of their citizenship. International and nongovernmental organizations can provide aid, but if they are to offer a better alternative, they must surmount a critical flaw: a constitutional inability to answer to the people they profess to serve.

Perhaps this problem can be overcome. But recent experience, such as the global campaign against HIV/AIDS, is not encouraging. For all the talk of "lessons learned," in some cases by the very same institutions that managed population control campaigns, they have been making many of the same mistakes. Once again, declaring a "crisis," even a national security threat, appeared necessary to attract media attention, scientific research, and humanitarian aid. Exponential growth in earmarked funds added to the pressure for "crash programs." Staff were hired away from already struggling health programs to work on this single problem. Standardized approaches to the "global AIDS crisis" override merely local concerns. After all, HIV, like "overpopulation," threatens everyone, including people in Geneva and New York, whereas "there are no anti-dysentery activists."[13]

It is commendable to feel a sense of urgency about helping people with AIDS, or an oppressive government, or an unwanted pregnancy. Perhaps declaring a global crisis once again really can "make poverty history." But the problems begin when those with money and power view "dialogue" as just a way to tell other people what is good for them. As much as they may scoff at the charge, the spirit of empire lives on when people are unaccountable to those they claim to serve. All that historians can do is remind them of where all this can lead. Little good can come from schemes to improve the human condition that start with such an impoverished sense of the past.

It might appear that the history of population control, at least, is over. Certainly the global movement to shape demographic trends now lies dormant. All over the world there has been a shift in the locus of control in how societies reproduce themselves, whether locally, nationally, or globally. Individuals are deciding for themselves, with or without anyone's help or permission. They are insisting on their right to choose where they live, and an ever-growing number are managing to get their way—nearly two hundred million having emigrated to another country, according to UN estimates. They are also choosing how many children they will have, and even what kind of children they will be.[14]

This trend may seem to be leading to the ultimate realization of reproductive rights. But in a world in which some own private jets while asylum-seekers are imprisoned, we must not be sanguine. Many parents already realize, for instance, that their choices between work and family, however difficult, would be impossible without the aid of poor women who have no choice but to leave their own families behind. Other women still cannot get contraception or use it without fear of violence. Abortion is the last line in women's defense of their bodily autonomy, and the attacks never cease. But in Asia, sex selection is making women a minority, subject to child marriage and sexual violence. In all these ways, we live with the legacies of policies and programs to shape populations, even if we are not ourselves living examples of "family planning."

Some have declared a new population crisis, only this time it is because the projections seem too low rather than too high, and we are told that we should fear too many elderly rather than too many children. Now most pronounced in Europe and Japan, the "aging" of populations may proceed much more rapidly in countries where fertility fell the fastest, such as China and Mexico, this time without benefit of a social safety net. Worldwide, *in vitro* fertilization and international adoption are increasingly in demand. For those who have put off childbearing and struggle to conceive, they can seem like miracles. But medical technology and the economics of the "baby business" have also brought us preimplantation genetic diagnosis, the outsourcing of surrogacy, and a global market for sperm, eggs, embryos, and infants. If privileged people are permitted to pick and choose, and make themselves a breed apart, how can future generations possibly fight prejudice or promote more equal opportunities?

There is therefore the prospect of a new age of population control. If history is any guide, projections of population decline may lead to increasingly coercive policies, even crash programs. The legacy of earlier "policy trade-offs," in which women's education, employment, and reproductive health care were promoted as methods to drive down birth rates, will make this more difficult to resist. After all, if these were not inalienable rights, but means to some other end, on what grounds can one now oppose governments that reverse these policies to prevent a population "implosion"?

Even without top-down coercion, we may already be witnessing something no less pernicious: the privatization of population control. It is governmentality without government, in which people police themselves, unconsciously reproducing and reinforcing inequality with every generation. The process is well under way in India and China because of sex-selective abortions. In the United States, so often a bellwether for population politics, it is more subtle. Parents increasingly experience genetic "counseling" and solicitous concern for fetal health as social pressure to have perfect children, even if the standards for perfection are constantly changing. In everyday conversation, people ascribe a whole range of behaviors to good or bad genes, faithfully reciting a eugenic catechism without the faintest idea of where it comes from or where it can lead. The cumulative effect of individual choices may be to make patriarchy permanent in the largest nations and endow the most privileged with genetic advantages they will lord over the world.

These problems *are* global in scope and transnational in nature. Can we devise means of addressing them without creating new forms of unaccountable power? There is no ready-made formula. But if there is to be some system in which different kinds of agencies, both public and private, cooperate to help people and also answer to them, it must have a heart and a brain and run on more than money. It must start with a commitment to reproductive freedom, not just fear of the future, and proceed with a sense of history. And it must be both pro-life *and* pro-choice, combining forces to oppose population control of any kind.

It may seem naive to think that challenges like coercive pro-natalism and genetic "enhancement" might bring about a peace process in this bitterest of culture wars. But what is the alternative? Those who consider themselves pro-life must eventually realize that making people breed at any price cheapens all of our lives. And those who consider themselves pro-choice would be in a stronger position if they were at the forefront in opposing *all* manipulative and coercive policies designed to control populations. There are some encouraging signs. Family planning groups are beginning to speak out in defense of Chinese dissidents who protest the one-child policy. Some pro-lifers have recognized that promoting access to contraception is the best way to reduce the incidence of abortion. And when one economist argued that abortion had reduced the crime rate—an eerie echo of population control arguments from the 1960s—the only person who rose to the bait was immediately condemned by all sides.[15]

A new agenda that can renew and revive the cause of reproductive freedom will require much more, beginning with a greater effort to find common ground by pro-life and pro-choice people of good faith. We must work together to ensure that everyone has access both to birth control and child care without being coerced to have more or fewer children. We might agree that society has an interest in potential life, to be balanced against the rights of the mother, and together fight sex-selective abortions worldwide. Both sides could also join in recognizing international adoption—now anarchic and inequitable—as ripe for advocacy and reform. We can demand that infertility treatment become part of comprehensive health care for all, in Africa no less than the United States. If we are to permit new technologies to select out predispositions for health problems, or even "enhance" future generations, these choices too must be given to everyone equally. And we might also recognize that, no less than capital, goods, and ideas, people should be able to move about the globe and seek opportunities in societies wise enough to accept them. In all these ways, we might begin to make real the ideal of the global family.

The struggle against population control has shown that it is never enough merely to insist on choice. Choices can be conditioned by default or design in ways that lead to new kinds of oppression. And the defense of life can also become an idol, a symbol devoid of substance, if the effect is to drive people to breed. Reproductive freedom is a cause that can and must stand on its own, now more than ever. But it can only take flight if it is animated by a vision of social justice in which every one of us is conceived in liberty and created equal.

Notes

Archives and Interviews

Acknowledgments

Index

NOTES

Abbreviations

AN	Archives Nationales, Paris
APS	American Philosophical Society, Philadelphia
CAC	Centre des Archives Contemporaines, Fontainebleau
CAOM	Centre des Archives d'Outre-mer, Aix-en-Provence
CMAC	Contemporary Medical Archives Center, Wellcome Institute for the History of Medicine, London
FAOA	Food and Agriculture Organization Archives, Rome
FFA	Ford Foundation Archives, New York City
FRUS	*Foreign Relations of the United States* (Washington, DC: U.S. Government Printing Office)
H&R	Harper & Row, Publisher Records, Manuscript Collections, Columbia University Library
ILOA	International Labour Organization Archives, Geneva
IOR	India Office Records, British Library, London
IPPFA	International Planned Parenthood Federation Archives, London
memcon	memorandum of conversation
MSP	Margaret Sanger Papers Microfilm Edition
NCCB Papers	National Conference of Catholic Bishops, USCCBA
NIHFW	National Institute of Health and Family Welfare, Documentation Centre, New Delhi
RAC	Rockefeller Archive Center, Tarrytown, NY
SA/EUG	Eugenics Society Papers
SA/FPA	Family Planning Association Papers

SSC	Sophia Smith Collection, Smith College
UKNA	National Archives, Kew, UK
UNESCOA	UNESCO Archives, Paris
UNRC	United Nations Archives and Records Centre, New York City
UPA	University Publications of America
USCC Papers	U.S. Catholic Conference Papers, USCCBA
USCCBA	U.S. Conference of Catholic Bishops Archives, Washington, DC
USNA	U.S. National Archives, College Park
VSA	Vatican Secret Archives, Rome
WBGA	World Bank Group Archives, Washington, DC
WHOA	World Health Organization Archives, Geneva

PREFACE

The Preface refers to the following sources: Paul Ehrlich, *The Population Bomb* (New York: Ballantine Books, 1968), prologue, 132; Paul R. Ehrlich and Anne H. Ehrlich, "Population Control and Genocide," in *Global Ecology: Readings toward a Rational Strategy for Man,* ed. John P. Holdren and Paul R. Ehrlich (New York: Harcourt Brace Jovanovich, 1971), 157–159; Paul Kennedy, *The Rise and Fall of the Great Powers: Economic Change and Military Conflict from 1500 to 2000* (New York: Random House, 1987); Matthew Connelly and Paul Kennedy, "Must It Be the Rest against the West?" *Atlantic Monthly,* December 1994, 61–84.

INTRODUCTION

1. Thomas Robert Malthus, *An Essay on the Principle of Population: Text, Sources and Background, Criticism* (New York: Norton, 1976), 70.

2. Robert William Fogel, *The Escape from Hunger and Premature Death, 1700–2100: Europe, America, and the Third World* (Cambridge: Cambridge University Press, 2004), 8–9, 13–15; Malthus, *An Essay on the Principle of Population,* 6th ed. (1826), book 3, chap. 2, para. 20, www.econlib.org/library/Malthus/malPlong.html.

3. Robert J. Richards, *The Meaning of Evolution: The Morphological Construction and Ideological Reconstruction of Darwin's Theory* (Chicago: University of Chicago Press, 1992), 82–84.

4. Fogel, *The Escape from Hunger,* 20–22, 40.

5. Edward O. Wilson, *Consilience: The Unity of Knowledge* (New York: Knopf, 1998), 11, 273.

6. Geoffrey Barraclough, *An Introduction to Contemporary History* (New York: Pelican, 1967), 92; Eric Hobsbawm, *The Age of Extremes: A History of the*

World, 1914–1991 (New York: Vintage, 1996), 345–346; William H. McNeill, *Population and Politics since 1750* (Charlottesville: University Press of Virginia, 1990), 49; Francis Fukuyama, *Our Posthuman Future: Consequences of the Biotechnology Revolution* (New York: Farrar, Straus and Giroux, 2002), 15.

7. "Observations Concerning the Increase of Mankind, Peopling of Countries, Etc.," in *Benjamin Franklin: Representative Selections, with Introduction, Bibliography, and Notes,* ed. Frank Luther Mott and Chester E. Jorgenson (New York: American Book Co., 1936), 221–223.

8. Journal excerpt, November 3, 1914, in *The Selected Papers of Margaret Sanger,* vol. 1, *The Woman Rebel, 1900–1928,* ed. Esther Katz (Urbana: University of Illinois Press, 2003), 98.

9. *Casti Connubii,* Encyclical of Pope Pius XI on Christian Marriage, December 31, 1930, www.vatican.va/holy_father/pius_xi/encyclicals/index.htm.

I. POPULATIONS OUT OF CONTROL

1. R. L. Dugdale, *The Jukes: A Study in Crime, Pauperism, Disease, and Heredity, with Further Studies of Criminals* (New York: G. P. Putnam's Sons, 1877), and review in the *New York Times,* July 29, 1877.

2. Roger Manvell, *The Trial of Annie Besant and Charles Bradlaugh* (London: Elek/Pemberton, 1976), 48–51.

3. Ibid., 86–92; see also "High Court of Justice," *The Times of London,* June 20, 1877.

4. Manvell, *Trial of Annie Besant,* 153; "Mrs. Annie Besant's Appeal," *The Malthusian,* May 1, 1879; Michael Teitelbaum, *The British Fertility Decline: Demographic Transition in the Crucible of the Industrial Revolution* (Princeton: Princeton University Press, 1984), 207–210.

5. Simon Szreter, *Fertility, Class and Gender in Britain, 1860–1940* (Cambridge: Cambridge University Press, 1996), 360–366; Peter Fryer, *The Birth Controllers* (London: Secker and Warburg, 1965), 175–176.

6. Manvell, *Trial of Annie Besant,* 137–138; D. V. Glass, *Population Policies and Movements in Europe* (1940; repr., London: Frank Cass and Co., 1967), 148–152; Angus McLaren, *Sexuality and Social Order: The Debate over the Fertility of Women and Workers in France, 1770–1920* (New York: Holmes and Meier, 1983), 18–25.

7. Ansley J. Coale and Roy Treadway, "A Summary of the Changing Distribution of Overall Fertility, Marital Fertility, and the Proportion Married in the Provinces of Europe," and Massimo Livi-Bacci, "Social Group Forerunners of Fertility Control in Europe," both in *The Decline of Fertility in Europe,* ed. Ansley J. Coale and Susan Watkins (Princeton: Princeton University Press, 1986), 39, 183–189. On new contraception and abortion, see Norman E. Himes, *Medical History*

of Contraception (New York: Gamut Press, 1963), 249–253; Angus McLaren, *A History of Contraception: From Antiquity to the Present Day* (Oxford: Blackwell, 1990), 184–191; John Peel, "The Manufacture and Retailing of Contraceptives in England," *Population Studies* 17, no. 2 (1963): 115–121; James Marchant, *The Declining Birth-Rate: Its Causes and Effects* (London: Chapman and Hall, 1916), 27–28.

8. Francine Van de Walle, "Infant Mortality and the European Demographic Transition," in Coale and Watkins, *The Decline of Fertility*, 211–215; Seth Koven and Sonya Michel, "Womanly Duties: Maternalist Politics and the Origins of Welfare States in France, Germany, Great Britain, and the United States, 1880–1920," *American Historical Review* 95 (1990): 1099.

9. *Rerum Novarum*, www.vatican.va/holy_father/leo_xiii/encyclicals.

10. Peter Gay, *Education of the Senses*, vol. 1, *The Bourgeois Experience* (New York: Oxford University Press, 1984), 236–238; Robert Woods, *The Demography of Victorian England and Wales* (Cambridge: Cambridge University Press, 2000), 248, 264; Szreter, *Fertility, Class and Gender*, 250. On typhoid, see Wesley W. Spink, *Infectious Diseases: Prevention and Treatment in the Nineteenth and Twentieth Centuries* (Minneapolis: University of Minnesota Press, 1978), 242.

11. Thomas McKeown's explanation in *The Modern Rise of Population* (New York: Academic Press, 1976) has not held up in all its particulars, but the relative unimportance of medicine in early mortality declines is generally accepted.

12. Simon Szreter, "The Idea of Demographic Transition and the Study of Fertility Change: A Critical Intellectual History," *Population and Development Review* 19 (1993): 659–701.

13. Van de Walle, "Infant Mortality," 228–231; Deborah Dwork, *War Is Good for Babies and Other Young Children: A History of the Child Welfare Movement in England, 1898–1918* (London: Tavistock, 1987), 4–7; Allan Sharlin, "Urban-Rural Differences in Fertility in Europe during the Demographic Transition," in Coale and Watkins, *The Decline of Fertility*, 243–244, 249–251.

14. On wealth flows, see John C. Caldwell, *Theory of Fertility Decline* (London: Academic Press, 1982). On cultural diffusion, see J. A. Banks, *Victorian Values, Secularism and the Size of Families* (Boston: Routledge and Kegan Paul, 1981), 107–108; but compare Szreter, *Fertility, Class, and Gender*, 478–481. On education, see T. Paul Schultz, "Demand for Children in Low Income Countries," *Handbook of Population and Family Economics* 1 (1997): 380–384.

15. Margaret Sanger, *Woman and the New Race* (New York: Brentano's, 1920).

16. Ian Hacking, *The Taming of Chance* (Cambridge: Cambridge University Press, 1990); Dwork, *War Is Good for Babies*, 9; Koven and Michel, "Womanly Duties," 1099–1100, 1105; Karen Offen, "Depopulation, Nationalism, and Feminism in Fin-de-Siècle France," *American Historical Review* 89 (1984): 648–676;

Michael S. Teitelbaum and Jay M. Winter, *Fear of Population Decline* (Orlando, FL: Academic Press, 1985), 28–30.

17. Rogers Brubaker, *Citizenship and Nationhood in France and Germany* (Cambridge: Harvard University Press, 1992), 64–70; Leo Lucassen, "Eternal Vagrants? State Formation, Migration, and Travelling Groups in Western Europe, 1350–1914," in *Migration, Migration History, History: Old Paradigms and New Perspectives,* ed. Jan Lucassen and Leo Lucassen (Bern: Peter Lang, 1997), 243–251; Frank Caestecker, "The Changing Modalities of Regulation in International Migration within Continental Europe, 1870–1940," in *Regulation of Migration: International Experiences,* ed. Anita Böcker et al. (Amsterdam: Het Spinhuis, 1998), 74–76, 80–81; Ernest Gellner, *Nations and Nationalism* (Ithaca: Cornell University Press, 1983), 1–2, 49; Benedict Anderson, *Imagined Communities: Reflections on the Origin and Spread of Nationalism,* rev. ed. (London: Verso, 1991), 80–87.

18. Quoted in Gérard Noiriel, *The French Melting Pot: Immigration, Citizenship, and National Identity,* trans. Geoffroy de Laforcade (Minneapolis: University of Minnesota Press, 1996), 52 (emphasis in original).

19. James C. Scott, *Seeing like a State: How Certain Schemes to Improve the Human Condition Have Failed* (New Haven: Yale University Press, 1998), 51, 81; John Torpey, *The Invention of the Passport: Surveillance, Citizenship and the State* (Cambridge: Cambridge University Press, 2000), 16–17; Hacking, *Taming of Chance,* 182–184.

20. Noiriel, *The French Melting Pot,* 48–49; Hacking, *Taming of Chance,* 2, 115–125, 194–195. On Farr, see Gay, *Education of the Senses,* 1:233–234. On calculating worth, see Paul Weindling, *Health, Race and German Politics between National Unification and Nazism, 1870–1945* (Cambridge: Cambridge University Press, 1989), 188.

21. Adam McKeown, *Melancholy Order: Asian Migration and the Globalization of Borders, 1834–1937* (New York: Columbia University Press, forthcoming); James Marchant, *Birth Rate and Empire* (London: Williams and Norgate, 1917), 17–18; C. Knick Harley, "Ocean Freight Rates and Productivity, 1740–1913: The Primacy of Mechanical Invention Reaffirmed," *Journal of Economic History* 48 (1988): 851–876; David Held et al., *Global Transformations: Politics, Economics and Culture* (Stanford: Stanford University Press, 1999), 291–292; Walter Nugent, *Crossings: The Great Transatlantic Migrations, 1870–1914* (Bloomington: Indiana University Press, 1992), 31, 42–43.

22. Ellen Ross, *Love and Toil: Motherhood in Outcast London, 1870–1918* (New York: Oxford University Press, 1993), 30–41; Held et al., *Global Transformations,* 155–157, 168–171, 193–198.

23. John T. Noonan, *Contraception: A History of Its Treatment by the Catholic Theologians and Canonists* (Cambridge: Harvard University Press, 1966), 417–419; Brubaker, *Citizenship and Nationhood,* 103–106.

24. Rosanna Ledbetter, *A History of the Malthusian League, 1877–1927* (Columbus: Ohio State University Press, 1976), 49–51, 62, 69–72, 149–152, 172–182; Deborah Anne Barrett, "Reproducing Persons as a Global Concern: The Making of an Institution" (Ph.D. diss., Stanford, 1995), 119–122. On French Malthusians, see Offen, "Depopulation, Nationalism, and Feminism," 661–663; McLaren, *Sexuality and Social Order,* 95; William H. Schneider, *Quality and Quantity: The Quest for Biological Regeneration in Twentieth-Century France* (New York: Cambridge University Press, 1990), 35–37; Francis Ronsin, *La Grève des Ventres: Propagande néo-malthusienne et baisse de la natalité française, 19e–20e siècles* (Paris: Aubier Montaigne, 1980), 48–50.

25. Mike Davis, *Late Victorian Holocausts: El Niño Famines and the Making of the Third World* (New York: Verso, 2001), 7; Arup Maharatna, *The Demography of Famines: An Indian Historical Perspective* (Delhi: Oxford University Press, 1996), 14–16.

26. Charles Drysdale, "The Life and Writings of T. R. Malthus," *The Malthusian,* June 1879, 37–38. On demography in Asia, see Nigel Crook, "On the Comparative Historical Perspective," in *India's Historical Demography: Studies in Famine, Disease and Society,* ed. Tim Dyson (London: Curzon Press, 1989), 286–287; James Z. Lee and Wang Feng, *One Quarter of Humanity: Malthusian Mythology and Chinese Realities, 1700–2000* (Cambridge: Harvard University Press, 1999), 88; S. B. Hanley and A. P. Wolf, eds., *Family and Population in East Asian History* (Stanford: Stanford University Press, 1985); S. B. Hanley and K. Yamamura, *Economic and Demographic Change in Pre-Industrial Japan, 1600–1968* (Princeton: Princeton University Press, 1977).

27. Davis, *Late Victorian Holocausts,* 71–79, 95–96, 106–108, 114. This discussion of public health and wealth differentials is indebted to Davis along with the earlier work of Ira Klein, Kenneth Pomeranz, and Sheldon Watts.

28. David Arnold quoting Gilberto Freyre, in *Famine: Social Crisis and Historical Change* (Oxford: Blackwell, 1988), 13–14, 93–94; see also Davis, *Late Victorian Holocausts,* 86–90.

29. Davis, *Late Victorian Holocausts,* 50–51. On Baring, see Sheldon Watts, *Epidemics and History: Disease, Power and Imperialism* (New Haven: Yale University Press, 1997), 203. On Malthusian influence, see also S. Ambirajan, "Malthusian Population Theory and Indian Famine Policy in the Nineteenth Century," *Population Studies* 30, no. 1 (1976): 5–14; John Caldwell, "Malthus and the Less Developed World: The Pivotal Role of India," *Population and Development Review* 24 (1998): 679–693.

30. Watts, *Epidemics and History,* 201–202; Davis, *Late Victorian Holocausts,* 38–40; David Arnold, "Cholera Mortality in British India, 1817–1947," in Dyson, *India's Historical Demography,* 268–270.

31. Ira Klein, "Imperialism, Ecology, and Disease: Cholera in India, 1850–

1950," *Indian Economic and Social History Review* 31, no. 4 (1994): 496–499, 505–510.

32. Sheldon Watts, "British Development Policies and Malaria in India 1897–c. 1929," *Past and Present,* no. 165 (November 1999): 141–144, 148–154; Leela Visaria and Pravin Visaria, "Population (1757–1947)," in *The Cambridge Economic History of India,* ed. Dharma Kumar (Cambridge: Cambridge University Press, 1983), 480–481; Ira Klein, "Death in India, 1871–1921," *Journal of Asian Studies* 32 (1973): 645–646.

33. Ira Klein, "Malaria and Mortality in Bengal, 1840–1921," *Indian Economic and Social History Review* 9, no. 2 (June 1972): 139–142. On cholera, see Klein, "Imperialism, Ecology, and Disease," 510, 516. On improvements and enclaves, see Radhika Ramasubban, *Public Health and Medical Research in India: Their Origins under the Impact of British Colonial Policy* (Stockholm: Swedish Agency for Research Cooperation with Developing Countries, 1982). On irrigation and malaria, see Watts, *Epidemics and History,* 205; and Watts, "British Development Policies," 170–174.

34. Davis, *Late Victorian Holocausts,* 26–27, 111–112; David Hardiman, "Usury, Dearth and Famine in Western India," *Past and Present,* no. 152 (August 1996): 133, 145–146. On shipping rates, see Kenneth Pomeranz and Steven Topik, *The World That Trade Created: Society, Culture, and World Economy, 1400– the Present* (Armonk, NY: M. E. Sharpe, 1999), 72. On Madras, see B. M. Bhatia, *Famines in India, 1850–1945* (Bombay: Konark, 1963), 94.

35. Watts, *Epidemics and History,* 203; Pomeranz and Topik, *The World That Trade Created,* 72–73; Davis, *Late Victorian Holocausts,* 33.

36. Pomeranz and Topik, *The World That Trade Created,* 120; Jörg Fisch, "Internationalizing Civilization by Dissolving International Society: The Status of Non-European Territories in Nineteenth-Century International Law," in *The Mechanics of Internationalism: Culture, Society, and Politics from the 1840s to the First World War,* ed. Martin H. Geyer and Johannes Paulmann (Oxford: Oxford University Press, 2001), 251.

37. On white deluge, see C. A. Bayly, *The Birth of the Modern World, 1780– 1914: Global Connections and Comparisons* (Oxford: Blackwell, 2004), 439–444.

38. Adam McKeown, "Conceptualizing Chinese Diasporas, 1842 to 1949," *Journal of Asian Studies* 58 (1999): 319–325. On inalienable right, see Lucy E. Salyer, *Laws Harsh as Tigers: Chinese Immigrants and the Shaping of Modern Immigration Law* (Chapel Hill: University of North Carolina Press, 1995), 7; see also Aristide R. Zolberg, "The Great Wall against China: Responses to the First Immigration Crisis, 1885–1925," in Lucassen and Lucassen, *Migration,* 299–302.

39. John Higham, *Strangers in the Land: Patterns of American Nativism, 1860–1925* (New York: Atheneum, 1963), 49. On coolies, see Donna Gabaccia, "The 'Yellow Peril' and the 'Chinese of Europe': Global Perspectives on Race and

Labor, 1815–1930," in Lucassen and Lucassen, *Migration,* 182–183. On attacks, see Alexander Saxton, *The Indispensable Enemy: Labor and the Anti-Chinese Movement in California* (Berkeley: University of California Press, 1971), 9–10, 76–77, 104.

40. Saxton, *Indispensable Enemy,* 104–106, 113ff.

41. Stuart Creighton Miller, *The Unwelcome Immigrant: The American Image of the Chinese, 1785–1882* (Berkeley: University of California Press, 1974), 189; M. J. Dee, "Chinese Immigration," *North American Review* 126, no. 262 (1878): 517–518; McKeown, "Conceptualizing Chinese Diasporas," 9–10; Zolberg, "The Great Wall," 303; Held et al., *Global Transformations,* 294–295; Adam McKeown, "Global Migration, 1846–1940," *Journal of World History* 15 (2004): 315–317.

42. Zolberg, "The Great Wall," 291–292. On border closings, see Salyer, *Laws Harsh as Tigers,* 1, 4–7; Gabaccia, "The 'Yellow Peril,'" 192–193; Robert A. Huttenback, *Racism and Empire: White Settlers and Colonial Immigrants in the British Self-Governing Colonies, 1830–1910* (Ithaca: Cornell University Press, 1976), 279–282. On scapegoating, see Saxton, *Indispensable Enemy,* 264.

43. Richard Hofstadter, *Social Darwinism in American Thought,* rev. ed. (Boston: Beacon Press, 1955), 38–42; Mike Hawkins, *Social Darwinism in European and American Thought, 1860–1945: Nature as Model and Nature as Threat* (Cambridge: Cambridge University Press, 1997), 85–86; *Report of the Special Joint Committee to Investigate Chinese* (Washington, DC: Government Printing Office, 1877), v; Dee, "Chinese Immigration," 510, 513–514, 520–521, 523–526.

44. William F. Wu, *The Yellow Peril: Chinese Americans in American Fiction, 1850–1940* (Hamden, CT: Archon Books, 1982), 30–46; John C. G. Röhl, *Wilhelm II: The Kaiser's Personal Monarchy, 1888–1900,* trans. Sheila de Bellaigue (Cambridge: Cambridge University Press, 2004), 753–755, 909–910; Heinz Gollwitzer, *Die gelbe Gefahr: Geschichte eines Schlagworts—Studien zum imperialistischen Denken* (Göttingen: Vandenhoeck und Ruprecht, 1962), 29–30, 50ff., 87–91, 131ff., 175–183; Arthur Herman, *The Idea of Decline in Western History* (New York: Free Press, 1997), 172–174.

45. Salyer, *Laws Harsh as Tigers.* Owen quoted in Saxton, *Indispensable Enemy,* 268, and see also 201–213.

46. On Prussia, see Zolberg, "The Great Wall," 311–314; and Jack Wertheimer, *Unwelcome Strangers: East European Jews in Imperial Germany* (New York: Oxford University Press, 1987), 43–49, 54–58. On France, see Brubaker, *Citizenship and Nationhood,* 101–102; Leo Lucassen, "The Great War and the Origins of Migration Control in Western Europe and the United States (1880–1920)," in Böcker et al., *Regulation of Migration,* 51; and Noiriel, *The French Melting Pot,* 54. On the United Kingdom, see Salyer, *Laws Harsh as Tigers,* 7; and Bernard Gainer, *The Alien Invasion: The Origins of the Aliens Act of 1905* (London: Heinemann, 1972), 97–99, 112. On Webb, see Daniel J. Kevles, *In the Name of Eugenics: Genet-*

ics and the Uses of Human Heredity (New York: Knopf, 1985), 74. On the spread of American-style racialist thinking, see Bayly, *Birth of the Modern World,* 189–190, 226–227.

47. On the lack of accountability, see McKeown, *Melancholy Order.* On solidarity and Gompers, see Lucassen, "The Great War," 46–54; and Saxton, *Indispensable Enemy,* 276–277. On frugality, see Richmond Mayo-Smith, "Control of Immigration: I," *Political Science Quarterly* 3, no. 1 (March 1888): 73, 75; and Gabaccia, "The 'Yellow Peril,'" 178. Henry Cabot Lodge, "Lynch Law and Unrestricted Immigration," *North American Review,* no. 152 (1891): 607–609, 612.

48. Mayo-Smith, "Control of Immigration," 73, 75; Keith Fitzgerald, *The Face of the Nation: Immigration, the State, and the National Identity* (Stanford: Stanford University Press, 1996), 114.

49. Walker's articles are reprinted in Walker, *Discussions in Economics and Statistics,* ed. Davis Dewey (New York: Henry Holt and Co., 1899), 1:420–422, 424–425, 438–439, 447.

50. Dennis Hodgson, "The Ideological Origins of the Population Association of America," *Population and Development Review* 17 (1991): 5; Kevles, *In the Name of Eugenics,* 33. On the United Kingdom and Germany, see Hawkins, *Social Darwinism,* 94–96, 111–113, 129–131; Sheila Faith Weiss, *Race Hygiene and National Efficiency: The Eugenics of Wilhelm Schallmayer* (Berkeley: University of California Press, 1987), 34–36. On Davenport, see Margaret Sanger, ed., *Proceedings of the World Population Conference* (London: Edward Arnold and Co., 1927), 343–344.

51. Charles E. Woodruff, "Ravenstein on Global Carrying Capacity," reprinted in *Population and Development Review* 16 (1990): 153–162; Edmund Morris, *The Rise of Theodore Roosevelt* (New York: Modern Library, 1979, 2001), 482; Gollwitzer, *Die gelbe Gefahr,* 50ff.; Benjamin Kidd, *The Control of the Tropics* (London: MacMillan, 1898), 29–31. On Ross, see David Arnold, "The Place of 'the Tropics' in Western Medical Ideas since 1750," *Tropical Medicine and International Health* 2, no. 4 (April 1997): 305, 310. Charles E. Woodruff, *The Effects of Tropical Light on White Men* (New York: Rebman Co., 1905), 261–282, 320.

52. Frederick Jackson Turner, *The Frontier in American History* (New York: Holt, Rinehart and Winston, 1920, 1962), 1–38; Louis S. Warren, "Buffalo Bill Meets Dracula: William F. Cody, Bram Stoker, and the Frontiers of Racial Decay," *American Historical Review* 107 (2002): 1124–1257; Ann Laura Stoler, "Sexual Affronts and Racial Frontiers: European Identities and the Cultural Politics of Exclusion in Colonial Southeast Asia," in *Tensions of Empire: Colonial Cultures in a Bourgeois World,* ed. Frederick Cooper and Ann Laura Stoler (Berkeley: University of California Press, 1997), 199–200.

53. On degeneration, see Daniel Pick, *Faces of Degeneration: A European Disorder, c. 1848–c. 1918* (New York: Cambridge University Press, 1989); and Herman, *The Idea of Decline.* On Lodge, see Higham, *Strangers in the Land,* 142.

54. Gollwitzer, *Die gelbe Gefahr,* 13–14; Ratzel, *Die Chinesische Auswanderung: Ein Beitrag zur Cultur- und Handelsgeographie* (Breslau: J. U. Kern, 1876); Ratzel, *Der Lebensraum: Eine Biogeographische Studie* (Tübingen: H. Laupp, 1901). My thanks to Cyrus Schayegh for pointing out this connection.

55. Benedict Anderson, *Imagined Communities: Reflections on the Origin and Spread of Nationalism,* rev. ed. (London: Verso, 1983, 1991), 164–170. On India, see Bernard S. Cohn, *An Anthropologist among Historians and Other Essays* (Delhi: Oxford University Press, 1987), 247–250; and Nicholas B. Dirks, *Castes of Mind: Colonialism and the Making of Modern India* (Princeton: Princeton University Press, 2001), 205–207, 210–224. On the spread of social Darwinism, see Bayly, *Birth of the Modern World,* 317, 323.

56. Zolberg, "The Great Wall," 291ff., 315; Avner Offer, *The First World War: An Agrarian Interpretation* (Oxford: Clarendon Press, 1989), 185–194; Robert A. Huttenback, *Racism and Empire: White Settlers and Colored Immigrants in the British Self-Governing Colonies 1830–1910* (Ithaca: Cornell University Press, 1976), 188–189. Britain's concern not to offend its Japanese ally was also a factor.

57. Edward Alsworth Ross, *Changing America: Studies in Contemporary Society* (New York: The Century, 1912), 46–48; and Ross, *The Old World in the New* (New York: The Century, 1914), preface.

58. Torpey, *Invention of the Passport,* 70, 75–77.

59. "The Lambeth Conference: Encyclical Letter," *The Times of London,* August 8, 1908, 6–7; Noonan, *Contraception,* 419–421.

60. Besant, "Malthus, Darwin, and Spencer," *The Malthusian,* March 1, 1879, 16; Donald K. Pickens, *Eugenics and the Progressives* (Nashville: Vanderbilt University Press, 1968), 44–45; Linda Gordon, *Woman's Body, Woman's Right: Birth Control in America,* rev. ed. (New York: Penguin Books, 1974, 1990), 118–120; Mark H. Haller, *Eugenics: Hereditarian Attitudes in American Thought* (New Brunswick: Rutgers University Press, 1963), 36ff.; Sandra J. Peart and David M. Levy, *The "Vanity of the Philosopher": From Equality to Hierarchy in Postclassical Economics* (Ann Arbor: University of Michigan Press, 2005), 224; Marouf Arif Hasian Jr., *The Rhetoric of Eugenics in Anglo-American Thought* (Athens: University of Georgia Press, 1996), 101–102; Alfred Ploetz, "Neo-Malthusianism and Race Hygiene," in *Problems in Eugenics: The First International Eugenics Congress,* vol. 2 (London: Eugenics Education Society, 1913), 189.

61. Madison Grant, *The Passing of the Great Race: Or, The Racial Basis of European History* (New York: Charles Scribner's Sons, 1916), 46–47.

62. Kevles, *In the Name of Eugenics,* 72; Prescott Hall, "Immigration Restriction and World Eugenics," *Journal of Heredity* 10, no. 3 (1919): 126.

63. On "embrace," see John Torpey, *Invention of the Passport,* 10–11. On Comstock, see Andrea Tone, *Devices and Desires: A History of Contraceptives in*

America (New York: Hill and Wang, 2001), 35–40, 44–45. On nonenforcement, see Noiriel, *The French Melting Pot,* 48–49; and Haller, *Eugenics,* 135–136. On labor and eugenics, see Lucassen, "The Great War," 46, 58; Fitzgerald, *Face of the Nation,* 117. On defining normalcy, see Hacking, *The Taming of Chance,* 3, 6. On Gladstone, see Martin Brookes, *Extreme Measures: The Dark Visions and Bright Ideas of Francis Galton* (New York: Bloomsbury, 2004), 238.

64. Matthew Connelly, "Seeing Beyond the State: The Population Control Movement and the Problem of Sovereignty," *Past and Present* 193, no. 1 (2006): 193–233. Regarding "bottom-up state building" more generally, see Bayly, *Birth of the Modern World,* 243, 248.

2. To Inherit the Earth

1. Ludendorff, *The General Staff and Its Problems,* vol. 1, trans. F. A. Holt (London: Hutchinson and Co., 1920), 201–203, 240 (emphasis in original).

2. National Council of Public Morals, Commission of Inquiry into the Declining Birth-Rate, *Problems of Population and Parenthood* (London: Chapman and Hall, 1920), 73. On Clemenceau, see Michael S. Teitelbaum and Jay M. Winter, *The Fear of Population Decline* (Orlando, FL: Academic Press, 1985), 36. On Russia, see Mark B. Adams, "Eugenics in Russia, 1900–1940," in *The Wellborn Science: Eugenics in Germany, France, Brazil, and Russia,* ed. Mark B. Adams (New York: Oxford University Press, 1990), 176.

3. John T. Noonan, *Contraception: A History of Its Treatment by the Catholic Theologians and Canonists* (Cambridge: Harvard University Press, 1966), 422; James Cardinal Gibbons, "Pastoral Letter of the Archbishop and Bishops of the United States," Vatican Secret Archives, Rome (hereafter VSA), Arch. Deleg. Stati Uniti, Sezione II (Stati Uniti), 209A.

4. Stephen Jay Gould, *The Mismeasure of Man,* rev. ed. (New York: Norton, 1986), 224–226; Richard A. Soloway, *Demography and Degeneration: Eugenics and the Declining Birthrate in Twentieth Century Britain* (Chapel Hill: University of North Carolina Press, 1990), 140–146; William H. Schneider, *Quality and Quantity: The Quest for Biological Regeneration in Twentieth-Century France* (New York: Cambridge University Press, 1990), 130–131.

5. Deborah Dwork, *War Is Good for Babies and Other Young Children: A History of the Infant and Child Welfare Movement in England, 1898–1918* (London: Tavistock, 1987), 208–216. On assisted emigration, see Soloway, *Demography and Degeneration,* 152–153; and A. M. Carr-Saunders, *Population* (London: Oxford University Press, 1925), 88. On other states, see Atina Grossmann, *Reforming Sex: The German Movement for Birth Control and Abortion Reform, 1920–1950* (New York: Oxford University Press, 1995), 8–9; Francis Ronsin, *La Grève des*

Ventres: Propagande néo-malthusienne et baisse de la natalité en France, 19e–20e siècles (Paris: Aubier Montaigne, 1980), 145–148; Victoria de Grazia, *How Fascism Ruled Women: Italy, 1922–1945* (Berkeley: University of California Press, 1992), 43–45.

6. "Extracts from the letter of James Cardinal Gibbons, May 5, 1919," VSA, A.E.S., periodo IV, Stati Uniti, 172 PO, fasc. 17; Administrative Committee of the National Catholic Welfare Council, "Report to His Holiness Pope Pius XI," 1922, VSA, A.E.S., periodo IV, Stati Uniti, 172 PO, fasc. 15; "Le Conferenze generali dell'Episcopato," December 1925, Arch. Deleg. Stati Uniti, Sezione II (Stati Uniti), 209/5.

7. Mark H. Haller, *Eugenics: Hereditarian Attitudes in American Thought* (New Brunswick: Rutgers University Press, 1963), 139–141. On migration, see John Higham, *Strangers in the Land: Patterns of American Nativism, 1860–1925* (New York: Atheneum, 1955, 1963), 316–324; Henry Chilton to William Philips, September 19, 1923, U.S. National Archives, Archives 1, Washington, DC, RG 59, Passport Office, Decimal File, 1910–1949, box 320, 130 (Armenia to Italy), file on "Hindus." On bishops, see Administrative Committee of the National Catholic Welfare Council, "Report to His Holiness Pope Pius XI," 1922.

8. "China," Vera Houghton, July 12, 1950, International Planned Parenthood Federation Archives, London (hereafter IPPFA), series B, reel 149, frames 2399–2409; Kingsley Davis, *The Population of India and Pakistan* (New York: Russell and Russell, 1951, 1968), 237; "Report of the City High Mortality Committee," July 25, 1927, British Library, India Office Records (hereafter IOR), V/26/810/2; Wendell Cleland, *The Population Problem in Egypt: A Study of Population Trends and Conditions in Modern Egypt* (Lancaster, PA: Science Press, 1936), 52–54; Yuehtsen Juliette Chung, *The Struggle for National Survival: Eugenics in Sino-Japanese Contexts* (New York: Routledge, 2002), 70–71; Peter M. Beattie, *The Tribute of Blood: Army, Honor, Race, and Nation in Brazil, 1864–1945* (Durham, NC: Duke University Press, 2001), 254; Nancy Stepan, *Beginnings of Brazilian Science: Oswaldo Cruz, Medical Research and Policy, 1890–1920* (New York: Science History Publications, 1976), 88–91, 112–115, 137–144.

9. *Buck v. Bell*, 274 U.S. 200 (1927), laws.findlaw.com/us/274/200.htm. On Darwin, see "Want More Babies in Best Families," *New York Times*, September 25, 1921, 16. On Davenport, see Allan Chase, *The Legacy of Malthus: The Social Costs of the New Scientific Racism* (New York: Knopf, 1977), 161–162. On social bodies, see David G. Horn, *Social Bodies: Science, Reproduction, and Italian Modernity* (Princeton: Princeton University Press, 1994), 3–5.

10. Rosika Schwimmer, "Birth Control or War," in *Problems of Overpopulation*, vol. 2, *The Sixth International Neo-Malthusian and Birth Control Conference*, ed. Margaret Sanger (New York: American Birth Control League, 1926), 136–137; *Casti Connubii*, Encyclical of Pope Pius XI on Christian Marriage, December 31,

1930, www.vatican.va/holy_father/pius_xi/encyclicals/index.htm; Sanger, *Women and the New Race* (New York: Brentano's, 1920), 94, and see also 151ff.

11. Harold Cox, *The Problem of Population* (New York: G. P. Putnam's Sons, 1923), 85; E. A. Ross, *Standing Room Only?* (New York: The Century, 1927), 93–98, 341.

12. Margaret Sanger, *An Autobiography* (New York: W. W. Norton, 1938), 89–90, 107–108; and Sanger, *My Fight for Birth Control* (New York: Farrar and Rinehart, 1931), 83.

13. David M. Kennedy, *Birth Control in America: The Career of Margaret Sanger* (New Haven: Yale University Press, 1970), 22–26, 79–87. On Japan and Shanghai, see Sanger to Juliet Rublee, n.d., Margaret Sanger Papers Microfilm Edition (hereafter MSP): Collected Documents Series, University Publications of America (hereafter UPA), reel 2, "Brief Impressions of Japan," diary entries ca. April 10 and May 1, 1922, MSP, Smith College Collections, UPA, reel 70, frames 110–120, 147–155.

14. Sanger, *My Fight for Birth Control,* 109–116; Sanger to S. Adolphus Knopf, November 11, 1924, MSP, Collected Documents Series, UPA, reel 3, frames 220–221.

15. Journal excerpt, November 3, 1914, in *The Selected Papers of Margaret Sanger,* vol. 1, *The Woman Rebel, 1900–1928,* ed. Esther Katz (Urbana: University of Illinois Press, 2003), 98. On her determination to travel, see Ellen Chesler, *Woman of Valor: Margaret Sanger and the Birth Control Movement in America* (New York: Simon and Schuster, 1992), 449–450. Sanger suffered chronic ailments of the sinuses and gallbladder.

16. James Reed, *From Private Vice to Public Virtue: The Birth Control Movement and American Society* (New York: Basic Books, 1978), 95; Chung, *Struggle for National Survival,* 133.

17. June Rose, *Marie Stopes and the Sexual Revolution* (London: Faber and Faber, 1992), 111, 144–145; Helen M. Hopper, *A New Woman of Japan: A Political Biography of Katô Shidzue* (Boulder, CO: Westview Press, 1996), 20–22, 56–58; Doris H. Linder, *Crusader for Sex Education: Elise Ottesen-Jensen (1886–1973) in Scandinavia and on the International Scene* (Lanham, MD: University Press of America, 1996), 68–70, 109–111.

18. Sunniva Engh, "Population Control in the 20th Century: Scandinavian Aid to the Indian Family Planning Programme" (Ph.D. diss., University of Oxford, 2005), 73; *Birth Control News,* no. 1, May 1922; Sanger to Dennett, December 16, 1921, MSP, Smith College Collections, UPA, reel 2, frames 77–78.

19. Marouf Arif Hasian Jr., *The Rhetoric of Eugenics in Anglo-American Thought* (Athens: University of Georgia Press, 1996), 101–103; David J. Garrow, *Liberty and Sexuality: The Right to Privacy and the Making of Roe v. Wade* (Berkeley: University of California, 1998), 17.

20. G. M. Chambers to Humphrey Roe, May 3, 1922, British Library, London, Stopes Papers, add. 58645, Eugenics Society; Davenport to Sanger, October 10 and October 21, 1921, MSP, Collected Documents Series, UPA, reel 1, frames 846, 861–862; Leslie Woodcock Tentler, *Catholics and Contraception: An American History* (Ithaca: Cornell University Press, 2004), 53–54.

21. "Want More Babies in Best Families," and "Eugenists Dread Tainted Aliens," *New York Times,* September 25, 1921, 16, 25; Barry Alan Mehler, "A History of the American Eugenics Society, 1921–1940" (Ph.D. diss., University of Illinois at Urbana-Champaign, 1988), 36–49. For Johnson and Laughlin, see Higham, *Strangers in the Land,* 313–314; Laughlin to Gosney, August 7, 1923, California Institute of Technology, Institute Archives, Pasadena, E. S. Gosney Papers and Records of the Human Betterment Foundation, box 7, folder 7.2; April 28, 1926, testimony in "The Eugenical Aspects of Deportation," in *The Committee on Immigration and Naturalization* (Washington, DC: U.S. Government Printing Office, 1928), 46.

22. Fumasoni-Biondi to Gasparri, March 27, 1925, VSA, Arch. Deleg. Stati Uniti, Sezione II (Stati Uniti), 224/2; "Minutes of the Seventh Annual Meeting of the American Hierarchy," September 1925, VSA, Arch. Deleg. Stati Uniti, Sezione II (Stati Uniti), 209/5; Sharon Mara Leon, "Beyond Birth Control: Catholic Responses to the Eugenics Movement in the United States, 1900–1950" (Ph.D. diss., University of Minnesota, 2004), 60, 70–77; Edwin Black, *War against the Weak: Eugenics and America's Campaign to Create a Master Race* (New York: Four Walls Eight Windows, 2003), 138–139.

23. Edward M. East, *Mankind at the Crossroads* (New York: Charles Scribner's Sons, 1923), 299; Kennedy, *Birth Control in America,* 198.

24. Robert J. Richards, *Darwin and the Emergence of Evolutionary Theories of Mind and Behavior* (Chicago: University of Chicago Press, 1989), 514–516.

25. Lothrop Stoddard, *The Rising Tide of Color against White World-Supremacy* (New York: Charles Scribner's Sons, 1920), 8–9.

26. Ibid., 259–260, 270; U.S. House Committee on Immigration and Naturalization, *Restriction of Immigration,* 68th Cong., 1st sess., 1924, 608–620.

27. East, *Mankind at the Crossroads,* 120, 347. See also O. E. Baker, "The Potential Supply of Wheat," *Economic Geography* 1 (March 1925): 17, 51.

28. Harold Wright, *Population* (New York: Harcourt, Brace, and Co., 1923), 123, 142, 176; Carr-Saunders, *Population,* 82–83; Cox, *The Problem of Population,* 85; C. P. Blacker, *Birth Control and the State: A Plea and Forecast* (London: Kegan Paul, 1926), 77–82.

29. Charles Richet, *La Sélection humaine* (Paris: Librairie Félix Alcan, 1919), 166; Georges Dequidt and Georges Forestier, "Les Aspects sanitaires du Problème de l'Immigration en France," *Revue d'Hygiene* 48 (1926): 1022.

30. Paul Demeny, "Ravenstein on Global Carrying Capacity," *Population and*

Development Review 16 (1990): 153–154; Philip M. Hauser, "World Population: Retrospect and Prospect," *Rapid Population Growth: Consequences and Policy Implications,* vol. 2 (Baltimore: Johns Hopkins University Press, 1971), 106.

31. Ross to Sanger, February 12, 1925, MSP, Collected Documents Series, UPA, reel 3, frame 366; see also Ross to Sanger, October 25, 1921, MSP, Collected Documents Series, UPA, reel 1, frame 866; Ross, *Standing Room Only,* 296.

32. Davis, *The Population of India and Pakistan,* 35–36. On public health, see Ira Klein, "Death in India, 1871–1921," *Journal of Asian Studies* 32 (1973): 657; Klein, "Imperialism, Ecology, and Disease: Cholera in India, 1850–1950," *Indian Economic and Social History Review* 31, no. 4 (1994): 501; East, *Mankind at the Crossroads,* 88–90; Wright, *Population,* 66.

33. Arup Maharatna, *The Demography of Famines: An Indian Historical Perspective* (Delhi: Oxford University Press, 1996), 15, 63–67; Ian D. Mills, "Influenza in India during 1918–19," in *India's Historical Demography: Studies in Famine, Disease and Society,* ed. Tim Dyson (London: Curzon Press, 1989), 256; Klein, "Death in India," 643, 656; "Annual Report of the Public Health Commissioner with the Government of India" for the years 1920–1929, esp. 1921, pp. 61–65, 1924, pp. 1–3, and 1926, pp. 207–211, IOR, V/24/3659.

34. Katherine Mayo, *Mother India* (New York: Harcourt, Brace and Co., 1927); Mrinalini Sinha, *Specters of Mother India: The Global Restructuring of an Empire* (Durham, NC: Duke University Press, 2006), 73–81.

35. Kabita Ray, *History of Public Health: Colonial Bengal, 1921–1947* (Calcutta: K. P. Bagchi and Co., 1998), 113–120; Sir Ronald Ross, "Malaria-Control in Malaya and Assam: A Visit of Inspection, 1926–7," June 13, 1927, IOR, L/E/7, file 1514; "Report of the Malaria Commission on Its Study Tour in India," 1930, IOR, L/E/7, file 1509; Ross, *Standing Room Only,* 296 (my emphasis).

36. *Buck v. Bell,* 274 U.S. 200 (1927); "The American Eugenics Society," dated "1927–2," Bentley Library, Ann Arbor, Michigan, C. C. Little Papers, box 7, folder 8.

37. East, *Mankind at the Crossroads,* 111–112; Pearl, "Some Eugenic Aspects of the Problem of Population," in *Eugenics in Race and State,* ed. Charles B. Davenport et al. (Baltimore: Williams and Wilkins, 1923), 214; Pearl, "The First International Eugenics Congress," *Science* 36, no. 926 (1912): 395; Pearl, *The Biology of Population Growth* (New York: Knopf, 1925), 171.

38. Pearl to East, May 7, 1925, American Philosophical Society, Philadelphia (hereafter APS), Raymond Pearl Papers, series I, Correspondence, Da-Ea, East, E., file #5; Elazar Barkan, *The Retreat of Scientific Racism: Changing Concepts of Race in Britain and the United States between the World Wars* (Cambridge: Cambridge University Press, 1992), 210–220.

39. Pearl, "The Biology of Superiority," *American Mercury* 2 (1927): 257–266. Garland Allen is particularly insightful on this point—see "Old Wine in New

Bottles: From Eugenics to Population Control in the Work of Raymond Pearl," in *The Expansion of American Biology,* ed. Jane Maienschein, Keith R. Benson, and Ronald Rainger (New Brunswick: Rutgers University Press, 1991): 232, 242–244, 256.

40. East, *Mankind at the Crossroads,* 299, 303; Laura Briggs, *Reproducing Empire: Race, Sex, Science, and U.S. Imperialism in Puerto Rico* (Berkeley: University of California Press, 2002), 105; Dickinson to Davis, November 11, 1926, Rockefeller Archive Center, Tarrytown, NY, Bureau of Social Hygiene Papers, series 3, box 7, "Committee on Maternal Health 1926–28."

41. Nancy Leys Stepan, *"The Hour of Eugenics": Race, Gender, and Nation in Latin America* (Ithaca: Cornell University Press, 1991), 174–176; Chung, *Struggle for National Survival,* 14, 27, 49, 53, 90–91, 102–103; Sumiko Otsubo, "Eugenics in Imperial Japan: Some Ironies of Modernity, 1883–1945" (Ph.D. diss., Ohio State University, 1998), 215–218; Chandrasekhar to Sanger, March 27, 1952, IPPFA, series B, reel 717, frames 1491–1492.

42. Minutes of the Permanent International Eugenics Committee, October 18, 1919, Contemporary Medical Archives Center, Wellcome Institute for the History of Medicine, London (hereafter CMAC), Eugenics Society Papers (hereafter SA/EUG), box 33, D.109. On the Indian Eugenics Society, see Little to Lady Chambers, October 20, 1921, and Darwin to Govaertz, December 13, 1921, CMAC, SA/EUG, box 48, E.8. On excluding non-Europeans, see also Stefan Kühl, *Die Internationale der Rassisten: Aufstieg und Niedergang der internationalen Bewegung für Eugenik und Rassenhygiene im 20. Jahrhundert* (Frankfurt: Campus, 1997), 73–74.

43. Stepan, *"The Hour of Eugenics,"* 174–182. Ramos's address is in APS, Charles Davenport Papers, Pan-American Conference, folder 4.

44. Stepan, *"The Hour of Eugenics,"* 177, 181; Mehler, "A History of the American Eugenics Society," 215–216; Otsubo, "Eugenics in Imperial Japan," 256–258; Kühl, *Die Internationale der Rassisten,* 86.

45. Kühl, *Die Internationale der Rassisten,* 13–15. On whitening, see Chung, *Struggle for National Survival,* 13, 61–64; Stepan, *"The Hour of Eugenics,"* 137–138, 147–148, 151; Beattie, *The Tribute of Blood,* 257–258; Thomas E. Skidmore, "Racial Ideas and Social Policy in Brazil, 1870–1940," in *The Idea of Race in Latin America, 1870–1940,* ed. Richard Graham (Austin: University of Texas Press, 1990), 7–36. On miscegenation, see Francesco Cassata, *Molti, Sani, e Forti: L'Eugenetica in Italia* (Torino: Bollati Boringhieri, 2006), 155–161; Ross to Sister Clement, April 24, 1913, State Historical Society of Wisconsin, Edward A. Ross Papers, Microfilm Edition, 1982, reel 7, frame 275; Gini to Darwin, August 1, 1919, and Darwin to Gini, May 7, 1920, CMAC, SA/EUG, box 11, C.123, Corrado Gini; "Minutes of the Meeting of the General Committee of the Second

International Eugenics Congress," April 10, 1920, CMAC, SA/EUG, box 33, D.109, International Eugenics Conferences.

46. Leon Whitney to C. C. Little, April 18, 1928, Bentley Library, Ann Arbor, MI, C. C. Little Papers, box 7, folder 8; Chung, *Struggle for National Survival,* 111–112.

47. Davenport to Hodson, June 26, 1928, APS, Charles B. Davenport Papers, International Federation of Eugenic Organizations, C. B. S. Hodson, folder 1, 1928, and membership files. On the Third Congress and Latin America, see Stepan, *"The Hour of Eugenics,"* 114–122, 128–133, 163–169; Deborah Anne Barrett, "Reproducing Persons as a Global Concern: The Making of an Institution" (Ph.D. diss., Stanford University, 1995), 156ff.

48. Dorothy Bocker, "Clinical Research Notes," attached to Sanger to Pearl, May 24, 1923, APS, Raymond Pearl Papers, series I, Correspondence, Ru-San, Sanger file #2. On courting eugenicists, see Sanger to Fosdick, October 19, 1925, MSP, Collected Documents Series, UPA, reel 3, frame 675; Dennis Hodgson, "The Ideological Origins of the Population Association of America," *Population and Development Review* 17 (1991): 15–16. On applying for parenthood, see Kennedy, *Birth Control in America,* 117. On the AMA, see Pearl to Sanger, July 30, 1924, APS, Raymond Pearl Papers, series I, Correspondence, Sanger file number 4; Pearl to Sanger, March 12, 1925, MSP, Collected Documents Series, UPA, reel 3, frame 415.

49. Diary Entries from October 6 to October 24, 1924, MSP, Smith College Collections, UPA, reel 70, frames 204ff.

50. Phadke to Sanger, August 20, 1925, frames 616–618; R. D. Karve to Sanger, October 9, 1925, frames 657–658; Sanger to Karve and Sanger to Phadke, November 4, 1925, frames 690–691; Sanger to Tagore, August 2, 1925, frame 610; all in MSP, Collected Documents Series, Subseries 1 (Correspondence), UPA, reel 3.

51. Margaret Sanger, ed., *International Aspects of Birth Control,* vol. 1, *The Sixth International Neo-Malthusian and Birth Control Conference* (New York: American Birth Control League, 1925), 173–240.

52. Pearl, "The Differential Birth Rate," and Cox, "A League of Low Birth-Rate Nations," in *Problems of Overpopulation,* ed. Margaret Sanger, vol. 2, *The Sixth International Neo-Malthusian and Birth Control Conference* (New York: American Birth Control League, 1925), 28–29, 152; Will Durant, "Message to Sixth International Neo-Malthusian and Birth Control Conference," MSP, Smith College Collections, UPA, reel 2, frame 51 (Sanger chose not to include Durant's message in the printed proceedings).

53. Taraknath Das, "The Population Problem in India," in Sanger, ed., *Religious and Ethical Aspects of Birth Control,* vol. 4, *The Sixth International Neo-*

Malthusian and Birth Control Conference (New York: American Birth Control League, 1926), 195–198; Dublin, "The Excesses of Birth Control," in Sanger, *The Problems of Overpopulation,* 180–181.

54. Sanger editorial, May 13, 1925, in Katz, *Selected Papers of Margaret Sanger,* 423–428; East to Sanger, May 15, 1925, MSP, Collected Documents Series, Subseries 1 (Correspondence), UPA, reel 3, frame 498. East originally wrote "a secondary aspect," but crossed that out.

55. On the parenthood license, see "Mrs. Sanger Answers Dr. Dublin," March 31, 1925, MSP, Collected Documents Series, Subseries 1 (Correspondence), UPA, reel 14, frame 239. On Catholic lobbying, see "Memorandum on the Question of Birth Control," September 1933, VSA, Arch. Deleg. Stati Uniti, Sezione II (Stati Uniti), Posiz. 345a. On her leave from the ABCL and Geneva, see Kennedy, *Birth Control in America,* 102–103; and Chesler, *Woman of Valor,* 236.

56. Merry del Val to Fumasoni-Biondi, January 5, 1927, and "Memorandum on Prevalence of Contraceptive Practices among Catholics of the United States," ca. March 1927, both in VSA, Arch. Deleg. Stati Uniti, Sezione II (Stati Uniti), Posiz. 345a

57. Yoshihisa Tak Matsusaka, *The Making of Japanese Manchuria, 1904–1932* (Cambridge: Harvard University Press, 2001), 272–273. On Libyan colonization as response to humiliation of emigrants, see Claudio G. Segre, *Fourth Shore: The Italian Colonization of Libya* (Chicago: University of Chicago Press, 1974), 21–22. On Gini and Mussolini, see Cassata, *Molti, Sani, e Forti,* 144–151; and Horn, *Social Bodies,* 46–47, 79–80.

58. On Sweden and France, see Margaret Sanger, ed., *Proceedings of the World Population Conference* (London: Edward Arnold and Co., 1927), 100, 280; Dequidt and Forestier, "Les Aspects sanitaires," 1003, 1008, 1014, 1018–1021, 1039; Patrick Weil, *Qu'est-ce qu'un Français? Histoire de la nationalité française depuis la révolution* (Paris: Grasset, 2002), 76–79. On Italy, see Horn, *Social Bodies,* 52, 136–137.

59. Skidmore, "Racial Ideas and Social Policy," 23–24; Alan Knight, "Racism, Revolution, and Indigenismo: Mexico, 1910–1940," in Graham, *The Idea of Race,* 96–97; Laughlin, "The Eugenical Aspects of Deportation," 24.

60. "Reunion du Bureau de la Commission d'Emigration," May 7, 1921, International Labour Organization Archives, Geneva (hereafter ILOA), A. Closed Series, E 100/E; Thomas, "International Migration and Its Control," in Sanger, *Proceedings of the World Population Conference,* 270. On Thomas's concerns about immigration debates in France, see his forward to Marcel Paon, *L'Immigration en France* (Paris: Payot, 1926).

61. On Thomas, see minutes, February 17, 1927, ILOA, B, Continuing Series, PO 1000/1/1; and Richard Symonds and Michael Carder, *The United Nations and the Population Question, 1945–1970* (New York: McGraw-Hill, 1973), 14.

On Geneva and How-Martyn, see Sanger, *An Autobiography* (New York: W. W. Norton, 1938), 170, 382.

62. Sanger to Little, March 16, 1927, "The World Population Conference: Preliminary Notice," ca. March 1927, Pearl to Juliet Barrett Rublee, April 16, 1927, and Sanger to Little, May 2, 1927, all in Bentley Library, Ann Arbor, MI, C. C. Little Papers, box 6, folder 26; Tait to the Deputy Director, September 5, 1927, ILOA, B, Continuing Series, PO 1000/1/1; Sanger, *Autobiography,* 379.

63. Symonds and Carder, *The United Nations,* 12–14. On Drummond, see James Barros, *Office without Power: Secretary-General Sir Eric Drummond, 1919–1933* (Oxford: Clarendon Press, 1979), 25–26.

64. Sanger, *Autobiography,* 385–386; "Miscellaneous Notes and Comments by M.S. from Geneva conference in 1927 to 1957," MSP, Smith College Collections, UPA, reel 70, frames 270, 330, 335.

65. How-Martyn to Sanger, n.d., MSP, Smith College Collections, UPA, reel 3, frames 979–980.

66. The papers appear in Sanger, *Proceedings of the World Population Conference.*

67. Thomas to Varlez, August 1, 1927, and Thomas to Legouis, August 6, 1927, both in ILOA, B, Continuing Series, PO 1000/1/1; Thomas, "International Migration," 257–260.

68. Thomas, "International Migration," 262–263.

69. Ibid., 263–265.

70. Henry Pratt Fairchild, *Immigration: A World Movement and Its American Significance* (New York: MacMillan, 1926), 500; Mallet, Opening and Closing Addresses, in Sanger, *Proceedings of the World Population Conference,* 18–19, 356. On belated recognition, see C. P. Blacker, "The International Planned Parenthood Federation: Aspects of Its History," IPPFA, series B, reel 664, frames 418–419; Chesler, *Woman of Valor,* 259.

71. Thomas to Varlez, September 12, 1927, ILOA, B, Continuing Series, PO 1000/1/1. Sanger to Pearl, October 27, 1927, Pearl to Sanger, November 10, 1927, and Pearl to Sanger, January 2, 1928, all in APS, Pearl Papers, Sanger file.

72. Henri Brenier, "Le Congrès de la population mondiale a Genève et la conspiration néo-malthusienne," October 10, 1927, *Correspondant,* 12–13, 16. The controversy is extensively documented in the World Health Organization Archives, Geneva, League Archives, 1928–1932, section 8A, 5638, Maternal Mortality, and, League Archives, 1933–1946, section 8f, 2363/895; Symonds and Carder, *The United Nations,* 25–28. On the Vatican campaign, see Catholic University of American Archives, NCWC, Collection 10, box 116, folder 31; Burke to Cicognani, October 30, 1933, VSA, Arch. Deleg. Stati Uniti, Sezione II (Stati Uniti), Posiz. 345a; Ministero degli Affari Esteri, January 17, 1933, VSA, Arch. Nunz. Italia, b.8, fasc. 1/2762.

73. Ferenczi to Maurette, September 10, 1930, ILOA, B, Continuing Series, PO 2/4/01/2; East to Pearl, October 9, 1928, December 5, 1928, and January 29, 1929, APS, Pearl Papers, series 1, Correspondence, East, files 7–8.

74. The Pearl-Wilson feud is detailed in Sharon E. Kingsland, *Modeling Nature: Episodes in the History of Population Ecology* (Chicago: University of Chicago Press, 1985), 87–91; and Richard Hankinson, "The U.S. Contribution to the Creation of the IUSSP," Population Association of America Archives, Silver Spring, MD, box 4, folder 62.

75. Wilson to Gini, May 2 and July 3, 1930, Population Association of America Archives, Silver Spring, MD, box 4, folder 62.

76. Hankinson, "The U.S. Contribution."

77. Varley to Thomas, April 29, 1927, ILOA, B, Continuing Series, PO 1000/1/1. On Sanger's faith in science, see Margaret Sanger, *The Pivot of Civilization* (New York: Brentano's, 1922), 220–222, 242.

78. Sanger, *Proceedings of the World Population Conference,* 85, 99, 102.

79. McGuire to Moore, August 17, 1929, VSA, Arch. Deleg. Stati Uniti, Sezione II (Stati Uniti), Posiz. 345a. The study, intended for confidential use by the hierarchy, is not in the files. On the secretive McGuire, who also helped set up the Georgetown School of Foreign Service, see www.tboyle.net/University/SFS_Founding.html.

80. Matthew Connelly, "To Inherit the Earth: Imagining World Population, from the Yellow Peril to the Population Bomb," *Journal of Global History* 1 (2006): 306–307. On worst enemies, see Sanger, *Proceedings of the World Population Conference,* 360.

81. Thomas, "International Migration," 269; Huxley, "Too Many People Seen as Real World Danger," *New York Times,* October 16, 1927, 6.

82. How-Martyn to Sanger, September 26, 1927, MSP, Smith College Collections, UPA, reel 3, frames 933–934.

3. POPULATIONS AT WAR

1. Margaret Sanger and Hannah M. Stone, eds., *The Practice of Contraception: An International Symposium and Survey—Proceedings of the Seventh International Birth Control Conference* (Baltimore: Williams and Wilkins, 1931), xiii.

2. Atina Grossmann, *Reforming Sex: The German Movement for Birth Control and Abortion Reform, 1920–1950* (New York: Oxford University Press, 1995), 38–44; "Fifth Congress of the World League for Sexual Reform on Scientific Basis," 1932, Sophia Smith Collection, Smith College (hereafter SSC), PPFA I, series 7: Subject Files, box 83, file 10.

3. Robert J. Richards, *Darwin and the Emergence of Evolutionary Theories of Mind and Behavior* (Chicago: University of Chicago Press, 1987), 514–516. On

Pearl, see "Fondation, Oeuvre et Historique de l'Union," Archives Nationales, Paris, Georges Mauco papers, AP577, box 5.

4. *Atti del Congresso internazionale per gli studi sulla popolazione (Roma, 7–10 Settembre, 1931) 1: Sezione di Storia*, p. 37, and *X: Sezione di Metodologia*, p. 292 (Roma: Istituto Poligrafico dello Stato, 1933–34).

5. Henri Decugis, *Le Destin des Races Blanches* (Paris: Librairie Félix Alcan, 1935); Enid Charles, *The Twilight of Parenthood* (New York: W. W. Norton, 1934); Friedrich Burgdörfer, *Sterben die Weissen Völker? Die Zukunft der weissen und farbigen Völker im Lichte der Biologische Statistik* (Munich: Georg D. W. Callwey Verlag, 1934); Yasumaru Shimojo, "An Inquiry concerning the Numerical Evolution of Population," and J. Van Gelderen, "The Numerical Evolution of Population with Particular Reference to the Population of Java," *Atti del Congresso internazionale, 1: Sezione di Storia*, 287, 312.

6. Mark Mazower, *Dark Continent: Europe's Twentieth Century* (New York: Vintage Books, 2000), 76–103.

7. Louise Young, *Japan's Total Empire: Manchuria and the Culture of Wartime Imperialism* (Berkeley: University of California Press, 1998), 326–328; Sabine Frühstück, *Colonizing Sex: Sexology and Social Control in Modern Japan* (Berkeley: University of California Press, 2003), 157–163. On war of populations, see Yuehtsen Juliette Chung, *The Struggle for National Survival: Eugenics in Sino-Japanese Contexts* (New York: Routledge, 2002), 141–143. On settlement, see John W. Dower, *War without Mercy: Race and Power in the Pacific War* (New York: Pantheon, 1986), 271–278.

8. Victoria de Grazia, *How Fascism Ruled Women: Italy, 1922–1945* (Berkeley: University of California Press, 1992), 53–56; David G. Horn, *Social Bodies: Science, Reproduction, and Italian Modernity* (Princeton: Princeton University Press, 1994), 59–60, 79–88. On colonization, see Claudio G. Segre, *Fourth Shore: The Italian Colonization of Libya* (Chicago: University of Chicago Press, 1974), 102–107. For examples of Mussolini's obsessive interest in national demographic statistics, see Archivio Centrale dello Stato, Rome, Segreteria Particolare del Duce, CO 1170–1171.

9. David L. Hoffman, *Stalinist Values: The Cultural Norms of Soviet Modernity, 1917–1941* (Ithaca: Cornell University Press, 2003), 98–102; Amir Weiner, "Nature, Nurture, and Memory in a Socialist Utopia: Delineating the Soviet Socio-Ethnic Body in the Age of Socialism," *American Historical Review* 104 (1999): 1114–1155. On extraction, see Peter Holquist, "To Count, to Extract, and to Exterminate: Population Statistics and Population Politics in Late Imperial and Soviet Russia," in *A State of Nations: Empire and Nation-Making in the Age of Lenin and Stalin*, ed. Ronald Grigor Suny and Terry Martin (New York: Oxford University Press, 2001), 133. On deportations, see Terry Martin, "Stalinist Forced Relocation Policies: Patterns, Causes, Consequences," in *Demography and National Security,*

ed. Myron Weiner and Sharon Stanton Russell (New York: Berghahn Books, 2001), 309, 315, 321–322.

10. Paul Weindling, *Health, Race, and German Politics between National Unification and Nazism, 1870–1945* (New York: Cambridge University Press, 1989), 489, 494–497. On Hitler, see Hermann Rauschning, *The Voice of Destruction* (New York: G. P. Putnam's Sons, 1940), 136–138; and Hervé Le Bras, *Le Sol et le Sang* (La Tour d'Aigues: Editions de l'Aube, 1991), 52–53.

11. Gunnar Broberg and Nils Roll-Hansen, eds., *Eugenics and the Welfare State: Sterilization Policy in Denmark, Sweden, Norway, and Finland* (East Lansing: Michigan State University Press, 1996). On Mexico and Brazil, see Nancy Leys Stepan, *The Hour of Eugenics: Race, Gender, and Nation in Latin America* (Ithaca: Cornell University Press, 1991), 125–133, 165–166. On France, see Francis Ronsin, *La Grève des Ventres: Propagande néo-malthusienne et baisse de la natalité française, 19e–20e siècles* (Paris: Aubier Montaigne, 1980), 210–211; William H. Schneider, *Quality and Quantity: The Quest for Biological Regeneration in Twentieth Century France* (New York: Cambridge University Press, 1990), chaps. 7–9; Francine Muel-Dreyfus, *Vichy and the Eternal Feminine: A Contribution to a Political Sociology of Gender,* trans. Kathleen A. Johnson (Durham, NC: Duke University Press, 2001), 284; Susan Pedersen, *Family, Dependence, and the Origins of the Welfare State: Britain and France, 1914–1945* (New York: Cambridge University Press, 1993), 401–408.

12. Michael S. Teitelbaum and Jay M. Winter, *The Fear of Population Decline* (Orlando, FL: Academic Press, 1985), 60–62; Allan Carlson, "The Roles of Alva and Gunnar Myrdal in the Development of a Social Democratic Response to Europe's 'Population Crisis'" (Ph.D. diss., Ohio University, 1990), 447–452.

13. Linda Gordon, *Woman's Body, Woman's Right: Birth Control in America,* rev. ed. (1974; New York: Penguin Books, 1990), 337–343.

14. On effect of the Depression, see Schneider, *Quality and Quantity,* 186; Richard A. Soloway, *Demography and Degeneration: Eugenics and the Declining Birthrate in Twentieth Century Britain* (Chapel Hill: University of North Carolina Press, 1990), 193–194. On the United States, see Adele Clarke, *Disciplining Reproduction: Modernity, American Life Sciences, and "The Problems of Sex"* (Berkeley: University of California Press, 1998), 173; Andrea Tone, *Devices and Desires: A History of Contraceptives in America* (New York: Hill and Wang, 2001), 151.

15. Ellen Chesler, *Woman of Valor: Margaret Sanger and the Birth Control Movement in America* (New York: Simon and Schuster, 1992), 342–343; Imré Ferenczi, "La crise des migrations et de la natalité," December 1, 1931, International Labor Organization Archives, Geneva, B. PO 2/01. On Keynes and Dublin, see Michael Teitelbaum and Jay Winter, *The Fear of Population Decline* (Orlando, FL: Academic Press, 1985), 60; David M. Kennedy, *Birth Control in America: The Career of Margaret Sanger* (New Haven: Yale University Press, 1970), 237–238.

16. Topping memorandum, "New Birth Control Magazine," March 27, 1931, Rockefeller Archive Center, Tarrytown, NY (hereafter RAC), Bureau of Social Hygiene Papers, series 3, box 7, file 166.

17. On deferring to Fairchild, see James Reed, *From Private Vice to Public Virtue: The Birth Control Movement and American Society since 1830* (New York: Basic Books, 1978), 204. On the conference, see "Preliminary Conference on a Population Association for the United States," December 15, 1930; "Second Conference of the Population Association of America," May 7, 1931; Topping memo to Harrison, June 6, 1931, all in RAC, Bureau of Social Hygiene Papers, series 3, box 9, "Population Conference 1931"; "Population Association of America, Minutes of Second Annual Meeting," May 12, 1933, RAC, Bureau of Social Hygiene Papers, series 3, box 9, "Population 1931–33."

18. Anders S. Lunde, "The Beginning of the Population Association of America," *Population Index* 47, no. 3 (1981): 481; Frank Notestein, "Reminiscences: The Role of Foundations, of the Population Association of America, Princeton University and of the United Nations in Fostering American Interest in Population Problems," *Milbank Memorial Fund Quarterly* 49, no. 4 (1971): 70–71.

19. Chesler, *Woman of Valor,* 355–356; Soloway, *Demography and Degeneration,* 193; "The Activities of the Birth Control International Information Centre," ca. March 1934, International Planned Parenthood Federation Archives, London (hereafter IPPFA), series B, reel 214, frames 1012–1018.

20. *Casti Connubii,* December 31, 1930, www.vatican.va/holy_father/pius_xi/encyclicals/index.htm.

21. Pacelli memorandum on relations with Germany, January 1936, and accompanying note, Pacelli to Diego von Bergen, Vatican Secret Archives, Rome (hereafter VSA), A.E.S., periodo IV, Germania, 692 P.O., fasc. 260; "Ancora dell'eugenetica et la morale cattolica," *Osservatore Romano,* October 6, 1933.

22. Pizzardo to Viollet, December 14, 1936, VSA, A.E.S., periodo IV, Francia, 767 P.O., fasc. 322; "A Statement on the Present Crisis," Administrative Committee of the NCWC, June 1, 1933, A.E.S., periodo IV, Stati Uniti, 230 P.O., fasc. 54.

23. How-Martyn to *New Generation,* July 1, 1932, IPPFA, series A, reel 10, Malthusian League, "The New Generation"; Horder in Michael Fielding, ed., *Birth Control in Asia: A Report of a Conference Held at the London School of Hygiene and Tropical Medicine November 24–25, 1933* (London: Birth Control International Information Center, 1935), 11; "An Appeal for Equality of Knowledge," ca. January 1931, IPPFA, series B, reel 214, frames 1012–1018.

24. Sanger to How-Martyn, March 4, 1936, The Margaret Sanger Papers Microfilm Edition (hereafter MSP), Collected Documents Series, University Publications of America (hereafter UPA), reel 6, frames 80–85, and reel 5, frames 866–867; Grossmann, *Reforming Sex,* 51–56; Janice R. MacKinnon and Stephen R.

MacKinnon, *Agnes Smedley: The Life and Times of an American Radical* (Berkeley: University of California Press, 1988), 114–115, 138–146; Chesler, *Woman of Valor,* 358–361.

25. Marjorie Martin to How-Martyn, February 26, 1932, and How-Martyn to Marjorie Martin, November 9, 1932, IPPFA, series B, reel 214, frames 480 and 556–557; "Either Convert Me or Be Converted: Gandhiji's Debate with a British Lady," *Sunday Times* (Madras), February 10, 1935.

26. B. L. Raina, *Planning Family in India: Prevedic Times to Early 1950s* (New Delhi: Commonwealth, 1990); James Z. Lee and Wang Feng, *One Quarter of Humanity: Malthusian Mythology and Chinese Realities, 1700–2000* (Cambridge: Harvard University Press, 1999).

27. On Mysore, see Sanjam Ahluwalia, "Controlling Births, Policing Sexuality: A History of Birth Control in Colonial India, 1877–1946" (Ph.D. diss., University of Cincinnati, 2000), 29; Frank Dikötter, "Race Culture: Recent Perspectives on the History of Eugenics," *American Historical Review* 103 (1998): 467–478; How-Martyn to Subha Rao, January 12, 1930, IPPFA, series B, reel 710, frames 1961–1962; and "Birth Control Clinics," June 11, 1930, IPPFA, series B, reel 710, frame 1869. On Yang, see Chung, *Struggle for National Survival,* 124–126. On the United States, see Topping, "Neglect of Birth Control and Eugenics at White House Conference, November 19–22, 1930," January 6, 1931, RAC, Bureau of Social Hygiene Papers, series 3, box 9, "Birth Control 1931–32."

28. Chung, *Struggle for National Survival,* 111, 119–120. On Hiratsuka Raichō and Myrdal, see Sumiko Otsubo, "Eugenics in Imperial Japan: Some Ironies of Modernity, 1883–1945" (Ph.D. diss., Ohio State University, 1998), 143–210.

29. Ahluwalia, "Controlling Births, Policing Sexualities," 95–97. On car and driver, see "Brief Impressions of Japan," Sanger diary entry ca. April 10, 1922, MSP, Smith College Collections, UPA, reel 70, frames 110–120.

30. How-Martyn to Sanger, February 6, 1935, MSP, Smith College Collections, UPA, reel 9, frames 616–620. On Indian cities, see Bernard S. Cohn, *An Anthropologist among Historians and Other Essays* (Delhi: Oxford University Press, 1987), 234–235. On Ehrlich, see Ehrlich, *The Population Bomb* (New York: Ballantine Books, 1968), 15–16.

31. Chesler, *Woman of Valor,* 560; Wattal, *The Population Problem in India: A Census Study* (Bombay: Bennet, Coleman and Co., 1934); Ranadive, *Population Problem of India* (Calcutta: Longmans, Green, and Co., 1930); Mukerjee (variant spelling), *Food Planning for Four Hundred Millions* (London: Macmillan and Co., 1938). On the journal *Marriage Hygiene,* see Barbara N. Ramusack, "Embattled Advocates: The Debates on Birth Control in India, 1920–1940," *Journal of Women's History* 1, no. 2 (1989): 45.

32. David Arnold, "Colonial Medicine in Transition: Medical Research in India, 1910–1947," *South Asia Research* 14, no. 1 (Spring 1994): 20; John Megaw, "Population and Health in India: The Real Problem," *Journal of the East India Association* 25, no. 2 (1934): 153–154.

33. Kingsley Davis, *The Population of India and Pakistan* (1951; New York: Russell and Russell, 1968), 27, 35–36; Ira Klein, "Population Growth and Mortality, Part II," *Indian Economic and Social History Review* 27, no. 1 (1990): 33–65; Sumit Guha, "Mortality in Early 20th Century India: A Preliminary Enquiry," *Indian Economic and Social History Review* 28, no. 4 (1991): 385–387. On public health officers, see Arnold, "Colonial Medicine," 18. On persistent poor health, see "Note on the Problem of Malnutrition in India," ca. 1935, British Library, London, India Office Records (hereafter IOR), L/E/9/265; *Report of the Health Survey and Development Committee,* vol. 1, *Survey* (Delhi: Government of India Press, 1946), 45–49.

34. Roger Jeffery, *The Politics of Health in India* (Berkeley: University of California Press, 1988), 68–73, 92–102; Arnold, "Colonial Medicine," 25–30.

35. Mrinalini Sinha, *Specters of Mother India: The Global Restructuring of an Empire* (Durham, NC: Duke University Press, 2006), 152ff.

36. This and subsequent quotations are from "Resolution *Re* Steps to Check the Increase in Population," March 18, 1935, *Council of State Debates: Official Report,* 1936, vol. 1, no. 15, IOR, V/9/244.

37. Megaw, "Population and Health in India," 161, 164, 167–168.

38. Ramusack, "Embattled Advocates," 43–44; Ahluwalia, "Controlling Births," 154–158.

39. Ahluwalia, "Controlling Births," 36–38, 41–43, 46–47, 53–55; and see also Fielding, *Birth Control in Asia,* 21–22.

40. Ahluwalia, "Controlling Births," 88–89.

41. Megaw, "Population and Health in India," 157; Wendell Cleland, *The Population Problem in Egypt: A Study of Population Trends and Conditions in Modern Egypt* (Lancaster, PA: Science Press, 1936), 109–111.

42. On the role of the BCIIC organizing conference, see Susanne Klausen, "From Imperial Feminist 'Rescue Work' to Population Control: Edith How-Martyn, Margaret Sanger and the Birth Control International Information Centre, 1930–1940," paper presented at conference on "Gender, Race, and Abolitionism," March 2006, London. Fielding, *Birth Control in Asia,* 52–53, 80; "Mrs. How-Martyn in Poona," *The Mahratta,* March 10, 1935; "Social Reformer" letters to the *Times of India,* January 2 and January 25, 1937.

43. Fielding, *Birth Control in Asia,* 65.

44. Ibid., 73, 92.

45. On the Committee on Maternal Health, see Dickinson to Davis, Novem-

ber 11, 1926, RAC, Bureau of Social Hygiene Papers, series 3, box 7, "Committee on Maternal Health 1926–28," and Topping memo, October 18, 1932, RAC, Bureau of Social Hygiene Papers, series 3, box 9, "Birth Control 1931–32." On the future of birth control and Voge, see Clarke, *Disciplining Reproduction,* 175–176, 187–188, 198–199. On vaccine studies, see Judith Richter, "Beyond Control: About Antifertility 'Vaccines,' Pregnancy Epidemics, and Abuse," in *Power and Decision: The Social Control of Reproduction,* ed. Gita Sen and Rachel C. Snow (Cambridge: Harvard University Press, 1994), 209–211. For Blacker appeal, see memo to the Eugenics Society Council, January 1, 1935, Contemporary Medical Archives Center, Wellcome Institute for the History of Medicine, London (hereafter CMAC), C. P. Blacker Papers, box 10, B.8, "Memos for Eugenics Society by CPB."

46. Sanger letter to How-Martyn, January 10, 1932, MSP, Collected Documents Series, UPA, reel 5, frames 314–321. On the Council, see "Meeting of the Committees of the American Birth Control League and the American Eugenics Society," November 3, 1933; "Conference on Coordination Called by the Population Association of America," February 15, 1934; and "Population Association of America: Minutes of the Third Annual Meeting," May 11, 1934; all in Population Association of America Archives, Silver Spring, MD, box 7, folder 126.

47. "Conference on Coordination," February 15, 1934, both in Population Association of America Archives, Silver Spring, MD, box 7, folder 126; "Memorandum on the Question of Birth Control," September 1933, VSA, Arch. Deleg. Stati Uniti, Sezione II (Stati Uniti), Posiz. 345a.

48. Cicognani to Pacelli, June 15, 1933, VSA, A.E.S., periodo IV, Stati Uniti, 230 P.O., fasc. 53. On Sanger and Roosevelt, see Kennedy, *Birth Control in America,* 224–240; and Chesler, *Woman of Valor,* 339–340, 352–353.

49. Sanger to How-Martyn, April 1, 1935, MSP, Library of Congress Microfilm Collection, reel 17, frames 734–735; "Memorandum on the Question of Birth Control," September 1933; "The Activities of the Birth Control International Information Center," ca. March 1, 1934, IPPFA, series B, reel 214, frames 1012–1018. On Nehru, see Raina, *Planning Family in India,* 151. On the Middle East tour, see How-Martyn's reports to Sanger, February 14 and 21, 1934, in MSP, Library of Congress Microfilm Collection, reel 20, frames 396–407. On her India tour, see BCIIC *Newsletter* no. 4, June 1935, CMAC, Family Planning Association Papers (hereafter SA/FPA), box 335.

50. How-Martyn to Marjorie Martin, March 31, 1933, IPPFA, series B, reel 214, frame 452; "Indian Dinner in Honour of Edith How-Martyn," November 19, 1934, IPPFA, series B, reel 214, frames 783–786; Sanger to Milbank, August 19, 1935, MSP, Library of Congress Microfilm Collection, reel 61, frame 707B.

51. Sanger to Blacker, November 14, 1935, CMAC, Eugenics Society Papers

(hereafter SA/EUG), box 26, D.14; "Statement by Margaret Sanger before Her Birth Control Campaign in the East," MSP, Library of Congress Microfilm Collection, reel 130, frames 587–589. Blacker endorsed her application, though the Eugenics Society made only a modest contribution to the BCIIC.

52. "News from Margaret Sanger," ca. November 1935, MSP, Collected Documents Series, subseries 1 (Correspondence), UPA, reel 5, frames 984–987; Huxley tribute, October 17, 1935, IPPFA, series B, reel 214, frames 321–327; Margaret Sanger, *An Autobiography* (New York: W. W. Norton, 1938), 463–464.

53. Sanger, *An Autobiography,* 465–466; Chesler, *Woman of Valor,* 361–364, 561.

54. How-Martyn to Sanger, February 6, 1935, MSP, Smith College Collections, UPA, reel 9, frames 616–620; "Either Convert Me or Be Converted"; Ahluwalia, "Controlling Births," 115–125; Cousins to Sanger, February 11, 1935, MSP, Library of Congress Microfilm Collection, reel 17, frame 718.

55. Margaret Sanger, "Address at Union of East and West," May 21, 1936, MSP, Smith College Collections, UPA, reel 71, frames 835–851; Ramusack, "Embattled Advocates," 50; Ahluwalia, "Controlling Births," 113, 118–121.

56. Sanger carefully transcribed this passage from Gandhi's writings into her journal, MSP, Smith College Collections, UPA, reel 70, frame 402.

57. Sanger to Newfield, December 4, 1935, MSP, Library of Congress Microfilm Collection, reel 15, frame 976; Chesler, *Woman of Valor,* 363–364.

58. How-Martyn to Gerda Guy, March 12, 1936, IPPFA, series B, reel 214, frames 865–869; Sanger to Kabanova, September 18, 1936, MSP, Library of Congress Microfilm Collection, reel 20, frames 704–705; and Sanger to S. W. Lee, October 21, 1936, MSP, Smith College Collections, UPA, reel 11, frames 721–722. On DeVilbiss, see Chesler, *Woman of Valor,* 379–380; and Ahluwalia, "Controlling Births," 105. On clinics, see How-Martyn, "Birth Control in India: Third Tour 1936–37," April 1937, IOR, Private Papers, European Manuscripts, Eileen Palmer Collection, D1182, file 5. On subsequent animal testing, see Sanger to Pillay, January 24, 1937, MSP, Library of Congress Microfilm Collection, reel 12, frames 831–832.

59. How-Martyn to Newfield, August 30, 1935, Harry Guy to How-Martyn, November 30, 1935, Gerda Guy to Sanger, December 11, 1935, and Harry Guy to Sanger, December 11, 1935, IPPFA, series B, reel 214, frames 801–802, 846–851, 889–890, 903; Sanger to How-Martyn, March 4, 1936, MSP, Collected Documents Series, UPA, reel 6, frames 80–85.

60. Reed, *From Private Vice to Public Virtue,* 218–222; Harry Guy to Sanger, December 11, 1935, IPPFA, series B, reel 214, frame 906; Sanger and Harry Guy to Kaufman, November 7, 1935, MSP, Library of Congress Microfilm Collection, reel 15, frame 940.

61. Sanger to Elmhirst, January 7, 1936, MSP, Library of Congress Microfilm Collection, reel 15, frame 1027; Sanger to Harry Guy, January 30, 1936, IPPFA, series B, reel 214, frames 830–831.

62. Newfield to Sanger, January 6, 1936, MSP, Library of Congress Microfilm Collection, reel 15, frame 1022; Hodson to Stopes, December 11, 1923, British Library, Stopes Papers, add. 58645, Eugenics Society.

63. "Recommendations as to the future relation of the National Birth Control Association to the Eugenics Society," April 21, 1937, CMAC, SA/EUG, box 27, D.23; Soloway, *Demography and Degeneration,* 209–214.

64. Allan Carlson, *The Swedish Experiment in Family Politics: The Myrdals and the Interwar Population Crisis* (New Brunswick: Transaction, 1990), 81–95; Sondra R. Herman, "Dialogue: Children, Feminism, and Power: Alva Myrdal and Swedish Reform, 1929–1956," *Journal of Women's History* 4, no. 2 (1992): 92–93.

65. Carlson, *The Swedish Experiment,* 90–91; Gunnar Broberg and Mattias Tydén, "Eugenics in Sweden: Efficient Care," in *Eugenics and the Welfare State: Sterilization Policy in Denmark, Sweden, Norway, and Finland,* ed. Gunnar Broberg and Nils Roll-Hansen (East Lansing: Michigan State University Press, 1996), 104–105, 108–121, 128–130.

66. Carlson, "Roles of Alva and Gunnar Myrdal," 447–452; Pedersen, *Origins of the Welfare State,* 320–321; "The Agricultural and the Health Problems," 1934, Food and Agriculture Organization Archives, Rome, RG 3.1, series D1, "Notes and Comments by Mr. F. L. McDougall."

67. *Final Report of the Mixed Committee of the League of Nations on the Relation of Nutrition to Health, Agriculture and Economic Policy* (Geneva: League of Nations, 1937). For the report's background and impact, see Amy Staples, "Constructing International Identity: The World Bank, Food and Agriculture Organization, and World Health Organization, 1945–1965" (Ph.D. diss., Ohio State University, 1998), 178–180; and Sean Turnell, *Eminence Grise: The Political Economy of F. L. McDougall,* Macquarie Economics Research Papers no. 13 (Sydney: Department of Economics, Macquarie University, 1998), 13–14.

68. Tomkins to Turner, November 14, 1935, IOR, L/E/9/265; Economic Advisory Council, Committee on Nutrition in the Colonial Empire, First Report—Part 1, "Nutrition in the Colonial Empire," July 25, 1939, IOR, L/E/9/266; and see also Cleland, *Population Problem in Egypt,* 79ff.

69. Marie E. Kopp, "Legal and Medical Aspects of Eugenic Sterilization in Germany," *American Sociological Review* 1 (1936): 761–770; Blacker to Bramwell, February 7, 1936, CMAC, SA/EUG, box 11, C.37.

70. Daniel J. Kevles, *In the Name of Eugenics: Genetics and the Uses of Human Heredity* (New York: Knopf, 1985), 170–173; Soloway, *Demography and Degeneration,* 195–198.

71. Frederick Osborn, "Development of a Eugenic Philosophy," *American*

Sociological Review 2, no. 3 (June 1937): 389–391; Mark H. Haller, *Eugenics: Hereditarian Attitudes in American Thought* (New Brunswick: Rutgers University Press, 1963), 174–176; Barry Alan Mehler, "A History of the American Eugenics Society, 1921–1940" (Ph.D. diss., University of Illinois at Urbana-Champaign, 1988), 271, 288–295; Carlson, "Roles of Alva and Gunnar Myrdal," 474–475.

72. Osborn to Taeuber, March 23, 1936, Population Association of America Archives, Silver Spring, MD, box 4A, folder 74; Haller, *Eugenics,* 174; Kevles, *In the Name of Eugenics,* 172–173; Soloway, *Demography and Degeneration,* 196–198.

73. Kevles, *In the Name of Eugenics,* 173–174; Soloway, *Demography and Degeneration,* 199–200; Mehler, "American Eugenics Society," 284–285; Stefan Kühl, *The Nazi Connection: Eugenics, American Racism, and German National Socialism* (New York: Oxford University Press, 1994), 75; Blacker to Pearson, July 2, 1931, CMAC, SA/EUG, box 20, C.268; Alva Myrdal, "The Swedish Approach to Population Policies," *Journal of Heredity* 30, no. 3 (1939): 113.

74. Shaughnessy to Cicognani, February 13, 1935, Shaughnessy to Burke, February 13, 1935, and Cicognani to Sbarretti, February 24, 1935, VSA, Arch. Deleg. Stati Uniti, Sezione II (Stati Uniti), Posiz. 345b.

75. Sbarretti to Cicognani, April 25, 1936, and Cicognani circular, May 23, 1936, VSA, Arch. Deleg. Stati Uniti, Sezione II (Stati Uniti), Posiz. 345b; Noll to Hayes, December 16, 1935, VSA, Arch. Deleg. Stati Uniti, Sezione II (Stati Uniti), Posiz. 345a.

76. Youngs Rubber Corporation v. C. I. Lee and Co., Inc., et al., 45 F. 2d 103 (1930); United States v. One Package of Japanese Pessaries, 86 F. 2d 737 (1936); Chesler, *Woman of Valor,* 373–374.

77. Kennedy, *Birth Control in America,* 242–243, 256–257; Chesler, *Woman of Valor,* 391–394.

78. Chesler, *Woman of Valor,* 391–394; Fairchild quoted in Clarke, *Disciplining Reproduction,* 184–185.

79. "Minutes of the Joint Committee," July 8, 1937, CMAC, SA/EUG, box 27, D.23; "The Family Planning Association," March 21, 1940, CMAC, SA/FPA, box 337, "1935–1949 Eugenics Society."

80. Luckily for historians, Olive Johnson retrieved as much as she could carry—Hawarden to Sanger, November 18, 1936, MSP, Collected Documents Series, UPA, reel 6, frames 220–222; Chance to Gerda Guy, March 8, 1937, IPPFA, series B, reel 214, frame 338; How-Martyn to Johnson, April 18, 1937, Margaret Sanger Papers Project, New York University, How-Martyn file; Sanger correspondence with Guy and Newfield, April–July 1938, MSP, Library of Congress Microfilm Collection, reel 16; Sanger to Harry and Gerda Guy, September 9, 1938, MSP, Smith College Collections, UPA, reel 15, frames 619–620.

81. Sanger to Hart, May 18, 1937, MSP, Library of Congress Microfilm Collection, reel 8, frame 1011; Helen M. Hopper, *A New Woman of Japan: A Political*

Biography of Katō Shidzue (Boulder, CO: Westview Press, 1996), 80–93; Chung, *Struggle for National Survival,* 144–147; Arata Ishimoto to Sanger, January 2, 1938, and Shidzue Ishimoto to Sanger, January 11, 1938, MSP, Smith College Collections, UPA, reel 14, frames 137–142, 187–189.

82. "Interview with Dr. Pillay in Bombay," December 2, 1938, IOR, European Manuscripts, D1182, Eileen Palmer Collection, file 7; Meetings of the International Sub-committee of the FPA, July 12, 1938, October 1, 1938, and June 26, 1941, SA/FPA, box 227; 31st Meeting of the BCIC, February 23, 1939, SA/FPA, box 333, "BCIC 1929–1939."

83. On Gini, see correspondence in Archivio Centrale dello Stato, Rome, Presidenza del Consiglio dei Ministri, 1940–1943, f.I.I.16.3.5.27.000-7; Pitt-Rivers, "Report to the President and Executive Committee," CMAC, SA/EUG, box 34, D.110; "Activité de l'Union internationale pour l'Etude scientifique de la Population de 1937 à 1953," Archives Nationales, Paris, Georges Mauco papers, AP577, box 5; Georges Mauco, "L'Immigration étrangère en France et le problème des réfugiés," *L'Ethnie Française* 6 (March 1942): 6–15.

84. Chance to Blacker, July 14, 1944, CMAC, C. P. Blacker Papers, box 10, B.10; Olive Johnson to Sanger, June 24, 1941, MSP, Smith College Collections, UPA, reel 19, frames 460–461; Sanger to Houghton, February 27, 1953, IPPFA, series B, reel 717, frames 1437–1438.

85. Grossmann, *Reforming Sex,* 145–153, 161–165; Abraham Stone, "The Stockholm Conference on Sex Education, Family Planning and Marriage Counseling," *Human Fertility* 11, no. 3 (September 1946): 93–94; Benno Muller-Hill, *Murderous Science: Elimination by Scientific Selection of Jews, Gypsies, and Others, Germany 1933–1945,* trans. George R. Fraser (New York: Oxford University Press, 1988).

86. Grossmann, *Reforming Sex,* 167, 170–171, 180; Doris H. Linder, *Crusader for Sex Education: Elise Ottesen-Jensen (1886–1973) in Scandinavia and on the International Scene* (Lanham, MD: University Press of America, 1996), 161–170.

87. Irene B. Taeuber and George W. Barclay, "Korea and the Koreans in the Northeast Asian Region," *Population Index* 16 (1950): 282; Henri Ulmer, "Actes du Congrès International de la Population, Paris, Août 1937," Centre des Archives d'Outre-mer, Aix-en-Provence (hereafter CAOM), Fonds ministériels, Affaires économiques, box 49 (1/affeco/49); Albert de Pouvourville, "L'Angoissante Prolificité de l'Asie: 'En Indochine, on naît trop et on ne meurt pas assez,'" ca. May–June 1936, CAOM, Fonds ministériels, Affaires économiques, box 49 (1/affeco/49), "Colonies, Affaires Economiques, Statistiques"; Le Conseil Supérieur de la Colonisation, séance inaugurale, June 30, 1938, CAOM, Fonds ministériels, Agence FOM, Indochine, Inde (7), (*Fm, agefom170/36); Graffeuil to Guernut, April 1, 1938, Commission d'enquête dans les territoires d'Outre-Mer (commission Guernut) 1937–1938, box 23; Henri Brenier, *Le Problème de la population*

dans les colonies françaises (Lyon: Chronique sociale de France, 1930), 26–27; "West India Royal Commission, 1938–39," National Archives, Kew, UK, CO 318/445/5.

88. Phyllis Page, "Conference with Dr. Ernest Gruening," October 26, 1936, and "Report on Washington Trip," January 18, 1937, SSC, PPFA I, series 7, Subject Files, box 62, folder 3; Annette B. Ramírez de Arellano and Conrad Seipp, *Colonialism, Catholicism, and Contraception: A History of Birth Control in Puerto Rico* (Chapel Hill: University of North Carolina Press, 1983), 49–60.

89. Henri Ulmer, "Sondages concernant les familles, la fecondité, la mortalité infantile," *Congrès International de la Population, Paris 1937,* vol. 6, *Démographie de la France d'Outre-mer* (Paris: Hermann et Cie, 1938), 120–127; T. Smolski, "L'Etat de la Population," ca. March 1937, CAOM, Fonds Ministériels, Affaires Economiques (FM77), box 49; Economic Advisory Council, Committee on Nutrition in the Colonial Empire, First Report—Part 1, "Nutrition in the Colonial Empire," July 25, 1939, IOR, L/E/9/266; Taeuber and Barclay, "Korea and the Koreans," 282; Commission des Reformes Musulmanes, 17e Séance, February 17, 1944, CAOM, 7CAB/17.

90. G. Mesnard, "La régression relative des Européens en Algérie," *Congrès International de la Population, Paris 1937,* 6:11–14.

4. Birth of the Third World

1. Department of State, Office of Intelligence Research, "World Population Estimates," March 1, 1947, U.S. National Archives, College Park, MD (hereafter USNA), RG 59, Records of the Bureau of Intelligence and Research, OIR Report No. 4192; Irene B. Taeuber, *The Population of Japan* (Princeton: Princeton University Press, 1958), 341. On fertility rates, see United Nations, *Demographic Yearbook, 1948* (Lake Success, NY: United Nations, 1949), 260–265. On infant mortality, see Frank McDougall to directors of divisions, "Food and Population," October 3, 1952, Food and Agriculture Organization Archives, Rome (hereafter FAOA), RG 3.1, series C1, "F. L. McDougall Memoranda to Director-General"; and UN Secretariat, Department of Social Affairs, Population Division, "The Situation and Recent Trends of Mortality in the World," *Population Bulletin,* no. 6 (1963): 16, 39.

2. Richard Law, John Boyd Orr, Alexander Kunosi, et al., *Freedom from Want of Food* (London: Lincolns-Prager, 1944), 31–32; Sauvy, "La Population dans le Monde et le Développement économique," June 13, 1960, Archives de la Ministère des Affaires Etrangères, Paris, Série Asie Océanie 1944, Dossiers Généraux, ECAFE, Juin-Août 1960, dossier 441.

3. Edmund Russell, *War and Nature: Fighting Humans and Insects with Chemicals from World War I to Silent Spring* (New York: Cambridge University

Press, 2001), 112–113, 124–136; UN Secretariat, Department of Social Affairs, Population Division, "The Situation and Recent Trends of Mortality in the World," *Population Bulletin,* no. 6 (1962): 10, 15, 35, 40, 50, 82; report by superintendent, Malaria Eradication Campaign, ca. January 1949, World Health Organization Archives, Geneva (hereafter WHOA), Records of the Central Registry, first generation of files (WHO.1), 453-4-29.

4. "Special Message to the Congress," March 24, 1952, *Public Papers of the Presidents of the United States: Harry S. Truman, 1952–53* (Washington, DC: U.S. Government Printing Office, 1966), 209.

5. Gunnar Broberg and Nils Roll-Hansen, eds., *Eugenics and the Welfare State: Sterilization Policy in Denmark, Sweden, Norway, and Finland* (East Lansing: Michigan State University Press, 1996), 61, 180–181; Clarence J. Gamble, "Human Sterilization and Public Understanding," *Eugenics Review* 45, no. 3 (1953): 166–167; Yuehtsen Juliette Chung, *The Struggle for National Survival: Eugenics in Sino-Japanese Contexts* (New York: Routledge, 2002), 173; Paul Weindling, *Health, Race, and German Politics between National Unification and Nazism, 1870–1945* (New York: Cambridge University Press, 1989), 564–567.

6. On the persistence of eugenics, see Bonnie Mass, *Population Target: The Political Economy of Population Control in Latin America* (Toronto: Women's Press, 1976); and Germaine Greer, *Sex and Destiny: The Politics of Human Fertility* (London: Secker and Warburg, 1984). On the importance of specificity, see Linda Gordon, *The Moral Property of Women: A History of Birth Control Politics in America* (Urbana: University of Illinois Press, 2002), 281–282, 415.

7. Peter Novick, *The Holocaust in American Life* (Boston: Houghton Mifflin, 1999), 63–66; "Address by Margaret Sanger," presented to the thirtieth annual meeting of the Planned Parenthood Federation of America, October 25, 1950, Contemporary Medical Archives Center, Wellcome Institute for the History of Medicine, London (hereafter CMAC), Eugenics Society Papers (hereafter SA/EUG), box 22, C.304.

8. "Working Party on the Relation between Differential Fertility of Different Groups and the Development of Intelligence in New Generations," February 1–4, 1954, and "Differential Fertility and Intelligence," July 13, 1954, both in UNESCO Archives, Paris (hereafter UNESCOA), 312.1 A 53, "Fertility Studies"; and see also related correspondence in 312: 308 A 06 (44) "54."

9. Mark Mazower, *Dark Continent: Europe's Twentieth Century* (New York: Knopf, 1998), 217–218; Taeuber, *Population of Japan,* 345–346, 369. On France, see Compte Rendu of the Commission Interministérielle de l'Immigration, January 11, 1946, Centre des Archives Contemporaines, Fontainebleau (hereafter CAC), versement 860269, **art. 8; De Gaulle to Pierre-Henri Teitgen, June 12, 1945, CAC, versement 860269, Haut Comité de la Population et de la Famille 1945–1970, art. 3.

10. "Point Four: Cooperative Program for Aid in the Development of Economically Underdeveloped Areas" (Washington, DC: Department of State, 1949), 9–10.

11. Rockefeller Foundation press release, July 21, 1913, Rockefeller Archive Center, Tarrytown, NY (hereafter RAC), RG 3.1, series 908, box 11, "Program and Policy, 1909–1927"; and see also Ilana Löwy and Patrick Zylberman, "Medicine as a Social Instrument: Rockefeller Foundation, 1913–45," *Studies in History and Philosophy of Biological and Biomedical Sciences* 31, no. 3 (2000): 367–369; Amy Staples, "Constructing International Identity: The World Bank, Food and Agriculture Organization, and World Health Organization, 1945–1965" (Ph.D. diss., Ohio State University, 1998), 387–389.

12. "Point Four: Cooperative Program for Aid"; Kennan, *Memoirs: 1925–1950* (Boston: Little, Brown, 1967), 559. On Churchill, see Christopher Thorne, *Allies of a Kind: The United States, Britain and the War against Japan, 1941–1945* (London: Hamish Hamilton, 1978), 7–9, 191. On Hitler, see Milan Hauner, "Did Hitler Want a World Dominion?" *Journal of Contemporary History* 13, no. 1 (1978): 25. On Roosevelt, see Gary R. Hess, *Vietnam and the United States: Origins and Legacy of War* (Boston: Twayne, 1990), 29.

13. David J. Garrow, *Liberty and Sexuality: The Right to Privacy and the Making of Roe v. Wade* (Berkeley: University of California Press, 1998), 107–108; John Dower, *War without Mercy: Race and Power in the Pacific War* (New York: Pantheon, 1986), 160; Aldous Huxley, "The Double Crisis: Part I," *World Review,* November 1948, 14.

14. Julian Huxley memo to Trygve Lie, March 30, 1948, UNESCOA, inactive correspondence files, 312 A 06 (45) "54"; E. Caine minute to file, National Archives, Kew, UK (hereafter UKNA), CO 927/10; Auguste Rencurel, "Rapport sur la création d'industries nouvelles capable d'absorber un plus grand nombre de travailleurs musulmans," ca. February 1944, Centre des Archives d'Outre-mer, Aix-en-Provence (hereafter CAOM), 7CAB/17; Sauvy, introduction to *Le Problème Démographique Nord-Africain,* by Louis Chevalier (Paris: Presses universitaires de France, 1947), 7.

15. "Compte Rendu de la Conférence du 30 Mars 1946 sur le Problème de l'Emploi de la Main d'Oeuvre Algérienne en France," March 30, 1946, Archives Nationales, Paris (hereafter AN), AJ80, Commissaire Général au Plan, box 75.

16. "Procès-verbal de la réunion de la Commission Interministérielle de l'Immigration," November 6, 1946, AN, AJ80, Commissaire Général au Plan, box 75; Siegfried, "Une Grande Aventure," *Le Figaro,* March 17, 1951; Boverat, "Gravité de la Discordance entre les Politiques de Population," May 20, 1952, CAC, versement 860269, Haut Comité de la Population et de la famille, 1945–1970; W. D. Sweaney, "Population Problems in the West Indies," March 8, 1949, UKNA, CO 318/511/9.

17. T. H. Davey, "The Growth of Tropical Populations," ca. March 1948, "Extracts from Minutes of C.A.M.C. 443rd Meeting," March 23, 1948, and accompanying minutes to file, UKNA, CO 859/154/6.

18. Frank Notestein, "Problems of Policy in Relation to Areas of Heavy Population Pressure," in *Demographic Studies of Selected Areas of Rapid Growth* (New York: Milbank Memorial Fund, 1944), 148–149, 153–157.

19. Dudley Kirk, "Population Changes and the Postwar World," *American Sociological Review* 9, no. 1 (1944): 32, 35; Dennis Hodgson, "Demography as Social Science and Policy Science," *Population and Development Review* 9 (1983): 7–12; Simon Szreter, "The Idea of Demographic Transition and the Study of Fertility Change: A Critical Intellectual History," *Population and Development Review* 19 (1993): 670–671.

20. "The World Demographic Transition," *Annals of the American Academy of Political and Social Science* 237 (January 1945): 7–8, emphasis in original; Susan Greenhalgh, "The Social Construction of Population Science: An Intellectual, Institutional, and Political History of Twentieth-Century Demography," *Comparative Studies in Society and History* 38, no. 1 (1996): 37–38.

21. Irene Taeuber and Hope T. Eldridge, "Some Demographic Aspects of the Changing Role of Women," *Annals of the American Academy of Political and Social Science* 251 (1947): 33; Notestein, "Problems of Policy," 151, 153; McDougall, "Notes for Director-General," April 22, 1946, FAOA, RG 3.1, series C1, F. L. McDougall memoranda to director-general; Huxley memo to Trygve Lie, March 30, 1948.

22. Orr, "The Choice Ahead: One World or None," December 14, 1946, FAOA; Orr to Scrutton, September 2, 1947, FAOA, RG 1.1, series A2, Lord John Orr Outgoing Letters; Huxley memo to Trygve Lie, March 30, 1948.

23. René Brunet, "Rapport présenté par la deuxième Commission à l'Assemblée," October 1, 1937, International Labour Organization Archives, Geneva, B, Continuing Series, PO 2/1/01.

24. Loveday to A. M. Carr-Saunders, April 19, 1939, LON, Registry Files, 1933–1946, box 4467, section 10A/36640/35637; and Loveday to Sean Lester, September 2, 1941, LON, Registry Files, 1933–1946, box R4469, section 10A/41207/41207; Notestein, "Demography in the United States: A Partial Account of the Development of the Field," *Population and Development Review* 8 (1982): 664–665.

25. John Cleland, "Demographic Data Collection in Less Developed Countries," *Population Studies* 50, no. 3 (1996): 435–436; Ping-ti Ho, *Studies on the Population of China* (Cambridge: Harvard University Press, 1959), 65–97; Robert Kuczynksi, "Memorandum on Colonial Population and Vital Statistics," ca. September 1943, and record of first meeting of Colonial Demography Research Group, September 24, 1943, UKNA, CO 927/10; T. Smolski, "L'Etat de la Pop-

ulation," ca. March 1937, CAOM, Fonds Ministériels, Affaires Economiques (FM77), box 49; Government of India, Department of Health, Education, and Lands, "Report of the Population Data Committee," 1945, British Library, India Office Records, V/26/810/1; "2nd Conference of Colonial Government Statisticians," ca. March 1953, UKNA, CO 1034/10.

26. Notestein, "Reminiscences: The Role of Foundations, of the Population Association of America, Princeton University and of the United Nations in Fostering American Interest in Population Problems," *Milbank Memorial Fund Quarterly* 49, no. 4 (1971): 78. This is an interpretation of what Notestein only implied in this article.

27. "Report of the Committee to Consider Proposals for the Work of the Demographic Commission of the Social and Economic Council," June 4, 1946, Population Association of America Archives, Silver Spring, MD, box 6, folder 13. Richard Symonds and Michael Carder provide an excellent account in *The United Nations and the Population Question, 1945–1970* (London: Sussex University Press, 1973), 40ff.

28. F. A. E. Crew et al., "Social Biology and Population Improvement," *Nature* 144, no. 3646 (1939): 521; Julian Huxley, *UNESCO: Its Purpose and Its Philosophy* (UNESCO, 1946), available from unesdoc.unesco.org/ulis/index.html; Julian Huxley, *Memories II* (London: George Allen and Unwin, 1973), 15–16.

29. Amy Staples, "Constructing International Identity: The World Bank, Food and Agriculture Organization, and World Health Organization, 1945–1965" (Ph.D. diss., Ohio State, 1998), 189–192.

30. Ibid., 211–229; John Boyd Orr, *As I Recall* (London: MacGibbon and Kee, 1966), 202.

31. Julian Huxley memo to Trygve Lie, March 30, 1948.

32. Arnaldo to Laves, April 24, 1948, Laugier to Huxley, April 21, 1948, and Arnaldo to Huxley, May 15, 1948, all in UNESCOA, inactive correspondence files, 312 A 06 (45) "54"; Philip Hauser to Kotschnig, April 19, 1949, University of Chicago Library, Special Collections Research Center, Philip M. Hauser Papers (hereafter Hauser Papers), box 30, folder 1.

33. John R. Wilmoth and Patrick Ball, "Arguments and Action in the Life of a Social Problem," *Social Problems* 42, no. 3 (1995): 323, 326–329; Fairfield Osborn, *Our Plundered Planet* (London: Faber and Faber, 1948), 29–30, 38, 44–45; William Vogt, *Road to Survival* (New York: William Sloane Associates, 1948).

34. Osborn, *Our Plundered Planet*, 3, 17, 80–84; Ronald Rainger, *An Agenda for Antiquity: Henry Fairfield Osborn and Vertebrate Paleontology at the American Museum of Natural History, 1890–1935* (Tuscaloosa: University of Alabama Press, 1991), 245.

35. Vogt, *Road to Survival*, xvi, 14, 48, 77, 257, 280–283.

36. "World Population Authority Named Director of Planned Parenthood,"

PPFA press release, May 18, 1951, International Planned Parenthood Federation Archives, London (hereafter IPPFA), series B, reel 718, frame 1743. For a similar analysis of the two books, see John H. Perkins, *Geopolitics and the Green Revolution: Wheat, Genes, and the Cold War* (New York: Oxford University Press, 1997), 135–136.

37. Huxley, "Double Crisis: Part I," 12–14. UNESCO promoted the piece. See W. E. Williams to Aldous Huxley, August 19, 1948, UNESCOA, inactive correspondence files, 312.

38. Robert C. Cook, *Human Fertility: The Modern Dilemma* (New York: William Sloane Assoc., 1951), 5.

39. Vogt, *Road to Survival,* 34, 59–69; Huxley, "Double Crisis: Part I," 11–13.

40. "International Congress on Sex," ca. August 1946, CMAC, SA/EUG, box 22, C.304; Doris H. Linder, *Crusader for Sex Education: Elise Ottesen-Jensen (1886–1973) in Scandinavia and on the International Scene* (Lanham, MD: University Press of America, 1996), 122–127, 171; Dagmar Herzog, *Sex after Fascism: Memory and Morality in Twentieth-Century Germany* (Princeton: Princeton University Press, 2005), 122–127.

41. Barbara Evans, *Freedom to Choose: The Life and Work of Dr. Helena Wright, Pioneer of Contraception* (London: The Bodley Head, 1984), 207–208; Linder, *Crusader for Sex Education,* 172–176.

42. Linder, *Crusader for Sex Education,* 176–179; Horder, "Presidential Address"; Orr, "World Resources," in *Proceedings of the International Congress on Population and World Resources in Relation to the Family* (London: H. K. Lewis and Co., 1948), 4–5, 9; Needham to Huxley, August 31, 1948, and S. Chandrasekhar to A. Brodersen, August 31, 1948, UNESCOA, 312: 620.91 A 06 (41–4) "49," Population and Resources—meeting 49—UK.

43. Huxley memo to Trygve Lie, March 30, 1948; Needham to Huxley, August 31, 1948; Needham, "Concluding Address," in *Proceedings of the International Congress,* 230, 232.

44. Needham, "Concluding Address," in *Proceedings of the International Congress,* 230, 232; *Proceedings of the International Congress,* 237–241.

45. Houghton, "A Review of the Past Eight Months," June 16, 1950, IPPFA, series B, reel 117, frames 1124–1128. The debate is documented in voluminous correspondence among Sanger, Ottesen-Jensen, Helen Cohen, and Helena Wright in the same file.

46. Ellen Chesler, *Woman of Valor: Margaret Sanger and the Birth Control Movement in America* (New York: Simon and Schuster, 1992), 416–418; Evans, *Freedom to Choose,* 208–209.

47. Vogt, *Road to Survival,* 282; *Proceedings of the International Congress,* 21–22, 82–83, 136–140.

48. Notestein quoted in Szreter, "Idea of Demographic Transition," 671. The

following discussion of Japan owes much to an extremely rich dissertation by Deborah Oakley, "The Development of Population Policy in Japan, 1945–1952, and American Participation" (Ph.D. diss., University of Michigan, 1977), here 278. On Rockefeller, see also John Ensor Harr and Peter J. Johnson, *The Rockefeller Century: Three Generations of America's Greatest Family* (New York: Scribner's, 1988), 462.

49. John W. Dower, *Embracing Defeat: Japan in the Wake of World War II* (New York: W. W. Norton, 1999), 48–54, 92–104; Taeuber, *Population of Japan,* 343–346, 355–356, 369–370; Deborah Oakley, "American-Japanese Interaction in the Development of Population Policy in Japan, 1945–1952," *Population and Development Review* 4 (1978): 619.

50. Oakley, "Population Policy in Japan," 114–131, 152–154; Tiana Norgren, "Abortion before Birth Control: The Interest Group Politics behind Postwar Japanese Reproduction Policy," *Journal of Japanese Studies* 24, no. 1 (1998): 67; Taeuber, *Population of Japan,* 372; John Sharpless, "Population Science, Private Foundations, and Development Aid: The Transformation of Demographic Knowledge in the United States, 1945–1965," in *International Development and the Social Sciences,* ed. Frederick Cooper and Randall Packard (Berkeley: University of California Press, 1997), 180.

51. Oakley, "Population Policy in Japan," 172–175; Oakley, "American-Japanese Interaction," 623.

52. Helen M. Hopper, *A New Woman of Japan: A Political Biography of Katô Shidzue* (Boulder, CO: Westview Press, 1996), 224–225; Oakley, "Population Policy in Japan," 226–235.

53. This account of the meeting is based on Hopper, *A New Woman of Japan,* 225–227.

54. Ibid.; Oakley, "Population Policy in Japan," 220, 235–236.

55. Daniel B. Luten memoranda of October 7 and 18, 1948, Hoover Institution Archives, Palo Alto, Daniel B. Luten Papers (hereafter Luten Papers), box 2; see also Deborah Oakley, "American-Japanese Interaction," 630–631.

56. Marshall C. Balfour, Roger F. Evans, Frank W. Notestein, and Irene B. Taeuber, *Public Health and Demography in the Far East* (New York: Rockefeller Foundation, 1950), 117–118; Notestein and W. Henry Lawrence memcon, October 5, 1948, RAC, Population Council Papers, RG IV3B4.2, General File Series, box 1, folder 3; Szreter, "Idea of Demographic Transition," 672–674; Hodgson, "Demography as Social Science," 12–13.

57. Balfour et al., "Public Health and Demography in the Far East," 83, 97, 118–119.

58. Luten memoranda for the record, October 8 and December 17, 1948, Luten Papers, box 2.

59. Luten memorandum for record, December 9, 1948, Luten Papers, box 2.

Thompson's principal works on Japan were *Danger Spots in World Population* (New York: Knopf, 1929) and *Population and Peace in the Pacific* (Chicago: University of Chicago Press, 1946).

60. Oakley, "Population Policy in Japan," 207–210.

61. Ibid., 220, 238, 421–427.

62. Thompson interview with MacArthur, April 9, 1949, Luten Papers, box 2; Warren Thompson, "Population," *American Journal of Sociology* 34, no. 6 (1929): 959–975.

63. Luten to Thompson, August 24, 1949; Luten to Whelpton, November 17 and December 19, 1949; Maurie Roche to Ed Ackerman, January 30, 1950; all in Luten Papers, box 1. Oakley, "Population Policy in Japan," 212–213, 228, 248–260.

64. Oakley, "Population Policy in Japan," 260–265; MacArthur to Roy Howard, February 23, 1950, Luten Papers, box 2.

65. Johnson and Harr, *The Rockefeller Century,* 465–466.

66. Norgren, "Abortion before Birth Control," 69–72; Oakley, "Population Policy in Japan," 240–244. Oakley argues that Japan had an explicit policy to reduce growth. In fact, officials went out of their way to deny there was any such policy. See, for instance, World Health Organization, *Official Records,* no. 42 (Geneva, 1952), 239.

67. Radhakamal Mukherjee, *Population (Report of the Sub-Committee), National Planning Committee Series* (Bombay: Vora and Co., 1947), 64–67, 87–88, 129–131, 134–135 (emphasis added).

68. Shah introduction and resolutions, both in Mukherjee, *Population,* 6–7, 144–145.

69. Chung, *The Struggle for National Survival,* 160–162; Vera Houghton, "China," July 12, 1950, IPPFA, series B, reel 149, frames 2511–2512; H. Yuan Tien, *China's Population Struggle: Demographic Decisions of the People's Republic, 1949–1969* (Columbus: Ohio State University Press, 1973), 177–179.

70. "Population Commission, Fifth Session, Provisional Summary Record of the Fifty-Fourth Meeting," May 23, 1950, Hauser Papers, box 30, folder 1; UN Secretariat, Department of Social Affairs, Population Division, "The 1960 World Population Census Programme," *Population Bulletin,* no. 5 (July 1956): 2.

71. R. A. Gopalaswami, *Census of India, 1951,* vol. 1, pt. I-A-Report (New Delhi: Government of India Press, 1953), vi.

72. UN Secretariat, Department of Social Affairs, Population Division, "The Past and Future Growth of World Population—A Long Range View," *Population Bulletin,* no. 1 (1951): 1–12; Symonds and Carder, *The United Nations and the Population Question,* 34, 78; Caldwell and Caldwell, *Limiting Population Growth,* 22–23.

73. Department of Social Affairs, United Nations, "Preliminary Report on the World Social Situation" (New York: United Nations, 1952), 41; Frank

McDougall to directors of divisions, "Food and Population," October 3, 1952, FAOA, RG 3.1, series C1, F. L. McDougall Memoranda to Director-General.

74. Sauvy, "Le 'Faux Problème' de la Population mondiale," originally published in *Population* 4 (1949): 447–462; reprinted in *Population and Development Review* 16 (1990): 759–774.

75. Sauvy, "Le 'Faux Problème,'" 764–765.

76. On the death threats, see Sauvy, "Le Population dans le Monde." For the debate, see Symonds and Carder, *The United Nations and the Population Question,* 71–72; and "Population Commission, Fifth Session, Committee of the Whole, Provisional Summary Record of the Second Meeting," May 25, 1950, as well as the undated, untitled record of Durand's comments, both in Hauser Papers, box 30, folder 1.

77. Symonds and Carder, *The United Nations and the Population Question,* 82.

78. "Note pour Monsieur Alphand," June 28, 1950, and "Suggestions pour la définition d'une politique française de l'Immigration," July 22, 1950, both in CAC, versement 770623, art. 69.

79. Myrdal to director-general, May 15, 1951, UNESCOA, 312 A 06(45) 54. On the results, see "Document Presented to the Regional Committee for South-East Asia at Its Fourth Session," November 28, 1951, WHO, *Official Records,* no. 40 (Geneva, 1952), 135–136; and C. Chandrasekaran, "Cultural Patterns in Relation to Family Planning in India," in *The Third International Conference on Planned Parenthood: Report of the Proceedings* (Bombay: Family Planning Association of India, 1952), 75–76.

80. Myrdal to Notestein, May 30, 1951, and Notestein to Myrdal, June 18, 1951, UNESCOA, 312.1 A 53, Fertility Studies.

81. J. P. Gupta, N. K. Sinha, and Amita Bardhan, *Evolution of Family Welfare Programme in India* (New Delhi: National Institute of Health and Family Welfare, 1992), 2:22; Caldwell and Caldwell, *Limiting Population Growth,* 40–41; B. L. Raina, *Population Policy* (Delhi: B. R. Publishing, 1988), 5–7, and "Recommendations of the Committee . . . on Population Growth and Family Planning," April 14, 1951, reprinted as appendix A2 in Gupta, Sinha, and Bardhan, *Evolution of Family Welfare Programme.*

82. On the Health Ministry position, see Vera Houghton, "Note of Interview with Sharya Pandit," April 4, 1951, IPPFA, series B, reel 156, frames 1681–1683; and Douglas Ensminger Oral History, pt. B1, November 1, 1971, Ford Foundation Archives, New York City. The draft policy is quoted in S. Chandrasekhar, "Demographic Disarmament for India: A Plea for Family Planning," November 30, 1951, Planning Commission Archives, Delhi.

83. Symonds and Carder, *The United Nations and the Population Question,* 61–62; Berkeley to Myrdal, December 11, 1951, UNESCOA, inactive correspondence files, 312.

84. J. de Romer, "L'Organisation mondiale de la Santé et les Catholiques," March 20, 1950, and "Réunion des Organisations catholiques internationales," April 29, 1950, Catholic University of American Archives, Washington, DC, NCWC, Collection 10, box 172, folder 32; S. de Lestapis, "L'Église catholique et les problèmes de population: Textes pontificaux récents," *Population* 7 (1952): 302–303.

85. Staples, "Constructing International Identity," 371–375.

86. World Health Organization, *Official Records,* no. 40 (Geneva, 1952), 141–143; "Document Presented to the Regional Committee for South-East Asia at Its Fourth Session," November 28, 1951.

87. Board quoted in Symonds and Carder, *The United Nations and the Population Question,* 63–64 (emphasis added). George Cadbury to Chisholm, January 23, 1952; Chisholm to Cadbury, February 2, 1952; Paul Henshaw to Chisholm, January 10, 1951; and Chisholm to Henshaw, January 21, 1952; all in WHOA, Records of the Central Registry, second generation of files (WHO.2), GH 12.

88. J. de Romer circular, April 2, 1952, Catholic University of America Archives, Washington, DC, NCWC, Collection 10, box 172, folder 32; WHO, *Official Records,* no. 42 (Geneva, 1952), 61, 90–91, 98–99, 111, 204–206.

89. WHO, *Official Records,* no. 42, pp. 206–208, 230–231. On Evang and eugenics, see Nils Roll-Hansen, "Norwegian Eugenics: Sterilization as Social Reform," in Broberg and Roll-Hansen, *Eugenics and the Welfare State,* 188, 264.

90. WHO, *Official Records,* no. 42, pp. 232–236.

91. Ibid., 237–240.

92. Ibid., 131, 240–242.

93. Symonds and Carder, *The United Nations and the Population Question,* 66; Myrdal to Torres Bodet, November 17, 1952, and April 13, 1953, and "Suggested Topics," July 14, 1953, all in UNESCOA, 312 A 06(45) 54.

94. Berkeley to Maheu, ca. May 1955, UNESCOA, 312 UN/A 02, "U.N. Population Commission"; Davis, "Confidential Report on the Eighth Session of the Population Commission," March 25, 1955, USNA, RG 59, Central Decimal Files, 340.15.

95. Myrdal to Luther Evans, September 16, 1954, UNESCOA, Paris, inactive correspondence files, 312 A 06 (45) "54"; Wilmoth and Ball, "Arguments and Action," 323.

96. "Trois mondes, une planète," *L'Observateur,* August 14, 1952.

97. Ibid.

5. THE POPULATION ESTABLISHMENT

1. John Ensor Harr and Peter J. Johnson, *The Rockefeller Conscience: An American Family in Public and in Private* (New York: Scribner's, 1991), 33–34;

John Ensor Harr and Peter J. Johnson, *The Rockefeller Century* (New York: Scribner's, 1988), 367–369, 552.

2. John H. Perkins, *Geopolitics and the Green Revolution: Wheat, Genes, and the Cold War* (New York: Oxford University Press, 1997), 133–135; Harr and Johnson, *Rockefeller Century,* 498–500.

3. This and subsequent quotations are from the verbatim transcripts of the conference "National Academy of Sciences: Conference on Population Problems," June 20–22, 1952, Rockefeller Archive Center, Tarrytown, NY (hereafter RAC), RG 5, John D. Rockefeller 3rd Papers, series 1, subseries 5, box 85, folders 720–723.

4. On Rockefeller's failed effort to desegregate Colonial Williamsburg, see Harr and Johnson, *Rockefeller Century,* 494–495.

5. "Excerpt from Remarks by Dr. Irene Taeuber regarding her trip to the Orient," January 16, 1954, RAC, Population Council Papers, RG IV3B4.2, General File Series, box 1, folder 3.

6. Harr and Johnson, *The Rockefeller Conscience,* 34–35. The records of the biweekly lunches are sketchy, but for examples see RAC, Population Council Papers, RG IV3B4.2, General File Series, box 32, folder 465.

7. John and Pat Caldwell, *Limiting Population Growth and the Ford Foundation Contribution* (London: Frances Pinter, 1986), 44; Osborn application to Ford Foundation, February 10, 1954, Ford Foundation Archives, New York City (hereafter FFA), 1953 Grant Files, PA 54–20, "The Population Council"; Bernard Berelson, "Summary Report of Discussions of Informal Consultative Group," April 18–19, 1953, FFA, 1953 General Correspondence, C-1165A, "Population."

8. For first draft, see "Proposed Establishment of Population Council," October 7, 1952, draft, RAC, RG 5, John D. Rockefeller 3rd Papers, series 1, subseries 5, box 81, folder 674 (I am grateful to Nancy Davenport for this document). For final draft, and continuing support for eugenics, see "Proposed Establishment of Population Council," November 20, 1952, and Osborn, "Population Council: Philosophy of the Rational Control of Family Life," June 9, 1954, RAC, RG 5, John D. Rockefeller 3rd Papers, series 1, subseries 5, box 82, folders 683–684; Maria Mesner, "Engineering Global Population: The Reasoning and Policies of the Population Council," and Edmund Ramsden, "Between Quality and Quantity: The Population Research Council and the Politics of 'Science-making' in Eugenics and Demography, 1952–1965," both in *Rockefeller Archive Center Research Reports Online,* no. 1 (April 2001), archive.rockefeller.edu/publications. For Osborn and Notestein views of "qualitative" issues, see Armes-Osborn memcon, November 3, 1954, and Osborn to Wood, May 11, 1955, Social Welfare History Archives, University of Minnesota, Minneapolis, Papers of the Association for Voluntary Sterilization, supp. 1, box 25, "Osborn, Frederick"; Notestein to Caryl

Haskins, April 17, 1959, RAC, Population Council Papers, RG IV3B4.2, General File Series, box 34, folder 489.

9. Taeuber to Osborn, March 18, 1956, RAC, RG IV3B4.2, Population Council, General Files, box 1, folder 7. On the link, see Taeuber and Hope T. Eldridge, "Some Demographic Aspects of the Changing Role of Women," *Annals of the American Academy of Political and Social Science* 251 (1947): 33; Marshall C. Balfour, Roger F. Evans, Frank W. Notestein, and Irene B. Taeuber, *Public Health and Demography in the Far East* (New York: Rockefeller Foundation, 1950), 83, 116; Notestein, "Gaps in the Existing Knowledge of the Relationships between Population Trends and Economic and Social Conditions," ca. June 1954, UNESCO Archives, Paris, 312 A 06(45) 54; "Conference on Study of Motivation Relevant to Fertility Control," May 29, 1959, RAC, RG 5, John D. Rockefeller 3rd Papers, series 1, subseries 5, box 82, folder 680.

10. "Minutes of a Meeting of the Subcommittee on the Quality Aspects of Population," March 13, 1954, RAC, Population Council Papers, RG IV3B4.2, General File Series, box 32, folder 470.

11. "This Christmas," *Commonweal* 53 (December 29, 1950): 291; W. S. and E. S. Woytinsky, *World Population and Production: Trend and Outlook* (New York: Twentieth Century Fund, 1953), 254–256 (the authors were actually skeptical of this view, considering projections for Soviet population growth to be overstated); Edmund Russell, *War and Nature: Fighting Humans and Insects with Chemicals from World War I to Silent Spring* (New York: Cambridge University Press, 2001), 187–188. Regarding typing of Soviets, see Matthew Connelly, "Taking Off the Cold War Lens: Visions of North–South Conflict during the Algerian War for Independence," *American Historical Review* 105 (2000): 753–754. On Eisenhower strategy, see Marc Trachtenberg, *A Constructed Peace: The Making of the European Settlement, 1945–1963* (Princeton: Princeton University Press, 1999), 160–162.

12. Kingsley Davis, "The Amazing Decline of Mortality in Underdeveloped Areas," *American Economic Review* 46, no. 2 (May 1956): 305–318; Dennis Hodgson, "Demography as Social Science and Policy Science," *Population and Development Review* 9 (1983): 12–13, 16–19; Frank W. Notestein, "The Economics of Population and Food Supplies: Economic Problems of Population Change," in *Proceedings of the Eighth International Conference of Agricultural Economists* (New York: Oxford University Press, 1953), 24–25; Davis, "World Population Trends and American Policy," November 1956, RAC, Rockefeller Brothers Fund, RG V4C, Special Studies Project, box 21, folder 237.

13. Donald T. Critchlow, *Intended Consequences: Birth Control, Abortion, and the Federal Government in Modern America* (New York: Oxford University Press, 1999), 30–33 (emphasis added).

14. Dudley Kirk, "Population Changes and the Postwar World," *American Sociological Review* 9, no. 1 (1944): 32; Van der Tak interview with Kirk, April 29,

1989, Population Association of America Archives, Silver Spring, MD, box 4A, folder 65; Frederick Osborn, "Notes on Ad Hoc Meeting," March 7, 1956, RAC, Population Council Papers, RG IV3B4.2, General File Series, box 1, folder 7.

15. Davis, "World Population Trends and American Policy," November 1956; Blacker to E. F. G. Haig, February 6, 1957, Contemporary Medical Archives Center, Wellcome Institute for the History of Medicine, London (hereafter CMAC), C. P. Blacker Papers, box 11, "Blacker: Correspondence, Eugenics Society, IPPF, January–June 1957."

16. Blacker to Dorothy Brush, April 13, 1956, CMAC, C. P. Blacker Papers, box 11, "Blacker: Correspondence, Eugenics Society, IPPF"; "Eugenics Society: Memorandum by Certain Members of the Council of the Society," January 18, 1960, CMAC, C. P. Blacker Papers, box 11, "Blacker: Eugenics Society, 1960"; Frederick Osborn, "Galton and Mid-Century Eugenics," *Eugenics Review* 48, no. 1 (1956): 16–17.

17. Francesco Cassata, *Molti, Sani, e Forti: L'Eugenetica in Italia* (Torino: Bollati Boringhieri, 2006), 315–319; Leslie Woodcock Tentler, *Catholics and Contraception: An American History* (Ithaca: Cornell University Press, 2004), 184–186; Frederick Osborn, "Galton and Mid-Century Eugenics," 21–22; Osborn, "The Contribution of Planned Parenthood to the Future of America," *Eugenics Review* 45, no. 3 (1953): 157–158. Osborn, "Talk with Father William Gibbons," November 16, 1956; "Concluding Statement by Chairman," April 4, 1957; and Osborn to members of Ad Hoc committee, April 15, 1957; all in RAC, Population Council Papers, RG IV3B4.2, General File Series, box 2, folders 12–13.

18. Molly Ladd-Taylor, "Eugenics, Sterilization and Modern Marriage in the USA: The Strange Career of Paul Popenoe," *Gender and History* 13 (August 2001): 298–327; Hardin, *Biology: Its Human Implications* (San Francisco: W. H. Freeman, 1949): 613–615.

19. Osborn, "Galton and Mid-Century Eugenics," 17, 21; Harr and Johnson, *The Rockefeller Conscience,* 72; Critchlow, *Intended Consequences,* 32–33; Moore, Clayton, and Barton to William Iliff, November 4, 1957, World Bank Group Archives, Washington, DC, Central Files 1947–1968, General Files—Projects and Studies, box 35, "Population," vol. 1 (emphasis added); "National Academy of Sciences: Conference on Population Problems," June 21, 1952.

20. Sanger to Gamble, April 5, 1940, Margaret Sanger Papers, Microfilm Edition, Collected Documents Series, University Publications of America, reel 7; "Draft Reply to Mrs. Sanger's Letter," January 9, 1950, Archives of the International Planned Parenthood Federation, London (hereafter IPPFA), series B, reel 117, frames 1208–1210; Beryl Suitters, *Be Brave and Angry: Chronicles of the International Planned Parenthood Federation* (London: International Planned Parenthood Federation, 1973), 42; Doris H. Linder, *Crusader for Sex Education: Elise Ottesen-Jensen (1886–1973) in Scandinavia and on the International Scene*

(Lanham, MD: University Press of America, 1996), 177; Houghton to Sanger, June 11, 1951, IPPFA, series B, reel 717, frames 1625–1627.

21. Sanger to Houghton, January 25, 1951, IPPFA, series B, reel 717, frames 1643–1644; Vembu to Sanger, February 21, 1952, and Rama Rau to Watumull, March 11, 1952, IPPFA, series B, reel 171, frames 1509, 1511–1512; Dhanvanthi Rama Rau, *An Inheritance: The Memoirs of Dhanvanthi Rama Rau* (New York: Harper and Row, 1977), 255–256; Suitters, *Be Brave and Angry,* 47–48.

22. *The Third International Conference on Planned Parenthood: Report of the Proceedings* (Bombay: Family Planning Association of India, 1952), 2–4. On Cariappa, see B. L. Raina, *Planning Family in India: Prevedic Times to Early 1950s* (New Delhi: Commonwealth, 1990), 112–113.

23. Radhakrishnan, "Inaugural Address," in *The Third International Conference,* 10–13; Linder, *Crusader for Sex Education,* 185.

24. Houghton to Pyke, September 16, 1951, and Blacker to Sanger, October 17, 1951, IPPFA, series B, reel 717, frames 1006–1009, 1545–1546; Linder, *Crusader for Sex Education,* 184; "Concluding Plenary Session," in *The Third International Conference,* 202–203; C. P. Blacker, "The International Planned Parenthood Federation: Aspects of Its History," *Eugenics Review* 56, no. 3 (1964): 139.

25. *The First Five Year Plan* (New Delhi: Planning Commission, 1953), 522–524; "Family Planning in India: A Review of the Progress in Family Planning Program April 1956–November 1958," National Institute of Health and Family Welfare, Documentation Centre, New Delhi (hereafter NIHFW), 204/83 IND; K. Srinivasan, *Regulating Reproduction in India's Population: Efforts, Results, and Recommendations* (New Delhi: Sage, 1995), 30. For India's influence, see Caldwell, *Limiting Population Growth,* 25; Harr and Johnson, *The Rockefeller Conscience,* 40–41. On the state of demography, see "Population Presentation to the Trustees," February 1953, FFA, 1953 General Correspondence, C-1165A, "Population."

26. Nielsen-Davis memcon, January 28, 1953; Nielsen-Whelpton memcon, January 28, 1953; and Thompson, "A Study . . . to Reduce the International Tensions," ca. March 1953; all in FFA, 1953 General Correspondence, C-1165A, "Population."

27. Oscar Harkavy, *Curbing Population Growth: An Insider's Perspective on the Population Movement* (New York: Plenum Press, 1995), 10–13; "Population Presentation to the Trustees"; Nielsen, "Population Discussion at Trustees Meeting," July 1, 1953, FFA, 1953 General Correspondence, C-1165A, "Population." For the first major grant to the Council, see "Research and Training in the Field of Population," February 19, 1954, FFA, 1953 Grant Files, PA 54–20, Population Council. For Rockefeller support, see "Grant in Aid to Harvard University," October 27, 1953, and subsequent grants in RAC, Rockefeller Foundation, RG 1.2, Projects, 200 United States, Harvard University—Indian Population 1953–1955, box 45, folder 369; Harr and Johnson, *The Rockefeller Conscience,* 42–43.

28. On "pump-priming," see Nielsen-Whelpton memcon. On India, see Srinivasan, *Regulating Reproduction,* 31–32; Ministry of Health circular to all state governments, February 27, 1954, in J. P. Gupta, N. K. Sinha, and Amita Bardhan, *Evolution of Family Welfare Programme in India,* vol. 1 (New Delhi: National Institute of Health and Family Welfare, 1992), appendix A4.

29. For examples of hyping, see Frank W. Notestein, "Policy of the Indian Government on Family Limitation," *Population Index* 17, no. 4 (1951): 254–263; and Hodgson, "Demography as Social Science," 19–20. On the establishment, see Bonnie Mass, *Population Target: The Political Economy of Population Control in Latin America* (Toronto: Women's Press, 1976), 33–34; and Julian L. Simon, "The Population Establishment, Corruption, and Reform," in *Population Policy: Contemporary Issues,* ed. Godfrey Roberts (Westport, CT: Praeger, 1990), 39.

30. On Indonesia, see Nielsen, "Draft Recommendations for Foundation Activity," May 28, 1953, FFA, 1953 General Correspondence, C-1165A, "Population." On fellowships, see Docket Item, "The Population Council," March 22, 1957; "Agenda, Meeting of Board of Trustees," October 19, 1960, RAC, RG IV3B4.2, Population Council, General File, box 36, folder 507; Caldwell, *Limiting Population Growth,* 46–47. On expert network, see John Sharpless, "The Rockefeller Foundation, the Population Council, and the Groundwork for New Population Policies," *Rockefeller Archive Center Newsletter* (Fall 1993): 3.

31. Caldwell, *Limiting Population Growth,* 44; Whelpton, "Report on People and Institutions in India," April 1954, RAC, Population Council Papers, RG IV3B4.2, General File Series, box 32, folder 470.

32. "Grant in Aid to Harvard University"; John B. Wyon and John E. Gordon, *The Khanna Study: Population Problems in the Rural Punjab* (Cambridge: Harvard University Press, 1971); Roger Revelle, "Application for Research Grant," June 22, 1967, RAC, Rockefeller Foundation, RG 1.2, Projects, series 200A, Harvard University—Population Studies in India, box 132, file 1170.

33. Wyon and Gordon, *The Khanna Study,* 295, 321, 329, 332; Gordon to Robert Morison, October 3, 1955, and "Harvard University Population Study," January 27, 1956, both in RAC, Rockefeller Foundation, RG 1.2, Projects, 200 United States, Harvard University—Indian Population 1953–1955, box 45, folder 369.

34. Taylor circular letter, December 1953, RAC, Rockefeller Foundation, RG 1.2, Projects, 200 United States, Harvard University—Indian Population 1953–1955, box 45, folder 369; Wyon and Gordon, *The Khanna Study,* 12, 15, 94, 130–131, 138, 154–155, 193–194; Mahmood Mamdani, *The Myth of Population Control: Family, Caste and Class in an Indian Village* (New York: Monthly Review Press, 1972), 158–160.

35. Balfour memo, dated April 3, 1956, RAC, Rockefeller Foundation, RG 1.2, Projects, 200 United States, Harvard University—Indian Population 1953–1955, box 45, folder 370; Wyon and Gordon, *The Khanna Study,* 195, 301; Sohan

Singh, John E. Gordon, and John B. Wyon, "Medical Care in Fatal Illnesses of a Rural Punjab Population: Some Social, Biological, and Cultural Factors and Their Ecological Implications," *Indian Journal of Medical Research* 50 (1962): 869, 874–875.

36. Mamdani, *Myth of Population Control,* 47–48, 66–104, 146–147; Amrit Kaur to Wyon and Taylor, December 5, 1955, and Rusk to Bugher, March 22, 1957, RAC, Rockefeller Foundation, RG 1.2, Projects, 200 United States, Harvard University—Indian Population 1953–1955, box 45, folders 369–371. On legibility, see James C. Scott, *Seeing like a State: How Certain Schemes to Improve the Human Condition Have Failed* (New Haven: Yale University Press, 1998).

37. Gamble to Blacker, May 5, 1955, IPPFA, series B, reel 716, frames 2103–2105; Doone Williams and Greer Williams, *Every Child a Wanted Child: Clarence James Gamble, M.D. and His Work in the Birth Control Movement,* ed. Emily P. Flint (Boston: Harvard University Press, 1978), 215–216; Yoshio Koya, *Pioneering in Family Planning: A Collection of the Papers on the Family Planning Programs and Research Conducted in Japan* (Tokyo: Japan Medical, 1963); Koya to Gamble, July 7, 1956, Francis A. Countway Library of Medicine, Boston, Clarence Gamble Papers, H MS C23, box 95, folder 1556 (Japan). For this last document, and my understanding of Gamble's work more generally, I am indebted to Nancy Davenport.

38. Williams and Williams, *Every Child a Wanted Child,* 220–222, 236–237, 242–243, 289; Laura Briggs, *Reproducing Empire: Race, Sex, Science, and U.S. Imperialism in Puerto Rico* (Berkeley: University of California Press, 2002), 125. See also James Reed, *From Private Vice to Public Virtue: The Birth Control Movement and American Society* (New York: Basic Books, 1978), 227; and Philip R. Reilly, *The Surgical Solution: A History of Involuntary Sterilization in the United States* (Baltimore: Johns Hopkins University Press, 1991), 133–134. On "natives," see Barbara Cadbury to Butcher, June 4, 1954, IPPFA, series B, reel 149, frames 2207–2208.

39. Adele Clarke, *Disciplining Reproduction: Modernity, American Life Sciences, and "The Problems of Sex"* (Berkeley: University of California Press, 1998), chap. 6. For the Sanger quote, see Bernard Asbell, *The Pill: A Biography of the Drug That Changed the World* (New York: Random House, 1995), 9; and see also Andrea Tone, *Devices and Desires: A History of Contraceptives in America* (New York: Hill and Wang, 2001), 212.

40. Reed, *From Private Vice to Public Virtue,* 323–324, 330, 332–333; Lara V. Marks, *Sexual Chemistry: A History of the Contraceptive Pill* (New Haven: Yale University Press, 2001), 55–56.

41. Reed, *From Private Vice to Public Virtue,* 334–340, 355–359; Ellen Chesler, *Woman of Valor: Margaret Sanger and the Birth Control Movement in America* (New York: Simon and Schuster, 1992), 432–433.

42. Paul Vaughan, *The Pill on Trial* (New York: Coward-McCann, 1970), 46–47.

43. Minutes, "Advisory Council on Scientific Policy," February 1, 1956, National Archives, Kew, UK (hereafter UKNA), CO 859/666. On the dubious trade-off, see Betsy Hartmann, *Reproductive Rights and Wrongs: The Global Politics of Population Control Surveillance,* rev. ed. (Boston: South End Press, 1995), 172–175. On the pill, see Marks, *Sexual Chemistry,* 104–105, 113; and Chesler, *Woman of Valor,* 45.

44. Chesler, *Woman of Valor,* 440–443; David J. Garrow, *Liberty and Sexuality: The Right to Privacy and the Making of Roe v. Wade* (Berkeley: University of California Press, 1998), 217–218; Loretta McLaughlin, *The Pill, John Rock, and the Church: The Biography of a Revolution* (Boston: Little, Brown, 1982), 155–164.

45. Marks, *Sexual Chemistry,* 77–79, 88; Fifth Meeting of the World Population and Planned Parenthood Appeal Committee, November 10, 1960, and receipts from British Drug Houses Ltd., Boots Pure Drug Co., and Beecham Group Ltd., IPPFA, series B, reel 715, frames 2285, 2316, 2318, 2420–2421. On Bombay, see Linder, *Crusader for Sex Education,* 187.

46. Suitters, *Be Brave and Angry,* 67–74; C. P. Blacker, "Family Planning and Eugenic Movements," *Eugenics Review* 47, no. 4 (1956): 231.

47. Blacker to Vogt, December 12, 1951, IPPFA, series B, reel 718, frames 1731–1734; Houghton to Barbara Cadbury, July 6, 1954, IPPFA, series B, reel 149, frames 2204–2206; Blacker to Sanger, August 8, 1955, IPPFA, series B, reel 716, frames 2078–2080. For objections, see Blacker to Stone, March 4, 1955; Hanna Rizk to Blacker, October 1, 1955; and Peers, "Confidential Reports on Miss Edith M. Gates," October 15, 1955; all in IPPFA, series B, reel 716, frames 2106–2108, 2049–2050, 2051–2052; and Blacker to Bird, October 3, 1956, IPPFA, series B, reel 717, frame 259. For Gamble's role assisting the new associations, see Williams and Williams, *Every Child a Wanted Child,* 240–241, 291, 299; and Osborn, "Notes on Ad Hoc Meeting," March 7, 1956.

48. Williams and Williams, *Every Child a Wanted Child,* 184, 289–290.

49. Amano to Gamble, January 26, 1956, Francis A. Countway Library of Medicine, Boston, Clarence Gamble Papers, H MS C23, box 95, folder 1555 (Japan), courtesy of Nancy Davenport.

50. Williams and Williams, *Every Child a Wanted Child,* 299–302, 305.

51. Osborn to Houghton, March 7, 1957, CMAC, C. P. Blacker Papers, box 11, "Blacker: Correspondence, Eugenics Society, IPPF, January–June 1957"; Betty Friedan, *The Feminine Mystique* (New York: W. W. Norton, 1963), 16–17; Reed, *From Private Vice to Public Virtue,* 304.

52. Dhanvanthi Rama Rau, *An Inheritance: The Memoirs of Dhanvanthi Rama Rau* (New York: Harper and Row, 1977), 272–275; Linder, *Crusader for Sex*

Education, 195; Ping-ti Ho, *Studies on the Population of China* (Cambridge: Harvard University Press, 1959), 87–94; H. Yuan Tien, *China's Population Struggle: Demographic Decisions of the People's Republic, 1949–1969* (Columbus: Ohio State University Press, 1973), 39–49, 175–176, 179–194, 354–356; Huan Hsiang, "China Faces the Problem," September 29, 1956, IPPFA, series B, reel 149, frames 2486–2494.

53. Tyrene White, *China's Longest Campaign: Birth Planning in the People's Republic, 1949–2005* (Ithaca: Cornell University Press, 2006), 38; Roderick MacFarquhar, Timothy Cheek, and Eugene Wu, eds., *The Secret Speeches of Chairman Mao: From the Hundred Flowers to the Great Leap Forward* (Cambridge: Harvard University Press, 1989), 159–160.

54. Tien, *China's Population Struggle,* 229–230.

55. Ibid., 187–188; Donald P. Warwick, *Bitter Pills: Population Policies and Their Implementation in Eight Developing Countries* (Cambridge: Cambridge University Press, 1982), 10; Saad Eddin Ibrahim, "State, Women, and Civil Society: An Evaluation of Egypt's Population Policy," in *Family, Gender, and Population in the Middle East,* ed. Carla Makhlouf Obermeyer (Cairo: American University in Cairo Press, 1995), 59–60. For the minister quote, see Vogt, draft IPPF leaflet, December 6, 1957, IPPFA, series B, reel 718, frames 893–900.

56. Richard Symonds and Michael Carder, *The United Nations and the Population Question, 1945–1970* (London: Sussex University Press, 1973), 77, 110; "Notes on Recent Developments in Moslem Countries Affecting Family Planning," ca. 1957, UKNA, CO/859/1026; Iverson to Howard, November 21, 1953, and Iverson, "Egypt, Report of Visit," November 20, 1953, FFA, 1953 General Correspondence, C-1165A, "Population." On Nasser's concerns, see Warwick, *Bitter Pills,* 10; and Bellivier, "Note au sujet du contrôle des naissances en Egypte," ca. December 1955, Archives de la Ministère des Affaires Etrangères, Paris, Mission de liaison algérien, Action extérieure, dossier 65, RAU. On the dam as an alternative, see Jon B. Alterman, *Egypt and American Foreign Assistance, 1952–1956: Hopes Dashed* (New York: Palgrave Macmillan, 2002), 101, 110–111, 132.

57. "Population Trends in British Colonial Territories and Policy in Regard to Population Control," January 20, 1956, UKNA, CO 859/1028; Margaret Chidell, "The Birth Control Clinics of the Family Planning Association of Hong Kong, 1951 to 1958," in *The Sixth International Conference on Planned Parenthood: Report of the Proceedings, 14–21 February 1959* (London: IPPF, 1959), 348.

58. Matthew Connelly, *A Diplomatic Revolution: Algeria's Fight for Independence and the Origins of the Post–Cold War Era* (New York: Oxford University Press, 2002), 11, 17–18, 20–21, 87–89; Fernand Boverat, "Le Surpeuplement accéléré de l'Algérie et ses conséquences tragiques," ca. 1956, Archives Nationales, Paris (hereafter AN), AJ80, Commissaire Général au Plan, box 149.

59. Compte Rendu, Haut Comité de la Population, January 27, 1956, April

11, 1956, and October 16, 1959, Centre des Archives Contemporaines, Fontainebleau, versement 860269, art. 1. On reforms targeting women in Algeria, see Marnia Lazreg, *The Eloquence of Silence: Algerian Women in Question* (New York: Routledge, 1994), 90–97.

60. K. Ram to V. K. B. Pillai, April 6, 1957, National Archives of India, New Delhi, Ministry of Health, PH Section II, file no. 39-4/45-H (I); Nehru to Huxley, January 14, 1955, IPPFA, series B, reel 156, frame 1879; Nehru to Chief Ministers, May 28, 1959, in *Jawaharlal Nehru Letters to Chief Ministers 1947–1964,* ed. G. Parthasarathi, vol. 5 (London: Oxford University Press, 1989), 256–257; *Second Five Year Plan* (New Delhi: Government of India, 1956), 7, 21–22.

61. Frank W. Notestein, "The Economics of Population and Food Supplies: Economic Problems of Population Change," in *Proceedings of the Eighth International Conference of Agricultural Economists* (New York: Oxford University Press, 1953), 18; Hodgson, "Demography as Social Science," 12–13, 18–19; Caldwell, *Limiting Population Growth,* 29. Particularly influential among these works, especially in India, was A. J. Coale and E. M. Hoover, eds., *Population Growth and Economic Development in Low Income Countries* (Princeton: Princeton University Press, 1958).

62. Gupta, Sinha, and Bardhan, *Family Welfare Programme in India,* 1:54–55, 98–99; "Report of the Family Planning Third Five Year Plan Committee," ca. 1960, NIHFW, Depository, 204 IND.

63. Directorate General of Health Services, Ministry of Health, "Family Planning in India: A Review of the Progress"; Gupta, Sinha, and Bardhan, *Family Welfare Programme,* 1:71.

64. "Report of the Family Planning Third Five Year Plan Committee." On incentives, see "Introduction to the Memorandum on Administrative Implementation of Family Planning Policy," in *The Sixth International Conference on Planned Parenthood,* 288; "Agenda for the Sixth Meeting of the Central Family Planning Board," February 1, 1959, IPPFA, series B, reel 156, frames 1805–1811; Gupta, Sinha, and Bardhan, *Family Welfare Programme,* 2:4.

65. "Family Planning in India: A Review of the Progress."

66. "Notes on Recent Developments in Moslem Countries Affecting Family Planning," ca. 1957, UKNA, CO/859/1026; Begum Saida Waheed, "Report of the Family Planning Association of Pakistan," in *Sixth International Conference on Planned Parenthood,* 363; Marshall C. Balfour, "Family Planning in Asia," *Population Studies* 15, no. 2 (1961): 105.

67. Suitters, *Be Brave and Angry,* 85–86, 163–164, 187; Rufus Day, "Summary of Field-Work Projects," n.d., IPPFA, series B, reel 664, frames 363–367; Linder, *Crusader for Sex Education,* 219–220, 236–237; Hannes Hyrenius and Ulla Åhs, *The Sweden-Ceylon Family Planning Project* (Göteborg: Demographic Institute, 1968); Symonds and Carder, *The United Nations,* 98.

68. NSC meeting, December 3, 1958, Dwight D. Eisenhower Library, Abilene, Ann Whitman File, NSC Series.

69. Phyllis Tilson Piotrow, *World Population Crisis: The United States Response* (New York: Praeger, 1973), 36–40; Peter J. Donaldson, *Nature against Us: The United States and the World Population Crisis, 1965–1980* (Chapel Hill: University of North Carolina, 1990), 23; Critchlow, *Intended Consequences,* 42–43.

70. "Composite Report of the President's Committee to Study the United States Military Assistance Program," August 17, 1959, U.S. National Archives, College Park, MD (hereafter USNA), RG 59, Central Files, 1967–1969, Soc 13-3, box 3124.

71. "Statement of Roman Catholic Bishops of U.S. on Birth Control," *New York Times,* November 26, 1959; Chesler, *Woman of Valor,* 454–455.

72. State Department circular, December 2, 1959, USNA, RG 59, Central Files, 1967–1969, Soc 13-3, box 3124; NSC Meetings, May 28 and August 18, 1959, Dwight D. Eisenhower Library, Abilene, Ann Whitman File, NSC Series.

73. Piotrow, *World Population Crisis,* 48; John R. Wilmoth and Patrick Ball, "Arguments and Action in the Life of a Social Problem," *Social Problems* 42, no. 3 (1995): 323; Krock, "The Worst Bomb of All Is the Population Bomb because It Produces an Explosion of People," *New York Times,* October 3, 1959; and Krock, "Politicians Ignore the Earth's Chief Problem but Others Give Attention to 'Population Bomb,'" *New York Times,* November 14, 1959.

74. "Eugenics Society: Memorandum by Certain Members of the Council of the Society," January 18, 1960, CMAC, C. P. Blacker Papers, box 11, "Blacker: Eugenics Society, 1960"; Blacker to Griessemer, October 17, 1959, IPPFA, series B, reel 719, frames 1036–1039 (emphasis in original). On Notestein's testimony, see "Health, Economic Development and Population Growth," RAC, RG 5, John D. Rockefeller 3rd Papers, series 1, subseries 5, Population Interests, box 82, folder 680; John Sharpless, "Population Science, Private Foundations, and Development Aid: The Transformation of Demographic Knowledge in the United States, 1945–1965," in *International Development and the Social Sciences,* ed. Frederick Cooper and Randall Packard (Berkeley: University of California Press, 1997), 193.

75. Harrison Brown, *The Challenge of Man's Future: An Inquiry Concerning the Condition of Man during the Years That Lie Ahead* (New York: Viking, 1954), 102–106, 232, 260–264; World Population Emergency Campaign, "A Statement of Purpose," March 20, 1960, "The World Population Emergency Campaign Reports on the Population Explosion . . . The Problem of Our Time," November 1961, "Grants Made by the World Population Emergency Campaign," 1962, and "Family Planning International Campaign: 12th Meeting of Steering Committee," January 26, 1965, all in IPPFA, series B, reel 715, frame 1069–1079, 1195–1207, 2223 (emphasis in original).

76. McPeak to Osborn, June 16, 1958, FFA, 1953 Grant Files, PA 54–20,

"The Population Council"; Carroll to Overseas Representatives, April 1, 1959, and Hill to Ensminger, April 15, 1959, both in FFA, 1953 Grant Files, PA59-482, "Government of India"; "Policy Paper: Population" March 28, 1963, FFA, reports, file 002856; Barnett, "Ford Foundation Interest in Population Problem," April 9, 1961, USNA, RG 59, Decimal Files, 800.401; Minutes of the Meeting of the Management and Planning Committee, June 13, 1962, IPPFA, 1.4.1.1.3.1.

77. Agnew, to FPA branches, June 27, 1960, IPPFA, series B, reel 715; Linder, *Crusader for Sex Education,* 233; Minutes of the Meeting of the Management and Planning Committee, May 12–13, 1960, IPPFA, 1.4.1.1.3.

78. Minutes of the Meeting of the Management and Planning Committee, May 12–13, 1960; Linder, *Crusader for Sex Education,* 234; Day to Burdick, January 16, 1961, IPPFA, series B, reel 719, frame 1027.

79. Pyke to Day, December 6, 1962, IPPFA, series B, reel 719, frames 876–877; Blacker to Griessemer, September 21, 1960, IPPFA, series B, reel 715, frame 1186.

80. "Birth Control Urged," *New York Times,* November 18, 1960; Arthur Conte, "Rapport d'information sur l'aide aux pays sous-développés," June 26, 1959, AN, Georges Bidault Papers, 457AP, dossier 180, Notes Politiques reçus et divers, 1958–1961. For India and Pakistan, see Chagla, "Text of Address," May 11, 1961, IPPFA, series B, reel 715, frames 2131–2137; and Commission économique pour l'Asie et Extrême-Orient, Procès-Verbaux officiels, seizième session, 9–21 Mars 1960, 223rd meeting, Archives de la Ministère des Affaires Etrangères, Paris, Asie Oceanie 1956–1967, Dossiers généraux, ECAFE, Juin–Août 1960, dossier 441.

81. "The Political Impact of Recent Population Trends," *Foreign Affairs* 36, no. 2 (1958): 295.

82. Cleveland, "Reflections on the 'Revolution of Rising Expectations,'" July 9, 1950; I am grateful to David Ellwood for this document. Davis, "The Political Impact of Recent Population Trends," 296.

83. "The World Population Emergency Campaign Reports."

84. Caldwell, *Limiting Population Growth,* 23–24.

85. Davis, "Confidential Report on the Tenth Session of the Population Commission of the United Nations," 1959, USNA, RG 59, Central Decimal Files, 340.15; Symonds and Carder, *The United Nations,* 79, 122–123; Halvor Gille, "What the Asian Censuses Reveal," *Far Eastern Economic Review* 32, no. 13 (1961): 636–637; Caldwell, *Limiting Population Growth,* 23–24.

86. Raina, "Family Planning Program: Report for 1962–63," 46, NIHFW, 204/83 RAI; India, Planning Commission, *Third Five Year Plan* (Delhi: Government of India, 1961), 25, 72; Central Bureau of Health Intelligence, *Annual Report of the Directorate General of Health Services: 1962* (New Delhi: Government of India, 1962), 151.

87. Raina, "Family Planning Program: Report for 1962–63," 33–34; Central Bureau of Health Intelligence, *Annual Report of the Directorate General of Health Services: 1960* (New Delhi: Government of India, 1960), 181; "Population Control and Family Planning: Report of the Indian Parliamentary and Scientific Committee," Planning Commission Archives, New Delhi; Gupta, Sinha, and Bardhan, *Family Welfare Programme,* 2:6.

88. Srinivasan, *Regulating Reproduction in India's Population,* 33–34; Chagla, "Text of Address."

89. Eccles, "Text of Address," May 12, 1961, IPPFA, series B, reel 715, frames 2110–2116; Paul Vaughan, *The Pill on Trial,* 25.

90. Cadbury, "The Role of the International Planned Parenthood Federation," IPPFA, series B, reel 715, frames 1104–1109.

91. Chesler, *Woman of Valor,* 457–458; "Population Boom Termed Perilous," *New York Times,* May 12, 1961; Critchlow, *Intended Consequences,* 119.

6. Controlling Nations

1. Dag Malm and W. McIlquham, "Explanatory Memorandum," August 18, 1961, National Archives, Kew, UK (hereafter UKNA), Overseas Development 10/32; Doris H. Linder, *Crusader for Sex Education: Elise Ottesen-Jensen (1886–1973) in Scandinavia and on the International Scene* (Lanham, MD: University Press of America, 1996), 132–133, 235; Richard Symonds and Michael Carder, *The United Nations and the Population Question* (London: Sussex University Press, 1973), 115–117; "Item 84: Population Growth and Economic Development," D. A. Burns to W. Bentley, October 7, 1961, UKNA, FO 371/161053.

2. UN General Assembly, Sixteenth Session, Official Records, Summary Record of the Second Committee, 430–431; Symonds and Carder, *The United Nations,* 117–118.

3. UN General Assembly, Seventeenth Session, Official Records, Summary Record of the Second Committee, 443–444, 449, 491.

4. Ibid., 451–452.

5. Ibid., 454, 459.

6. UN General Assembly, Seventeenth Session, Official Records, Summary Record of the Plenary Meeting, 1171–1178; Symonds and Carder, *The United Nations,* 121–122.

7. Ball to Rusk, June 30, 1961, U.S. National Archives, College Park, MD (hereafter USNA), RG 59, Decimal Files, 800.401; Phyllis Tilson Piotrow, *World Population Crisis: The United States Response* (New York: Praeger, 1973), 55–56, 59–60, 63, 73.

8. McGhee to Ball, August 30, 1961, USNA, RG 59, Records of the Policy Planning Staff 1957–1961, Lot 67D548; UN General Assembly, Sixteenth Ses-

sion, Official Records, Summary Record of the Second Committee, 458–459; UN General Assembly, Seventeenth Session, Official Records, Summary Record of the Plenary Meeting, 1178–1179; Piotrow, *World Population Crisis,* 60–61, 67.

9. Piotrow, *World Population Crisis,* 73–74.

10. Ball to Rusk, June 30, 1961; Fifth Meeting of the Management and Planning Committee, October 22, 1963, International Planned Parenthood Federation Archives, London (hereafter IPPFA), 1.4.1.1.3, "IPPF Management and Planning Committee"; Piotrow, *World Population Crisis,* 68.

11. Piotrow, *World Population Crisis,* 68–69, 72–75, 78; Guttmacher, "Notes on December 10, 1963 Washington, DC Meetings," Rockefeller Archive Center, Tarrytown, NY (hereafter RAC), RG IV3B4.5, Population Council, Organization File, box 107.

12. Second Meeting of the Management and Planning Committee, April 6, 1961, IPPFA, 1.4.1.1.3, "IPPF Management and Planning Committee." On Nehru, see Nehru to Vijaya Lakshmi Pandit, November 12, 1952, Nehru Memorial Museum and Library, Delhi, Vijaya Lakshmi Pandit Papers, file 47; Nehru, July 2, and July 20, 1959, minutes to file, and Rusk to Deshmukh, December 16, 1958, all in National Archives of India, New Delhi, Ministry of External Affairs, American Division, file no. 67(4)-AMS/58.

13. Douglas Ensminger Oral History, November 1, 1971, and October 28, 1972, Ford Foundation Archives, New York City (hereafter FFA), B.1, A.38. Kathleen McCarthy, "The Ford Foundation's Population Program in India, Pakistan and Bangladesh, 1959–1981," 1985, FFA, report no. 011011, pp. 14, 100.

14. Meredith Minkler, "Consultants or Colleagues: The Role of U.S. Population Advisors in India," *Population and Development Review* 3 (1977): 413; Ensminger Oral History, October 28, 1972, FFA, A.38; Naid minute to file, July 1, 1958, National Archives of India, New Delhi, Ministry of External Affairs, American Division, file no. 67(4)-AMS/58.

15. "Agenda, Meeting of Board of Trustees," October 19, 1960, RAC, RG IV3B4.2, Population Council, General File, box 36, folder 507; McCarthy, "Ford Foundation's Population Program," 21–22; Radhika Ramasubban and Bhanwar Singh Rishyasringa, "From Population Control to Reproductive Health and Sexual Rights: Fifty Years of the Ford Foundation Program in India," 2001, unpublished manuscript in FFA, 22–24; Ensminger Oral History, November 1, 1971, and October 28, 1972, FFA, B.1, A.38.

16. Lyle Saunders, "Next Steps," September 27, 1968, FFA, file 001405.

17. Clipping from *Daily Herald,* November 24, 1931, Contemporary Medical Archives Center, Wellcome Institute for the History of Medicine, London (hereafter CMAC), SA/FPA, box 333, BCIC 1929–1939; Huxley, "World Population," *Scientific American* 194, no. 3 (1956): 66, 69; Chagla, "Text of Address," May 11, 1961, IPPFA, series B, reel 715, frames 2131–2137.

18. Linder, *Crusader for Sex Education,* 236–237; Notestein Diary Notes, March 1–8, 1963, RAC, RG IV3B4.2, Population Council, General File, box 32, folder 462.

19. Warren O. Nelson and Alan F. Guttmacher, "Introduction," and Christopher Tietze, "Intra-Uterine Contraceptive Rings: History and Statistical Appraisal," in *Intra-Uterine Contraceptive Devices: Proceedings of the Conference April 30–May 1, 1962, New York City, International Congress Series No. 54,* ed. Christopher Tietze and Sarah Lewit (Amsterdam: Excerpta Medica, 1962), 7, 10, 13–14; Guttmacher to Pyke, August 22, 1962, IPPFA, series B, reel 718, frame 1049; James Reed, *From Private Vice to Public Virtue: The Birth Control Movement and American Society* (New York: Basic Books, 1978), 275–276.

20. Reed, *From Private Vice to Public Virtue,* 305–306; Tietze and Lewit, *Intra-Uterine Contraceptive Devices,* 122.

21. Tietze and Lewit, *Intra-Uterine Contraceptive Devices,* 122–125.

22. Ibid., 123; Reed, *From Private Vice to Public Virtue,* 307.

23. Tietze and Lewit, "Intra-uterine Contraception: Effectiveness and Acceptability," in *Intra-Uterine Contraception: Proceedings of the Second International Conference, October 2–3, 1964, International Congress Series,* ed. S. J. Segal, A. L. Southam, and K. D. Shafer (Amsterdam: Excerpta Medica, 1964), 98–110, and discussion, 118. On Guttmacher, see Tietze and Lewit, *Intra-Uterine Contraceptive Devices,* 122.

24. Tietze to Segal, September 16, 1964; John C. Cobb to Tietze, August 19, 1964; and Tietze to Cobb, September 29, 1964; all in RAC, RG IV3B4.4b, Population Council, National Committee on Maternal Health, box 94, folder 1764.

25. C. H. Lee and P. Y. Wei, "Control of Fertility in Taichung, Taiwan, by Using Intra-Uterine Contraceptive Devices: Medical Follow-Up"; Han Su Shin, "The Intra-Uterine Device Program in Korea"; Robert E. Hall, "Comparative Merits of the Spiral, Loop, and Bow: A Survey of 11,833 Woman-Months of Experience at Sloane Hospital"; and "Discussion," all in Segal, Southam, and Shafer, *Intra-Uterine Contraception,* 23–27, 45–51, 66–75, 119, and 121.

26. IPPF press release, October 15, 1964, Harper & Row Publisher Records, Manuscript Collections, Columbia University Library, series 2, International Planned Parenthood, box 303, "I."

27. Barbara Evans interview with Joan Swingler and Penny Kane, November 23, 1982, CMAC, PP/PRE/J.1/27, box 21; Barbara Evans, *Freedom to Choose: The Life and Work of Dr. Helena Wright, Pioneer of Contraception* (London: The Bodley Head, 1984), 220, 224, 231; John and Pat Caldwell, *Limiting Population Growth and the Ford Foundation Contribution* (London: Frances Pinter, 1986), 120–121.

28. Segal, Southam, and Shafer, *Intra-Uterine Contraception,* 217. On Mrs. Searle, see Elizabeth Siegel Watkins, *On the Pill: A Social History of Oral Contracep-*

tives, 1950–1970 (Baltimore: Johns Hopkins University Press, 1998), 70. For IPPF recommendations, see press release, October 15, 1964.

29. Notestein Diary Notes, March 8, 1963, RAC, RG IV3B4.2, Population Council, General File, box 32, folder 462.

30. Ibid., March 3, 1963; John Ensor Harr and Peter J. Johnson, *The Rockefeller Conscience: An American Family in Public and in Private* (New York: Scribner's, 1991), 164. On loop investment, see Reed, *From Private Vice to Public Virtue,* 307.

31. "Conversation with Sam Keeny," March 16, 1965, RAC, RG IV3B4.8, Population Council, Foreign Correspondence, box 48; Notestein Diary Notes, September 8–10, 1964, RAC, RG IV3B4.2, Population Council, General File, box 32, folder 462.

32. Notestein Diary Notes, September 8–10, 1964; "Monthly Progress Reports" in FFA, 1953 Grant Files, PA 54–20, "The Population Council"; Ronald Freedman and John Y. Takeshita, *Family Planning in Taiwan: An Experiment in Social Change* (Princeton: Princeton University Press, 1969). On Taiwan and South Korea as examples, see "Conversation with Sam Keeny," March 16, 1965.

33. "Policy Paper: Population," ca. March 1963, FFA, file 002856; David Hopper, "The Population Council: A Review," May, 1975, FFA, Papers of David Bell, series V, box 33, folder 783; Fifth Meeting of the Management and Planning Committee, October 22, 1963, IPPFA, 1.4.1.1.3, "IPPF Management and Planning Committee"; Beryl Suitters, *Be Brave and Angry: Chronicles of the International Planned Parenthood Federation* (London: IPPF, 1973), 246, 260.

34. "Policy Items to Be Raised in the Report," June 1962, and Third Meeting of the Management and Planning Committee, June 13, 1962, both in IPPFA, 1.4.1.1.3.1, "IPPF Management and Planning Committee."

35. Ian Dowbiggin, "'A Rational Coalition': Euthanasia, Eugenics, and Birth Control in America, 1940–1970," *Journal of Policy History* 14, no. 3 (2002): 226–231; "Proceedings of the First International Conference on Voluntary Sterilization," April 16, 1964, National Institute of Health and Family Welfare, Documentation Centre, New Delhi (hereafter NIHFW), 247 INT.

36. "Proceedings of the First International Conference on Voluntary Sterilization."

37. Ibid.

38. Ibid. James F. Donnelly to Wylda B. Cowles, January 23, 1964, RAC, RG IV3B4.5, Population Council, Organization File, box 107, folder 1992. On Davis, see Nicole J. Grant, *The Selling of Contraception: The Dalkon Shield Case, Sexuality, and Women's Autonomy* (Columbus: Ohio State University Press, 1992), 31; Andrea Tone, *Devices and Desires: A History of Contraceptives in America* (New York: Hill and Wang, 2001), 258–259. On Madras, see "Introduction to the Memorandum

on Administrative Implementation of Family Planning Policy," in IPPF, *The Sixth International Conference on Planned Parenthood: Report of the Proceedings, 14–21 February 1959* (London: IPPF, 1959), 288.

39. AVS press release, January 8, 1968, United Nations Archives and Records Centre, New York City (hereafter UNRC), S[eries]-0291-[box] 0009, Chef de Cabinet, unnumbered files, 1958–1973, India Family Planning, September 1964–March 1970.

40. H. W. Brands, *The Wages of Globalism: Lyndon Johnson and the Limits of American Power* (New York: Oxford University Press, 1995), 131–133; Paul Y. Hammond, *LBJ and the Presidential Management of Foreign Relations* (Austin: University of Texas Press, 1992), 63–67, 73–74.

41. Piotrow, *World Population Crisis,* 78; Bundy to Valenti, March 7, 1965, and Valenti to Johnson, March 9, 1965, Lyndon Baines Johnson Library, Austin (hereafter LBJ Library), LBJ Papers, Welfare, EX WE 11/22/1963, box 1.

42. Komer to Bundy, April 27, 1965, and Enke, "Lower Birth Rates—Some Economic Aspects," February 12, 1965, both in LBJ Library, LBJ Papers, National Security File, Files of Robert W. Komer, box 48, "Population Control 1965–March 1966."

43. Enke, "Lower Birth Rates—Some Economic Aspects."

44. Ibid.

45. Ibid.

46. Regarding discounting, see Warren C. Robinson and David E. Horlacher, "Evaluating the Economic Benefits of Fertility Reduction," *Studies in Family Planning* 1, no. 39 (1969): 4–8.

47. Enke, "Lower Birth Rates—Some Economic Aspects."

48. Margaret Sanger, "Address of Welcome," in *Problems of Overpopulation,* vol. 1, *The Sixth International Neo-Malthusian and Birth Control Conference,* ed. Margaret Sanger (New York: American Birth Control League, 1925), 7.

49. Komer to LBJ, April 27, 1965, LBJ Library, LBJ Papers, National Security File, Files of Robert W. Komer, box 48, "Population Control 1965–March 1966" (emphasis in original).

50. Komer to LBJ, April 27, 1965, *Foreign Relations of the United States, 1964–1968,* vol. 9 (Washington, DC: U.S. Government Printing Office, 1997) (hereafter *FRUS* with year and volume), 95–96; Komer to Moyers, April 28, 1965, and Komer to Bundy, August 4, 1965, both in LBJ Library, LBJ Papers, National Security File, Files of Robert W. Komer, box 48, "Population Control 1965–March 1966."

51. "Address in San Francisco," June 25, 1965, *Public Papers of the Presidents of the United States: Lyndon B. Johnson, 1965,* book 2 (Washington, DC: U.S. Government Printing Office, 1966), 705. On the short leash, see Brands, *The Wages of*

Globalism, 133–134; Hammond, *LBJ,* 74. Neither mentions India's population policies as a factor in these negotiations.

52. Hammond, *LBJ,* 74–75.

53. On the World Bank, see ibid., 77–79; George D. Woods to Narasimhan, January 4, 1965, UNRC, S[eries]-0291-[box] 0009, Chef de Cabinet, unnumbered files, 1958–1973, India Family Planning, September 1964–March 1970; Florence Mahoney to Douglas Cater, July 17, 1965, LBJ Library, LBJ Papers, Welfare, EX WE 11/22/1963, box 1; Berelson memo to files, September 14, 1964, RAC, RG IV3B4.5, Population Council, General File, box 29, folder 425. On the UN, see Julia Henderson to Narasimhan, September 14, 1964, S[eries]-0175-[box] 0627-06, Family Planning—India (210-1A). For the Ford–Planning Commission report, see "Evaluation of the Family Planning Programme, Reports of Assessment Teams and the Panel of Consultants," June 25, 1965, NIHFW, Depository, 06/213.8/IND.

54. Reuben Hill, "Comments on Programs in India," October 18, 1965, FFA, report no. 003684. On problems with the program, see "Evaluation of the Family Planning Programme, Reports of Assessment Teams," June 25, 1965; "Report on Demographic Training and Research in India under the Guidance of the Demographic Advisory Committee," ca. March 1965, National Archives of India, New Delhi, Ministry of External Affairs, U.I. Section, file no. UI/3532-01/65; J. P. Gupta, N. K. Sinha, and Amita Bardhan, *Evolution of Family Welfare Programme in India* (New Delhi: National Institute of Health and Family Welfare, 1992), 1:119–120 and 2:8.

55. Hill, "Comments on Programs in India," October 18, 1965; "Evaluation of the Family Planning Programme, Reports of Assessment Teams," June 25, 1965; "Conversation with Sam Keeny," March 16, 1965.

56. Deverell to Henderson, April 12, 1965, UNRC, S[eries]-0175-[box] 0627-06, Family Planning—India (210-1A); United Nations Advisory Mission, "Report on the Family Planning Programme in India," February 20, 1966, NIHFW, 204 UNI.

57. "Evaluation of the Family Planning Programme, Reports of Assessment Teams," June 25, 1965; "United Nations Family Planning Mission to India: Main Draft Recommendations," n.d., but ca. December 1965, UNRC, S[eries]-0291-[box] 0009, Chef de Cabinet, Unnumbered files, 1958–1973, India Family Planning, September 1964–March 1970; Committee on the Status of Women in India, *Towards Equality* (Delhi: Government of India, 1974), 317–318.

58. Berelson memo to files, September 14, 1964. On Nayar and IUDs, see McCarthy, "Ford Foundation's Population Program," 15; Oscar Harkavy, *Curbing Population Growth: An Insider's Perspective on the Population Movement* (New York: Plenum Press, 1995), 154. On the studies, see United Nations Advisory Mission,

"Report on the Family Planning Programme in India," February 20, 1966. On the Ford Foundation and medical research, see "Conversation with Sam Keeny," March 16, 1965.

59. "Technical Assistance Authorization," August 10, 1965, RAC, RG IV3B4.2, Population Council, General File, box 34, folder 490; Barbara Ehrenreich, Mark Dowie, and Stephen Minkin, "The Charge: Gynocide," November/December 1979, www.motherjones.com; United Nations Advisory Mission, "Report on the Family Planning Programme in India," February 20, 1966; "Unedited Transcript of Statement made by Sir Colville Deverell," April 9, 1965; "Evaluation of the Family Planning Programme, Reports of Assessment Teams," June 25, 1965.

60. I. P. M. Cargill, "Efforts to Influence Recipient Performance: Case Study of India," in *The World Bank Group, Multilateral Aid, and the 1970s*, ed. John P. Lewis and Ishan Kapur (Lexington, MA: Lexington Books, 1973), 94; Komer to Bundy, September 13, 1965, and Bowles to Komer, September 21, 1965, *FRUS, South Asia, 1964–1968*, 25:393, 420.

61. Mitra to B. Mukherjee, October 22, 1965, Nehru Memorial Museum and Library, Delhi, Asok Mitra Papers, National Planning Commission, box 152 (provisional box number).

62. Ibid.

63. "Evaluation of the Family Planning Programme, Reports of Assessment Teams," June 25, 1965; "Summary Record of the First Meeting of the Central Family Planning Council," December 31, 1965, attached to "Central Family Planning Council 2nd Meeting Agenda," June 27, 1966, NIHFW, Depository, 204 IND. On "the enemy," see "The Problem—Some Broad Conclusions," n.d., but ca. April 1965, UNRC, S[eries]-0175-[box] 0627-06, Family Planning—India (210-1A).

64. "Address of Doctor Sushila Nayar," December 31, 1965, attached to "Central Family Planning Council 2nd Meeting Agenda," June 27, 1966, NIHFW, Depository, 204 IND. Budget figures are from the Central Bureau of Health Intelligence, "Annual Report of the Directorate General of Health Services, 1964–66" (New Delhi, Government of India), 262, and "Progress of Family Planning Programme in India," March 1970, NIHFW, 204/83 IND.

65. Mukherjee to State Health Ministers, November 10, 1965; "Proceedings" of meetings on "The Physiological Approach to Fertility Control," September 18, 1952, FFA, 1953 General Correspondence, C-1165, "Population"; "Unedited Transcript of Statement made by Sir Colville Deverell," April 9, 1965.

66. W. Parker Mauldin, "Retention of IUD's," August 1966, RAC, RG IV3B4.4b, Population Council, National Committee on Maternal Health, box 95, folder 1773; Freedman and Takeshita, *Family Planning in Taiwan*, 249–250.

67. Commissioner, Agricultural and Rural Development, to all Block Development and Panchayat Officers, January 29, 1965, Ministry of Health and Family

Welfare, Department of Family Welfare, Archives, New Delhi; B. Mukherjee, "Note for the Cabinet Committee on Family Planning Programme," December 4, 1965, Nehru Memorial Museum and Library, Asok Mitra Papers, National Planning Commission, box 152 (provisional box number); "Address of Doctor Sushila Nayar," December 31, 1965. On the plant, see Asok Mitra, "Proposals for Implementation of the Family Planning Program," December 2, 1965, Nehru Memorial Museum and Library, Asok Mitra Papers, National Planning Commission, box 152 (provisional box number).

68. Freeman to Johnson, November 26, 1965, Komer to Johnson, December 6, 1965, and LBJ-Subramaniam, memcon, December 20, 1965, all in *FRUS, South Asia, 1964–1968,* 25:476–479, 484–486, 516–518.

69. Komer to LBJ, January 4, 1966, Rostow to LBJ, January 6, 1966, and Komer to LBJ, January 18, 1966, all in *FRUS, South Asia, 1964–1968,* 25:524, 526–527, 541–542.

70. J. Anthony Lukas, "Nehru's 'Munshi' Comes Out of Nehru's Shadow," *New York Times Magazine,* November 28, 1965. On ancestral home and radios, see Gandhi to Narasimhan, September 13, 1964, and November 4, 1965, UNRC, S[eries]-0291-[box] 0009, Chef de Cabinet, unnumbered files, 1958–1973, India Family Planning, September 1964–March 1970. On Gandhi and the IUD, see Berelson memo to files, September 14, 1964.

71. Bowles to LBJ, January 20, 1966, *FRUS, South Asia, 1964–1968,* 25:546–548.

72. Joseph A. Califano, *The Triumph and Tragedy of Lyndon Johnson: The White House Years* (College Station: Texas A&M University Press, 2000), 154–155; Califano, *Inside: A Public and Private Life* (New York: Public Affairs, 2004), 172–173 (emphasis in original).

73. Komer to LBJ, March 10, March 21, March 23, and March 27, 1966, and Gaud to LBJ, March 25, 1966, all in LBJ Library, LBJ Papers, National Security File, Files of Robert W. Komer, box 133, Prime Minister Gandhi Visit Papers (emphasis in original).

74. Lewis to Komer, March 21, 1966; Rusk to LBJ, March 26, 1966; "Visit of Prime Minister Indira Gandhi of India: Strategy Paper," March 21, 1966, all in LBJ Library, LBJ Papers, National Security File, Files of Robert W. Komer, box 133, Prime Minister Gandhi Visit Papers (emphasis in original).

75. Memcon, Gandhi-LBJ, March 28, 1966, *FRUS, South Asia, 1964–1968,* 25:596–598; "Special Message to the Congress," March 30, 1966, *Public Papers of the Presidents of the United States: Lyndon B. Johnson,* 1966, book 1 (Washington, DC: U.S. Government Printing Office, 1967), 369.

76. Mukherjee Committee Report, April 16, 1966; Govind Narain minutes to file, April 20 and April 21, 1966, Ministry of Health and Family Welfare, Department of Family Welfare, Archives, New Delhi.

77. Mukherjee Committee Report, April 16, 1966; United Nations Advisory Mission, "Report on the Family Planning Programme in India," February 20, 1966 (emphasis in original).

78. Ibid.; Narain minute to file, April 20, 1966.

79. Ministry of Health and Family Planning, "A Danger Signal," in RAC, Rockefeller Foundation, RG 1.2 Projects, series 200A, Harvard University—Population Studies in India, box 132, file 1173; memcon Johnson-Mehta, May 4, 1966, *FRUS,* South Asia, 1964–1968, 25:637–638.

80. Adaline Satterthwaite to Tietze, June 21, 1966, and Guttmacher to Notestein, July 26, 1966, RAC, RG IV3B4.4b, Population Council, National Committee on Maternal Health, box 95, folder 1773; Mauldin, "Retention of IUD's."

81. Guttmacher to Notestein, July 26, 1966; Mauldin, "Retention of IUD's"; "Considerations of Suggestions Made by the State Governments and Other Members of the Central Family Planning Council," attached to "Central Family Planning Council 2nd Meeting Agenda," June 27, 1966, NIHFW, Depository, 204 IND; "India's Family Planning Programme: A Brief Analysis," Ford Foundation Report, April 1970, National Institute for Health and Family Welfare, Delhi, Depository, 291 FOR.

82. Narain, "Note for the Cabinet Committee on Family Planning," June 7, 1966, Department of Family Welfare, Ministry of Health and Family Welfare, Archives; Ramachandran to B. P. Patel, March 30, 1970, Department of Family Welfare, Ministry of Health and Family Welfare, Archives, file number 1-1/71-PLY; Gupta, Sinha, and Bardhan, *Family Welfare Programme in India,* 1:31.

83. M. E. Khan and C. V. S Prasad, *Fertility Control in India: A Critical Evaluation of the Role of Incentives in Promoting Sterilization among Industrial Workers* (New Delhi: Manohar, 1980), 1.

84. Jean Drèze, "Famine Prevention in India," in *Famine Prevention,* vol. 2 of *The Political Economy of Hunger,* ed. Jean Drèze and Amartya Sen (Oxford: Clarendon Press, 1990), 57.

85. "Statement of the President upon Signing the Food for Peace Act of 1966," November 12, 1966, *Public Papers of the Presidents of the United States: Lyndon B. Johnson,* 1966, book 2, 608; Piotrow, *World Population Crisis,* 117, 127; R. N. Madhok, "Note for the Committee of Cabinet . . . Suggestions for Incentives," September 15, 1967, Department of Family Welfare, Ministry of Health and Family Welfare, Archives, New Delhi, file number V 13011/4/75; Drèze, "Famine Prevention in India," 57.

86. These figures, as well as those for Madhya Pradesh and Uttar Pradesh, are from the following sources: Mukherjee, "Note for the Cabinet Committee on Family Planning," March 13, 1966, Department of Family Welfare, Ministry of Health and Family Welfare, Archives, New Delhi; Ramachandran to B. P. Patel,

March 30, 1970, Department of Family Welfare, Ministry of Health and Family Welfare, Archives, New Delhi, file number 1-1/71-PLY; Gupta, Sinha, and Bardhan, *Family Welfare Programme in India,* 2:74.

87. Ramachandran to B. P. Patel, March 30, 1970, Department of Family Welfare, Ministry of Health and Family Welfare, Archives, New Delhi, file number 1-1/71-PLY; Gupta, Sinha, and Bardhan, *Family Welfare Programme in India,* 2:74.

88. R. E. Elder Jr., *Development Administration in a North Indian State: The Family Planning Program in Uttar Pradesh* (Chapel Hill: Carolina Population Center, University of North Carolina, 1972), 21–22, 39, 94–100, 106–107, 118–120.

89. "Dept. of F. P.: Govt. of Maharashtra Proposals," October 7, 1967, Department of Family Welfare, Ministry of Health and Family Welfare, Archives, New Delhi, file number V 13011/4/75; Programme Evaluation Organization, "Family Planning Programme, An Evaluation: Maharashtra," and "Family Planning Programme, An Evaluation: Punjab," 1971, Planning Commission Archives, New Delhi; "Statement Showing the Decisions Taken," May 15, 1967, Department of Family Welfare, Ministry of Health and Family Welfare, Archives, New Delhi, file number V 13011/4/75.

90. Elder, *Development Administration,* 122–125; "Family Planning Programme, An Evaluation: Maharashtra."

91. Elder, *Development Administration,* 32, 121, 129; R. N. Madhok, "Note for the Committee of Cabinet . . . Suggestions for Incentives," September 15, 1967; "Family Planning Programme, An Evaluation: Maharashtra"; "Family Planning Programme, An Evaluation: Punjab."

92. Johnson to Gandhi, January 16, 1967, and Johnson to Gandhi, May 9, 1967, both in *FRUS, South Asia, 1964–1968,* 25:808–809, 852.

93. Madhok, "Note for the Committee . . . Pilot Project for Gift of Transistor Radio," September 15, 1967, Department of Family Welfare, Ministry of Health and Family Welfare, Archives, New Delhi, file number V 13011/4/75; Mitra, "Implementation of the Family Planning Program," March 3, 1966, Asok Mitra Papers, National Planning Commission, box 152 (provisional box number).

94. Madhok, "Note for the Committee of Cabinet . . . Suggestions for Incentives," September 15, 1967; V. P. Naik to Indira Gandhi, June 27, 1967, both in Department of Family Welfare, Ministry of Health and Family Welfare, Archives, New Delhi, file number V 13011/4/75.

95. V. P. Kataria, "Family Planning Program of India—An Appraisal," ca. 1975, in "Some Studies of Indian Demographic Trends," Nehru Memorial Museum and Library, Asok Mitra Papers, National Planning Commission, box 181 (provisional box number), "Meetings, Conferences."

96. V. P. Naik to Indira Gandhi, June 27, 1967.

97. "Summary Proceedings of the Central Family Planning Council," June 27, 1966, NIHFW, 204 IND; "Minutes of the Meeting Held under the Chair-

manship of Mir Mushtaq Ahmed," November 23, 1966, National Archives of India, New Delhi, Ministry of Home Affairs, Delhi Section, file 11/29/68-Delhi; Elder, *Development Administration,* 29–31, 110–116.

98. K. N. Srinistava, "Note for the Committee of the Cabinet . . . Critical Analysis of the Family Planning Programme," March 29, 1967; "Minutes of the Meeting of the Cabinet Committee on Family Planning," March 31, 1967, both in Department of Family Welfare, Ministry of Health and Family Welfare, Archives, New Delhi, file number 4-4/67-C&C; "Note for the Committee of the Cabinet . . . Suggestions for Incentives and Disincentives," April 2, 1968, Department of Family Welfare, Ministry of Health and Family Welfare, Archives, New Delhi, file number V 13011/4/75; Elder, *Development Administration,* 141–142.

99. Joseph Lelyveld, "India: Is Sterilization the Answer?" *New York Times,* August 6, 1967.

100. "Meeting of the Cabinet Committee on Family Planning," November 25, 1967, and "Extract from File . . . of Department of Social Welfare," November 3, 1967, both in Department of Family Welfare, Ministry of Health and Family Welfare, Archives, New Delhi, file numbers V 16011/3/82 and V 13011/4/75.

101. "Statement Prepared for Presentation before the Subcommittee on Foreign Aid Expenditures," June 1965, IPPFA, series B, reel 148, frames 2064–2069.

102. Draper to Rusk, August 22 and October 16, 1967, USNA, Nixon Papers, NSC Files, Subject Files, box 369, "Population Crisis, 1969–1970"; "Conference on Direction and Support of Research," May 15, 1968, LBJ Library, LBJ Papers, Files of S. Douglass Cater, box 66.

103. "Expenditures on Population," ca. October 1966, RAC, RG IV3B4.2, Population Council, General File, box 36, folder 526.

104. Donald T. Critchlow, *Intended Consequences: Birth Control, Abortion, and the Federal Government in Modern America* (New York: Oxford University Press, 1999), 66; Piotrow, *World Population Crisis,* 118, 123–126, 130–138, 151.

105. Draper to Canfield, March 11, 1968, Harper & Row Publisher Records, Manuscript Collections, Columbia University Library, series 2, International Planned Parenthood, box 301, "A"; Rettie to Fox, March 22, 1966, CMAC, Family Planning Association Papers, box 305, 1955–1966, IPPF.

106. William Gaud, "Statement," February 1, 1968; David Hopper, "The Population Council: A Review," May, 1975; "The Foundation's Work on Population," September 1970, FFA, file 001976.

107. Symonds and Carder, *The United Nations,* 146–149; "Additional Financing of the Expanded United Nations Population Program," June 12, 1967, UNRC, S[eries]-0512-[box] 0038, Office of the Secretary-General, Records of U Thant, 1961–1971, DAG-1/5.2.1.6-4.

108. Notestein Diary Notes, March 1–12, September 8–10, 1964, RAC, RG IV3B4.2, Population Council, General File, box 32, folder 462.

109. Guttmacher to Deverell, March 10, 1967, and Guttmacher circular letter, May 17, 1967, IPPFA, series B, reel 148, frames 1609–1610, 1628; "The Ford Foundation: Travel Policies," May 15, 1970, FFA, Population Office Files, box 8142; Ensminger Oral History, November 6, 1972, FFA, A.39.

110. Ensminger Oral History, May 10, October 4, and October 25, 1972, FFA, A.18, A.42, and A.38; G. R. Christie, "Critical Appraisal of IPPF Policies and Opportunities," January 6, 1964, IPPFA, series B, reel 642, frames 2102–2107.

111. Taylor circular letter, December 1953, RAC, Rockefeller Foundation, RG 1.2, Projects, 200 United States, Harvard University—Indian Population 1953–1955, box 45, folder 369.

112. "Conversation with Sam Keeny," March 16, 1965; Richmond K. Anderson, Ansley Coale, Lyle Saunders, and Howard Taylor, "Report to the Ministry of Economic Planning and Development, Republic of Kenya, on Population Growth and Family Planning in Kenya," August 1965, RAC, RG IV3B4.2, Population Council, Acc. II, George Zeidenstein, box 7.

113. Sam Keeny, Parker Mauldin, George Brown, and Louis Hellman, "Report to the Ministry of Health, Government of Iran, on Population Growth and Family Planning in Iran," August 1966, and Notestein to M. Shahgholi, August 13, 1966, RAC, RG IV3B4.2, Population Council, Acc. II, George Zeidenstein, box 7. For a classic analysis of cut-and-paste "development," see James Ferguson, *The Anti-Politics Machine: "Development," Depoliticization, and Bureaucratic Power in Lesotho* (Cambridge: Cambridge University Press, 1990).

114. Chris Scott and Susheela Singh, "Problems of Data Collection in the World Fertility Survey," in *World Fertility Survey Conference 1980: Record of Proceedings* (London, 1980), 32, 72–73.

115. David I. Kertzer and Tom Fricke, "Toward an Anthropological Demography," in *Anthropological Demography: Toward a New Synthesis,* ed. Kertzer and Fricke (Chicago: University of Chicago Press, 1997), 18–19; McCarthy, "Ford Foundation's Population Program," 31; Minkler, "Consultants or Colleagues," 411.

116. Kantner, "Population, Policy, and Political Atavism," *Demography* 19, no. 4 (1982): 430.

7. Beyond Family Planning

1. "India's Urban Tensions," February 1976, National Archives of India, New Delhi, Ministry of Home Affairs, file 1/18/70-RP, vol. 5; Bernard Berelson, "Beyond Family Planning," *Studies in Family Planning* 1, no. 38 (1969): 1–16.

2. W. Parker Mauldin, "Patterns of Fertility Decline in Developing Countries, 1950–1975," *Studies in Family Planning* 9 (1978): 75–84. On increasing en-

rollment, see Jeremi Suri, *Power and Protest: Global Revolution and the Rise of Détente* (Cambridge: Harvard University Press, 2003), 92–93, 269. On Johnson, see Gaud to Johnson, February 1, 1968, and "Excerpts—News Briefing," July 16, 1968, both in LBJ Library, Austin, TX, LBJ Papers, Files of S. Douglass Cater, box 66. On McNamara, see Harkavy memo to Bell, September 21, 1970, Ford Foundation Archives, New York City (hereafter FFA), Papers of David Bell, series 5, box 32, folder 772. On Gandhi, see Haksar to Gandhi, December 2, 1970, Nehru Memorial Museum and Library, New Delhi, P. N. Haksar Papers, Subject Files, file 48.

3. "Report on Meeting between UNFPA Staff and Representatives of Non-UN Organisations," June 4, 1970, National Archives, Kew, UK (hereafter UKNA), OD 62/27; Deverell to Winfield Best, July 12, 1966, Harper & Row Publisher Records, Manuscript Collections, Columbia University Library, series 2, International Planned Parenthood, box 302, "Deverell." On family planning accelerating preexisting trends, see Friedman to McNamara, April 29, 1968, World Bank Group Archives, Washington, DC (hereafter WBGA), Central Files 1947–1968, General Files—Projects and Studies, box 36, "Population," vol. 4; and Dudley Kirk, "Prospects for Reducing Natality in the Underdeveloped World," *Annals of the American Academy of Political and Social Science* 369 (1967): 58.

4. "Proclamation of Tehran," May 13, 1968, reprinted in *American Journal of International Law* 63 (1969): 674–677; "Elements of Population Strategy," August 1970, FFA, Report no. 007610.

5. Kingsley Davis, "Population Policy: Will Current Programs Succeed?" *Science* 158 (November 10, 1967): 730–739.

6. "Off Record Note re Kingsley Davis from Norm Ryder," May 11, 1988, Population Association of America Archives, Silver Spring, MD, box 2a, folder 35; Davis, "Population Policy," 736; Paul Ehrlich, *The Population Bomb* (New York: Ballantine Books, 1968), prologue, 132; David Dempsey, "Dr. Guttmacher Is the Evangelist of Birth Control," *New York Times,* February 9, 1969.

7. Bell handwritten notes from Lake Como meeting, April 10, 1968, FFA, Papers of David Bell, series 5, box 34, folder 787.

8. Peter J. Donaldson, *Nature Against Us: The United States and the World Population Crisis, 1965–1980* (Chapel Hill: University of North Carolina Press, 1990), 48–49; Donald P. Warwick, *Bitter Pills: Population Policies and Their Implementation in Eight Developing Countries* (Cambridge: Cambridge University Press, 1982), 48; Donald T. Critchlow, *Intended Consequences: Birth Control, Abortion, and the Federal Government in Modern America* (New York: Oxford University Press, 1999), 186; Gabriel to Ravenholt, July 29, 1976, Ravenholt papers provided to author.

9. Draper to Moynihan, February 15, 1969, U.S. National Archives, College Park, MD (hereafter USNA), RG 59, Central Files, 1967–1969, Soc 13-3,

box 3121; "The Ford Foundation's Activities in Population," August 1968, FFA, file 001405.

10. Halvor Gille, *World Fertility Survey: Major Findings and Implications* (Voorburg: International Statistical Institute, 1984), 1–5; Gary L. Lewis, "The Contraceptive Prevalence Study: Content and Status," *Population Index* 49, no. 2 (1983): 189–198.

11. Linda E. Atkinson, Richard Lincoln, and Jacqueline D. Forrest, "Worldwide Trends in Funding for Contraceptive Research and Evaluation," *Family Planning Perspectives* 17 (September–October 1985): 204; Deborah Maine, "Depo: The Debate Continues," *Family Planning Perspectives* 10, no. 6 (1978): 342–345; Irving Sivin et al., "Norplant: Reversible Implant Contraception," *Studies in Family Planning* 11, no. 7/8 (1980): 227–235.

12. Lant H. Pritchett, "Desired Fertility and the Impact of Population Policies," *Population and Development Review* 20 (1994): 37.

13. Jason L. Finkle and Barbara B. Crane, "The World Health Organization and the Population Issue: Organizational Values in the United Nations," *Population and Development Review* 2 (1976): 373–374.

14. Dudley Kirk, "Natality in the Developing Countries: Recent Trends and Prospects," in *Fertility and Family Planning: A World View,* ed. S. J. Behrman Jr., Leslie Corsa, and Ronald Freedman (Ann Arbor: University of Michigan Press, 1969), 87–88.

15. "Secretary-General's Report," Appendix H to Tenth Meeting of the Management and Planning Committee, November 9, 1967, IPPF Archives, London (hereafter IPPFA), 1.4.1.1.3, "IPPF Management and Planning Committee." Notestein is quoted in Critchlow, *Intended Consequences,* 177. On Noonan and Zeidenstein, see "Minutes of the Directors Meeting," February 24, 1976, Rockefeller Archive Center, Tarrytown, NY (hereafter RAC), RG IV3B4.2, Population Council, Acc. II, George Zeidenstein, box 2, "Officers Meetings, July 1975–June 1976"; Mauldin memo to files, June 7, 1974, RAC, RG IV3B4.2, Population Council, Acc. II, George Zeidenstein, box 3, "Joan Dunlop."

16. Sunniva Engh, "Population Control in the 20th Century: Scandinavian Aid to the Indian Family Planning Programme" (Ph.D. diss., University of Oxford, 2005). For an overview of the law program, see Luke T. Lee, "Law, Human Rights, and Population Policy," in *Population Policy: Contemporary Issues,* ed. Godfrey Roberts (Westport, CT: Praeger, 1990), 1–20. On abortion and sterilization, see Lester R. Brown and Kathleen Newland, "Abortion Liberalization: A Worldwide Trend," *Futurist,* June 1976, 140–143; "The Future Program of the Population Council on Sterilization and Abortion," June 2, 1976, RAC, RG IV3B4.2, Population Council, Acc. II, George Zeidenstein, box 5, "Population Council Committee on Sterilization and Abortion"; Christopher Tietze, *Induced Abortion: A World Review,* 4th ed. (New York: Population Council, 1981), 7–17.

17. "Instant Abortion Pill," *Washington Post,* October 16, 1970, 52. On abortion and poor people, see "Report on Meeting between UNFPA Staff and Representatives of Non-UN Organisations," June 4, 1970. On sex determination, see Margaret Snyder, "A Summary of the Discussions," *Studies in Family Planning* 1, no. 23 (1967): 10; Lyle Saunders, "Workshop on Contraceptive Specifications," August 13, 1969, FFA, report no. 10088. On interest in sex determination, see also Allan C. Barnes, "Contraceptive Development," March 1973, FFA, file 002854. On paying for abortions, see Bell handwritten notes, May 10, 1973, FFA, David Bell Papers, series 5, box 34, file 796. On menstrual regulation kits, see Ravenholt, "Taking Contraceptives to the World's Poor: Creation of USAID's Population/Family Planning Program, 1965–80," www.ravenholt.com; author's interview with Ravenholt, January 8, 2005.

18. Berelson, "Beyond Family Planning"; Hoffman to Knapp, March 28, 1969, WBGA, Central Files 1969–1971, IBRD—IDA Administration, Liaison, and Policy Files, box 145, "Population-1969-I."

19. Hoffman to Knapp, March 28, 1969; Douglas Ensminger Oral History, November 1, 1971, FFA, pt. A27; "Indian Family Planning: Is Failure Inevitable," FFA, report no. 006866. On "natural history," see "Beyond FP," attached to "The Ford Foundation's Activities in Population," August 1968, FFA, file 001405. On researching involuntary methods, see Saunders to Files, August 13, 1969, FFA, report no. 010088.

20. Draper to Owen, May 5, 1969, Harper & Row Publisher Records, Manuscript Collections, Columbia University Library, series 2, International Planned Parenthood, box 304, "A.I.D. 1969"; David Dempsey, "Dr. Guttmacher Is the Evangelist of Birth Control," *New York Times,* February 9, 1969; Cooper to Scowcroft, November 26, 1973, USNA, Nixon Papers, NSC Institutional Files, Study Memorandums, NSSM 200, box H-204.

21. "Re Compulsory Sterilization and All That in India," October 26, 1976, FFA, David Bell Papers, series 5, box 32, folder 772; Kang to Files, December 1, 1976, WBGA, Records of the Health Services Development Sector, Liaison with International and Other Organizations—UNFPA—vol 5. On Saunders and Freedman, see "Agenda Questions: Session III: Program Organization and Administration Continued," April 8, 1968, FFA, David Bell Papers, series 5, box 34, folder 787.

22. Annan to Aurelius, June 20, 1974, and Annan to Crowley, October 3, 1972, FAO Archives, Rome, RG 16, PA 1/1, UNFPA-UNDP. On coercive use of food aid, see Betsy Hartmann, *Reproductive Rights and Wrongs: The Global Politics of Population Control,* rev. ed. (Boston: South End Press, 1994), 227–228, 230–231. On ethics of community incentives, see "Agenda Questions: Session III: Program Organization and Administration Continued," April 8, 1968.

23. Kanagaratnam to Files, August 9, 1975, WBGA, Credit 0300—Indone-

sia—Population Project—Correspondence, vol. 9. On Singapore, see "Report of the Second Ministerial Conference and Third Official Meeting of the Intergovernmental Coordinating Committee, South East Asia Regional Cooperation in Population and Family Planning," May 1973, National Institute of Health and Family Welfare, Documentation Centre, New Delhi, 204 MIN.

24. Memcon, LBJ-Subramaniam, December 20, 1965, *Foreign Relations of the United States,* South Asia, 1964–1968, vol. 25 (hereafter *FRUS* with year and volume), 516–518.

25. Oscar Harkavy, Frederick S. Jaffe, and Samuel W. Wishik, "Family Planning and Public Policy: Who Is Misleading Whom?" *Science* 165 (July 25, 1969): 367–373.

26. William Ryan, *Blaming the Victim,* rev. ed. (New York: Vintage, 1976), 89–116; Phyllis Tilson Piotrow, *World Population Crisis: The United States Response* (New York: Praeger, 1973), 88; Eisenhower quoted in Critchlow, *Intended Consequences,* 245; Garrett Hardin, "The Tragedy of the Commons," *Science* 162 (December 13, 1968): 1243–1248.

27. Carl Degler, *In Search of Human Nature: The Decline and Revival of Darwinism in American Social Thought* (New York: Oxford University Press, 1991), chap. 9; Burdick to Moore, August 20, 1968, Francis A. Countway Library of Medicine, Center for the History of Medicine, Boston, Alan Guttmacher Papers, box 5, folder 18.

28. Dempsey, "Dr. Guttmacher Is the Evangelist of Birth Control"; John R. Rague to Hugh Moore, November 15, 1968, Social Welfare History Archives, University of Minnesota, Minneapolis, Papers of the Association for Voluntary Sterilization, SW 15.1, box 17, "Hugh Moore Correspondence"; Jaffe to Guttmacher et al., October 2, 1969, Sophia Smith Collection, Smith College (hereafter SSC), Planned Parenthood Federation of America Papers 2, Acc. #90S-33, box 4, folder 5.61.

29. "Special Message to the Congress: America's Unfinished Business, Urban and Rural Poverty," March 14, 1967, *The Public Papers of the Presidents of the United States,* www.hti.umich.edu:80/p/ppotpus.

30. Lodge, "The Restriction of Immigration," *North American Review* 152 (January 1891): 34; Walker, *Discussions in Economics and Statistics* (New York: Henry Holt and Co., 1899), 2:447–448; and see also John Higham, *Strangers in the Land: Patterns of American Nativism, 1860–1925* (1962; New York: Atheneum, 1973), 40. On Moynihan, see "The Negro Family: The Case for National Action," March 1965, dol.gov/oasam/programs/history/webid-meynihan.htm [*sic*]; and see also Daniel Patrick Moynihan, "A Family Policy for the Nation," *America* 113 (September 18, 1965): 280–283. For the reception, and Moynihan's reaction, see Godfrey Hodgson, *The Gentleman from New York: Daniel Patrick Moynihan* (Boston: Houghton Mifflin, 2000), 113–119.

31. On pathologies, see esp. the sections "The Roots of the Problem" and "The Tangle of Pathology." On sterilization legislation, see Philip R. Reilly, *The Surgical Solution: A History of Involuntary Sterilization in the United States* (Baltimore: Johns Hopkins University Press, 1991), 160–161. Willie quoted in Thomas B. Littlewood, *The Politics of Population Control* (Notre Dame: University of Notre Dame Press, 1977), 49–50.

32. Moynihan to MacPherson, September 22, 1966, LBJ Library, LBJ Papers, Welfare, EX WE 11/22/1963, box 1; Charles Schultze to Califano, June 10, 1967, LBJ Library, LBJ Papers, Welfare, EX WE 2/16/1967, box 2; Gulick to Draper, May 15, 1968, Harper & Row Publisher Records, Manuscript Collections, Columbia University Library, series 2, Planned Parenthood P, box 298, "Population Crisis Committee 1968–1970."

33. Herbert L. Smith, S. Philip Morgan, Tanya Koropeckyj-Cox, "A Decomposition of Trends in the Nonmarital Fertility Ratios of Blacks and Whites in the United States, 1960–1992," *Demography* 33 (May 1996): 141–142; Frederick S. Jaffe, "Low-Income Families: Fertility Changes in the 1960s," *Family Planning Perspectives* 4 (1972): 44–45; Kevin P. Phillips, "Census Hints Race Shift," *Washington Post,* October 17, 1970, A11.

34. J. Mayone Stycos, "Opinion, Ideology, and Population Problems: Some Sources of Domestic and Foreign Opposition to Birth Control," in National Academy of Sciences Office of the Foreign Secretary, *Rapid Population Growth: Consequences and Policy Implications* (Baltimore: NAS, 1971), 553, 555–556.

35. Critchlow, *Intended Consequences,* 106–109; Deverell to Canfield, April 5, 1966, Harper & Row Publisher Records, Manuscript Collections, Columbia University Library, series 2, International Planned Parenthood, box 304, "O."

36. Piotrow, *World Population Crisis,* 141. On the 1967 bill, and backlash, see also Critchlow, *Intended Consequences,* 89–93, 109–111; and David R. Seidman, "Alternative Modes of Delivering Family Planning Services," *Studies in Family Planning* 1 (April 1970): 7.

37. Nicole J. Grant, *The Selling of Contraception: The Dalkon Shield Case, Sexuality, and Women's Autonomy* (Columbus: Ohio State University Press, 1992), 31, 45, 51; Barbara Ehrenreich, Mark Dowie, and Stephen Minkin, "The Charge: Gynocide," November/December 1979, www.motherjones.com.

38. Critchlow, *Intended Consequences,* 86, 104–106; "The Ford Foundation's Activities in Population," August 1968, FFA, file 001405.

39. K. Parker to Solly Zuckerman, April 8, 1968, UKNA, CAB 168/240; Francis Ronsin, *La Grève des Ventres: Propagande néo-malthusienne et baisse de la natalité française, 19e–20e siècles* (Paris: Aubier-Montaigne, 1980), 216–220.

40. Stycos, "Opinion, Ideology, and Population Problems," 551–553. On the conference, see "Family Planning vs. Population Control," December 1971, SSC,

Planned Parenthood Federation of America (PPFA II) Records, Acc. #90S-11, box 1, "Negros." On Bond, see Littlewood, *Politics of Population Control,* 60.

41. Michael Flamm, *Law and Order: Street Crime, Civil Disorder, and the Crisis of Liberalism in the 1960s* (New York: Columbia University Press, 2005), 173; Nixon-Peterson-Kissinger memcon, September 2, 1969, *FRUS,* 1969–1976, 4:280–285.

42. NSC meeting, December 3, 1958, Dwight D. Eisenhower Library, Abilene, Ann Whitman File, NSC Series; Nixon-Peterson-Kissinger memcon, September 2, 1969.

43. Moynihan to Rogers, March 28, 1969, USNA, RG 59, Central Files 1967–1969, Soc 13-3, box 3124; Cole to Rogers et al., Nixon Papers, NSC Files, Subject Files, box 369, "Population Crisis, 1969–1970." On the commission, see Finn to Moynihan, April 1, 1969, USNA, Nixon Papers, WHCF, Subject Files, FG 99, "Committee on Population and Family Planning," box 1. On HEW, see Rockefeller to LBJ, May 3, 1968, and Berelson to Lee, May 29, 1968, both in LBJ Library, LBJ Papers, Welfare, EX WE 2/16/1967, box 2; and Nixon, "Special Message to the Congress on Problems of Population Growth," July 18, 1969, *Public Papers of the Presidents of the United States: Richard Nixon, 1969* (Washington, DC: U.S. Government Printing Office, 1971), 521–530. On involuntarism, see Littlewood, *Politics of Population Control,* 56–57.

44. Draper to Annabelle Arnold, June 2, 1972, SSC, Planned Parenthood Federation of America (PPFA II) Records, box 89, "Victor Bostrum Fund"; The Victor Fund for the International Planned Parenthood Federation, report no. 5 (Spring 1967): 2, 23.

45. Gandhi-LBJ memcon, March 28, 1966, *FRUS,* South Asia, 1964–1968, 25:597; "Remarks to American and Korean Servicemen at Camp Stanley," November 1, 1966, *The Public Papers of the Presidents of the United States,* www.hti.umich.edu:80/p/ppotpus; Claxton to Rusk, November 2, 1966, Declassified Documents Reference System, CDROM Id: 1995110103408, fiche#: 1995-281; Cater to Johnson, November 29, 1966, LBJ Library, LBJ Papers, Welfare, EX WE 11/22/1963, box 1.

46. William Clark to John Nicholls, February 21, 1966, UKNA, FO 953/2496; memcon, Head-Toumanoff, April 7, 1971, USNA, RG 59, Central Files, 1970–1973, Soc 13-3, box 3037.

47. William Vogt, *People! Challenge to Survival* (New York: William Sloane Associates, 1960), 50–51.

48. Maggie Black, *A Cause for Our Times: Oxfam, the First 50 Years* (Oxford: Oxford University Press, 1992), 116; Charline Dekens, "Oxfam and Family Planning, 1964–1994" (M.A. diss., Oxford Brookes University, 2005), 38, 48.

49. Critchlow, *Intended Consequences,* 151–152.

50. Deborah Shapley, *Promise and Power: The Life and Times of Robert McNamara* (New York: Little, Brown, 1993), 480–481.

51. Swingler to M. Peto, February 13, 1964, and Shirley Wren-Lewis to Peto, April 21 and May 5, 1964, IPPFA, series B, reel 400, frames 2017, 2021, 2027–2028; Leland C. Devinney memo, October 29, 1957, RAC, Rockefeller Foundation, RG 1.2, Projects, 200 United States, Harvard University—Indian Population 1953–1955, box 45, folder 371.

52. Bruce D. Carlson, "Report on Asian Trip, January 24–March 4, 1967," School of Oriental and African Studies, Archives and Manuscript Collections, Christian Aid, Organisation, section 1, box 2, file 8, "Family Planning, 1967–."

53. Author's interview with Ehrlich, April 10, 2007; Betty Ballantine to Ehrlich, May 1, 1968, Paul and Anne Ehrlich Papers, Stanford University Archives, SC 223, series 6, folder 52; Ehrlich, *The Population Bomb* (New York: Ballantine Books, 1968).

54. Ehrlich, *The Population Bomb,* 15–16 (emphasis in original).

55. Mahmood Mamdani, *The Myth of Population Control: Family, Caste, and Class in an Indian Village* (New York: Monthly Review Press, 1972), 1–2; Ehrlich, *The Population Bomb,* 22–23, 37, 40.

56. Ehrlich, *The Population Bomb,* 133, 158–166.

57. See January 15, 1970, Ballantine Books press release along with Ehrlich correspondence in Paul and Anne Ehrlich Papers, Stanford University Archives, SC 223, series 6, box 2, folder 52. On sales and ZPG, see also ZPG monthly reports and Ehrlich correspondence in Paul and Anne Ehrlich Papers, Stanford University Archives, SC 223, series 4, box 4, folder 7. On congressional interest, see correspondence in Ehrlich Papers, SC 223, series 6, box 2, folders 57–58. For sample letters, see Ehrlich, *The Population Bomb,* 179–180.

58. McCloskey article for the *Stanford Alumni Almanac* in Stanford University Archives, Paul and Anne Ehrlich Papers, SC 223, series 6, box 2, folders 57–58; Ehrlich, *The Population Bomb,* 138; Daddario quoted in *Congressional Record—House,* December 1968, p. 4682.

59. Randall Hansen, "The Kenyan Asians, British Politics, and the Commonwealth Immigrants Act, 1968," *Historical Journal* 42 (September 1999): 809–834; "Mr. Powell Filled with Foreboding on Immigrants," *Times* (London), April 21, 1968. First published in *The Ecologist* in January 1972, "Blueprint" is available at www.theecologist.info/page32.html.

60. Matthew Connelly, *A Diplomatic Revolution: Algeria's Fight for Independence and the Origins of the Post–Cold War Era* (New York: Oxford University Press, 2002), 179, 269–270; Réunion du Haut Comité de la Population et de la Famille, December 12, 1967, Archives Nationales, Archives Privée, Mauco Papers, 577, box 3; "Note: Les objectifs d'une politique de la natalité" and "Les Prestations Familiales," n.d., both in Centres des Archives Contemporaines, Fontainebleau,

CAC/12, Santé et Protection Sociale, Direction de la Population et des Migrations, versement 760133, art. 13. On European coordination, see "Note: Eighth Meeting of the Standing Committee on Employment," December 4, 1975, Council of the European Communities Archive, file 31928, see esp. the annexes.

61. Keith Fitzgerald, *The Face of the Nation: Immigration, the State, and the National Identity* (Stanford: Stanford University Press, 1996); Betty K. Koed, "The Politics of Reform: Policymakers and the Immigration Act of 1965" (Ph.D. diss., University of California, Santa Barbara, 1999); Davis, "Sociological Aspects of Genetic Control," in *Genetics and the Future of Man,* ed. J. D. Roslansky (Amsterdam: North Holland, 1966), 173–204; Hardin, "Tragedy of the Commons," p. 1247.

62. "Conference on Study of Motivation Relevant to Fertility Control," RAC, RG 5, John D. Rockefeller 3rd Papers, series 1, subseries 5, Population Interests, box 82, folder 680.

63. "Report to the Secretary-General of the United Nations from the Review Committee of the UNFPA," October 1972, UNRC, S-0290-0017-05, "UNDP— Report of the Michanek Committee." On the veterinary approach, see Harry E. Wilhelm, "Notes on the Ford Foundation's Activities in Population," January 31, 1968, FFA, Papers of David Bell, series 5, box 32, folder 771.

64. Bell handwritten notes from Lake Como meeting, April 8, 1968, FFA, Papers of David Bell, series 5, box 34, folder 787; President's Council Meeting, March 3, 1969, WBGA, 03–04, Office of the President, Records of President McNamara, series 02, President's Council Minutes, box 1. On education, see David Hopper, George Brown, and Wendy Marson, "An Overview of Agency Activities," May 1973, FFA, file 002854.

65. Taylor to Huw Jones, October 29, 1971, WBGA, Credit 0300—IND— Negotiations 1969 IV; Ravenholt to Zahra, March 13, 1973, WHO Archives, Geneva, P13/372/6 (emphasis in original); Metcalfe to Kangas, April 21, 1975, Ravenholt papers provided to author.

66. "Summary Comments: Bellagio IV Meeting on Population: Ulvshale, Denmark," June 1977, FFA, report no. 010333.

67. Richard Reynolds, "Some Observations on the Indian Family Planning Program," April 1969, FFA, file 007634. The IPPF archives contain a vast collection of family planning posters as well as the 1967 Disney film *Family Planning,* which presented the same theme.

68. Anna L. Southam to Lyle Saunders, July 31, 1969, FFA, International Division/Population Papers of Lyle Saunders, series 1, box 1, folder 10; Dennis N. De Tray, "Child Quality and the Demand for Children," *Journal of Political Economy* 81, pt. 2 (March–April 1973): S-71. On Singapore, see Suitters, *Be Brave and Angry,* 380.

69. Davis, "Population Policy," 738–739.

70. Betty Friedan, *The Feminine Mystique* (New York: W. W. Norton, 1963),

16–17, 31–32; Judith Blake, "Population Policy for Americans: Is the Government Being Misled?" *Science* 164 (1969): 522–529.

71. Notestein, "Population as a Factor in National Power," September 1, 1970, Seeley G. Mudd Manuscript Library, Princeton, Notestein Papers, box 5, National War College file.

72. Gwatkin to Bell, July 11, 1977, FFA, report no. 010333; Reynolds, "Some Observations on the Indian Family Planning Program"; Nancy E. Williamson, "Sex Preferences, Sex Control, and the Status of Women," *Signs* 1 (1976): 853–854, 860–862.

73. Robert McClory, *Turning Point* (New York: Crossroad, 1995), 55.

74. Ibid., 62–63, 72–74, 86–92, 102–107; Leslie Woodcock Tentler, *Catholics and Contraception: An American History* (Ithaca: Cornell University Press, 2004), chaps. 4 and 5.

75. "Report on the 4th Session of the Commission," March 25, 1965, University of Notre Dame Archives, Crowley Papers, box 5.

76. Robert Blair Kaiser, *The Politics of Sex and Religion* (Kansas City: Sheed and Ward, 1985), 170–177, 248–263; Noonan, "The Process by Which the Position of the Church on Contraception Has Been Shaped," University of Notre Dame Archives, Crowley Papers, box 7, folder 8. On Paul VI, see McClory, *Turning Point,* 130–132; Peter Hebblethwaite, *Paul VI: The First Modern Pope* (New York: Paulist Press, 1993), 472.

77. *Humanae Vitae,* July 25, 1968, www.vatican.va/holy_father/paul_vi/encyclicals; Kaiser, *Politics of Sex and Religion,* 197.

78. McClory, *Turning Point,* 140–146; Kaiser, *Politics of Sex and Religion,* 197–198; O'Brien to Moore, August 6, 1968, University of Notre Dame Archives, John A. O'Brien Papers (COBR), box 4, folder 32.

79. Suzanne Staggenborg, *The Pro-Choice Movement: Organization and Activism in the Abortion Conflict* (New York: Oxford University Press, 1991), 19–20; Cynthia Gorney, *Articles of Faith: A Frontline History of the Abortion Wars* (1998; New York: Simon and Schuster, 2000), 82–84. On Hardin's influence, see David J. Garrow, *Liberty and Sexuality: The Right to Privacy and the Making of Roe v. Wade* (Berkeley: University of California Press, 1998), 293–298.

80. Barbara Seaman, *The Doctors' Case against the Pill,* 25th ed. (1969; Alameda, CA: Hunter House, 1995), 223–224; Elizabeth Siegel Watkins, *On the Pill: A Social History of Oral Contraceptives, 1950–1970* (Baltimore: Johns Hopkins University Press, 1998), 112–117.

81. Watkins, *On the Pill,* 108–110.

82. Harkavy, "Informal Notes on Bellagio Population Conference Discussion," June 24, 1971, FFA, report no. 009549.

83. Grant, *The Selling of Contraception,* 65–68. On sterilization, see Littlewood, *Politics of Population Control,* 107–111; Jael Silliman et al., *Undivided*

Rights: Women of Color Organize for Reproductive Justice (Cambridge, MA: South End Press, 2004), 111–112; Nancy Ordover, *American Eugenics: Race, Queer Anatomy, and the Science of Nationalism* (Minneapolis: University of Minnesota Press, 2003), 165–170.

84. Peter Novick, *The Holocaust in American Life* (Boston: Houghton Mifflin, 1999), 19–20; Samuel Moyn, *A Holocaust Controversy: The Treblinka Affair in Postwar France* (Waltham, MA: Brandeis University Press), 2005, preface; Pieter Lagrou, "Victims of Genocide and National Memory: Belgium, France, and the Netherlands 1945–1965," *Past and Present* 154 (February 1997): 217–219; Robert G. Moeller, "War Stories: The Search for a Usable Past in the Federal Republic of Germany," *American Historical Review* 101 (1996): 1034–1035. On eugenic training, see Paul A. Lombardo and Gregory M. Dorr, "Eugenics, Medical Education, and the Public Health Service: Another Perspective on the Tuskegee Syphilis Experiment," *Bulletin of the History of Medicine* 80, no. 2 (2006): 291–316.

85. Woods to State, October 17, 1972, USNA, RG 59, Central Files, 1970–1973, Soc 13-3, box 3037; Greer, *Sex and Destiny*, 360–361.

86. Berelson, "Beyond Family Planning," 2; Joel N. Shurkin, *Broken Genius: The Rise and Fall of William Shockley, Creator of the Electronic Age* (New York: Macmillan, 2006), 80–81, 193–194, 200–206, 223–225; "Recommendations with Respect to the Behavioral and Social Aspects of Human Genetics," in NAS Letter to Members, vol. 1, no. 6, February 1971, Kingsley Davis Papers, Hoover Institution Archives, Palo Alto, box 16.

87. Davis, "Population Policy and the Theory of Reproductive Motivation," 1974, Kingsley Davis Papers, Hoover Institution Archives, Palo Alto, box 3; Susan Greenhalgh, "The Social Construction of Population Science: An Intellectual, Institutional, and Political History of Twentieth-Century Demography," *Comparative Studies in Society and History* 38 (1996): 44, 50–51.

88. Kreitner to Stephenson, April 30, 1970, WBGA, Bank General Records, Central Files Station 1972–1974, Operations Policy Files, box 42, "OP—Population and Nutrition—V"; Julian J. Simon, *A Life against the Grain: The Autobiography of an Unconventional Economist* (New Brunswick, NJ: Transaction, 2002), 239–248; Ester Boserup, *The Conditions of Agricultural Growth: The Economics of Agrarian Change under Population Pressure* (Chicago: Aldine, 1965); R. A. Easterlin, "Effects of Population Growth on Economic Development," *Annals of the American Academy of Political and Social Science* 369 (1967): 98–108; S. Kuznets, "Population and Economic Growth," *Proceedings of the American Philosophical Society* 111 (1967): 170–193; John Tierney, "Betting the Planet," *New York Times Magazine*, December 2, 1990, 6; Paul R. Ehrlich and Anne H. Ehrlich, *Betrayal of Science and Reason: How Anti-Environmental Rhetoric Threatens Our Future* (Washington, DC: Island Press, 1996), 100–101.

89. Mamdani, *Myth of Population Control*, 147.

90. Julian L. Simon, *The Ultimate Resource 2* (Princeton: Princeton University Press, 1996), 566–567; Ed Regis, "The Doomslayer," *Wired*, February 1997; Paul R. Ehrlich and Anne H. Ehrlich, "Population Control and Genocide," in *Global Ecology: Readings toward a Rational Strategy for Man*, ed. John P. Holdren and Paul R. Ehrlich (New York: Harcourt Brace Jovanovich, 1971), 157–159.

8. A SYSTEM WITHOUT A BRAIN

1. Cadbury to Kleinwort, January 17, 1969, Harper & Row Publisher Records, Manuscript Collections, Columbia University Library (hereafter H&R), series 2, International Planned Parenthood, box 303, "Fund Raising."

2. "Declaration of Population," *Studies in Family Planning* 1, no. 16 (1967): 1; "Declaration on Population: The World Leaders' Statement," *Studies in Family Planning* 1, no. 26 (1968): 1; Dorothy L. Nortman and Ellen Hofstatter, *Population and Family Planning Programs: A Factbook*, 9th ed. (New York: Population Council, 1978), 19–30; Beryl Suitters, *Be Brave and Angry: Chronicles of the International Planned Parenthood Federation* (London: IPPF, 1973), 362–374.

3. Day to Canfield, January 25, 1968, H&R, series 2, International Planned Parenthood, box 301, "Budget."

4. John A. O'Brien, "Birth Control and the Catholic Conscience," December 23, 1968, *Reader's Digest*. On decentralized strategy, see Bernardin to Luigi Raimondi, July 18, 1969, U.S. Catholic Conference Papers (hereafter USCC Papers), U.S. Conference of Catholic Bishops Archives, Washington, DC (hereafter USCCBA), box 79, "Developmental Committees: Social Development: Family Life: Birth Control, 1968–69." On emotional fire, see "Comment on Right to Life Program," April 30, 1967, USCC Papers, box 79, "Developmental Committees: Social Development: Family Life: Birth Control, 1968–69." On going farther, see McHugh to Bernardin, November 10, 1971, USCC Papers, box 140, "U.S. Government: Poverty: Family Planning: 1971–72."

5. The phrase "system without a brain" is from Richard Symonds and Michael Carder, who used it to denote UN development aid more generally, in their *The United Nations and the Population Question* (London: Sussex University Press, 1973), 192.

6. Dorothy L. Nortman and Ellen Hofstatter, *Population and Family Planning Programs: A Compendium of Data through 1978*, 10th ed. (New York: Population Council, 1980), 37.

7. Interview with J. Joseph Speidel, October 10–11, 2002, Population and Reproductive Rights Oral History Project, Sophia Smith Collection, Smith College (hereafter SSC). On earmarking, see Phyllis Tilson Piotrow, *World Population Crisis: The United States Response* (New York: Praeger, 1973), 146–150. On the USAID budget, see "Joint Report to the President by the Secretary of State and

the Administrator of the Agency for International Development," November 14, 1969, U.S. National Archives, College Park, MD (hereafter USNA), Nixon Papers, NSC Files, Subject Files, box 369, "Population Crisis, 1969–1970"; Bernstein to Williams, February 10, 1971, USNA, RG 286, USAID, Office of the Administrator, Office of the Executive Secretariat, Subject Files 1970–71, "BUD, FY 71, December 70–June 71."

8. Ravenholt interview with author, January 8, 2005; Ravenholt, "Taking Contraceptives to the World's Poor," www.ravenholt.com; Stephen S. Rosenfeld, "Do Family Planning Programs Work?" *Washington Post,* February 27, 1976, A25.

9. Draper to Deverell, January 19, 1968, H&R, series 2, International Planned Parenthood, box 303, "India"; Deverell to Draper, January 26, 1968, H&R, series 2, International Planned Parenthood, box 302, "Draper."

10. Lubin to Deverell, June 18, 1968, IPPF Archives, London (hereafter IPPFA), series B, reel 285, frames 1175–1176.

11. Peter J. Donaldson, *Nature Against Us: The United States and the World Population Crisis, 1965–1980* (Chapel Hill: University of North Carolina Press, 1990), 73. The quote "what any of us dreamed" from Lubin to Deverell, November 26, 1968, IPPFA, series B, reel 285, frame 1172. On Deverell resignation threat, see Deverell to Lubin, July 3, 1968, IPPFA, series B, reel 285, frames 900–901. On "sudden access," see Canfield to Cadbury (quoting Cadbury's memo), February 15, 1968, H&R, series 2, International Planned Parenthood, box 301, "Cadbury."

12. Avabai B. Wadia, *The Light Is Ours: Memoirs and Movements* (London: IPPF, 2001), 257–258; Canfield to Cadbury, February 15, 1968.

13. Draper to Dorsey, December 28, 1971, SSC, Planned Parenthood Federation of America (PPFA II) Records, box 89, "Victor Bostrum Fund"; Draper to Canfield, March 11, 1968, H&R, series 2, International Planned Parenthood, box 301, "A."

14. Draper to Canfield, March 11, 1968; Day to Canfield, January 25, 1968. On the new initiative, see Draper to Medawar, July 7, 1969, H&R, series 2, Planned Parenthood P, box 302, "Draper." On the richest families, see Draper to Owen, December 22, 1969, H&R, series 2, International Planned Parenthood, box 303, "Fund Raising."

15. William O. Sweeney to Saunders, February 14, 1970, Ford Foundation Archives, New York City (hereafter FFA), International Division/Population Papers of Lyle Saunders, box 2, folder 67.

16. Rufus Day to Deverell, October 21, 1968, H&R, series 2, International Planned Parenthood, box 301, "Braestrup." "IPPF Estimated Cash Receipts for 1969," n.d., Contemporary Medical Archives Center, Wellcome Institute for the History of Medicine, London (hereafter CMAC), Alan S. Parkes Papers (hereafter PP/ASP), box 12.

17. Cadbury to Braestrup, September 20, 1968, H&R, series 2, International Planned Parenthood, box 301, "Braestrup"; Lubin, "Meeting of 23rd September with USAID in Washington," September 30, 1969, H&R, series 2, International Planned Parenthood, box 304, "A.I.D. 1969."

18. See evaluations in FFA, International Division/Population Papers of Lyle Saunders, series 1, box 1, folder 8: "IPPF Meeting—FF Sep. 22, 1969"; and see also "Agenda Questions: Session II: National Family Planning Programs: Organization and Administration," April 8, 1968, FFA, Papers of David Bell, series 5, box 34, folder 787.

19. Heaps to Harkavy, July 31, 1969, and Nagel to Harkavy, August 29, 1969, both in FFA, International Division/Population Papers of Lyle Saunders, series 1, box 1, folder 8; Bernbaum to State, November 8, 1968, USNA, RG 59, Central Files, 1967–1969, Soc 13-3, box 3124.

20. Saunders to Nagel, August 15, 1969, and Nagel to Harkavy, August 29, 1969, both in FFA, International Division/Population Papers of Lyle Saunders, series 1, box 1, folder 8.

21. Metcalfe to Ravenholt, May 16, 1972, IPPFA, series B, reel 186; Canfield to Cadbury, April 30, 1969, H&R, series 2, International Planned Parenthood, box 301, "A."

22. Hanson to Harkavy, August 22, 1969, FFA, International Division/Population Papers of Lyle Saunders, series 1, box 1, folder 8; D. Wolfers minute to file, May 13, 1970, National Archives, Kew, UK (hereafter UKNA), OD 62/20.

23. On the IFRP, Peter J. Donaldson interview with author, November 2, 2006.

24. "Meeting of Governing Council of UNDP," June 3, 1970, UKNA, OD 62/27; Harkavy to Bell, April 30, 1970, FFA, Papers of David Bell, series 5, box 32, folder 772.

25. "Additional Financing of the Expanded United Nations Population Program," June 12, 1967, and Draper to U Thant, April 24, 1968, both in United Nations Archives and Records Centre, New York City (hereafter UNRC), S-0512-0038, Office of the Secretary-General, Records of U Thant, 1961–1971, DAG-1/5.2.1.6-4.

26. Juan de Onis, "Major U.N. Effort to Curb Birth Rate Asked in Study," *New York Times,* May 25, 1969.

27. Narasimhan, "UNFPA," August 13, 1969, UNRC, S-0512-0038, Office of the Secretary-General, Records of U Thant, 1961–1971, DAG-1/5.2.1.6-4; Buffum to Department of State, January 8, 1970, USNA, RG 59, Central Files, 1970–1973, Soc 13-3, UN, box 3037.

28. Rockefeller memo, June 16, 1969, UNRC, S-0512-0038, Office of the Secretary-General, Records of U Thant, 1961–1971, DAG-1/5.2.1.6-4. On con-

sultation, see "Record of Meeting," January 14, 1970, D. F. Milton to Wilberforce, March 20, 1970, and Wilberforce to D. F. Milton, March 24, 1970, all in UKNA, OD 62/25. On funding NGOs, see "Record of Meeting," January 14, 1970; Records of Meeting of Advisory Board of UNFPA, January 26, 1970, UKNA, OD 62/26.

29. "Projet de création de poste d'un Commissaire des Nations Unies à la Population," n.d. but ca. October 1969, Centres des Archives Contemporaines, Fontainebleau, Fonds Alfred Sauvy, versement 20000115, art. 7; Symonds and Carder, *The United Nations,* 194–195.

30. Hoffman-Hart memcon, February 20, 1970; Hoffman to Hart, February 3, 1970, UKNA, OD 62/25. On Draper, see Robertson minute to Kerby, May 1, 1970, UKNA, OD 62/26, and Wolfers to Robertson, July 7, 1970, UKNA, OD 62/27.

31. Gille to Narasimhan, May 28, 1971, UNRC, S-0290-0019-05, "UNDP—UN Fund for Population Activities"; Bell to Goodman, September 8, 1971, World Bank Group Archives, Washington, DC (hereafter WBGA), Credit 0300—IND—Negotiations 1969—vol. 4; Buffum to Department of State, USNA, RG 59, Central Files, 1970–1973, Soc 13-3, UN, box 3037. On "Catholic and brown," see Hoffman-Hart memcon, February 20, 1970, UKNA, OD 62/25. On UNFPA's inadequacy, see "Note of Meeting," February 6, 1970, and Salas memcon with Leach, March 11, 1970, both in UKNA, OD 62/25.

32. Gille memcon, March 5, 1971, Food and Agriculture Organization Archives, Rome (hereafter FAOA), RG 12, ESD, File 75, PR 4/22; Mathieson minute to minister, December 31, 1970, UKNA, OD 62/28. On Norway, see Sunniva Engh, "Population Control in the 20th Century: Scandinavian Aid to the Indian Family Planning Programme" (Ph.D. diss., University of Oxford, 2005), 151–154. For other aid figures, see Stanley Johnson, *World Population—Turning the Tide: Three Decades of Progress* (London: Graham and Trotman / Martinus Nijhoff, 1994), 100.

33. Engh, "Population Control," 156–158; Margaret Wolfson, *Profiles in Population Assistance: A Comparative Review of the Principal Donor Agencies* (Paris: Development Center for the Organization for Economic Cooperation and Development, 1983).

34. Ravenholt interview with author, January 6, 2005; Engh, "Population Control," 100, 184.

35. Thomas Schindlmayr, "Explicating Donor Trends for Population Assistance," *Population Research and Policy Review* 23, no. 1 (2004): 34; Johnson, *World Population,* 100.

36. "United Nations Fund for Population Activities: Principles and Procedures," February 27, 1970, UKNA, OD 62/26. On Wahren, see "Report on Meet-

ing between UNFPA Staff and Representatives of Non-UN Organisations," June 4, 1970, UKNA, OD 62/27. On World Population Year, see Symonds and Carder, *The United Nations,* 182.

37. Symonds and Carder, *The United Nations,* 182; Narasimhan, "Note for the Record," December 10, 1969, UNRC, S-0512-0038, Office of the Secretary-General, Records of U Thant, 1961–1971, DAG-1/5.2.1.6-4.

38. "Report on Meeting Between UNFPA Staff," June 4, 1970; undated memo ca. January 1971, UKNA, OD 62/20. On Lindström, see Symonds and Carder, *The United Nations,* 194. On Lleras-Camargo, see "United Nations Fund for Population Activities Advisory Board Meeting," May 25, 1967, UKNA, OD 62/27.

39. Haksar to Gandhi, December, 2, 1970, Nehru Memorial Museum and Library, P. N. Haksar Papers, Subject Files, file no. 48. On consultants and USAID, see Meredith Minkler, "Consultants or Colleagues: The Role of U.S. Population Advisors in India," *Population and Development Review* 3 (1977): 413. On other donors, see Engh, "Population Control," 146–147.

40. Engh, "Population Control," 149, 159, 165, 184, 195–197, 201. This analysis of Scandinavian aid in family planning relies heavily on Engh's well-researched and insightful thesis.

41. Ibid., 94, 152–153, 183.

42. Ibid., 185–190, 281–284.

43. Ibid., 165–169; "UNFPA 1969–1972: Progress Report by the Executive Director," April 10, 1973, UNRC, S-0290-0019-08, "UNDP—UN Fund for Population Activities."

44. UN Advisory Mission, "An Evaluation of the Family Planning Programme of the Government of India," November 24, 1969, Planning Commission Archives, Delhi; Ensminger to Shah, April 20, 1969, FFA, report no. 010049; "Population Policy: Beyond Family Planning," March 16, 1970.

45. Charline Dekens, "Oxfam and Family Planning, 1964–1994" (M.A. diss., Oxford Brookes University, 2005), 39; "Letter from Mr. Lubin in Lahore," November 30, 1971, IPPFA, series B, reel 81, frames 1533–1535; "The Story of the Ernakulam Experiment in Family Planning" (Government of Kerala, 1971), 192; A. K. Nanda, *Vasectomy Camps: A Study* (New Delhi: National Institute of Family Planning, 1973), 27.

46. Viond M. Patel, "Campaign Approach to Family Planning: An Emergency Technique," April 1972, National Institute of Health and Family Welfare, Documentation Centre, New Delhi (hereafter NIHFW), Depository, 204 PAT; Nanda, *Vasectomy Camps,* 18. On the importance of local officials, see Marika Vicziany, "Coercion in a Soft State: The Family Planning Program of India: Part 2: The Sources of Coercion," *Pacific Affairs* 55, no. 4 (1982–83): 575–576. On Kerala, see "Fourth Plan Performance and Outlook: Introduction," Nehru Memo-

rial Museum and Library, Asok Mitra Papers (National Planning Commission), box 153 (provisional file no.); L. Phatak circular, August 5, 1972, Ministry of Health and Family Welfare, Department of Family Welfare, Archives, New Delhi, "Cabinet Committee Meeting—August 23, 1972"; "Proceedings of the Seminar on Population Problems," November 16–17, 1974, Trivandrum, Department of Health, Government of Kerala, NIHFW, 443.1/83 PRO.

47. Patel, "Campaign Approach to Family Planning"; Rockefeller, "Personal Observations," April 6–8, 1970, FFA, Papers of David Bell, series 5, box 34, folder 793.

48. Ravenholt, "Taking Contraceptives to the World's Poor"; Rague to Moore, February 1, 1968, Papers of the AVS, Social Welfare History Archives, University of Minnesota, Minneapolis, SW 15.1, box 36, "Ravenholt, Dr. R.T."; Draper to McNamara, August 2, 1968, WBGA, Central Files 1947–1968, General Files—Projects and Studies, box 36, "Population—Vol. 5."

49. "Would You Be Sterilized for a Sari and a Bag of Rice?" n.d. but ca. August 1974, FFA, Population Office Files, box 6992; Deverell to Canfield, April 5, 1966, H&R, series 2, International Planned Parenthood, box 304, "O."

50. Author's interview with Adrienne Germain, November 10, 2006; author's interview with Nafis Sadik, November 9, 2006; "Secretary-General's Report," April 25, 1975, IPPFA, 1.4.1.1.3, "IPPF Management and Planning Committee," "Minutes of the 26th Meeting."

51. Author's interview with Adrienne Germain, November 10, 2006.

52. Rockefeller, "Personal Observations," April 6–8, 1970.

53. Engh, "Population Control," 173; "India" project description, December 12, 1979, WBGA, Records of the Health Services Development Sector, Liaison with International and Other Organizations, UNFPA—vol. 6, box 15.

54. Matheson to Asst. Director-General, January 10, 1972, UNESCO Archives, Paris (hereafter UNESCOA), 312 A 06 USNA "72." On UNFPA as umbrella, see Kanagaratnam to Bernard Bell, May 14, 1971, WBGA, Credit 0300—IND—Negotiations 1969 III. On Pakistan, see Carder, "Report on a Mission to Pakistan," ca. June 1973, UNRC, S-0175-1618, 322/1 Paki (260-1).

55. McNamara to Suharto, August 14, 1970, and Kanagaratnam to files, January 26, 1971, WBGA, Credit 0300—IND—Negotiations 1969 II. On massive projects, see Kanagaratnam to Bernard Bell, May 14, 1971, and Tolbert to Kanagaratnam, November 4, 1971, WBGA, Credit 0300—IND—Negotiations 1969 IV.

56. Candau to McNamara, December 3, 1971, WBGA, Credit 0300—IND—Negotiations 1969 IV; Salas to de Seynes, January 30, 1972, and de Seynes to Salas, February 8, 1972, both in UNRC, S-0290-0019-05, UNDP—UN Fund for Population Activities.

57. Draper, "Aide-Memoire on Conversation with the Secretary-General,"

April 4, 1972, UNRC, S-0513-0233, Office of the Secretary-General, Records of Kurt Waldheim, DAG-1/5.3.3.4.3.

58. De Seynes to Waldheim, March 23, 1972, and de Seynes to Waldheim, April 6, 1972, both in UNRC, S-0513-0233, Office of the Secretary-General, Records of Kurt Waldheim, DAG-1/5.3.3.4.3 (emphasis in original).

59. Alisbah to de Silva, April 18, 1972, WBGA, Country Operational Files, Credit 0300—Indonesia—Population Project, vol. 3; UN General Assembly Resolution 2815, 26th session, January 7, 1972, UNRC, S-0290-0017-06, UNDP—Report of the Review Committee.

60. Berelson to Gille et al., May 16, 1972, IPPFA, series B, reel 222, frame 358; "Main Points Discussed," June 22–23, 1972, IPPFA, series B, reel 222, frames 338–343.

61. Isam Nazer, "Recent Developments in Family Planning in the Near East," November 14, 1968, IPPFA, attached to Minutes of the Thirteenth Meeting of the Management and Planning Committee, 1.4.1.1.3, "IPPF Management and Planning Committee." On Haiti, see Stephen Douglas, "Mission Report," 27 November–2 December 1982, UNRC, Q400-R021-SU24 UNFPA Global and Interregional Project Files, box 9, HAI 1982–1983.

62. Misha D. Jezernik, "Travel Report," December 1971, FAOA, RG 12, ESH, File 140, PA 2/4; "Human Environment," June 14, 1972, in *India: The Speeches and Reminiscences of Indira Gandhi* (London: Hodder and Stoughton, 1975), 195.

63. Donald T. Critchlow, *Intended Consequences: Birth Control, Abortion, and the Federal Government in Modern America* (New York: Oxford University Press, 1999), 62.

64. "Statement about the Report of the Commission on Population Growth and the American Future," May 5, 1972, *Public Papers of the Presidents of the United States: Richard Nixon, 1972* (Washington, DC: U.S. Government Printing Office, 1974), 576–577; Bernardin to Cooke, March 24, 1972, USCC Papers, USCCBA, box 140, "U.S. Government: Poverty: Family Planning: 1971–72"; Nixon to Cooke, April 20, 1972, Reagan Presidential Library, Mariam Bell Files, OA 17964, "Abortion-General (10)."

65. "Main Points Discussed," June 22–23, 1972 (emphasis added).

66. Ibid.

67. Michanek Report, October 1972.

68. Narasimhan Note for the Record, November 10, 1972, UNRC, S-0290-0019-05, UNDP—UN Fund for Population Activities; Michanek to Review Committee, November 11, 1972, and Keppel Note to Files, November 29, 1972, both in UNRC, S-0290-0017-06, UNDP—Report of the Review Committee.

69. Bodart, "Contribution pour l'Entretien entre le Directeur Général et Monsieur Rafael Salas," March 25, 1983, UNESCOA, 312.12 A 8 UNFPA/SS, pt.

13; Henderson, "Relations with Other International Bodies," ca. April 1971, CMAC, PP/ASP, box 12 (emphasis in original). On Ravenholt, see "The Use of Loans for External Financing of Population Projects," December 5, 1972, WBGA, Bank General Records, Central Files Station 1972–1974, Operations Policy Files, box 43, "Staff Reports and Papers 2"; and Ravenholt to Zahra, March 13, 1973, World Health Organization Archives, Geneva (hereafter WHOA), P13/372/6.

70. Gatuiria to Stedman, August 5, 1972, IPPFA, series B, reel 128, frames 2129–2130; Higgs and Schulte, "Report on Duty Travel," January 1973, FAOA, RG 16, PA 1/1, UNFPA-UNDP.

71. Donald P. Warwick, *Bitter Pills: Population Policies and Their Implementation in Eight Developing Countries* (Cambridge: Cambridge University Press, 1982), 21–22; Draper memo, March 22, 1972, IPPFA, series B, reel 469, frames 1609–1610.

72. "Note to Episcopal Conferences regarding the World Population Year and Conference," October 18, 1973, National Conference of Catholic Bishops Papers, USCCBA, box 62, "Pro-Life Activities," file on "Ad-Hoc Committee: Pro-Life Activities, 1974, Jan–June."

73. Fazzi to Bellerive, March 15, 1973, WHOA, P12/80/3; Director, UN Liaison Office to Bellerive, April 12, 1973, WHOA, P13/86/2(B); "Note to Episcopal Conferences," October 18, 1973; Cadbury, "Special Representative's Report," April 1974, IPPFA, 1.4.1.1.3, "IPPF Management and Planning Committee."

74. "Family Planning—In Service Training," July 8, 1971, WBGA, Credit 0300—IND—Negotiations 1969—III.

75. De Silva to Jones, June 23, 1972, WBGA, Credit 0300—IND—vol. 3. On Banjar system, see Tom Reese to Ravenholt, October 5, 1976, and July 1, 1977, Brackett to Ravenholt, July 28, 1975, and American Embassy Islamabad to State, April 1976, all from Ravenholt Papers; Bernard Berelson and Jonathan Lieberson, "Government Efforts to Influence Fertility: The Ethical Issues," *Population and Development Review* 5 (1979): 588.

76. Paul Demeny, "Observations on Population Policy and the Population Program in Bangladesh," *Population and Development Review* 1 (1975): 308.

77. Wolfson, "Profiles in Population Assistance," 173; Ravenholt, "Use of Surgical Laparoscopy for Fertility Management Overseas," www.ravenholt.com; Barbara Ehrenreich, Mark Dowie, and Stephen Minkin, "The Charge: Gynocide," *Mother Jones,* November/December 1979, www.motherjones.com/news/feature/1979/11/ehrenreich.html. Regarding Mexico, Ravenholt interview with author, January 8, 2005. On AVS subgrantees, see "International Project, AVS, Annual Report, April 1973," Papers of the AVS, Social Welfare History Archives, University of Minnesota, Minneapolis, SW 15.2, box 4.

78. "Letter from Mr. Lubin in Lahore," November 30, 1971; Ravenholt, "Taking Contraceptives to the World's Poor," www.ravenholt.com. On making

abortion easy, see "Minutes of the Directors Meeting," March 9, 1976, Rockefeller Archive Center, Tarrytown, NY (hereafter RAC), RG IV3B4.2, Population Council, Acc. II, George Zeidenstein, box 2, Officers Meetings, July 1975–June 1976; Officer's Meeting: Minutes, July 13, 1976, RAC, RG IV3B4.2, Population Council, Acc. 85-41, box 1, "Officers' Meetings, 7/76–3/77."

79. Frank Notestein, "Reminiscences: The Role of Foundations, of the Population Association of America, Princeton University and of the United Nations in Fostering American Interest in Population Problems," *Milbank Memorial Fund Quarterly* 49, no. 4 (1971): 67–84.

80. Donaldson, *Nature Against Us,* 106; Warwick, *Bitter Pills,* 50.

81. Ravenholt interview with author, January 8, 2005; Thomas B. Littlewood, *The Politics of Population Control* (Notre Dame: University of Notre Dame Press, 1977), 96–106; Dan Morgan, "Passman Sought AID Fund for Friend," *Washington Post,* February 8, 1975.

82. Rebecca Sharpless interviews with Joan Dunlop, April 14–15, 2004, Population and Reproductive Rights Oral History Project, SSC.

83. Rebecca Sharpless interviews with Joan Dunlop, April 14–15, 2004, and Adrienne Germain, June 19–20, and September 25, 2003, Population and Reproductive Rights Oral History Project, SSC; Notestein, "Demography in the United States: A Partial Account of the Development of the Field," *Population and Development Review* 8 (1982): 686.

84. Berelson, "Excerpts from Status Report on Population Developments," March 1973, FFA, report no. 002854; Bell handwritten notes, May 10, 1973, FFA, Papers of David Bell, series 5, box 34, folder 796.

85. "Third Population Conference: Summary and Projections," May 12, 1973, FFA, report no. 010334; Bell to Harkavy, September 4, 1973, FFA, Papers of David Bell, series 5, box 33, folder 785.

86. Bell to Harkavy, September 4, 1973. On the nine hundred projects, see *The Population Debate: Dimensions and Perspectives* (New York: United Nations, 1974), 663. On Indonesia, see Jarrett Clinton to Ravenholt, January 23, 1975, Ravenholt papers. On Pakistan, see Bean-Lowenstein memcon, December 2, 1974, Ravenholt papers.

87. Anrudh Jain, "India," *Studies in Family Planning,* 6, no. 8 (1974): 253. On Gandhi and family planning officials, see Debabar Banerji, "Political Economy of Population Control in India," in *Poverty and Population Control,* ed. Lars Bondestam and Staffan Bergström (London: Academic Press, 1980), 85. On 59 percent of budget, see Committee on the Status of Women in India, *Towards Equality* (Delhi: Government of India, 1974), 330. On number of sterilizations, see "Family Planning Association of India Report, 1973–1974," NIHFW. On the period of reappraisal and the impact of the UNFPA grant more generally, see also

Marika Vicziany, "Coercion in a Soft State: The Family Planning Program of India: Part 1: The Myth of Voluntarism," *Pacific Affairs* 55, no. 3 (1982): 395–397.

88. "India Report of Mission on Needs Assessment for Population Assistance," February 9, 1979, Population Foundation of India Library, New Delhi. On sterilization payments, see Singh to all State Governments and Union Territories, April 30, 1974, and "A Note on Compensation Money," in J. P. Gupta, N. K. Sinha, and Amita Bardhan, *Evolution of Family Welfare Programme in India* (New Delhi: National Institute of Health and Family Welfare, 1992), reprinted as appendices H (V) 16 and 30, pp. 180, 188.

89. Unless otherwise noted, the following account of the conference is based on these sources: daily reports of *Planet,* the conferences' unofficial newspaper published by the IPPF, available in FFA, Population Office Files, box 6992; "The World Population Conference," WBGA, Central Files 1972–1974, Bank Administration and Policy, Central Files Station 1972–1974, Operations Policy Files, box 43, "Population and Nutrition—Staff Reports and Papers"; "Report on United Nations World Population Conference," FAOA, Rome; Claxton, "World Population Conference: Report of U.S. Delegation," and Weinberger to Kissinger, September 19, 1974, both in USNA, Nixon Papers, NSC Institutional Files, Study Memorandums, NSSM 200, box H-204. On botched abortions, see Gail Kligman, *The Politics of Duplicity: Controlling Reproduction in Ceausescu's Romania* (Berkeley: University of California Press, 1998), 56–59.

90. On setting an example, see Robert W. Barnett, "Population: Policy and Program," March 25, 1966, USNA, RG 59, Central Files, 1964–1966, Soc 13-3, box 3200. On the IPPF and the British, see A. J. Sobrero to Ferguson, November 4, 1966, H&R, series 2, International Planned Parenthood, box 304, "M."

91. Wahren, "FPAs at a Crossroads: Warnings from Bucharest," October 28, 1975, IPPFA, 1.4.1.1.1, "IPPF Governing Body," "31st October/2nd Nov. 1975 13th Meeting of the Governing Body." On UNDP, see Hoffman memcon with Hart, February 20, 1970, UKNA, 62/25. On lack of experience, see Hanson to Harkavy, August 22, 1969, FFA, International Division/Population Papers of Lyle Saunders, series 1, box 1, folder 8; Stone to State, July 9, 1970, USNA, RG 59, Central Files, 1970–1973, Soc 13-3, UN, box 3037. Regarding the Public Safety Division, author's interview with Steven Sinding, February 21, 2005. For population control as counterinsurgency, see, for instance, "Report of U.S. Counterinsurgency Objectives in Thailand, South Vietnam, and Cambodia," ca. January 1963, Declassified Documents Reference System, doc. no. CK3100478100.

92. "A Review of the Ford Foundation's Population Work in Asia," June 1973, FFA, New York, report no. 002976. On the World Bank, see "President's Council Meeting," September 16, 1974, WBGA, 03–04, Office of the President, Records of President McNamara, series 02, President's Council Minutes, box 1;

and see also Aart van de Laar, *The World Bank and the Poor* (Boston: Kluwer-Nijhoff, 1980). On Draper and Mexico, see "U.S., Mexico Negotiating Pact," *Wall Street Journal,* September 10, 1954; "Mexico Grants 21.2% Power Rate Rise; Amity toward Foreign Investment Seen," *New York Times,* September 20, 1954; Paul P. Kennedy, "Mexico Industry in Fight over U.S.," *New York Times,* May 1, 1955.

93. On food riots and cartels, see Claxton to Members of Inter-Agency Committee for the World Population Conference, December 5, 1973, USNA, Nixon Papers, NSC Institutional Files, Study Memorandums, NSSM 200, box H-204; Ingersoll to Ford, December 14, 1974, Declassified Documents Reference System, doc. no. CK3100290297. On "Holy War," see Taylor, "Population Growth as a Threat to World Peace," May 2, 1973, USNA, Nixon Papers, NSC Institutional Files, study memorandums, NSSM 200, box H-204.

94. Bernard Gwertzman, "Ford in Warning on U.S. Defenses," *New York Times,* August 13, 1974, 1, 73.

95. Malcolm W. Brown, "Expert Says World Has 27 Days' Food," *New York Times,* August 21, 1974, 2.

96. Nan Robertson, "Parley Gives Bucharest a Taste of Overpopulation," *New York Times,* August 24, 1974; Geoff Watts, "A Chronicle of Wasted Time," *World Medicine,* September 25, 1974, 17.

97. Erland Hofsten, "Bucharest and After," in Bondestam and Bergström, *Poverty and Population Control,* 216–217; Teitelbaum to Salas, August 26, 1974, FFA, Population Office Files, box 6992, World Population Year, Bucharest, 1974.

98. Interview with Joan Dunlop, August 10, 2007; Critchlow, *Intended Consequences,* 179–183. Rockefeller's speech, "Population Growth: The Role of the Developed World," is reprinted in *Population and Development Review* 4 (1978): 509–516.

99. Dusko Doder, "Population Conferees Reject 2 U.S. Efforts," *Washington Post,* August 28, 1974, A23; Notestein, "Uninhibited Notes on Bucharest," September 12, 1974, Seeley G. Mudd Manuscript Library, Princeton, (unprocessed) Papers of Frank Notestein.

100. For a good analysis of how the WPPA demoted international and nongovernmental organizations, see Bernard Berelson, "The World Population Plan of Action: Where Now?" *Population and Development Review* 1 (1975): 115–146. For the text itself, see "World Population Plan of Action," www.un.org/popin.

101. Notestein, "Uninhibited Notes"; "World Population Conference Report," October 31–November 2, 1975, IPPFA, 1.4.1.1.1, IPPF Governing Body, 13th Meeting of the Governing Body.

102. Teitelbaum, "Status of Women at the World Population Conference," October 4, 1974, FFA, Population Office Files, box 6992, "World Population Year, Bucharest, 1974."

103. Ibid.; Teitelbaum to Harkavy, November 30, 1973, FFA, Papers of David Bell, series 5, box 32, folder 772.

104. Dunlop to Rockefeller, April 7, 1976, RAC, RG IV3B4.2, Population Council, Acc. II, George Zeidenstein, box 3, "Personal and Confidential." On WIN, see "Women's International Network—For Whom?" March 24, 1975, Bobbye Ortiz Papers, Duke University Rare Book, Manuscript, and Special Collections Library, box 17, Subject File, Latin America 2/4 (I am grateful to Jocelyn Olcott for this document). On the IPPF panel, see Piepmeier to Persinger, February 17, 1975, SSC, International Women's Tribune, box 3; and "Family Planning Talk Raises Emotions," *Xilonen,* June 27, 1975, 3.

105. Potts quoted in Germaine Greer, *Sex and Destiny: The Politics of Human Fertility* (London: Secker and Warburg, 1984), 349. On Berelson, see "The 'New Look' of Bucharest, or 'Population in Context of Development,'" June 1975, FFA, Papers of David Bell, series 5, box 34, folder 797.

106. "India's Urban Tensions," February 1976, National Archives of India, New Delhi, Ministry of Home Affairs, file 1/18/70-RP, vol. 5.

107. Singh letter to all health ministers, lieutenant-governors, and chief commissioners, ca. August 1975, reprinted as appendix A23 in Gupta, Sinha, and Bardhan, *Family Welfare Programme in India,* 17–18. On crash programme and Bihar, see Shah Commission of Inquiry, *Third and Final Report* (New Delhi: Government of India Press, 1978), 153–154, 158. Along with the Shah report, the best accounts of population control during the emergency are Banerji, "Political Economy of Population Control"; Davidson R. Gwatkin, "Political Will and Family Planning: The Implications of India's Emergency Experience," *Population and Development Review* 5 (1979): 29–59; and Vicziany, "Coercion in a Soft State."

108. "Statement on Population Policy," April 16, 1976, Ministry of Health and Family Welfare, Department of Family Welfare, Archives, New Delhi, File V13011/3/75-Ply, vol. 3.

109. Katherine Frank, *Indira: The Life of Indira Nehru Gandhi* (New York: Houghton Mifflin, 2001), 391, 399; P. N. Dhar, *Indira Gandhi, the 'Emergency,' and Indian Democracy* (New Delhi: Oxford University Press, 2000), 304–305, 317–318.

110. Dhar, *Indira Gandhi,* 325–329. On Sanjay as doer, and PRB, see Pranay Gupte, *Mother India: A Political Biography of Indira Gandhi* (New York: Scribner's, 1992), 441–442, 445–446. On worst abuses, see Gwatkin, "Political Will and Family Planning," 38–42.

111. On migration control, see, for instance, "Research Project on Rural Youth and Population Problems," April 12, 1972, FAOA, RG 12, ESD, File 75, PA 4/1. Kingsley Davis, "Asia's Cities: Problems and Options," *Population and Development Review* 1 (1975): 83.

112. Emma Tarlo has documented how people were told that resettlement required that they produce a sterilization certificate, but she also makes clear that we will never know what, precisely, provoked the people of Delhi that day; see her *Unsettling Memories: Narratives of the Emergency in Delhi* (London: Hurst and Co., 2003), 37–39, 68–69.

113. Ibid., 39–41.

114. Ibid., 15. On Varanasi, see Frank, *Indira*, 403. On sterilization certificate, see Shah Commission, *Third and Final Report*, 195.

115. Gwatkin, "Political Will and Family Planning," 45–46. On Pipli, see Michael Henderson, *Experiment with Untruth: India under Emergency* (Delhi: Macmillan, 1977), 76. On "people pollution," see Henry Kamm, "Indian State Is Leader in Forced Sterilization," *New York Times*, August 13, 1976.

116. Huw Jones to I. Z. Husain, April 29, 1976, WBGA, Credit 0981—India—Population project (2)—Correspondence vol. 1.

117. On Bank position, see Messenger to Kanagaratnam, June 1, 1976, WBGA, Records of the Health Services Development Sector, Liaison with International and Other Organizations—UNFPA, vol. 5. On the IPPF position, see minutes of the 28th meeting, April 30, 1976, IPPFA, 1.4.1.1.3, "IPPF Management and Planning Committee." On FPAI during the Emergency, see "Secretary-General's Comments: India," June 21, 1976, IPPFA, series B, reel 389, frames 95–97; Avabai B. Wadia, *The Light Is Ours: Memoirs and Movements* (London: IPPF, 2001), 587–589; "FPAI 'Think Tank,'" September 20, 1976, IPPFA, series B, reel 479, frames 1828–1836.

118. "Discussion of Additional General Grant Support from USAID," November 2, 1976, IPPFA, 1.4.1.1.1, "IPPF Governing Body, 14th Meeting (Special Meeting) 5th–7th November"; "Indian Ocean Region: Report to the Management and Planning Committee," March 25, 1977, IPPFA, 1.4.1.1.3, "IPPF Management and Planning Committee," "Minutes of the 31st Meeting."

119. Tarlo, *Unsettling Memories*, 11.

120. Engh, "Population Control," 214–215; Shah Commission, *Third and Final Report*, 195.

121. Laurenti memo to files, December 8, 1976, and "Notes on Visit to India, November 6–12, 1976," both in WBGA, 03–04, Office of the President, Records of President McNamara, series 05, Contacts [Member Countries] files, box 8, "India (1976–1977)." On Sadik, see Kang to files, December 1, 1976, WBGA, Records of the Health Services Development Sector, Liaison with International and Other Organizations—UNFPA—vol. 5.

122. "VSC Coordination Meetings," July 20, 1976, IPPFA, series B, reel 266, frames 926–930. On WHO, see Ravenholt to Zahra, October 5, 1976, WHOA, Geneva, P13/372/6; Hammer, "Trip Report to New York and Washington," De-

cember 12, 1976, WHOA, P13/372/9. On laparoscopes, see Ravenholt, "Use of Surgical Laparoscopy."

123. "Report of the External Advisory Panel on Population," August 1976, WBGA, General Records, Central Files Station 1975–1977, Operations Policy Files, box 47. Considering the incentives and the disincentives, it is likely that there was some false reporting, but the risk of detention would have minimized the extent. See Gwatkin, "Political Will and Family Planning," 58.

124. Gupta, Sinha, and Bardhan, *Family Welfare Programme in India*, 81–82. On Banerji, see Meredith Minkler, "Thinking the Unthinkable: The Prospect of Compulsory Sterilization in India," ca. September 1976, FFA, Papers of David Bell, series 5, box 32, folder 772.

125. "Report of the External Advisory Panel on Population," August 1976.

126. Shah Commission, *Third and Final Report*, 170–206; see also Greer, *Sex and Destiny*, 357–358.

127. Dhar, *Indira Gandhi*, 324–325, 340–341, 343–344; Singh letters to all State Governments and Union Territories, November 11 and 18, 1976, reprinted as appendices H (II) 61 and 61a in Gupta, Sinha, and Bardhan, *Family Welfare Programme in India*, 96–97. On Maharashtra, see Shah Commission, *Third and Final Report*, 162. On overruling Sanjay, see Frank, *Indira*, 410.

128. Dhar, *Indira Gandhi*, 265.

129. Gwatkin, "Political Will and Family Planning," 58; Frank, *Indira*, 411.

130. Gwatkin, "Political Will and Family Planning," 58; Henderson, *Experiment with Untruth*, ix.

9. Reproducing Rights, Reproducing Health

1. *World Development Report 1984* (Washington, DC: World Bank, 1984), 148, 153.

2. "World Population Conference Report," October 31, 1975, IPPF Archives, London (hereafter IPPFA), 1.4.1.1.1, "31st October/2nd Nov. 1975 13th Meeting of the Governing Body"; "Integration of Family Planning and Socio-economic Development," November 5, 1976, IPPFA, 1.4.1.1.1, "IPPF Governing Body, 14th Meeting (Special Meeting) 5th–7th November"; Avabai B. Wadia, *The Light Is Ours: Memoirs and Movements* (London: IPPF, 2001), 309–313.

3. "Discussion on Additional General Grant Support," November 2, 1976, IPPFA, 1.4.1.1.1, "IPPF Governing Body, 14th Meeting (Special Meeting) 5th–7th November." On cuts at the Council, see "Minutes of the Senior Staff Meeting," November 26, 1974, and June 5, 1975, Rockefeller Archive Center, Tarrytown, NY (hereafter RAC), RG IV3B4.2, Population Council, Acc. II, George Zeidenstein, box 2, Officers Meetings July 1975–June 1976.

4. "The Population Council: A Review," May 1975, and Teitelbaum to Harkavy, April 25, 1975, Ford Foundation Archives, New York City (hereafter FFA), Papers of David Bell, series 5, box 33, folder 783.

5. Bell, "Notes for Phone Conversation," June 1, 1975, FFA, Papers of David Bell, series 5, box 33, folder 783; author's interview with Adrienne Germain, November 10, 2006; Maria Mesner, "Engineering Global Population: The Reasoning and Policies of the Population Council," *Rockefeller Archive Center Research Reports Online* (April 2001), archive.rockefeller.edu/publications/resrep/mesner.pdf.

6. Author's interview with Adrienne Germain, November 10, 2006; "Minutes of the Directors Meeting," March 9, 1976, RAC, RG IV3B4.2, Population Council, Acc. II, George Zeidenstein, box 2, "Officers Meetings, July 1975–June 1976."

7. "The Foundation's Strategy for Work on Population," June 1977, FFA, report no. 004137; "RF meeting," December 21, 1977, FFA, Papers of David Bell, series 5, box 34, folder 799.

8. "Minutes of the Thirty-First Meeting of the Management and Planning Committee," March 25, 1977, IPPFA, 1.4.1.1.3, "IPPF Management and Planning Committee."

9. Transcript of House Committee on International Relations Hearing, April 18, 1977, Reagan Presidential Library, Morton C. Blackwell Files, OA 12448; "Questions Submitted to AID Administrator Peter McPherson by Senator Jesse Helms," April 24, 1986, Reagan Presidential Library, Mariam Bell Files, OA 17964.

10. Sunniva Engh, "Population Control in the 20th Century: Scandinavian Aid to the Indian Family Planning Programme" (Ph.D. diss., University of Oxford, 2005), 229–256.

11. "CWL Wins Round 1 of 'Condom War,'" *Times Journal* (Philippines), July 9, 1975. On Muslim opposition, see Reese to Ravenholt, March 30, 1976, Ravenholt Papers; editorial, *Indonesian Observer*, July 1, 1976; "Condom from the U.S. and Sex from Indonesia," *Merdeka* (Jakarta), June 30, 1976,

12. Byroade to State, February 25, 1977, Ravenholt Papers; author's interview with Steven Sinding, February 21, 2005; Lois Wessel, "Reproductive Rights in Nicaragua: From the Sandinistas to the Government of Violeta Chamorro," *Feminist Studies* 17 (Autumn 1991): 538–539; Homa Hoodfar and Samad Assadpour, "The Politics of Population Policy in the Islamic Republic of Iran," *Studies in Family Planning* 31 (March 2000): 20.

13. Harkavy memo, April 22, 1975, and Bell handwritten notes, ca. June 1977, FFA, Papers of David Bell, series 5, box 34, folders 797, 798.

14. Deborah Shapley, *Promise and Power: The Life and Times of Robert*

McNamara (New York: Little, Brown, 1993), 674; Bell, "Summary Comments," June 9, 1977, FFA, Papers of David Bell, series 5, box 34, folder 798.

15. "Report of the Committee on Women's Roles and Population," November 7, 1975, RAC, RG IV3B4.2, Population Council, Acc. II, George Zeidenstein, box 3, "The Population Council, Vol. I. George Zeidenstein."

16. Dunlop to Rockefeller, April 6, 1976, RAC, RG IV3B4.2, Population Council, Acc. II, George Zeidenstein, box 3, "Personal and Confidential."

17. Author's interview with Nafis Sadik, November 9, 2006; Dunlop to Rockefeller, April 6, 1976; "Suggested Discussion Outline," June 2, 1977, FFA, Papers of David Bell, series 5, box 34, folder 798.

18. "The Unborn and the Born Again," *New Republic,* July 2, 1977, 5–6.

19. Jaffe to Dunlop, July 11, 1977, RAC, RG IV3B4.2, Population Council, Acc. II, George Zeidenstein, box 3, "Joan Dunlop."

20. "The President's News Conference," July 12, 1977, www.presidency.ucsb.edu; Peterson to Costanza, July 13, 1977, Jimmy Carter Presidential Library, Atlanta, WHCF, box WE6, file WE3.

21. Adrienne Germain, personal communication, August 15, 2007; Bell, "Summary Comments," June 9, 1977, FFA, 010333.

22. "Minutes of the Senior Staff Meeting," July 22, 1975, RAC, RG IV3B4.2, Population Council, Acc. II, George Zeidenstein, box 2, "Officers Meetings, July 1975–June 1976"; Bell to Harkavy, June 9, 1977, FFA, Papers of David Bell, series 5, box 34, folder 799; "RF meeting," December 21, 1977.

23. Stern to McNamara, April 22, 1977, World Bank Group Archives, Washington, DC (hereafter WBGA), 03–04, Office of the President, Records of President McNamara, series 05, Contacts [Member Countries] files, box 8, "India (1976–1977)"; "Operational Manual Statement," November 1977, and Lee to Baldwin, July 21, 1977, both in WBGA, Bank General records, Central Files Station 1975–1977, Operations Policy Files, box 47, "Population and Nutrition," vol. 2.

24. Ravenholt circular, October 5, 1977, Ravenholt Papers.

25. Ravenholt to Terry, June 5, 1975; Esmundo to Terry, August 4, 1975; Ravenholt to Zimmerly, January 14, 1976; Ravenholt to Kangas, January 15, 1976; Zimmerly to Ravenholt, February 18, 1976; all in Ravenholt Papers. Kang to Kanagaratnam, September 7, 1976, WBGA, Bank General Records, Central Files Station 1975–1977, Operations Policy Files, box 47, vol. 1.

26. Wheeler to Adler, November 25, 1976; Haynal to Zahidi, August 5, 1976; Ravenholt to Sinding, November 28, 1977; all in Ravenholt Papers.

27. Ravenholt to Esmundo, September 12, 1975, Ravenholt Papers; author's interview with Ravenholt, January 8, 2005. On Ravenholt's use of IPPF to distribute abortion kits, see Smith to Robbins, December 10, 1974, Sophia Smith Col-

lection, Smith College (hereafter SSC), Planned Parenthood Federation of America (PPFA II) Records, box 83. For example of audit, see "Audit Report—Abortion Activities Financed by the Pathfinder Fund," December 22, 1977, USAID, DEXS On-Line Database, DOCID/Order No: PD-AAX-678.

28. Author's interview with Ravenholt, January 8, 2005. On the revolt, see Shaw to Rausch, September 27, 1976, National Conference of Catholic Bishops Papers, U.S. Conference of Catholic Bishops Archives, Washington, DC, box 62, "Pro-Life Activities," file on "Ad-Hoc Committee: Pro-Life Activities, 1976, July–Sept." The Carter and Ford negotiations are extensively documented in this file.

29. Peter J. Donaldson, *Nature Against Us: The United States and the World Population Crisis, 1965–1980* (Chapel Hill: University of North Carolina Press, 1990), 89–90, 101; Paul Wagman, "U.S. Goal: Sterilize Millions of World's Women," *St. Louis Post-Dispatch,* April 22, 1977, 1, 6A.

30. Bill Peterson, "Battling the Birth Boom," *Washington Post,* May 18, 1978, A1; Donaldson, *Nature Against Us,* 85; Harold J. Logan, "Population Control Office Targeted in AID Reorganization," *Washington Post,* July 1, 1978, A6.

31. Logan, "Population Control Office Targeted."

32. The original cable is available under the title "Family Planning Program Success" at www.ravenholt.com. See also Ravenholt to Wheeler, March 30, 1976; Ravenholt to Niblock, March 15, 1977; Reese to Ravenholt, July 1, 1977; Ravenholt to Niblock, February 15, 1978; all in Ravenholt Papers.

33. Paul Demeny, "On the End of the Population Explosion," *Population and Development Review* 5 (1979): 150–152.

34. Ibid.; Reese to Ravenholt, July 1, 1977, Ravenholt Papers.

35. Taylor to Bell, March 8, 1979, FFA, Papers of David Bell, series 5, box 34, folder 800 (emphasis in original); "Minutes of the Fourth Meeting," April 26, 1979, IPPFA, 1.4.1.1.3, "Executive Committee of the Central Council, 26th/27th April 1979"; Agenda for Meeting on Health and Population in Developing Countries," April 1979, FFA, Papers of David Bell, series 5, box 34, folder 800.

36. Tyrene White, *China's Longest Campaign: Birth Planning in the People's Republic, 1949–2005* (Ithaca: Cornell University Press, 2006), 60–61; Susan Greenhalgh and Edwin A. Winckler, *Governing China's Population: From Leninist to Neoliberal Biopolitics* (Stanford, CA: Stanford University Press), 86–87.

37. Linda Matthews, "China's Saturation Approach to Birth Control," *International Herald Tribune,* September 14, 1977; "General View of China," October 12, 1977, IPPFA, series B, reel 493, frame 393.

38. Davis, "World Population Growth and United States Foreign Aid," March 31, 1977, Kingsley Davis Papers, Hoover Institution Archives, Palo Alto, box 3; Bernard Berelson and Jonathan Lieberson, "Government Efforts to Influence Fertility: The Ethical Issues," *Population and Development Review* 5 (1979): 591–596.

39. White, *China's Longest Campaign,* 53. On the possible USAID influence on China, author's interview with Ravenholt, January 8, 2005. On Song Jian, see Susan Greenhalgh, "Science, Modernity, and the Making of China's One-Child Policy," *Population and Development Review* 29 (2003): 170–171. Donella H. Meadows et al., *The Limits to Growth: A Report for the Club of Rome's Project on the Predicament of Mankind* (New York: Universe Books, 1972).

40. Susan Greenhalgh, "Missile Science, Population Science: The Origins of China's One-Child Policy," *China Quarterly* 182 (June 2005): 263; Greenhalgh, "Science, Modernity," 171–178.

41. Greenhalgh, "Science, Modernity," 183–186; White, *China's Longest Campaign,* 63–67; Joan Kaufman et al., "The Creation of Family Planning Service Stations in China," *International Family Planning Perspectives* 18 (1992): 18–23.

42. White, *China's Longest Campaign,* 86–89.

43. Ibid., 104–105.

44. Ibid., 120–121, 136, 173–193; Steven W. Mosher, *Broken Earth: The Rural Chinese* (New York: Free Press, 1983), 225–226.

45. Miro, "Report on Mission . . . October 29–November 19, 1979," and "Project Agreement," December 7, 1979, both in UN Archives and Records Centre, New York City (hereafter UNRC), S-0024-0011, China Demographic Training and Research, box 11, files 9, 6.

46. Ansley Coale, "Report to UNFPA on Trip to People's Republic of China," October 1980, and "Draft Proposal for the Meeting," November 17, 1981, both in UNRC, S-0024-0011, China Demographic Training and Research, box 11, files 1, 4.

47. Kane to Wahren, January 14, 1980, IPPFA, series B, reel 493, frame 365.

48. H. Yuan Tien, "Demography in China: From Zero to Now," *Population Index* 47 (Winter 1981): 686; Laurence Schneider, *Biology and Revolution in Twentieth-Century China* (Lanham, MD: Rowman and Littlefield, 2003), 225; Frank Dikötter, *Imperfect Conceptions: Medical Knowledge, Birth Defects and Eugenics in China* (New York: Columbia University Press, 1998), 122–123. On elimination measures, see Jay Mathews, "Breeding a New China," *Washington Post,* July 21, 1980, A1, A24.

49. Kane Trip Report, November 26–December 1, 1980, IPPFA, series B, reel 493, frames 285–291.

50. Kane to Li Xiuzhen, May 17, 1982, IPPFA, series B, reel 493, frame 1419. On Japan, see Aluvihare Trip Reports, November 17–24, 1980, and October 24–November 5, 1982, IPPFA, series B, reel 465, frames 1599, 1862–1864; Hammer report on UNDP Governing Council, June 29, 1981, World Health Organization Archives, Geneva (hereafter WHOA), P13/372/9. On stability in Asia, see Wahren Trip Report, March 16–20, 1982, and Stiernborg Trip Report, May

17–28, 1982, IPPFA, series B, reel 465, frames 1615–1624, 1627–1631. On Japanese reaction to criticism, see Lubin Trip Report, May 18–23, 1981, IPPFA, series B, reel 465, frames 1839–1840, and Aluvihare Trip Report, October 24–November 5, 1982.

51. Aluvihare Trip Report, March 2–7, 1981, IPPFA, series B, reel 493, frames 264–271. On the great deal of attention, see Aluvihare to Rivera, March 22, 1982, IPPFA, series B, reel 493, frames 1510–1511. On being treated differently, see Lubin to Wahren, December 10, 1982, IPPFA, series B, reel 493, frame 1469.

52. Michele Vink, "Abortion and Birth Control in Canton, China," *Wall Street Journal,* November 30, 1981, 26; "Killing of Baby Girls in China Reported," *Los Angeles Times,* November 21, 1982, 19.

53. Wahren, "Discussions with Chinese Group," December 15, 1982, IPPFA, series B, reel 493, frame 1466.

54. Wahren Trip Report, April 12–24, 1983, IPPFA, series B, reel 493, frames 220–225. It's not clear whether the IPPF provided such equipment, but by 1984 the IPPF was arranging for Chinese researchers interested in genetics and genetic counseling to study perinatal care (Aluvihare to Li Xiuzhen, July 26, 1984, IPPFA, series B, reel 493, frame 1290).

55. Susan Greenhalgh, "Controlling Births and Bodies in Village China," *American Ethnologist* 21 (1994): 8; White, *China's Longest Campaign,* 139–141; Wahren to members of the Executive Committee, April 11, 1983, IPPFA, series B, reel 493, frames 1349–1350.

56. White, *China's Longest Campaign,* 143–144. On explicit regulation, see Michael Weisskopf, "China Orders Sterilization for Parents," *Washington Post,* May 28, 1983, A1, A27. On lines of communication, see Aluvihare to Aird, July 22, 1983, IPPFA, series B, reel 493, frames 1368–1370. On ten million volunteers, see Wahren Trip Report, April 12–24, 1983.

57. White, *China's Longest Campaign,* 141–142. On government apparatus in India, see Working Group on Population Policy interim and final reports, along with Ministry of Health and Family Welfare letter to Chief Secretaries, March 3, 1983, in J. P. Gupta, N. K. Sinha, and Amita Bardhan, *Evolution of Family Welfare Programme in India* (New Delhi: National Institute of Health and Family Welfare, 1992), reprinted as appendices A42–A44, pp. 35–40. On performance figures and foreign aid, see Gupta, Sinha, and Bardhan, *Evolution of Family Welfare Programme in India,* 86, 110. On cabinet proposals, see minutes to file, "Future Strategy for Family Welfare Programme," V.13011/4/81-US(P), Department of Family Welfare Archives, Ministry of Health and Family Welfare, New Delhi.

58. Murray Gendell, "Stalls in the Fertility Decline in Costa Rica and South Korea," *International Family Planning Perspectives* 15 (1989): 18–19.

59. Sharon M. Lee, Gabriel Alvarez, and J. John Palen, "Fertility Decline and Pronatalist Policy in Singapore," *International Family Planning Perspectives* 17 (1991): 66–67.

60. Muskie to Bujumbura, June 4, 1980; Tyson to Baldi, June 17, 1980; Levin to Bennet, June 30, 1980; all in U.S. National Archives, College Park, MD, RG 286, USAID, HQ Office of the Administrator, Office of the Executive Secretary, Communications and Coordinating Staff, Classified Subject Files of Executive Officers 1979–80, file HLS-9, 2919–2922.

61. Sandberg to Files, January 15, 1980, and North to Files, January 24, 1980, both in WBGA, Bank General Records, Central Files Station 1975–1977, Operations Policy Files, box 49. On donor plan, see "Draft Appraisal Report," March 16, 1981, UNRC, Q400-R021-SU24, UNFPA Global and Interregional Project Files, box 9, KEN 1981–1982–1983. On release of loan, see Operations Evaluation Department, *Population and the World Bank: A Review of Activities and Impacts from Eight Case Studies* (Washington, DC: World Bank, 1992), 54.

62. On donor plan, see "Position Paper on the Population Control and Family Planning Programme in Bangladesh," March 1, 1983; on bystanders, see Keller to Ando, March 16, 1983; both in UNRC, Q400-R021-SU24, UNFPA Global and Interregional Project Files, box 8, BGD 1980–81–82–83.

63. See Betsy Hartmann's devastating account in *Reproductive Rights and Wrongs: The Global Politics of Population Control,* rev. ed. (Boston: South End Press, 1994), 221–233.

64. UN press release, September 30, 1983, IPPFA, series B, reel 681, frames 1946–1947.

65. Bernard D. Nossiter, "Population Prizes from U.N. Assailed," *New York Times,* June 24, 1983, A4. On Chinese publicity, see Michael Weisskopf, "China Orders Sterilization for Parents," *Washington Post,* May 28, 1983, A1, A27. On posters and backing, see Margaret Wolfson, "China Revisited," April 1982, IPPFA, series B, reel 493, frames 234–247. On Qian, see H. Yuan Tien, *China's Strategic Demographic Initiative* (Westport, CT: Praeger, 1991), 133.

66. Tien, *China's Strategic Demographic Initiative,* 123–129, 135–137; White, *China's Longest Campaign,* 151–152, 159–160, 163–164.

67. Minutes, Europe Regional Council, May, 1984, and McLaughlin to Aluvihare, June 15, 1984, IPPFA, series B, reel 493, frames 1283–1285 and 1307.

68. Reimert Ravenholt, "Foremost Achievements of USAID's Population Program, 1966–1979," www.ravenholt.com; Cynthia Gorney, *Articles of Faith: A Frontline History of the Abortion Wars* (New York: Simon and Schuster, 1998), 338–348, 358–362.

69. Gorney, *Articles of Faith,* 375; David J. Garrow, *Liberty and Sexuality: The Right to Privacy and the Making of Roe v. Wade* (Berkeley: University of California

Press, 1998), 638–642. On bishops, see Uhlmann and Barr to Harper, January 20, 1983, Reagan Presidential Library, Edwin Meese III Papers, OA 9447, "Abortion (1)."

70. U.S. Coalition for Life press release, January 21, 1983, Reagan Presidential Library, Morton Blackwell Papers, OA 12448, "Pro-Life."

71. "A United Pro-Life Movement," January 1984, National Library of Medicine, History of Medicine Division, Bethesda, MD, C. Everett Koop Papers, series 10, Subject Files, box 32, folder 17.

72. Marx to Mahony, March 16, 1989, Marx Papers, University of Notre Dame Archives (hereafter CMRX), box 1, folder 10. On checks, see Clemens to Marx, April 23, 1986; Marx to Gagnon, January 7, 1988; Sin to Marx, January 7, 1988; all CMRX, box 2, folder 8. On HLI campaign, see Marx to Fisher, November 17, 1987, CMRX, box 12, folder 5; and Marx to Jose Sanchez, December 10, 1985, CMRX, box 1, folder 19. On Trujillo, see Alberto Rizo and Laura Roper, "The Role of Sterilization in Colombia's Family Planning Program: A National Debate," *International Family Planning Perspectives* 12 (1986): 45.

73. "General Plan of Appeal to Catholics," n.d., Reagan Presidential Library, Morton Blackwell papers, OA 12450, "Catholic Strategy" (3 of 3).

74. Tad Szulc, *Pope John Paul II* (New York: Simon and Schuster, 1996), 272–275, 282–283. On Nicaragua, see Jeffrey Klaiber, *The Church, Dictatorships, and Democracy in Latin America* (New York: Orbis, 1998), 203–207. For Reagan response, see Reagan to Quarracino, February 13, 1984, Reagan Presidential Library, Edwin Meese III Papers, box OA 11844.

75. Lubin to Weerakoon, July 6, 1984, IPPFA, series B, reel 140, frame 1208.

76. See copious documentation—apart from some items still classified—Reagan Presidential Library, Executive Secretariat: NSC: Records: Subject File, box 82. On concrete assurances, see Buckley to Baker, July 12, 1984, Reagan Presidential Library, Executive Secretariat: NSC: Records: Subject File, box 82.

77. "'Overpopulation' Has McNamara, Buckley at Odds," *Boston Globe*, August 6, 1984, 1; "U.S. Policy Statement for the International Conference on Population," *Population and Development Review* 10 (1984): 576–577.

78. Jason L. Finkle and Barbara B. Crane, "Ideology and Politics at Mexico City: The United States at the 1984 International Conference on Population," *Population and Development Review* 11 (1985): 12; Hartmann, *Reproductive Rights and Wrongs,* 227–230; "Sino-American Cooperation in Reproductive Physiology and Family Planning Techniques," November 3, 1982, National Library of Medicine, History of Medicine Division, Bethesda, MD, C. Everett Koop Papers, series 2, Sequential Files, box 19, "People's Republic of China (2)."

79. Laura Lopez, "A Debate over Sovereign Rights," *Time,* August 20, 1984.

80. Nick Cater, "Family Planning Aid Cut," *Guardian,* December 31, 1984.

81. Finkle and Crane, "Ideology and Politics," 12; "Statement of M. Peter

McPherson," November 25, 1985, Reagan Presidential Library, Mariam Bell Files, OA 17964.

82. "Statement of M. Peter McPherson," November 25, 1985. On twenty million members, see "CFPA 10th Anniversary Celebrations," December 5, 1990," IPPFA, series B, reel 493, frame 841.

83. Susan Greenhalgh, "The Peasantization of the One-Child Policy in Shaanxi," in *Chinese Families in the Post-Mao Era,* ed. Deborah Davis and Stevan Harrell (Berkeley: University of California Press, 1993), 249–250; Greenhalgh and Winckler, *Governing China's Population,* 272–273.

84. Terence H. Hull, "Recent Trends in Sex Ratios at Birth in China," *Population and Development Review* 16 (1990): 74. On sex determination, see Chapter 7 and Ehrlich, *The Population Bomb* (New York: Ballantine Books, 1968), 139. On ultrasound imports, see Zeng Yi et al., "Causes and Implications of the Recent Increase in the Reported Sex Ratio at Birth in China," *Population and Development Review* 19 (1993): 291; M. Daly, "Abortion Row over Exports to China," *Sunday Herald* (Australia), September 16, 1990. On "hot items," see Christine A. Genzberger, *China Business: The Portable Encyclopedia for Doing Business with China* (San Rafael, CA: World Trade Press, 1994), 36. On World Bank loans, see, for example, Rural Health and Medical Education Project (P003424), Staff Appraisal Report, 1984/04/30, report no. 4864, provided by WBGA.

85. P. Nicholas and D. Kristof, "The Chosen Sex—A Special Report; Chinese Turn to Ultrasound, Scorning Baby Girls for Boys," *New York Times,* July 21, 1993, A1; T. Bangsberg, "GE Officials Tour China for Potential Investments," *Journal of Commerce,* September 4, 1992, 4A. Underreporting of female births and excess female infant mortality are also contributing causes to the shift in the sex ratio, but recent research suggests they are secondary; see Chu Junhong, "Prenatal Sex Determination and Sex-Selective Abortion in Rural Central China," *Population and Development Review* 27 (2001): 259–281; and see also Zeng et al., "Causes and Implications," 283.

86. Linda Feldman, "China, UN Alter Population Policy," *Christian Science Monitor,* January 3, 1989; Greenhalgh and Winckler, *Governing China's Population,* 225; White, *China's Longest Campaign,* 220–221, 225, 230.

87. Greenhalgh, "Controlling Births and Bodies in Village China," 23–24; Kay Ann Johnson and Amy Klatzkin, *Wanting a Daughter, Needing a Son: Abandonment, Adoption, and Orphanage Care in China* (St. Paul, MN: Yeong and Yeong, 2004), 75.

88. The best case for the UNFPA's positive role in China is made by Edwin Winckler, "Maximizing the Impact of Cairo on China," in *Where Human Rights Begin: Health, Sexuality, and Women in the New Millennium,* ed. Wendy Chavkin and Ellen Chesler (New Brunswick: Rutgers University Press, 2005), 204–234.

89. "Draft for Inclusion in CEC Chairman's Report," October 7, 1983,

IPPFA, series B, reel 642, frames 1562–1564. On women in UNFPA, see Laskin, "Report on UNDP Governing Council Meeting," June 6, 1988, IPPFA, series B, reel 685, frames 699–703. On the turn away from incentives, see Kupperstein Trip Report, January 18–February 6, 1987, IPPFA, series B, reel 671, frame 862.

90. "Background Note on Recent Developments," October 2, 1985, WHOA, P13/372/9. On Salas, see "Paying for Abortions," *Wall Street Journal*, April 9, 1984, 34; but compare Gille to Kang, July 3, 1980, WBGA, Records of the Health Services Development Sector, Liaison with International and Other Organizations, box 15, UNFPA—vol. 6. On the Council, see Sheldon J. Segal, *Under the Banyan Tree: A Population Scientist's Odyssey* (New York: Oxford University Press, 2003), xxvii.

91. Service to Cossey et al., December 10, 1986; "Task Force on Opposition," April 24, 1986; "Opposition to Family Planning and the IPPF—An Update," October 1986; all in Contemporary Medical Archives Center, Wellcome Institute for the History of Medicine, London, SA/FPA/C/B/2/7/2.

92. Barbara B. Crane, "The Transnational Politics of Abortion," *Population and Development Review* 20 (1994): 252; Paige Whaley Eager, *Global Population Policy: From Population Control to Reproductive Rights* (Aldershot, UK: Ashgate, 2004), 86, 107.

93. Gita Sen and Caren Grown, *Development, Crises, and Alternative Visions: Third World Women's Perspectives* (New York: Monthly Review Press, 1987), 50.

94. Adrienne Germain, interview with author, November 10, 2006, and personal communication, August 15, 2007.

95. Rebecca Sharpless interview with Adrienne Germain, June 19–20, September 25, 2003, Population and Reproductive Rights Oral History Project, SSC.

96. Rebecca Sharpless interview with Joan Dunlop, April 14–15, 2004, and with Adrienne Germain, June 20, 2003, Population and Reproductive Rights Oral History Project, SSC.

97. This account is based on Eager, *Global Population Policy*, 108–109; author's interview with Adrienne Germain, November 10, 2006; Sharpless interview with Adrienne Germain, June 19–20, September 25, 2003, Population and Reproductive Rights Oral History Project, SSC.

98. "Report of the World Conference to Review and Appraise the Achievements of the United Nations Decade for Women," July 1985, www.un.org/esa/gopher-data/conf/fwcw/nfls/nfls.en.

99. Aluvihare Trip report, May 24, 1987, IPPFA, series B, reel 685, frames 934–940; "Mission to UNFPA Headquarters," January 25, 1988, UNESCO Archives, Paris, 312.12 A 8 UNFPA/SS, pt. 17. For examples of new UNFPA projects, see UNRC, Q400-R013-SU20, UNFPA Project Files, box 39, RAF/88/P28, "Population, Women, and Development."

100. Author's interview with Nafis Sadik, November 9, 2006; Nicholas

Kristof, "A U.N. Agency May Leave China over Coercive Population Control," *New York Times,* May 14, 1993.

101. Author's interview with Nafis Sadik, November 9, 2006; Susan A. Cohen, "The Road from Rio to Cairo: Toward a Common Agenda," *International Family Planning Perspectives* 19 (1993): 62–63.

102. Dennis Hodgson and Susan Cotts Watkins, "Feminists and Neo-Malthusians: Past and Present Alliances," *Population and Development Review* 23 (1997): 503

103. Westoff, "What's the World's Priority Task? Finally, Population Control," and Chesler, "No, the First Priority Is Stop Coercing Women," both in *New York Times Magazine,* February 6, 1994.

104. Author's interview with Nafis Sadik, November 9, 2006.

105. This account is based on author's interview of Sadik, along with Carl Bernstein and Marco Politi, *His Holiness: John Paul II and the Hidden History of Our Time* (New York: Doubleday, 1996), 513–518.

106. Bernstein and Politi, *His Holiness,* 518–521; Szulc, *Pope John Paul II,* 464–465; "Woman's Dignity Must Be Respected," June 22, 1984, www.vatican.va/holy_father/john_paul_ii/audiences.

107. John Hooper, "Pope and Tehran Do Abortion Deal," *Guardian,* August 9, 1994, 20; John Lancaster, "Muslims Echo Pope's Rejection of U.N. Population Document," *Washington Post,* August 11, 1994; Laurie Goodstein, "Catholics and Muslims Find Common Moral Ground," *Washington Post,* August 19, 1994, A21. On Libya, see Jim Hoagland, "John Paul II: Two Devils," *Washington Post,* August 23, 1994, A19.

108. "Muslim Preachers in Egypt Condemn Population Conference," *New York Times,* August 22, 1994, A9; Robert Fisk, "'Holy Alliance' Tries to Wreck Birth-Control Conference," *Independent,* August 31, 1994, 8; author's interview with Nafis Sadik, November 9, 2006.

109. Author's interview with Ellen Chesler, December 2, 2006; Elisabeth Bumiller, *May You Be the Mother of a Hundred Sons: A Journey among the Women of India* (New York: Fawcett Columbine, 1990).

110. Barbara Crossette, "Women's Advocates Flocking to Cairo, Eager for Gains," *New York Times,* September 2, 1994, A3.

111. Ramon G. McLeod, "Abortion Fight Dominates Cairo Meeting," *San Francisco Chronicle,* September 6, 1994, A1; Brundtland, "Key-Note Address to the ICPD," September 5, 1994, "International Conference on Population and Development—statement by Albert Gore," September 5, 1994, and "Address by H. E. Mohtarama Benazir Bhutto," September 5, 1994, all in www.un.org/popin/icpd/conference/gov.

112. "Vatican Envoy Is Booed," *St. Louis Post-Dispatch,* September 7, 1994, 1A.

113. Gwynne Dyer, "It's About Power," *Gazette* (Montreal), September 7, 1994, B3.

114. Jyoti Shankar Singh, *Creating a New Consensus on Population: The International Conference on Population and Development* (London: Earthscan, 1998), 65–68; Ramon G. McLeod, "Abortion Fight Dominates Cairo Meeting"; Mark Nicholson, "Vatican Makes Waves at Cairo Conference," *Financial Times,* September 8, 1994, 4; Bob Hepburn, "How Vatican Stumbled on World Stage," *Toronto Star,* September 18, 1994, E5.

115. George Weigel, "What Really Happened at Cairo," *First Things* 50 (February 1995): 24–31.

116. K. Srinivasan, *Regulating Reproduction in India's Population: Efforts, Results, and Recommendations* (New Delhi: Sage, 1995), 60–62; White, *China's Longest Campaign,* 234–235; Rounaq Jahan, "Securing Maternal Health through Comprehensive Reproductive Health Services: Lessons from Bangladesh," *American Journal of Public Health* 97, no. 7 (July 2007): 1186–1190.

CONCLUSION

1. Peter J. Donaldson, *Nature Against Us: The United States and the World Population Crisis, 1965–1980* (Chapel Hill: University of North Carolina Press, 1990), 176.

2. See, for example, Werner Fornos, "U.S. Relinquishes Leadership on Global Population Control," *Seattle Post-Intelligencer,* June 21, 2005; Martha Campbell and Malcolm Potts, "The Pill Is Mightier than the Sword," *San Francisco Chronicle,* May 18, 2006.

3. Jianguo Liu et al., "Effects of Household Dynamics on Resource Consumption and Biodiversity," *Nature* 421 (2003): 530–533.

4. William Easterly, *The Elusive Quest for Growth: Economists' Adventures and Misadventures in the Tropics* (Cambridge: MIT Press, 2001), 91–92.

5. Nancy Birdsall, Allen C. Kelley, and Steven W. Sinding, *Population Matters: Demographic Change, Economic Growth, and Poverty in the Developing World* (New York: Oxford University Press, 2001). For an introduction to the debate, see the special issue of *Finance and Development* 43, no. 3 (2006).

6. The classic article on the subject is Lant H. Pritchett, "Desired Fertility and the Impact of Population Policies," *Population and Development Review* 20 (1994): 1–55.

7. UNFPA, "State of World Population 2006," www.unfpa.org/swp/2006/english/notes/indicators.html.

8. Pritchett, "Desired Fertility"; see also Pritchett, "The Impact of Population Policies," *Population and Development Review* 20 (1994): 611–630.

9. Pritchett, "Desired Fertility," 38; T. Paul Schultz, "Demand for

Children in Low Income Countries," *Handbook of Population and Family Economics* 1 (1997), 380–384, provides a good review. For further confirmation, see Jean Dreze and Mamta Murthi, "Fertility, Education, and Development: Evidence from India," *Population and Development Review* 27 (2001): 33–63.

10. On Myrdal: Nick Cullather, "Binay R. Sen, the FAO, and the 'Balance' of Food and Population," unpublished paper provided to author.

11. Michael Ignatieff, *Empire Lite: Nation Building in Bosnia, Kosovo, Afghanistan* (London: Vintage, 2003). What Foucault described as characteristic of the eighteenth century was far more relevant to the time in which he was writing; Foucault, *The History of Sexuality,* ed. and trans. Robert Hurley, vol. 1: *An Introduction* (New York: Pantheon, 1978), 7–8, 26, 140–147; Foucault, "Governmentality," in *The Foucault Effect: Studies in Governmentality,* ed. Graham Burchell, Colin Gordon, and Peter Miller (repr.; Chicago: University of Chicago Press, 1991), 90ff.

12. On Drysdale, see Deborah Anne Barrett, "Reproducing Persons as a Global Concern: The Making of an Institution" (Ph.D. diss., Stanford University, 1995), 113. On Gini, see 1921 note attached to "Le Ricerche del Comitato italiano," Archivio Centrale dello Stato, Rome, Corrado Gini (documentazione), Scatola 9, "Spedizione messico." On Laughlin, see Edwin Black, *War against the Weak: Eugenics and America's Campaign to Create a Master Race* (New York: Four Walls Eight Windows, 2003), 50. On Huxley, see F. A. E. Crew et al., "Social Biology and Population Improvement," *Nature* 144, no. 3646 (1939): 521.

13. Laurie Garrett, "What's Missing Is Political Will," *Toronto Star,* August 25, 2006; and Garrett, "The Challenge of Public Health," *Foreign Affairs,* January/February 2007, 14–38.

14. Population Division of the Department of Economic and Social Affairs of the United Nations Secretariat, "Trends in Total Migrant Stock: The 2005 Revision," *esa.un.org/migration.*

15. "White House Criticizes Bennett for Remarks on Race," *New York Times,* September 30, 2005.

ARCHIVES AND INTERVIEWS

Interviews

Ellen Chesler
Paul Demeny
Peter Donaldson
Joan Dunlop
Paul Ehrlich
Adrienne Germain
Nina Puri
Reimert Ravenholt
Nafis Sadik
Sheldon Segal
Steven Sinding
Michael Teitelbaum

Private Papers

Georges Bidault Papers, Archives Nationales, Paris
C. P. Blacker Papers, Contemporary Medical Archives Center, Wellcome Institute
    for the History of Medicine, London
Paul Blanshard Papers, Bentley Historical Library, Ann Arbor
Cass Canfield Papers, Harper & Row Publisher Records, Columbia University
    Rare Books and Manuscript Library
C. C. Cook Papers, Library of Congress
Patrick and Patricia Crowley Papers, University of Notre Dame Archives
Charles B. Davenport Papers, American Philosophical Society, Philadelphia

Kingsley Davis Papers, Hoover Institution Archives, Palo Alto
Paul and Anne Ehrlich Papers, Stanford University Archives
Havelock Ellis Papers, British Library, Manuscripts Collection, London
Clarence Gamble Papers, Francis A. Countway Library of Medicine, Boston
Corrado Gini Papers, Archivio Centrale dello Stato, Rome
Allan Guttmacher Papers, Francis A. Countway Library of Medicine, Boston
P. N. Haksar Papers, Nehru Memorial Museum and Library, New Delhi
Philip M. Hauser Papers, University of Chicago Library, Special Collections Re-
    search Center
John Harvey Kellogg Papers, Bentley Historical Library, Ann Arbor
C. Everett Koop Papers, National Library of Medicine, Bethesda
C. C. Little Papers, Bentley Historical Library, Ann Arbor
Daniel B. Luten Papers, Hoover Institution Archives, Palo Alto
Florence Mahoney Papers, National Library of Medicine, Bethesda
John Marshall Papers, University of Notre Dame Archives, South Bend, IN
Paul Marx Papers, University of Notre Dame Archives, South Bend, IN
Georges Mauco Papers, Archives Nationales, Paris
Asok Mitra Papers, Nehru Memorial Museum and Library, New Delhi
Hugh Moore Fund Collection, Seeley G. Mudd Manuscript Library, Princeton
Frank W. Notestein Papers, Seeley G. Mudd Manuscript Library, Princeton
John O'Brien Papers, University of Notre Dame Archives
William Fielding Ogburn Papers, University of Chicago Library, Special Collec-
    tions Research Center
Frederick Osborn Papers, American Philosophical Society, Philadelphia
Eileen Palmer Collection, India Office Records, Private Papers/European Manu-
    scripts, British Library, London
Lakshmi Pandit Papers, Nehru Memorial Museum and Library, New Delhi
Alan Sterling Parkes Papers, Contemporary Medical Archives Center, Wellcome
    Institute for the History of Medicine, London
Raymond Pearl Papers, American Philosophical Society, Philadelphia
Gregory Pincus Papers, Library of Congress
Reimert Ravenholt Papers, Seattle
John D. Rockefeller 3rd Papers, Rockefeller Archives Center, Tarrytown, NY
Alfred Sauvy Papers, Archives Nationales, Section Contemporaine, Fontainebleau
Marie Stopes Papers, British Library, Manuscripts Collection, London
George D. Woods Papers, Columbia University Rare Books and Manuscript Library

Nongovernmental Organization Archives

American Eugenics Society Papers, American Philosophical Society, Philadelphia
Association for Voluntary Sterilization Papers, Social Welfare History Archives,
    University of Minnesota, Minneapolis

Christian Aid Papers, School of Oriental and African Studies Library, London

Eugenics Society Papers, Contemporary Medical Archives Center, Wellcome Institute for the History of Medicine, London

Family Planning Association (of the UK) Papers, Contemporary Medical Archives Center, Wellcome Institute for the History of Medicine, London

Ford Foundation Archives, New York City

Human Betterment Foundation Papers, California Institute of Technology

International Planned Parenthood Federation Archives, London

International Women's Tribune Centre Papers, Sophia Smith Collection, Smith College

National Catholic Welfare Conference Papers, Catholic University Archives, Washington, DC

National Conference of Catholic Bishops Papers, U.S. Conference of Catholic Bishops Archives, Washington, DC

Planned Parenthood Federation of America Papers, Sophia Smith Collection, Smith College

Population Association of America Archives, Silver Spring, MD

Population Council Papers, Rockefeller Archives Center, Tarrytown, NY

Population Foundation of India Library, New Delhi

Rockefeller Brothers Fund, Rockefeller Archives Center, Tarrytown, NY

Rockefeller Foundation, Rockefeller Archives Center, Tarrytown, NY

Vatican Secret Archives, Rome

NATIONAL ARCHIVES

Archives Nationales, Paris

Archives Nationales, Section Contemporaine, Fontainebleau

Archives d'Outre-Mer, Aix-en-Provence

Archivio Centrale dello Stato, Rome

Jimmy Carter Presidential Library, Atlanta

Dwight D. Eisenhower Presidential Library, Abilene

Gerald R. Ford Presidential Library, Ann Arbor

India Office Records, British Library, London

Institut de Medicine Tropicale du Service de Santé des Armées, Aix-en-Provence

Institut National d'Etudes Démographiques, Paris

Lyndon B. Johnson Presidential Library, Austin

Ministry of Health and Family Welfare, Department of Family Welfare, New Delhi

National Archives, Kew, UK

National Archives, New Delhi

National Institute of Health and Family Welfare, Documentation Centre, New Delhi

Planning Commission Archives, New Delhi
Ronald Reagan Presidential Library, Simi Valley
United States National Archives, College Park

INTERNATIONAL ARCHIVES

Archives of the League of Nations, Geneva
Council of the European Communities Archives, Brussels
European Commission Archives, Brussels
Food and Agriculture Organisation Archives, Rome
International Labour Organisation Archives, Geneva
United Nations Archives and Records Centre, New York City
UN Educational, Scientific, and Cultural Organization Archives, Paris
World Bank Group Archives, Washington, DC
World Health Organization Archives, Geneva

ACKNOWLEDGMENTS

This book argues against the notion that our problems are caused by excess people. Completing it led me to ask many people for help. Hence the paradox: How can I possibly thank them all? On the acknowledgments page, at least, too many names really can amount to a horde, in which no one—not even the author—retains his or her individuality. Those acknowledged here must therefore stand in for still more people. In some cases they helped at the crucial early stages of this project, now so long ago I fear I do not remember them all, or when I was so preoccupied and needy that I scarcely realized how much help I was getting. All of the mistakes are my own, but these are the ones I regret most.

However voluminous, the notes say less than they should about what I owe to other scholars. I would not have known where to begin without them. Time and again, rather than leave the impression that some insight was my own, I wanted to write "as so and so argues"—so often that most readers would have found it distracting. But others who are familiar with these arguments will already realize that this whole book is part of a conversation with people like Alison Bashford, Ellen Chesler, Nicholas Cullather, Mike Davis, Linda Gordon, Ian Hacking, Betsy Hartmann, Stefan Kühl, Mahmood Mamdani, Adam McKeown, Alexander Saxton, James C. Scott, Sheldon Watts, and Aristide Zolberg. I also profited enormously from dissertations, especially those of Sanjam Ahluwalia, Charline Dekens, Sunniva Engh, Sharon Mara Leon, Deborah Oakley, and Amy Staples. Still others were both protagonists in this story and helped me to understand it. In addition to all those who let me interview them, I feel indebted to A. R. Nanda, Lant Pritchett, and Susan Watkins.

The Archives section lists many archives, but none of the people who opened them to me. Some of those who went far out of their way to point me in the right direction include Giuliano Fregoli at the FAO, Alan Divack and Anthony Maloney at the Ford Foundation, Remo Becci and Fiona Rolian at the ILO, Esther Katz at the Margaret Sanger Papers Project, Ken Rose at the Rockefeller Archive Center, Linnea Anderson at the Social Welfare History Archives, Neshantha Karunanayake at the United Nations, Mahmoud Ghander of UNESCO, Nancy Patterson at the U.S. Council of Catholic Bishops, Ineke Deserno at WHO, and Bertha Wilson at the World Bank.

Columbia University generously supported this research, and the Institute for the Humanities in Ann Arbor, the Center for Social Theory and Comparative History in Los Angeles, and the Woodrow Wilson International Center for Scholars in Washington provided intellectual community and comfortable surroundings when I paused to write up my findings. Here, there, and everywhere other historians read these drafts and helped me think harder, particularly Vicky de Grazia, Akira Iriye, Alice Kessler-Harris, Bill Leach, Mark Mazower, Ken Pomeranz, Emily Rosenberg, and Jay Winter. My editor, Joyce Seltzer, provided crucial midcourse correction and stuck with me to the end. My research assistant, Matt Christiansen, became my tutor in economics and statistics. Others assisted with translations or corrected my notes, including Nancy Davenport, Ed Fitzmaurice, Tim Jolis, Kerri Liming, and Cyrus Schayegh.

Still others who helped me remain anonymous—the reviewers of articles I submitted to academic journals. Revising and resubmitting helped immeasurably in refining my arguments. I am also grateful for receiving permission to reproduce parts of the articles here in revised form: "Population Control in India: Prologue to the Emergency Period," *Population and Development Review* 32 (November 2006): 629–667; "To Inherit the Earth: Imagining World Population, from the Yellow Peril to the Population Bomb," *Journal of Global History* 1 (November 2006): 299–319; "Seeing Beyond the State: The Population Control Movement and the Problem of Sovereignty," *Past & Present* 193 (December 2006): 197–233.

All along, old friends let me talk about all this, and at times it was all I could talk about. I am thinking particularly of Susan Ferber, Nathalie James, Martin Keady, Thomas LeBien, and David Thacher. And family members, because they are family, performed the no less important role of reminding me that I should also listen now and then. Thank you especially Patrick, Jeanellen, and Peter.

By the end, if only from lack of sleep, I sometimes dreamed that the muse of history was whispering in my ear. But now I realize that it was Sarah Kovner, urging me to wake up and get on with it. She was all of the above—historian, guide, critic, editor, confidante—and more. She helped in more ways than I can acknowledge, especially since she was the one who convinced me to close this book and open my eyes to the world beyond.

INDEX

Abortion: and Catholic Church, 11, 14–
15, 43, 48, 49, 147, 148, 269–270,
274, 278, 301, 336–337, 351, 352–
353, 364–366, 367, 368, 369, 377; vs.
contraception, 20, 77, 204, 265, 375,
381–382, 383; criminalization of, 48,
49, 51, 67, 79, 80, 111–112, 142, 243,
247, 272, 306–307, 310, 353, 363,
366, 367–368; and family planning,
240, 242–244, 252, 265, 367–368; and
IPPF, 243, 244, 336, 355, 359; legalization of, 11, 104, 111, 135, 136, 137,
139, 152, 179, 244, 263, 267, 269–
270, 353, 373; "menstrual regulation"
kits, 241, 244, 266, 306–307, 336;
Mexico City policy, 352, 353–355, 361;
Roe v. Wade, 269, 351, 352; as sex-selective, 244, 267, 356, 367, 381, 382–
383; and UNFPA, 244, 355. *See also
under individual countries*
Abrams, Frank, 169
Abzug, Bella, 363
Acheson, Dean, 213
Adoption, international, 382, 383
AES. *See* American Eugenics Society
Afghanistan, 371, 379

Africa: Belgian Congo, 32, 39, 125; contraception in, 86, 134; demographers
from, 301; family planning in, 197,
248, 302, 311, 316, 348–349, 354,
377; fertility rates in, 113, 127, 371;
French West Africa, 125; infant mortality in, 113; migration from, 260, 261;
mortality rates in, 143; population
growth in, 2, 5, 49, 57, 120, 121, 125,
190, 255. *See also* Algeria; Egypt;
Kenya; Morocco; Nigeria
African Americans: Black Panthers, 253;
Black Power, 13; fertility rates among,
107, 127, 208, 250–251, 274; and *The
Negro Family,* 249–250; racism regarding, 249–251, 253–254, 271, 272, 274;
syphilis experiment on, 271
Age structure, 343, 356, 372–373, 382
A. H. Robbins Corp., 252, 271
Aid for Families with Dependent Children
(AFDC), 250, 251
Algeria, 313; independence of, 261; popu-
lation growth in, 113, 114, 121, 181–
182, 196
Aluvihare, Bernard, 347
AMA. *See* American Medical Association